Welcome to the 73rd International Symposium on Molecular Spectroscopy
June 18-22, 2018
Urbana-Champaign, IL

On behalf of the Executive Committee, I extend a heartfelt welcome to all the attendees of the 73rd Symposium and welcome you to the University of Illinois at Urbana-Champaign.

The Symposium presents research in fundamental molecular spectroscopy and a wide variety of related fields and applications. The continued vitality and significance of spectroscopy is annually re-affirmed by the number of talks, their variety, and the fact that many are given by students. These presentations are the heart of the meeting and are documented by this Abstract Book. Equally important is the information flowing from informal exchanges and discussions. As organizers, we strive to provide an environment that facilitates both kinds of interactions.

The essence of the meeting lies in the scientific discussions and your personal experiences this week independent of the number of times that you have attended this meeting. It is our sincere hope that you will find this meeting informative and enjoyable both scientifically and personally, whether it is your first or 50th meeting. If we can help to enhance your experience, please do not hesitate to ask the Symposium staff or the Executive Committee.

Ben McCall
Symposium Chair

SCHEDULE OF TALKS

Monday (M) 1
Tuesday (T) 8
Wednesday (W) 20
Thursday (R) 32
Friday (F) 39

ABSTRACTS

Monday (M) 45
Tuesday (T) 79
Wednesday (W) 138
Thursday (R) 197
Friday (F) 228

AUTHOR INDEX 252

VENUE AND SPONSOR INFORMATION FOLLOWS AUTHOR INDEX

73rd INTERNATIONAL SYMPOSIUM ON MOLECULAR SPECTROSCOPY

International Advisory Committee
Tucker Carrington, Queen's University
Emilio Cocinero, Universidad del País Vasco
Richard Dawes, Missouri University of Science & Technology
Gary Douberly, University of Georgia, Chair-Elect
Jens-Uwe Grabow, Leibniz Universität Hannover
Kaori Kobayashi, University of Toyama
Helen Leung, Amherst College
Laurent Margules, Université de Lille
Terry Miller, Ohio State University
Takamasa Momose, University of British Columbia
Hiroyuki Sasada, Keio University
Tim Steimle, Arizona State University
Nick Walker, Newcastle University
Mathias Weber, University of Colorado and JILA
Susanna Widicus Weaver, Emory University
Yunjie Xu, University of Alberta, Chair
Shanshan Yu, Jet Propulsion Laboratory
Mingfei Zhou, Fudan University

Executive Committee
Ben McCall, Chair
Brian DeMarco
Dana Dlott
Gary Eden
Nick Glumac
Martin Gruebele
So Hirata
Leslie Looney
Josh Vura-Weis
Dave Woon

Please send correspondence to
Ben McCall
International Symposium on Molecular Spectroscopy
Department of Chemistry
600 S. Mathews Avenue
Urbana IL 61801 USA
e-mail: chair@isms.illinois.edu
http://isms.illinois.edu

Mini-Symposia

FAR-INFRARED SPECTROSCOPY

Organized by **Olivier Pirali** (SOLEIL Synchrotron) and **Shanshan Yu** (Caltech, Jet Propulsion Laboratory) Covers recent instrumental development, quantitative spectroscopy, reactive species, intermolecular interactions and biomolecules, and astrophysical lab/observation studies. Invited Speakers: **Brian Drouin** (Caltech, Jet Propulsion Laboratory), **Jean-François Lampin** (Lille University of Science and Technology), **Hiroyuki Ozeki** (Toho University), **Charles Schmuttenmaer** (Yale University), **Jean Vander Auwera** (Université Libre de Bruxelles)

FREQUENCY-COMB SPECTROSCOPY

Organized by **Aleksandra Foltynowicz** (Umeå University), **Nathalie Picqué** (Max-Planck Institute of Quantum Optics), and **Jun Ye** (JILA, University of Colorado). New frequency-comb-based experimental techniques and spectroscopy results, from gas-phase to condensed-phase, are showcased with an emphasis on intriguing applications that advance the frontiers of chemical physics. Invited Speakers: **Kjeld Eikema** (Vrije Universiteit, Amsterdam), **Paolo De Natale** (National Institute of Optics-CNR, Florence), **Greg Rieker** (University of Colorado, Boulder), **Takeshi Yasui** (University of Tokushima)

NEW WAYS OF UNDERSTANDING MOLECULAR SPECTRA

Organized by **Per Jensen** (University of Wuppertal) and **Stephan Schlemmer** (University of Cologne). Recent years have seen a number of new and alternative approaches to the interpretation of molecular spectra. They range very widely. We will showcase new and alternative approaches for interpreting molecular spectra, from heavily computational methods to those that are almost computer-free. We will bring together protagonists from these diverse fields in the hope of fostering discussion across discipline borders. Invited Speakers: **Renato Lemus** (National Autonomous University of Mexico, UNAM), **Brooks Pate** (University of Virginia), **Sergey Yurchenko** (University College London)

Picnic (Tuesday)

The Symposium picnic will be held on **Tuesday evening** at Ikenberry Commons. The cost of the picnic is included in your registration (at below cost to students), so that all may attend the event. The **Coblentz Society** is the host for refreshments for one hour. Please see your packet for additional details.

Sponsorship

We are pleased to acknowledge the many organizations that support the 73rd Symposium. Principal funding comes from the **Army Research Office** (ARO) and the **National Radio Astronomy Observatory** (NRAO). We are most grateful to ARO for their long-standing support. We also acknowledge the many efforts and contributions of **The University of Illinois** in hosting the meeting, including financial contributions from the Departments of Chemistry, Electrical and Computer Engineering, Astronomy, and Physics.

Our Corporate Sponsors are **Amplitude, Bristol Instruments, Elsevier/JMS, Ideal Vacuum Products, Journal of Physical Chemistry/ACS,** and **Quantel**. Please see the back of this book for their advertisements.

We are also pleased to acknowledge **IMRA, JASCO, Light Conversion USA, Lockheed Martin Laser and Sensor Sytems, M Squared Lasers, Menlo Systems, and Toptica Photonics** as Contributing Sponsors. Our sponsors will have exhibits at the Symposium and we encourage you to visit their displays.

Rao Prize

The three Rao Prizes for the most outstanding student talks at the 2017 meeting will be presented. The winners are **Bryce Bjork**, University of Colorado at Boulder; **Anna Huff**, University of Minnesota; and **Christopher Shingledecker**, University of Virginia. The Rao Prize was created by a group of spectroscopists who, as graduate students, benefited from the emphasis on graduate student participation, which has been a unique characteristic of the Symposium. This year three more Rao Prize winners will be selected.

The award is administered by a Prize Committee chaired by Gary Douberly, University of Georgia, and comprised of David Anderson (University of Wyoming); Brooks Pate (University of Virginia); Rebecca Peebles (Eastern Illinois University); Jennifer van Wijngaarden (University of Manitoba); and Tim Zwier, (Purdue University). Any questions or suggestions about the Prize should be addressed to the Committee. Anyone (especially post-docs) willing to serve on a panel of judges should contact Gary Douberly (douberly@uga.edu).

Miller Prize

The Miller Prize was created in honor of Professor Terry A. Miller, who served as chair of the International Symposium on Molecular Spectroscopy from 1992 to 2013. The Miller Prize for the best presentation given by a recent PhD at the 2017 meeting will be presented. The winner, **Leah Dodson** (University of Colorado, Boulder), will give a lecture on Thursday.

The Miller Prize winner and his or her co-authors will be invited to submit an article to the Journal of Molecular Spectroscopy based on the research in the prize-winning talk. After passing the normal review process, the article will appear in the Journal with a caption identifying the paper with the talk that received the Miller Prize.

The award is administered by a Prize Committee chaired by Mike Heaven, Emory University and comprised of Stephen Cooke (Purchase College SUNY); Richard Dawes (Missouri University of Science and Technology); Jinjun Liu (University of Louisville); Mark Marshall (Amherst College); Rebecca Peebles (Eastern Illinois University); Trevor Sears (Brookhaven National Lab); Tim Steimle (Arizona State University); Susanna Widicus Weaver (Emory University). Any questions or suggestions about the Prize should be addressed to the Committee. Anyone willing to serve on a panel of judges should contact Mike Heaven (mheaven@emory.edu).

Information

ACCOMMODATIONS

The check-in for dormitory accommodations is located in Bousfield Hall, 1214 South First Street, opens at noon on Saturday, June 16th at 3:00 PM. The desk will be closed from 2:30 – 6:30 AM on 6/17, 6/19, 6/20, 6/21& 6/22. The desk at Nugent Hall will be staff 24/7. The desk at Wassaja is unstaffed. Hotel information is listed on the ISMS website.

PARKING

Parking permits are for lot E14 (map @ end of book). Purchase parking as part of your check-in at the dorm. You may purchase meter hang-tags for parking near the meeting rooms at the registration desk.

REGISTRATION (SAME AS IN 2017!)

The registration desk is located in the Chemistry Library in Noyes Lab, and is open on Sunday from 4:00-6:00 PM, and Monday through Friday from 8:00 AM-4:30 PM. Refreshments will be available from 8:00 AM-4:30 PM.

CHEMISTRY LIBRARY

The Chemistry Library will be the home for our Registration desk and exhibitor space (plus coffee and donuts) again this year. The library has a few small conference rooms, and comfy chairs (and books!).

READY ROOM/STATION

We will have 2 desks in the Library with computers that you can use to test your powerpoint presentation. If you have any problems, the staff at the "Ready Station" (right next to registration) can assist you.

COMPUTER LAB (VizLab)

Noyes Lab 151 is a small computer lab with Apple computers that is available for your use during the meeting. Please look in your packet for an access code to enter the room. You can also use the PCs in the Chemistry Library. The VizLab will be unavailable from 1:00 – 4:30 PM M-F.

INTERNET ACCESS/Wi-Fi

Each attendee will receive a login and password to access campus WiFi (SSID: IllinoisNet) as a guest. This access should work in most locations through campus. Please read the Internet Acceptable Use Policy below.

AUDIO/VIDEO INFORMATION

Each session room is equipped with a laptop computer, onto which presentation files will be pre-loaded by Symposium staff. To submit your presentation file, you must go to the **Manage Presentations** link on our web site and follow the instructions. All files must be submitted by **11:59 PM CDT THE DAY BEFORE** your presentation session. All submitted files will be loaded onto the presentation computer one half-hour prior to the beginning of the session.

ACKNOWLEDGMENTS

The Symposium Chair wishes to acknowledge the hard work of numerous people who made this meeting possible. First and foremost is the Symposium Coordinator Birgit McCall, who has smoothly and single-handedly taken care of almost all of the electronic and logistical aspects of the meeting. Second are our symposium assistants, Charlie Markus, Philip Kocheril, Amber Rose, and Kori Sye, who have handled innumerable important details to ensure the sessions and exhibitions go well. I wish to acknowledge the hospitality of the Chemistry Department and the School of Chemical Sciences (as well as the School of Molecular and Cell Biology) in tolerating our takeover of their buildings.

DISCLAIMER

The views, opinions, and/or findings contained in this report are those of the authors and should not be construed as an official Department of the Army position, policy, or decision, unless so designated by other documentation.

LIABILITY

The Symposium fees DO NOT include provisions for the insurance of participants against personal injuries, sickness, theft, or property damage. Participants and companions are advised to obtain whatever insurance they consider necessary. The Symposium organizing committee, its sponsors, and individual committee members DO NOT assume any responsibility for loss, injury, sickness, or damages to persons or belongings, however caused. The statements and opinions stated during oral presentations or in written abstracts are solely the author's responsibilities and do not necessarily reflect the opinions of the organizers.

INTERNET ACCEPTABLE USE POLICY

Each attendee will receive a login and password to access campus WiFi (SSID: IllinoisNet) as a guest. Guest accounts are intended to support a broad range of communications. Professional and appropriate etiquette is required. Anonymous access and posting through guest accounts is forbidden. All users must accept that their identity may be associated with any content they provide while using the service. By accessing the campus WiFi network, you expressly acknowledge and agree to the following:

Use of the guest account service is at your sole risk and the entire risk as to satisfactory quality and performance is with you. You agree not to use the guest account intentionally or unintentionally to violate any applicable local, state, national or international law, including, but not limited to, any regulations having the force of law. To the extent not prohibited by law, in no event shall the university be liable for personal injury, or any incidental, special, indirect or consequential damages whatsoever, including, without limitation, damages for loss of profits, loss of data, business interruption or any other commercial damages or losses, arising out of or related to your use or inability to use the guest account, however caused, regardless of the theory of liability (contract, tort or otherwise) and even if the university has been advised of the possibility of such damages. The use of the guest account is subject, but not limited to, all University policies and regulations detailed at the Campus Administrative Manual (http://www.cam.illinois.edu). See the University's Web Privacy Notice (http://www.vpaa.uillinois.edu/policies/web_privacy.cfm) for all applicable laws and policies.

MA. Plenary
Monday, June 18, 2018 – 8:30 AM
Room: Foellinger Auditorium

Chair: Martin Gruebele, University of Illinois at Urbana-Champaign, Urbana, IL, USA

Welcome — 8:30
Susan Martinis, Vice Chancellor for Research
University of Illinois at Urbana-Champaign

MA01 8:40 – 9:20
UNDERSTANDING MOLECULES WITH NEW TOOLS, Jun Ye

MA02 9:25 – 10:05
ULTRAFAST VIBRONIC DYNAMICS OF FUNCTIONAL ORGANIC POLYMER MATERIALS: COHERENCE, CONFINEMENT, AND DISORDER, Irene Burghardt

Intermission

MA03 10:40 – 11:20
ELECTRONIC STRUCTURES OF MIXED METAL SUB-OXIDE CLUSTERS, Caroline Chick Jarrold

MA04 11:25 – 12:05
EXPLORATIONS OF INFRARED SPECTRA OF CRIEGEE INTERMEDIATES AND THEIR REACTIONS, Yuan-Pern Lee

MG. Mini-symposium: Frequency-Comb Spectroscopy
Monday, June 18, 2018 – 1:45 PM
Room: 116 Roger Adams Lab

Chair: Marissa L. Weichman, JILA, Boulder, CO, USA

MG01 *INVITED TALK* 1:45 – 2:15
PRECISION RAMSEY-COMB SPECTROSCOPY OF MOLECULAR HYDROGEN IN THE DEEP-UV, L.S. Dreissen, R.K. Altmann, C. Roth, M.G.J. Favier, J. Krauth, Edcel John Salumbides, Wim Ubachs, K.S.E. Eikema

MG02 2:19 – 2:34
PRECISION MEASUREMENT OF THE IONIZATION ENERGY OF METASTABLE He_2, Paul Jansen, Luca Semeria, Frederic Merkt

MG03 2:36 – 2:51
PRECISION SPECTRA OF $A\,^2\Sigma^+, v' = 0 \leftarrow X\,^2\Pi_{3/2}, v'' = 0, J'' = 3/2$ TRANSITIONS IN ^{16}OH AND ^{16}OD, Arthur Fast, John Furneaux, Samuel Meek

MG04 2:53 – 3:08
DETERMINATION OF ROVIBRATIONAL INTERVALS IN H_2^+ WITH SUB-MHZ ACCURACY, Maximilian Beyer, Nicolas Hoelsch, Frederic Merkt, Christian Jungen

MG05 3:10 – 3:25
DIRECT FREQUENCY-COMB-DRIVEN RAMAN TRANSITIONS IN THE TERAHERTZ RANGE, Cyrille Solaro, Steffen Meyer, Karin Fisher, Michael DePalatis, Michael Drewsen

Intermission

MG06 4:01 – 4:16
RELATIVE INTENSITY OF A CROSSOVER RESONANCE TO LAMB DIPS OBSERVED IN STARK SPECTROSCOPY OF METHANE II, Shoko Okuda, Hiroyuki Sasada

MG07 4:18 – 4:33
ACCURATE COMB-ASSISTED CAVITY RING DOWN SPECTROSCOPY OF MOLECULAR HYDROGEN, Samir Kassi, Tim Stoltmann, Magdalena Konefal, Alain Campargue

MG08 4:35 – 4:50
COMB-REFERENCED COHERENT RAMAN SPECTROSCOPY ON PURE H_2, Marco Lamperti, Lucile Rutkowski, Davide Gatti, Riccardo Gotti, Giulio Cerullo, Dario Polli, Marco Marangoni, Franck Thibault, Piotr Maslowski, Szymon Wojtewicz, Piotr Wcislo

MG09 4:52 – 5:07
SUB-DOPPLER SPECTROSCOPY OF THE ν_2 FUNDAMENTAL BAND AND FIRST HOT BAND OF THE H_3^+ CATION, Charles R. Markus, Philip A. Kocheril, Anne Marie Esposito, Alex W Schrader, Benjamin J. McCall

MG10 5:09 – 5:24
SUB-DOPPLER FREQUENCY METROLOGY IN HD FOR TEST OF FUNDAMENTAL PHYSICS, F.M.J. Cozijn, Edcel John Salumbides, K.S.E. Eikema, Wim Ubachs, Patrick Dupré

MH. Mini-symposium: New Ways of Understanding Molecular Spectra
Monday, June 18, 2018 – 1:45 PM
Room: 100 Noyes Laboratory

Chair: Stephan Schlemmer, I. Physikalisches Institut, Köln, Germany

MH01 *INVITED TALK* 1:45 – 2:15
ALGEBRAIC APPROACHES AND THEIR CONNECTION WITH PHASE SPACE METHODS: APPLICATIONS TO SPECTROSCOPY, Renato Lemus

MH02 2:19 – 2:34
VIBRATIONAL QUANTUM GRAPHS AND THEIR APPLICATION TO THE QUANTUM DYNAMICS OF CH_5^+, Csaba Fábri, Attila Császár

MH03 2:36 – 2:51
A VARIATIONAL METHOD FOR COMPUTING VIBRATIONAL SPECTRA OF MOLECULES WITH UP TO 18 ATOMS, Phillip Thomas, Tucker Carrington

MH04 2:53 – 3:08
A NONDIRECT PRODUCT DISCRETE VARIABLE REPRESENTATION-LIKE METHOD FOR CALCULATING VIBRATIONAL SPECTRA OF POLYATOMIC MOLECULES, Emil J Zak, Tucker Carrington

MH05 3:10 – 3:25
THE $\tilde{A}^2 E''$ STATE OF NO_3: NEW VIBRONIC CALCULATIONS, Bryan Changala, John F. Stanton

Intermission

MH06 4:01 – 4:16
MULTI-CHANNEL QUANTUM DEFECT THEORY CALCULATION OF VIBRATIONAL AUTOIONIZATION RESONANCE WIDTH OF $v=1$, $n^* \approx 14$ CaF RYDBERG STATE, Jun Jiang, Timothy J Barnum, Robert W Field

MH07 4:18 – 4:33
ENERGETICS AND COUPLINGS IN OLIGOACENE-BASED SINGLET FISSION: EFFICIENT AND ACCURATE DENSITY FUNCTIONAL THEORY TELLS THE STORY, Zhou Lin, Hikari Iwasaki, Hong-Zhou Ye, Troy Van Voorhis

MH08 4:35 – 4:50
COMPUTATIONAL SPECTROSCOPY OF NCS IN THE RENNER-DEGENERATE ELECTRONIC STATE $\tilde{X}\,^2\Pi$, Jens Freund, Sarah Caroll Galleguillos Kempf, Per Jensen, Umpei Nagashima, Tsuneo Hirano

MH09 4:52 – 5:07
RO-VIBRATIONALLY AVERAGED STRUCTURE OF $^2\Pi$ NCS: RE-INTERPRETATION OF THE B_0 VALUES, Tsuneo Hirano, Umpei Nagashima, Per Jensen

MH10 5:09 – 5:24
A COLLOCATION-BASED MULTI-CONFIGURATION TIME-DEPENDENT HARTREE METHOD FOR COMPUTING VIBRATIONAL SPECTRA, Robert Wodraszka, Tucker Carrington

MI. Linelists, Lineshapes, Collisions
Monday, June 18, 2018 – 1:45 PM
Room: 1024 Chemistry Annex

Chair: Matthew J. Cich, TOPTICA Photonics Inc., Farmington, NY, USA

MI01   1:45 – 2:00
VELOCITY-CHANGING COLLISIONS IN SUB-DOPPLER AND DOPPLER-BROADENED LINES, Trevor Sears, Gregory Hall, Sylvestre Twagirayezu

MI02   2:02 – 2:17
SUB-DOPPLER SPECTROSCOPY OF THE ν_3 BAND OF METHANE, Philip A. Kocheril, Charles R. Markus, Anne Marie Esposito, Alex W Schrader, Thomas S Dieter, Benjamin J. McCall

MI03   2:19 – 2:34
H_2 BROADENING IN THE ν_3 AND ν_4 BANDS OF CH_4 AT ROOM TEMPERATURE, Ehsan Gharib-Nezhad, Alan N Heays, James R Lyons, Michael R Line, Glenn Stark, Hans A Bechtel

MI04   2:36 – 2:51
COLLISION INDUCED ABSORPTION OF THE $a^1\Delta_g$-$X^3\Sigma^-_g$ BAND OF OXYGEN NEAR 1.27 μM BY CAVITY RING DOWN SPECTROSCOPY, Didier Mondelain, Alain Campargue, Samir Kassi

MI05   2:53 – 3:08
LOW-TEMPERATURE HIGH PRECISION MEASUREMENTS OF LINE MIXING and COLLISIONAL INDUCED ABSORPTION IN THE OXYGEN A-BAND, Erin M. Adkins, Mélanie Ghysels, David A. Long, Joseph T. Hodges

Intermission

MI06   3:44 – 3:59
HIGH PRECISION LINE PARAMETERS OF N_2O NEAR 1.5μm BY CAVITY RING-DOWN SPECTROSCOPY, Gu-Liang Liu, Peng Kang, Tian-Peng Hua, Shui-Ming Hu, An-Wen Liu

MI07   4:01 – 4:16
UPDATES AND CURRENT STATUS OF THE HITRAN APPLICATION PROGRAMMING INTERFACE (HAPI), Roman V Kochanov, Iouli E Gordon, Laurence S. Rothman, Yan Tan, Joshua Karns, Wyatt Matt, Christian Hill

MI08   4:18 – 4:33
NEW MEASUREMENTS OF THE WATER VAPOR ABSORPTION CROSS SECTION IN THE BLUE-VIOLET RANGE BY CAVITY-ENHANCED DIFFERENTIAL OPTICAL ABSORPTION SPECTROSCOPY, Randall Chiu, Oleg L. Polyansky, Rainer Volkamer

MI09   4:35 – 4:50
COMPLETE PHOTOABSORPTION LINELIST FOR CO AND ITS ISOTOPOLOGUES BETWEEN 101 AND 115 NM, Alan Heays, Jean Louis Lemaire, Michele Eidelsberg, Lisseth Gavilan, Glenn Stark, Steven Federman, James R Lyons, Wim Ubachs, Nelson de Oliveira

MI10   4:52 – 5:07
SHIFTS AND BROADENING IN THE CN $A\,^2\Pi - X\,^2\Sigma^+$ (2-0) BAND INDUCED BY ARGON COLLISIONS, James Lockhart, Trevor Sears, Gregory Hall

MI11   5:09 – 5:24
UNUSUAL POWER DEPENDENT PEAK SPLITTING IN REMPI SPECTRUM, Junggil Kim, Jean Sun Lim, Sang Kyu Kim

MJ. Radicals
Monday, June 18, 2018 – 1:45 PM
Room: 217 Noyes Laboratory

Chair: Melanie A.R. Reber, University of Georgia, Athens, GA, USA

MJ01   1:45 – 2:00
AN UPDATED LOOK AT THE INFRARED SPECTRUM OF FULVENALLENE AND FULVENALLENYL, Alaina R. Brown, Joseph T. Brice, Peter R. Franke, Gary E. Douberly

MJ02   2:02 – 2:17
INFRARED SPECTRA OF THE 1,1-DIMETHYLALLYL AND 1,2-DIMETHYLALLYL RADICALS ISOLATED IN SOLID *PARA*-HYDROGEN, Jay C. Amicangelo, Yuan-Pern Lee

MJ03   2:19 – 2:34
MATRIX-ISOLATION FTIR SPECTROSCOPY OF THE 1-BUTYN-3-YL RADICAL, Glenna J. Brown, Martha Ellis, Laura R. McCunn

MJ04   2:36 – 2:51
SUB-DOPPLER INFRARED SPECTROSCOPY OF JET COOLED BENZYl $C_6H_5CH_2$: A CLASSIC RESONANTLY STABILIZED ORGANIC HYDROCARBON RADICAL, Andrew Kortyna, Daniel Lesko, Preston G. Scrape, David Nesbitt

MJ05   2:53 – 3:08
ANALYSIS OF THE $\tilde{A} - \tilde{X}$ BANDS OF THE ETHYNYL RADICAL NEAR 1.48μm AND RE-EVALUATION OF $\tilde{X}$ STATE ENERGIES, Eisen C. Gross, Anh T. Le, Gregory Hall, Trevor Sears

MJ06   *Post-Deadline Abstract*   3:10 – 3:25
DECOMPOSITION OF VIBRONIC AND RENNER-TELLER STRUCTURE IN C_2H AND C_2D FROM ANION HIGH-RESOLUTION PHOTOELECTRON IMAGING, Stephen T Gibson, Benjamin A Laws

MJ07   3:27 – 3:42
OPTICAL DETECTION OF TWO NEW ISOMERS OF C_7H_7 IN A TOLUENE DISCHARGE, Meredith Ward, Jonathan Flores, Sederra D Ross, Damian L Kokkin, Neil J Reilly

Intermission

MJ08   4:18 – 4:33
SUB-DOPPLER INFRARED SPECTROSCOPY OF JET COOLED CH_2Br RADICAL: CH_2 STRETCH VIBRATIONS, Andrew Kortyna, Preston G. Scrape, Daniel Lesko, David Nesbitt

MJ09   4:35 – 4:50
LIF SPECTROSCOPY OF A $^1\Sigma$ SPECIES CONTAINING Si: LINEAR SiOSi ?, Masaru Fukushima, Takashi Ishiwata

MJ10   4:52 – 5:07
THERMAL DECOMPOSITION OF THE LIGNIN MODEL COMPOUNDS: SALICYLALDEHYDE AND CATECHOL, Thomas Ormond, Joshua H Baraban, Jessie P Porterfield, Adam M Scheer, Patrick Hemberger, Tyler Troy, Musahid Ahmed, Mark R Nimlos, David Robichaud, John W Daily, Barney Ellison

MJ11   5:09 – 5:24
CAVITY RING-DOWN SPECTROSCOPY OF 1-, 2- AND 3-METHYL ALLYL PEROXY RADICALS, Md Asmaul Reza, Hamzeh Telfah, Anam C. Paul, Jahangir Alam, Jinjun Liu

MJ12   5:26 – 5:41
OBSERVATION OF THE $\tilde{A} \leftarrow \tilde{X}$ ELECTRONIC TRANSITIONS OF TETRAHYDROPYRANYL AND TETRAHYDROFURANYL PEROXY RADICALS BY ROOM-TEMPERATURE CAVITY RING-DOWN SPECTROSCOPY, Hamzeh Telfah, Md Asmaul Reza, Anam C. Paul, Jinjun Liu

MK. Clusters/Complexes
Monday, June 18, 2018 – 1:45 PM
Room: B102 Chemical and Life Sciences

Chair: Susana Blanco, Universidad de Valladolid, Valladolid, Spain

MK01   1:45–2:00
TEMPERATURE DEPENDENCE OF HYDROGEN-BONDED STRUCTURES OF PHENOL CATION INVESTIGATED BY UV PHOTODISSOCIATION SPECTROSCOPY, Itaru Kurusu, Reona Yagi, Ryota Kato, Hikaru Sato, Masataka Orito, Yasutoshi Kasahara, Haruki Ishikawa

MK02   2:02–2:17
OBSERVATION OF THE IR-INDUCED ISOMERIZATION OF HYDRATED PHENOL CATIONS TRAPPED IN THE COLD ION TRAP, Hikaru Sato, Ryota Kato, Yasutoshi Kasahara, Haruki Ishikawa

MK03   2:19–2:34
FINE AND HYPERFINE STRUCTURE IN THE MID-INFRARED ABSORPTION SPECTRUM OF THE AR-NO COMPLEX, Chuanxi Duan

MK04   2:36–2:51
SPECTRA OF C_6H_6-Rg_n (n=1,2) IN THE 3 MIRCON INFRARED BAND SYSTEM OF BENZENE, A. J. Barclay, Bob McKellar, Nasser Moazzen-Ahmadi

MK05   2:53–3:08
CHARACTERIZATION OF OCS-HCCCH AND N2O-HCCCH DIMERS: THEORY AND EXPERIMENT, A. J. Barclay, Andrea Pietropolli Charmet, K. H. Michaelian, Nasser Moazzen-Ahmadi

Intermission

MK06   3:44–3:59
LASER SPECTROSCOPIC STUDY ON PHENOL-ETHYLDIMETHYLSILANE DIHYDROGEN-BONDED CLUSTER, Masaaki Uchida, Takutoshi Shimizu, Yasutoshi Kasahara, Haruki Ishikawa, Yoshiteru Matsumoto

MK07   4:01–4:16
IR SPECTROSCOPIC STUDY ON PHENOL-TRIETHYLSILANE DIHYDROGEN-BONDED CLUSTER IN THE ELECTRONIC EXCITED STATE, Takutoshi Shimizu, Masaaki Uchida, Yasutoshi Kasahara, Haruki Ishikawa, Yoshiteru Matsumoto

MK08   4:18–4:33
COMPOUND-MODEL MORPHED POTENTIAL FOR THE HYDROGEN BOND HCN–HF, Luis A. Rivera-Rivera, Robert R. Lucchese, John W. Bevan

MK09   4:35–4:50
SPECTRAL SHIFTS AND STRUCTURES OF O-H•••π HYDROGEN BONDED COMPLEXES BETWEEN CARBOXYLIC ACIDS AND BENZENE: A MATRIX ISOLATION INFRARED SPECTROSCOPIC STUDY, Pujarini Banerjee, Indrani Bhattacharya, Deb Pratim Mukhopadhyay, Tapas Chakraborty

MK10   4:52–5:07
INFRARED SPECTROSCOPY OF ALANINE AND ITS WATER CLUSTER ISOLATED IN SOLID PARAHYDROGEN, Brendan Moore, Shin Yi Toh, Ying-Tung Angel Wong, Pavle Djuricanin, Takamasa Momose

ML. Structure determination
Monday, June 18, 2018 – 1:45 PM
Room: 2079 Natural History

Chair: S. A. Cooke, Purchase College SUNY, Purchase, NY, USA

ML01   1:45 – 2:00
PUSHING THE LA-MB-FTMW TO THE LIMIT: CONTROLLED DOUBLE PULSE LASER ABLATION FOURIER TRANSFORM MICROWAVE SPECTROSCOPY, Iker León, Elena R. Alonso, Santiago Mata, José L. Alonso

ML02   2:02 – 2:17
A PICTURE OF THE GLY-PRO SEQUENCE OF COLLAGEN IS CAUGHT IN GAS PHASE, Iker León, Elena R. Alonso, Carlos Cabezas, Santiago Mata, José L. Alonso

ML03   2:19 – 2:34
USING HYPERFINE STRUCTURE TO QUANTIFY THE EFFECTS OF SUBSTITUTION IN 2-, 3-, AND 4-PICOLYLAMINE, Lindsey M McDivitt, Timothy J McMahon, Josiah R Bailey, Caleb B Shiery, Ryan G Bird

ML04   2:36 – 2:51
ELECTRON-WITHDRAWING EFFECTS ON THE MOLECULAR STRUCTURE OF 2- AND 3-NITROBENZONITRILE REVEALED BY BROADBAND MICROWAVE SPECTROSCOPY, Jack B. Graneek, Melanie Schnell

ML05   2:53 – 3:08
ROTATIONAL SPECTRUM OF METHYL PIVALATE, Nobuhiko Kuze, Yoshiyuki Kawashima

ML06   3:10 – 3:25
FTMW SPECTRA OF $CH_3OC(O)NCO$ AND $CH_3OC(O)N_3$, Shinichiro Watanabe, Yoshiyuki Kawashima, Nobuhiko Kuze

ML07   3:27 – 3:42
CARBON-13 STUDIES OF SULFUR-TERMINATED CARBON CHAINS: CHEMICAL BONDING, MOLECULAR STRUCTURES AND FORMATION PATHWAYS, Michael C McCarthy, Kelvin Lee

Intermission

ML08   *Post-Deadline Abstract*   4:18 – 4:33
THE CURIOUS(ER) CASE OF MASS 63: A THRILLING AND CONFUSING SEQUEL, Christopher C Blackstone, Andrei Sanov

ML09   4:35 – 4:50
CONFORMATION-SPECIFIC IR AND UV SPECTROSCOPY OF A MODEL SYNTHETIC FOLDAMER: β^3-ALA TRIPEPTIDE, Dewei Sun, Joshua L. Fischer, Karl N. Blodgett, Timothy S. Zwier

ML10   4:52 – 5:07
THE EFFECTS OF C-TERMINAL FLUOROPHORE CAPS ON THE STRUCTURE OF AMINOISOBUTYRIC ACID DIPEPTIDES, Joshua L. Fischer, Brayan R. Elvir, Karl N. Blodgett, Matthew A. Kubasik, Timothy S. Zwier

ML11   5:09 – 5:24
GAS PHASE INFRARED SPECTROSCOPY OF ISOMERIC BENZYL AND TROPYLIUM CATIONS, J. Philipp Wagner, David C McDonald II, Michael A Duncan

ML12   5:26 – 5:41
NMR, RAMAN AND DFT STUDY OF TAUTOMERIC EQUILIBRIUM AND SLOW SELF-ASSEMBLING IN LYOTROPIC CHROMONIC LIQUID CRYSTAL SSY AQUEOUS SOLUTIONS, Nomeda Rima Valeviciene, Vytautas Balevicius, Kestutis Aidas, Laurynas Dagys, Arunas Marsalka, Kristina Kristinaityte, Valeriy Pogorelov, Yelyzaveta Chernolevska, Yevhenii Vaskivskyi, Iryna Doroshenko, George Pitsevich, Uladzimir Sapeshka

TA. Mini-symposium: Frequency-Comb Spectroscopy
Tuesday, June 19, 2018 – 8:30 AM
Room: 116 Roger Adams Lab

Chair: Frans Harren, Radboud University, Nijmegen, Netherlands

TA01 *INVITED TALK* 8:30 – 9:00
DUAL THZ COMB SPECTROSCOPY, Takeshi Yasui

TA02 9:04 – 9:19
DOPPLER-LIMITED BROADBAND DUAL-COMB SPECTROSCOPY AT 3 μm, Zaijun Chen, Theodor W. Hänsch, Nathalie Picqué

TA03 9:21 – 9:36
MASSIVELY PARALLEL DETECTION OF TRACE MOLECULES AND ISOTOPOLOGUES WITH A SUBHARMONIC MID-IR DUAL COMB SYSTEM, Andrey Muraviev, Viktor O Smolski, Zachary E Loparo, Konstantin L Vodopyanov

TA04 9:38 – 9:53
DUAL-COMB SPECTROSCOPY USING QUANTUM AND INTERBAND CASCADE LASERS, Jonas Westberg, Lukasz A. Sterczewski, Gerard Wysocki

Intermission

TA05 10:29 – 10:44
FREQUENCY-AGILE COMBS: APPLICATIONS IN SPECTROSCOPY AND PHYSICAL METROLOGY, David A. Long, Adam J. Fleisher, Feng Zhou, Yiliang Bao, Jason J Gorman, Thomas W LeBrun, David F. Plusquellic, Joseph T. Hodges

TA06 10:46 – 11:01
DUAL-COMB SPECTROSCOPY USING A DUAL-COMB FIBER LASER BASED ON HYBRID PULSE FORMATION MECHANISMS, Ting Li, Xin Zhao, Zijun Yao, Yuehan Wu, Jie Chen, Zheng Zheng

TA07 *Post-Deadline Abstract* 11:03 – 11:18
DUAL COMB GENERATION FROM A SINGLE FIBER LASER CAVITY VIA SPECTRAL SUBDIVISION, Georg Winkler, Jakob Fellinger, Oliver H Heckl

TA08 11:20 – 11:35
PHASE-CONTROLLED DUAL-COMB COHERENT ANTI-STOKES RAMAN SPECTROSCOPIC IMAGING, Haoyun Wei, Kun Chen, Tao Wu, Yan LI

TA09 11:37 – 11:52
ATTENUATED TOTAL REFLECTANCE SPECTROSCOPY OF LIQUIDS USING A MID-IR DUAL FREQUENCY COMB SPECTROMETER, Daniel I. Herman, Gabriel Ycas, Eleanor Waxman, Fabrizio Giorgetta, Esther Baumann, Nathan R. Newbury, Ian Coddington

TB. Mini-symposium: New Ways of Understanding Molecular Spectra
Tuesday, June 19, 2018 – 8:30 AM
Room: 100 Noyes Laboratory

Chair: Renato Lemus, Universidad Nacional Autonoma de Mexico, Mexico City, CDMX, Mexico

TB01   8:30 – 8:45
NUMERICAL ANALYSIS OF VIBRONIC STRUCTURE OF THE SiCN $\tilde{X}\ ^2\Pi$ SYSTEM, Masaru Fukushima, Takashi Ishiwata

TB02   8:47 – 9:02
ARE LINEAR MOLECULES REALLY LINEAR?   I. THEORETICAL PREDICTIONS, Tsuneo Hirano, Umpei Nagashima, Per Jensen

TB03   9:04 – 9:19
ARE LINEAR MOLECULES REALLY LINEAR?   II. RE-INTERPRETATION OF EXPERIMENTAL B_0-VALUES., Tsuneo Hirano, Umpei Nagashima, Per Jensen

TB04   9:21 – 9:36
DIPOLE MOMENTS OF LINEAR MOLECULES: A COMPUTATIONAL MOLECULAR SPECTROSCOPY STUDY, Hui Li, Per Jensen, Tsuneo Hirano

Intermission

TB05   10:12 – 10:27
THE JAHN-TELLER EFFECT AS A TREATMENT OF MOLECULAR ANHARMONICITY, David S. Perry, Bishnu P. Thapaliya, Mahesh B. Dawadi

TB06   10:29 – 10:44
COMPARING EXPERIMENTAL AND CALCULATED SPECTRAL PARAMETERS FOR JAHN-TELLER ACTIVE MOLECULES. PART I, Ketan Sharma, Scott M. Garner, Terry A. Miller, John F. Stanton

TB07   10:46 – 11:01
COMPARING EXPERIMENTAL AND CALCULATED SPECTRAL PARAMETERS FOR JAHN-TELLER ACTIVE MOLECULES: PART II, Scott M. Garner, Ketan Sharma, Terry A. Miller, John F. Stanton

TB08   11:03 – 11:18
UNDERSTANDING QUANTUM YIELDS IN NAPHTHALENES AND BORON-DIPYRROMETHENES: TOWARDS A PREDICTION OF NON-RADIATIVE DECAY PATHWAYS IN ORGANIC OPTOELECTRONIC MATERIALS, Zhou Lin, Alexander W. Kohn, Troy Van Voorhis

TC. Chirality and stereochemistry
Tuesday, June 19, 2018 – 8:30 AM
Room: 1024 Chemistry Annex

Chair: Sonia Melandri, University of Bologna, Bologna, Italy

TC01   8:30 – 8:45
REACTION FLASK ANALYSIS OF THE ASYMMETRIC HYDROGENATION OF ARTEMISINIC ACID, Reilly E. Sonstrom, Justin L. Neill, B Frank Gupton, Yuan Yang, Luca Evangelisti, Brooks Pate

TC02   8:47 – 9:02
FAST CHIRAL MONITORING IN A CONTINUOUS PHARMACEUTICAL SYNTHESIS BY MOLECULAR ROTATIONAL RESONANCE SPECTROSCOPY, Justin L. Neill, Matt Muckle, Brooks Pate, Yuan Yang, B Frank Gupton

TC03   9:04 – 9:19
MICROWAVE SPECTRUM AND MOLECULAR STRUCTURE OF THE CHIRAL TAGGING CANDIDATE, 3,3-DIFLUORO-1,2-EPOXYPROPANE, AND ITS COMPLEX WITH THE ARGON ATOM, Helen O. Leung, Mark D. Marshall

TC04   9:21 – 9:36
EFFECTS OF CHIRALITY IN HOMODIMERS OF 3,3,3-TRIFLUORO-1,2-EPOXYPROPANE, Mark D. Marshall, Helen O. Leung, Nathan A. Seifert, Yunjie Xu, Wolfgang Jäger

TC05   9:38 – 9:53
CHIRAL ANALYSIS OF BIOLOGICALLY RELEVANT SAMPLES USING BROADBAND ROTATIONAL SPECTROSCOPY, María Mar Quesada-Moreno, Anna Krin, Melanie Schnell

TC06   9:55 – 10:10
ENANTIOMERIC EXCESS MEASUREMENTS OF ISOPULEGOL USING CHIRAL TAG SPECTROSCOPY, Kevin J Mayer, Caitlin Embly, Brooks Pate, Luca Evangelisti

Intermission

TC07   10:46 – 11:01
DETERMINATION OF ABSOLUTE CONFIGURATION IN CEDROL USING CHIRAL TAG ROTATIONAL SPECTROSCOPY, Luca Evangelisti, Brooks Pate, Taylor Smart, Channing West, Martin S. Holdren, Kevin J Mayer

TC08   11:03 – 11:18
ENANTIOMERIC EXCESS MEASUREMENTS USING MICROWAVE THREE-WAVE MIXING, Martin S. Holdren, Brooks Pate, Caitlin Embly, Arthur Wu, Kevin J Mayer, James Dittman, Patrick Buoniconti, Golara Haghtalab, Brianna Mitchell, Luca Evangelisti

TC09   11:20 – 11:35
CP-FTMW SPECTROSCOPY OF THE LOW ENERGY CONFORMERS OF TWO CHIRAL ALCOHOLS: MYRTENOL AND NOPOL, Galen Sedo, Frank E Marshall, G. S. Grubbs II

TC10   11:37 – 11:52
A COMPARATIVE STUDY OF CHIRAL ANALYSIS OF FENCHYL ALCOHOL USING NUCLEAR MAGNETIC RESONANCE, INFRARED, AND ROTATIONAL SPECTROSCOPY, Kevin J Mayer, Supraja Chittari, Alysa Modi, Eric Odermatt, Charles Spivey, Julian Stashower, Brooks Pate

TC11   11:54 – 12:09
DESIGNING CHIRAL TAGS TO IMPROVE ABSOLUTE CONFIGURATION DETERMINATION BY ROTATIONAL SPECTROSCOPY, Luca Evangelisti, Channing West, Mark D. Marshall, Helen O. Leung, Brooks Pate

TD. Dynamics and kinetics
Tuesday, June 19, 2018 – 8:30 AM
Room: 217 Noyes Laboratory

Chair: James Lockhart, Brookhaven National Laboratory, Long Island, NY, USA

TD01   8:30 – 8:45
STATE-TO-STATE ROTATIONAL RATE COEFFICIENTS FOR AMMONIA SELF COLLISIONS FROM PUMP-PROBE CHIRPED-PULSE EXPERIMENTS, Christian Endres, Paola Caselli, Stephan Schlemmer

TD02   8:47 – 9:02
WAVEGUIDE CP-FTMW SPECTROSCOPY OF ETHYL CYANOFORMATE: EXPLORING THE POSSIBILITY OF SEEING EXCHANGE-AVERAGED STATES, Steven Shipman, J. H. Westerfield

TD03   9:04 – 9:19
INFRARED SPECTROSCOPIC STUDIES OF ORTHO-PARA CONVERSION IN SOLID HYDROGEN CATALYZED BY HYDROGEN ATOMS, David T. Anderson, Morgan E. Balabanoff, Aaron I. Strom

TD04   9:21 – 9:36
INFRARED EMISSION FROM UV-IRRADIATED MIXTURES OF CH_2I_2 AND O_2 PROBED WITH A STEP-SCAN FTIR SPECTROMETER, Ting-Yu Chen, Yuan-Pern Lee

TD05   9:38 – 9:53
ULTRAFAST DYNAMICS OF DIBROMOCYCLOALKANES, Darya S. Budkina, Kanykey E. Karabaeva, R. Marshall Wilson, Alexander N Tarnovsky

Intermission

TD06   10:29 – 10:44
BEYOND VPTn-SCTST FOR CHEMICAL KINETICS: COMPLEX SCALING AND CURVILINEAR VIBRATIONAL SELF-CONSISTENT FIELD THEORY, Bryan Changala, John F. Stanton

TD07   10:46 – 11:01
MCTDH ROVIBRATIONAL STATES AND STATE-TO-STATE INELASTIC SCATTERING CALCULATIONS ON THE H_2O-H_2 SYSTEM, Steve Alexandre Ndengue, Richard Dawes, Yohann Scribano, Fabien Gatti

TD08   11:03 – 11:18
RELAXATION OF VIBRATIONAL, ROTATIONAL, AND CORIOLIS ENERGIES OF AN EXCITED NITROMETHANE MOLECULE IN ARGON BATH, Luis A. Rivera-Rivera, Albert F. Wagner

TD09   11:20 – 11:35
CONFORMATIONAL DYNAMICS OF THE CYTOCHROME P450CAM-PUTIDAREDOXIN COMPLEX PROBED VIA 2D IR SPECTROSCOPY, Sashary Ramos, Edward Basom, Megan Thielges

TD10   11:37 – 11:52
RESOLVING ULTRAFAST PHOTOCHEMISTRY OF COORDINATION COMPLEXES USING HIGH HARMONIC GENERATION XANES SPECTROSCOPY, Elizabeth S Ryland, Josh Vura-Weis, Kaili Zhang, Muffaddal Burhani, Max A Verkamp, Ming-Fu Lin, Kristin Benke, Michaela Carlson

TD11   11:54 – 12:09
FEMTOSECOND EXTREME ULTRAVIOLET SPECTROSCOPY OF SEMICONDUCTOR CARRIER DYNAMICS, Josh Vura-Weis, Max A Verkamp

TE. Clusters/Complexes
Tuesday, June 19, 2018 – 8:30 AM
Room: B102 Chemical and Life Sciences

Chair: Juan Carlos Lopez, Universidad de Valladolid, Valladolid, Spain

TE01   8:30 – 8:45
INFRARED PHOTODISSOCIATION SPECTROSCOPIC AND THEORETICAL STUDY OF BORON CARBONYL COMPLEXES, Jiaye Jin, Guanjun Wang, Mingfei Zhou

TE02   8:47 – 9:02
WITHDRAWN

TE03   9:04 – 9:19
SPIN-ORBIT STATE-SELECTIVE AUTODETACHMENT OF VIBRATIONALLY EXCITED CCP^-, G. Stephen Kocheril, Joseph Czekner, Ling Fung Cheung, Lai-Sheng Wang

TE04   9:21 – 9:36
INFRARED SPECTRA OF C_2H_4 DIMER AND TRIMER, A. J. Barclay, Koorosh Esteki, Bob McKellar, Nasser Moazzen-Ahmadi

TE05   9:38 – 9:53
GAS PHASE SPECTROSCOPY ON HYDRATED CLUSTERS OF OXYTOCIN BY ELECTROSPRAY IONIZATION / COLD ION TRAP TECHNIQUE, Mizuki Tabata, Masaaki Fujii

TE06   9:55 – 10:10
BLUE SHIFTED HYDROGEN BOND IN $CH/D_3CN\ldots HCCL_3$ COMPLEXES, Bedabyas Behera, Puspendu Kumar Das

Intermission

TE07   10:46 – 11:01
BORAZINE AND BENZENE: CHALK AND CHEESE. MATRIX ISOLATION INFRARED AND AB INITIO STUDIES., Kanupriya Verma, K S Viswanathan

TE08   11:03 – 11:18
INTERROGATION OF $MoO_yC_nH_n^-$ CHEMIFRAGMENTS ILLUMINATES $Mo-(\eta^2-ACETYLENE)$ INTERACTIONS WITHIN $Mo_xO_y^-$ AND ETHYLENE REACTIONS, Josey E Topolski, Richard N Schaugaard, Manisha Ray, Krishnan Raghavachari, Caroline Chick Jarrold

TE09   11:20 – 11:35
PROPYLBENZENE-$(H_2O)_n$ CLUSTERS: EFFECT OF THE ALKYL CHAIN ON THE π H-BOND, Piyush Mishra, Joshua L. Fischer, Edwin Sibert, Timothy S. Zwier

TE10   11:37 – 11:52
WEAKLY-BOUND COMPLEXES OF FURAN AND WATER AS INVESTIGATED BY MATRIX ISOLATION FTIR, Tyler G. Fuller, Schuyler P Lockwood, Josh Newby

TE11   11:54 – 12:09
CHARACTERIZATION OF A HYDROGEN PEROXIDE-BENZENE COMPLEX USING MATRIX ISOLATION INFRARED SPECTROSCOPY, Jay C. Amicangelo, Yudhishtara Payagala, Dylan Johnson, Catherine Kaiser

TF. Atmospheric science

Tuesday, June 19, 2018 – 8:30 AM
Room: 2079 Natural History

Chair: Peter F. Bernath, Old Dominion University, Norfolk, VA, USA

TF01   8:30 – 8:45
SPECTROSCOPIC DATABASES FOR THE VAMDC PORTAL: NEW TOOLS AND IMPROVEMENTS, Cyril Richard, Vincent Boudon, Nicolas Moreau, Marie-Lise Dubernet

TF02   8:47 – 9:02
GLOBAL ANALYSES OF SF_6 HIGH-RESOLUTION SPECTRA FOR THE VAMDC/SHeCaSDa DATABASE, Vincent Boudon, Hanzhang Ke, Cyril Richard, Mbaye Faye, Laurent Manceron

TF03   9:04 – 9:19
NEW DATA AND ANALYSIS FOR SF_6 ABSORPTION MODELLING IN THE 10 MICRON ATMOSPHERIC WINDOW, Mbaye Faye, Vincent Boudon, Michel Loete, Cyril Richard, P. Roy, Laurent Manceron

TF04   9:21 – 9:36
TRENDS IN ATMOSPHERIC HCL, HFC – 134A AND CHF_3 FROM THE ACE SATELLITE MISSION , Anton Madushanka Fernando, Peter F. Bernath, Chris Boone

Intermission

TF05   10:12 – 10:27
QUANTITATIVE INFRARED SPECTROSCOPY OF HALOGENATED SPECIES FOR ATMOSPHERIC REMOTE SENSING, Jeremy J. Harrison

TF06   10:29 – 10:44
ACCURATE LABORATORY DETERMINATION OF THE MID AND SHORT WAVE INFRARED WATER VAPOR SELF-CONTINUUM. NEW MEASUREMENTS AND TEST OF THE MT_CKD MODEL, Didier Mondelain, Alain Campargue, Roberto Grilli, Samir Kassi, Loïc Lechevallier, Lucile Richard, Daniele Romanini, Semyon Vasilchenko, Irene Ventrillard

TF07   10:46 – 11:01
A HIGHLY ACCURATE *AB INITIO* DIPOLE MOMENT SURFACE FOR WATER: TRANSITIONS EXTENDING INTO THE ULTRAVIOLET , Eamon K Conway, Aleksandra A. Kyuberis, Oleg L. Polyansky, Jonathan Tennyson

TF08   11:03 – 11:18
HIGH-RESOLUTION INFRARED SPECTROSCOPY OF ISOPRENE AND METHYL VINYL KETONE IN THE 10 μm REGION, Michael Cyrus Iranpour, Minh Nhat Tran, Marcus Vinicus Pereira, Tyler Hoadley, Jacob Stewart

TG. Mini-symposium: Frequency-Comb Spectroscopy
Tuesday, June 19, 2018 – 1:45 PM
Room: 116 Roger Adams Lab

Chair: Hairun Guo, École polytechnique fédérale de Lausanne, Lausanne, Switzerland

TG01 *INVITED TALK* 1:45 – 2:15
EXTENDING FREQUENCY COMB SPECTROSCOPY TO THE MID AND FAR INFRARED RANGE, Paolo De Natale

TG02 2:19 – 2:34
HIGH-PRECISION MID-IR MOLECULAR SPECTROSCOPY WITH TRACEABILITY TO PRIMARY FREQUENCY STANDARDS USING SUB-Hz FREQUENCY COMB-STABILIZED QCLS, Dang Bao An Tran, Rosa Santagata, Olivier Lopez, Sean Tokunaga, Michel Abgrall, Yann Le Coq, Rodolphe Le Targat, Won-Kyu Lee, Dan Xu, Paul-Eric Pottie, Anne Amy-Klein, Benoit Darquie

TG03 2:36 – 2:51
SINGLE-SHOT SUB-MICROSECOND SPECTROSCOPY OF THE BACTERIORHODOPSIN PHOTOCYCLE WITH QUANTUM CASCADE LASER FREQUENCY COMBS, Markus Mangold, Tilman Kottke, Jessica Klocke, Andreas Hugi, Pitt Allmendinger, Jerome Faist

TG04 2:53 – 3:08
WIDELY TUNABLE UV/VIS CAVITY-ENHANCED ULTRAFAST SPECTROSCOPY AND EXCITED STATE PROTON TRANSFER IN JET-COOLED MOLECULES AND CLUSTERS, Yuning Chen, Myles C Silfies, Thomas K Allison

Intermission

TG05 3:44 – 3:59
WIDE-BANDWIDTH COMB-ASSISTED SPECTROSCOPY IN THE FINGERPRINT REGION AND APPLICATION TO THE ν_1 FUNDAMENTAL BAND OF $^{14}N_2^{16}O$, Bidoor Alsaif, Marco Lamperti, Davide Gatti, Paolo Laporta, Martin Fermann, Aamir Farooq, Oleg Lyulin, Alain Campargue, Marco Marangoni

TG06 4:01 – 4:16
OPTICAL FEEDBACK STABILIZED LASER CAVITY RING DOWN SPECTROSCOPY: FROM SATURATED SPECTROSCOPY TO ISOTOPIC RATIO., Samir Kassi, Tim Stoltmann, Mathieu Daëron, Mathieu Casado, Amaelle Landais, Alain Campargue

TG07 4:18 – 4:33
FEED-FORWARD COHERENT LINK FROM A COMB TO A DIODE LASER : APPLICATION TO SATURATED CAVITY RING DOWN SPECTROSCOPY, Riccardo Gotti, Marco Prevedelli, Samir Kassi, Marco Marangoni, Daniele Romanini

TG08 4:35 – 4:50
BROADBAND CALIBRATION-FREE COMPLEX REFRACTIVE INDEX SPECTROSCOPY IN A CAVITY USING A COMB-BASED FOURIER TRANSFORM SPECTROMETER, Alexandra C Johanssson, Lucile Rutkowski, Anna Filipsson, Thomas Hausmaninger, Gang Zhao, Ove Axner, Aleksandra Foltynowicz

TG09 4:52 – 5:07
FOURIER-TRANSFORM COMPLEX REFRACTIVE INDEX SPECTROSCOPY AT Hz-LEVEL WITH OPTICAL FREQUENCY COMBS , Dominik Charczun, Grzegorz Kowzan, Akiko Nishiyama, Agata Cygan, Daniel Lisak, Ryszard S. Trawiński, Piotr Maslowski, Michael Debus, Philipp Huke, Dorota Tomaszewska, Grzegorz Soboń

TG10 5:09 – 5:24
BROADBAND CAVITY-ENHANCED MOLECULAR ABSORPTION AND DISPERSION SPECTROSCOPY WITH A FREQUENCY COMB-BASED VIPA SPECTROMETER, Grzegorz Kowzan, Dominik Charczun, Agata Cygan, Ryszard S. Trawiński, Daniel Lisak, Piotr Maslowski

TH. Mini-symposium: New Ways of Understanding Molecular Spectra
Tuesday, June 19, 2018 – 1:45 PM
Room: 100 Noyes Laboratory

Chair: Per Jensen, University of Wuppertal, Wuppertal, Germany

TH01 1:45 – 2:00
AUTOMATED ASSIGNMENT OF ROTATIONAL SPECTRA USING ARTIFICIAL NEURAL NETWORKS, Daniel P. Zaleski, Kirill Prozument

TH02 2:02 – 2:17
NEW APPROACHES TO DECODING ROTATIONAL SPECTRA: APPLICATIONS TO FLUOROETHYLENE MICRO-SOLVATION BY CO_2, Rebecca A. Peebles, Prashansa Kannangara, Sean A. Peebles, Brooks Pate

TH03 2:19 – 2:34
ATTEMPTS TO SOLVE O_2-CONTAINING VAN DER WAALS INTERACTIONS USING SPFIT AND SPCAT WITH MICROWAVE MEASUREMENT PRECISION: PROBLEMS, PITFALLS, AND SUCCESSES., Frank E Marshall, Nicole Moon, Amanda Jo Duerden, G. S. Grubbs II

TH04 2:36 – 2:51
STEPPING ACROSS THE DISSOCIATION THRESHOLD OF THE $I^-\cdot(H_2O)$ COMPLEX: RESONANCE ENHANCED TWO-COLOR IR-IR PHOTODISSOCIATION (R2PD), Nan Yang, Chinh H. Duong, Patrick J Kelleher, Justin J Talbot, Ryan P Steele, Mark Johnson

TH05 2:53 – 3:08
DECIPHERING THE EXCITED-STATE VIBRATIONAL SIGNATURES OF THE WATER-IODIDE BINARY COMPLEX THROUGH QUANTUM SIMULATIONS., Justin J Talbot, Ryan P Steele, Nan Yang, Chinh H. Duong, Patrick J Kelleher, Mark Johnson

TH06 3:10 – 3:25
FROM LINES TO STATES WITHOUT A MODEL, Stefan Brackertz, Stephan Schlemmer, Oskar Asvany

Intermission

TH07 *Journal of Molecular Spectroscopy Review Lecture* 3:58 – 4:31
THE ANALYSIS OF COMPLEX CHEMICAL MIXTURES BY BROADBAND ROTATIONAL SPECTROSCOPY, Brooks Pate, Justin L. Neill

TH08 4:35 – 4:50
INFRARED PHOTODISSOCIATION SPECTROSCOPY OF THE EXOTIC H_6^+ CATION IN THE GAS PHASE, David C McDonald II, J. Philipp Wagner, Michael A Duncan

TH09 4:52 – 5:07
SURFACE AND SPECTROSCOPIC PROPERTIES OF 1,8-DIAZAFLUOREN-9-ONE IN TITANIUM DIOXIDE THIN FILMS, Aneta Lewkowicz, Anna Synak, Michał Mońka, Piotr Bojarski, Karol Szczodrowski, Robert Bogdanowicz, Jakub Karczewski

TH10 5:09 – 5:24
DETERMINATION OF GLYCOL CONTAMINATION IN ENGINE OIL BY INFRARED AND UV-VIS SPECTROSCOPY, Torrey E. Holland, Robinson Karunanithy, Ali Mazin Abdul-Munaim, P Sivakumar, Dennis G. Watson

TI. Instrument/Technique Demonstration

Tuesday, June 19, 2018 – 1:45 PM
Room: 1024 Chemistry Annex

Chair: Elangannan Arunan, Indian Institute of Science, Bangalore, India

TI01 1:45 – 2:00
SUB-NANOMETER IMAGING OF ELECTRONICALLY EXCITED QUANTUM DOTS: STARK EFFECT, ORIENTATION DEPENDENCE AND ENERGY TRANSFER, Duc Nguyen, Martin Gruebele, Joseph Lyding

TI02 2:02 – 2:17
LASER ABLATION-RESONANCE ENHANCED PHOTOIONIZATION MASS SPECTROMETRY (LA-REPMS) OF PARTICLE-BASED ASSAYS TO IMPROVE EARLY DETECTION OF CANCER, Christopher Mandrell, Jessica C Jurak, P Sivakumar

TI03 2:19 – 2:34
LED-CAVITY ENHANCED ABSORPTION SPECTROSCOPY FOR SENSING THE ATMOSPHERE, Hongming Yi, Tao Wu, Eirc Fertein, Cécile Coeur, Weixiong Zhao, Guishi Wang, Xiaoming Gao, Weijun Zhang, Weidong Chen

TI04 2:36 – 2:51
COMPARISON OF CAVITY ENHANCED FARADAY ROTATION SPECTROSCOPY TECHNIQUES , Link Patrick, Jonas Westberg, Gerard Wysocki

TI05 2:53 – 3:08
APPLICATION OF COHERENT ANTI-STOKES RAMAN SCATTERING THERMOMETRY IN TURBULENT AND LAMINAR FLAMES, Aman Satija, Ziqiao Chang, Dong Han, Albyn Lowe, Levi Michael Thomas, Jay P Gore, Assaad R Masri, Robert P. Lucht

TI06 3:10 – 3:25
LASER-BASED MOLECULAR SPECTROSCOPY FOR MONITORING EMISSION IN ANIMAL FARMING, Michal Nikodem, Dorota Stachowiak

TI07 3:27 – 3:42
HIGH-RESOLUTION LINEAR SPECTROSCOPY ON A MICROMETRIC LAYER OF MOLECULAR VAPOR, Junior Lukusa Mudiayi, Benoit Darquie, Isabelle Maurin, Sean Tokunaga, Alexander Shelkovnikov, Jose Roberto Rios Leite, Daniel Bloch, Athanasios Laliotis

Intermission

TI08 4:18 – 4:33
DEVELOPMENT OF A HYBRID LASER-MASS SPECTROMETER WITH TWO INSTRUMENT ARMS: IRMPD AND HENDI COLD ION SPECTROSCOPIC EXPERIMENTS , Matthias Heger, Joseph Cheramy, Fan Xie, Zhihao Chen, Haolu Wang, Wolfgang Jäger, Yunjie Xu

TI09 4:35 – 4:50
AC STARK EFFECT OBSERVED IN A MICROWAVE-(SUB)MILLIMETERWAVE DOUBLE RESONANCE EXPERIMENT , Kevin Roenitz, Brian M Hays, Carson Reed Powers, Morgan N McCabe, Houston H Smith, Susanna L. Widicus Weaver, Steven Shipman

TI10 4:52 – 5:07
A USB - TO - W-BAND TRANSMITTER: MILLIMETER-WAVE MOLECULAR SPECTROSCOPY WITH CMOS TECHNOLOGY, Deacon J Nemchick, Brian Drouin, Adrian Tang, Yanghyo Kim, Gabriel Virbila, M.-C. Frank Chang

TI11 5:09 – 5:24
THE CONFORMER SPECIFIC ROOM-TEMPERATURE ROTATIONAL SPECTRUM OF ALLYL CHLORIDE UTILIZING STRONG FIELD COHERENCE BREAKING, Erika Riffe, Erika Johnson, Steven Shipman, Sean Fritz, Alicia O. Hernandez-Castillo, Timothy S. Zwier

TI12 5:26 – 5:41
A 6–18 GHZ DIRECT DIGITAL SYNTHESIS TUNABLE SEGMENTED CHIRPED PULSE FOURIER TRANSFORM MICROWAVE SPECTROMETER, Haley N. Scolati, Sommer L. Johansen, Anna L Pischer, Kyle N. Crabtree

TJ. Rotational structure/frequencies
Tuesday, June 19, 2018 – 1:45 PM
Room: 217 Noyes Laboratory

Chair: Ranil Gurusinghe, West-Ward Pharmaceuticals, Bedford, OH, USA

TJ01   1:45 – 2:00
HIGH SENSITIVITY CRDS OF CO_2 IN THE 1.74 μM TRANSPARENCY WINDOW. A VALIDATION TEST FOR THE SPECTROSCOPIC DATABASES, Peter Čermák, Ekaterina Karlovets, Didier Mondelain, Samir Kassi, Valery Perevalov, Alain Campargue

TJ02   2:02 – 2:17
LINE POSITIONS AND INTENSITIES FOR THE ν_3 BAND OF 5 ISOTOPOLOGUES OF GERMANE FOR PLANETARY APPLICATIONS, Vincent Boudon, Tigran Grigoryan, Florian Philipot, Cyril Richard, F. Kwabia Tchana, Laurent Manceron, Athena Rizopoulos, Jean Vander Auwera, Thérèse Encrenaz

TJ03   2:19 – 2:34
INVESTIGATION OF THIOKETENE ISOMERS: MICROWAVE SPECTROSCOPY AND FORMATION CHEMISTRY OF HCCSH, Kelvin Lee, Marie-Aline Martin-Drumel, Valerio Lattanzi, Brett A. McGuire, Michael C McCarthy

TJ04   2:36 – 2:51
THE MILLIMETER/SUBMILLIMETER-WAVE SPECTRUM OF F_2SO ($\tilde{X}^1A'$), John P Keogh, DeWayne T Halfen, Lucy M. Ziurys

TJ05   2:53 – 3:08
FLUORINATION EFFECT ON HYDROGEN BOND TOPOLOGIES IN WATER ADDUCTS OF FLUOROPYRIDINES, Juan Wang, Xiaolong Li, Gang Feng, Qian Gou

Intermission

TJ06   3:44 – 3:59
TIPPING THE BALANCE BETWEEN ELECTROSTATICS AND STERIC EFFECTS: THE MICROWAVE SPECTRA AND MOLECULAR STRUCTURES OF 2-CHLORO-1,1-DIFLUOROETHYLENE–ACETYLENE AND *CIS*-1,2-DIFLUOROETHYLENE–ACETYLENE, Helen O. Leung, Mark D. Marshall

TJ07   4:01 – 4:16
MICROWAVE SPECTRUM AND MOLECULAR STRUCTURE OF 2,3,3,3-TETRAFLUOROPROPENE–HYDROGEN CHLORIDE, Helen O. Leung, Mark D. Marshall, Miles A. Wronkovich

TJ08   4:18 – 4:33
THE CONFORMATIONS OF PROTEINOGENIC AMINO ACID GLUTAMINE: MORE ACCURACY IS URGENTLY NEEDED IN THEORETICAL CALCULATIONS, Iker León, Elena R. Alonso, Carlos Cabezas, Santiago Mata, José L. Alonso

TJ09   4:35 – 4:50
SULFUR HYDROGEN BONDING: A COMPARISON OF THE DIMERS AND MONOHYDRATES OF THENYL AND FURFURYL ALCOHOLS AND MERCAPTANS, Marcos Juanes, Rizalina Tama Saragi, Alberto Lesarri, Ruth Pinacho, José Emiliano Rubio, Lourdes Enriquez, Martin Jaraiz

TJ10   *Post-Deadline Abstract*   4:52 – 5:07
STRUCTURE DETERMINATION OF 5 MEMBERED SILANE RINGS USING MICROWAVE SPECTROSCOPY, Frank E Marshall, Amanda Jo Duerden, Nicole Moon, David Joseph Gillcrist, Ivan Sedlacek, Grier Jones, Theodore Carrigan-Broda, Gamil A Guirgis, G. S. Grubbs II

TK. Large amplitude motions, internal rotation
Tuesday, June 19, 2018 – 1:45 PM
Room: B102 Chemical and Life Sciences

Chair: Jens-Uwe Grabow, Gottfried-Wilhelm-Leibniz-Universität, Hannover, NI, Germany

TK01   1:45–2:00
THE THERMAL SELF-POLYMERIZATION OF METHYL METHACRYLATE — ROTATIONAL CHARACTERIZATION OF THE METHYL METHACRYLATE DIMER (IT'S NOT A COMPLEX!), Sven Herbers, Daniel A. Obenchain, Kevin G. Lengsfeld, Henning Kuper, Jens-Uwe Grabow, Jörg August Becker

TK02   2:02–2:17
TOWARDS THE DETECTION OF EXPLOSIVE TAGGANTS: MICROWAVE AND MILLIMETER-WAVE GAS PHASE SPECTROSCOPIES OF 3-NITROTOLUENE, Anthony Roucou, Isabelle Kleiner, Manuel Goubet, Sabath Bteich, Gaël Mouret, Robin Bocquet, Francis Hindle, W. Leo Meerts, Arnaud Cuisset

TK03   2:19–2:34
INTERNAL ROTATION ANALYSIS OF THE FTMW AND MILLIMETER WAVE SPECTRA OF FLUORAL (CF_3CHO), Celina Bermúdez, R. A. Motiyenko, L. Margulès, Carlos Cabezas, Yasuki Endo, J.-C. Guillemin

TK04   2:36–2:51
GLOBAL FIT OF O-FLUOROTOLUENE TORSIONAL STATES FROM WAVEGUIDE CP-FTMW SPECTROSCOPY, J. H. Westerfield, Steven Shipman

TK05   2:53–3:08
SPECTROSCOPIC CHARACTERIZATION OF THE ELUSIVE *GAUCHE*-ISOPRENE BY HIGH RESOLUTION MICROWAVE SPECTROSCOPY, Jessie P Porterfield, J. H. Westerfield, Bryan Changala, Thanh Lam Nguyen, Joshua H Baraban, Steven Shipman, Michael C McCarthy

TK06   3:10–3:25
HYPERFINE SPLITTINGS OF METHANOL IN THE FIRST EXCITED TORSIONAL STATE, Li-Hong Xu, Jon T. Hougen, G Yu Golubiatnikov, Sergey Belov, Alexander Lapinov, E. A. Alekseev, Igor Krapivin, L. Margulès, R. A. Motiyenko, Stephane Bailleux

TK07   3:27–3:42
FURTHER PROGRESS IN FITTING 13000 TORSION-WAGGING-ROTATIONAL MW AND IR $v_t = 0,1$ TRANSITIONS IN CH_3NH_2 USING THE HYBRID (TUNNELLING + INTERNAL ROTATION) PROGRAM, Isabelle Kleiner, Jon T. Hougen

Intermission

TK08   4:18–4:33
A NEW MULTI-STATE VIBRATION-TORSION-ROTATION FITTING PROGRAM FOR MOLECULES WITH A C_{3v} TOP AND C_s FRAME: APPLICATION TO THE ν_{10} BAND OF ACETALDEHYDE, V. Ilyushin, E. A. Alekseev, Olga Dorovskaya, L. Margulès, R. A. Motiyenko, Manuel Goubet, Olivier Pirali, Sigurd Bauerecker, Christof Maul, Christian Sydow, Georg Ch. Mellau, Isabelle Kleiner, Jon T. Hougen

TK09   4:35–4:50
FIRST RESULTS FOR ETHYLPHOSPHINE, $CH_3CH_2PH_2$, FROM AN EFFECTIVE ROTATIONAL HAMILTONIAN FOR TWO-ROTOR SYSTEMS WITH SYMMETRIC AND ASYMMETRIC INTERNAL ROTORS (LIKE ETHANOL), Peter Groner

TK10   4:52–5:07
TORSIONAL SPLITTING AND FOUR-FOLD BARRIER TO INTERNAL ROTATION: THE ROTATIONAL SPECTRA OF VINYLSULFUR PENTAFLUORIDE, W. Orellana, Susanna L. Stephens, Wallace C. Pringle, Stewart E. Novick, Peter Groner, S. A. Cooke

TK11   5:09–5:24
FURTHER STUDIES OF A FOUR-FOLD BARRIER TO INTERNAL ROTATION: THE ROTATIONAL SPECTRA OF PROPEN-1-YLSULFUR PENTAFLUORIDE AND BUTEN-1-YLSULFUR PENTAFLUORIDE, W. Orellana, Susanna L. Stephens, Stewart E. Novick, S. A. Cooke

TK12   5:26–5:41
THE ROTATIONAL STUDY OF THE VITAMINE B6 FORM PYRIDOXINE, Elena R. Alonso, Iker León, José L. Alonso

TL. Astronomy
Tuesday, June 19, 2018 – 1:45 PM
Room: 2079 Natural History

Chair: Jay A Kroll, University of Colorado at Boulder, Boulder, CO, USA

TL01 1:45 – 2:00
THE ROTATIONAL SPECTRUM OF PROTONATED ETHYL CYANIDE, Harshal Gupta, Kelvin Lee, Sven Thorwirth, Oskar Asvany, Stephan Schlemmer, Michael C McCarthy

TL02 2:02 – 2:17
TOWARDS LABORATORY LINE LISTS TO SEARCH FOR CH_3OD and $^{13}CH_3OD$ IN SPACE, Li-Hong Xu, Ronald M. Lees, Olena Zakharenko, Holger S. P. Müller, Frank Lewen, Stephan Schlemmer, Karl M. Menten

TL03 2:19 – 2:34
PREBIOTIC MOLECULES IN INTERSTELLAR SPACE: ROTATIONAL SPECTROSCOPY OF CYANOMETHANIMINE AND ETHANIMINE, Cristina Puzzarini, Lorenzo Spada, Mattia Melosso, Luca Dore, Vincenzo Barone

TL04 2:36 – 2:51
SEARCH FOR THE ROTATIONAL SPECTRUM OF THE β-CYANOVINYL RADICAL, Sommer L. Johansen, Kyle N. Crabtree

TL05 2:53 – 3:08
INDIRECT ROTATIONAL SPECTROSCOPY OF THE D_2H^+ MOLECULAR ION, Charles R. Markus, Philip A. Kocheril, Benjamin J. McCall

TL06 3:10 – 3:25
HIGH-RESOLUTION MICROWAVE SPECTROSCOPY OF RADIOACTIVE MOLECULES: MASS-INDEPENDENT STUDIES OF AlO, TiO, AND FeO, Alexander A. Breier, Björn Waßmuth, Thomas Büchling, Guido W Fuchs, Thomas Giesen, Jürgen Gauss

Intermission

TL07 4:01 – 4:16
THE PURE ROTATIONAL SPECTRUM OF THE T-SHAPED AlC_2 RADICAL ($\tilde{X}^2A_1$), DeWayne T Halfen, Lucy M. Ziurys

TL08 4:18 – 4:33
TOWARDS UNRAVELLING THE FORMATION OF ICE GRAINS: THE PHENANTHRENE-WATER COMPLEX, Donatella Loru, Sébastien Gruet, Amanda Steber, Cristobal Perez, Melanie Schnell

TL09 4:35 – 4:50
LABORATORY DETECTION OF VIBRATION-ROTATION TRANSITIONS OF $^{12}CH^+$ AND $^{13}CH^+$ AND IMPROVED MEASUREMENT OF THEIR ROTATIONAL TRANSITION FREQUENCIES, José Luis Doménech, Pavol Jusko, Stephan Schlemmer, Oskar Asvany

TL10 4:52 – 5:07
LABORATORY INVESTIGATION OF ASTRONOMICAL REACTIVE SPECIES: THE VIBRATIONAL SATELLITES OF c-C_3H_2 RE-VISITED, Marie-Aline Martin-Drumel, Bryan Changala, Harshal Gupta, J. H. Westerfield, Olivier Pirali, Sven Thorwirth, Joshua H Baraban, John F. Stanton, Michael C McCarthy

TL11 5:09 – 5:24
SPECTROSCOPY OF NEW IMINE ASTROPHYSICS TARGET: METHYLIMINO-ACETONITRILE ($CH_3N=CHCN$), L. Margulès, R. A. Motiyenko, J.-C. Guillemin

WA. Mini-symposium: Frequency-Comb Spectroscopy
Wednesday, June 20, 2018 – 8:30 AM
Room: 116 Roger Adams Lab

Chair: Masatoshi Misono, FUKUOKA UNIVERSITY, Fukuoka, Japan

WA01 *INVITED TALK* 8:30 – 9:00
DUAL FREQUENCY COMB METHANE LEAK DETECTION AT OPERATIONAL OIL AND GAS FACILITIES, Gregory B Rieker, Sean Coburn, Caroline Alden, Robert Wright, Alex Rybchuk, Kuldeep Prasad, Kevin C Cossel, Esther Baumann, Ian Coddington

WA02 9:04 – 9:19
DUAL-COMB SPECTROSCOPY OF GREENHOUSE GAS BASED ON AN ERBIUM DUAL-COMB FIBER LASER, Siyao Yin, Jie Chen, Xin Zhao, Ting Li, Zheng Zheng

WA03 9:21 – 9:36
DUAL FREQUENCY COMB SPECTROSCOPY FOR DEVELOPMENT AND TESTING OF HIGH PRESSURE, HIGH TEMPERATURE ABSORPTION MODELS, Ryan K. Cole, Paul James Schroeder, Anthony D. Draper, Matthew J. Cich, Brian Drouin, Gregory B Rieker

WA04 9:38 – 9:53
DYNAMIC REGIONAL AND CITY SCALE SENSING OF GHG'S USING A DUAL-COMB SPECTROMETER, Eleanor Waxman, Kevin C Cossel, Fabrizio Giorgetta, Gar-Wing Truong, Micheal Cermak, William C Swann, Daniel Hesselius, Gregory B Rieker, Nathan R. Newbury, Ian Coddington

Intermission

WA05 10:29 – 10:44
FREQUENCY COMB VERNIER SPECTROSCOPY OF METHANE IN THE MID-IR WITH TEMPORAL RETRIEVAL OF COMB LINES, James R Bounds, Feng Zhu, Alexander Kolomenskii, Hans A Schuessler, Paotai Lin, Junchao Zhou

WA06 10:46 – 11:01
DIRECT FREQUENCY COMB SPECTROSCOPY WITH AN 8.5 μm OPO, Kana Iwakuni, Thinh Quoc Bui, Justin Niedermeyer, Bryan Changala, Marissa L. Weichman, Takashi Sukegawa, Jun Ye

WA07 11:03 – 11:18
MID-INFRARED FREQUENCY COMB SPECTROSCOPY USING A VIRTUALLY IMAGED PHASED ARRAY, Adam J. Fleisher

WA08 *Post-Deadline Abstract* 11:20 – 11:35
HIGH-RESOLUTION AND ULTRA-BROADBAND DIRECT-COMB ABSOLUTE-SPECTROSCOPY BY MEANS OF THE SCANNING MICRO-CAVITY RESONATOR (SMART) TECHNIQUE, Alessio Gambetta, Edoardo Vicentini, Yuchen Wang, Nicola Coluccelli, Paolo Laporta, Gianluca Galzerano

WB. Mini-symposium: New Ways of Understanding Molecular Spectra
Wednesday, June 20, 2018 – 8:30 AM
Room: 100 Noyes Laboratory

Chair: Taylor Smart, University of Virginia, Charlottesville, Virginia, USA

WB01   *INVITED TALK*   8:30 – 9:00
MOLECULAR SPECTROSCOPY FROM FIRST PRINCIPLES, Sergei N. Yurchenko

WB02   9:04 – 9:19
SPECTRA AND ASSIGNMENTS OF HOT METHANE UP TO 1000 K IN THE 1–2 μm REGION, Andy Wong, Peter F. Bernath, Michael Rey, Andrei V. Nikitin, Vladimir Tyuterev

WB03   9:21 – 9:36
NEURAL NETWORK VS GUASSIAN PROCESS FITTING FOR REPRESENTING POTENTIAL ENERGY SURFACES, Sergei Manzhos, Tucker Carrington, Roman Krems, Rodrigo Hernandez

WB04   9:38 – 9:53
PolyMLR: AN ANALYTIC MODEL FOR POLYATOMIC POTENTIALS WITH FEWER UNPHYSICAL PARAMETERS. APPLICATION TO CO_2., Nikesh S. Dattani

WB05   9:55 – 10:10
VMS-ROT: A NEW MODULE OF THE VIRTUAL MULTIFREQUENCY SPECTROMETER FOR SIMULATION, INTERPRETATION, AND FITTING OF ROTATIONAL SPECTRA, Daniele Licari, Nicola Tasinato, Lorenzo Spada, Cristina Puzzarini, Vincenzo Barone

Intermission

WB06   10:46 – 11:01
PHOTOPHYSICS AND ELECTRONIC STRUCTURE STUDIES OF PROTONATION OF QUINOLINE, Hirdyesh Mishra

WB07   11:03 – 11:18
EXPERIMENTAL AND THEORETICAL INVESTIGATIONS OF THE THRESHOLD PHOTOELECTRON SPECTRUM OF THE CH_2 RADICAL, B. Gans, F. Holzmeier, L. H. Coudert, J.-C. Loison, G. A. Garcia, C. Alcaraz

WB08   11:20 – 11:35
$0.06\,\text{cm}^{-1}$ DISCREPANCY FOR $Li_2 \rightarrow 2Li$ AND $0.994\,\text{cm}^{-1}$ FOR $C \rightarrow C^+$ BETWEEN LABORATORY AND COMPUTER SPECTROMETERS., Nikesh S. Dattani

WC. Mini-symposium: Far-Infrared Spectroscopy
Wednesday, June 20, 2018 – 8:30 AM
Room: 1024 Chemistry Annex

Chair: Olivier Pirali, Synchrotron SOLEIL, Gif-sur-Yvette, France

WC01 *INVITED TALK* 8:30 – 9:00
LINE INTENSITIES AND BROADENING COEFFICIENTS FROM HIGH RESOLUTION FAR INFRARED SPECTRA, Jean Vander Auwera

WC02 9:04 – 9:19
HIGH RESOLUTION IR SPECTROSCOPY AND ANALYSIS OF THE BENDING DYAD OF RuO_4, Sébastien Reymond-Laruinaz, Mbaye Faye, Vincent Boudon, Denis Doizi, Laurent Manceron

WC03 9:21 – 9:36
HIGH RESOLUTION STUDY OF THE ν_2 AND ν_5 ROVIBRATIONAL FUNDAMENTAL BANDS OF THIONYL CHLORIDE : INTERPLAY OF AN EVOLUTIONARY ALGORITHM AND A LINE-BY-LINE ANALYSIS , Anthony Roucou, Guillaume Dhont, Arnaud Cuisset, Marie-Aline Martin-Drumel, Sven Thorwirth, Daniele Fontanari, W. Leo Meerts

WC04 9:38 – 9:53
FAR-INFRARED AND MICROWAVE SPECTROSCOPY OF $HCOOCH_3$, Kaori Kobayashi, Ryo Ohyama, Nobukimi Ohashi, Dennis W. Tokaryk, Brant E. Billinghurst

Intermission

WC05 10:29 – 10:44
ON THE IMPORTANCE OF FAR-INFRARED SPECTROSCOPY FOR NON-POLAR SPHERICAL-TOP MOLECULES, Vincent Boudon, Olivier Pirali, Laurent Manceron, Mbaye Faye

WC06 10:46 – 11:01
IMPROVED FAR-INFRARED AMMONIA INTENSITY FROM EMPIRICAL HAMILTONIAN MODEL, John Pearson, Shanshan Yu, Keeyoon Sung, Jeniveve Pearson, Brian Drouin, Olivier Pirali

WC07 11:03 – 11:18
THE JET-COOLED HIGH-RESOLUTION FAR-IR SPECTRUM OF FORMIC ACID CYCLIC DIMER, Sabath Bteich, Manuel Goubet, Therese R. Huet, Olivier Pirali, Pascale Soulard, Pierre Asselin, Robert Georges

WC08 11:20 – 11:35
THE STRUCTURE OF *gauche*-BUTADIENE: INSIGHTS FROM THE CENTIMETER, MILLIMETER, AND FIR-INFRARED HIGH RESOLUTION SPECTRA, Marie-Aline Martin-Drumel, Joshua H Baraban, Bryan Changala, Matthew Nava, Jessie P Porterfield, Barney Ellison, Olivier Pirali, John F. Stanton, Michael C McCarthy

WC09 11:37 – 11:52
IMPROVE THE PREDICTION ACCURACY OF ISOTOPOLOGUE MICROWAVE SPECTRA BY COMBINING AMES-296K SO_2 IR LISTS WITH EXPERIMENTAL MODELS: A BENCHMARK STUDY, Xinchuan Huang, David Schwenke, Timothy Lee

WD. Radicals

Wednesday, June 20, 2018 – 8:30 AM
Room: 217 Noyes Laboratory

Chair: Neil J Reilly, University of Massachusetts Boston, Boston, MA, USA

WD01   8:30 – 8:45
SUB-DOPPLER INFRARED SPECTROSCOPY OF JET COOLED HCCL DIRADICAL: THE CH STRETCH AND VIBRATIONAL COUPLING IN THE GROUND ELECTRONIC STATE, Andrew Kortyna, Preston G. Scrape, Daniel Lesko, David Nesbitt

WD02   8:47 – 9:02
2C-R4WM SPECTROSCOPY OF JET COOLED NO_3 (II), Masaru Fukushima, Takashi Ishiwata

WD03   9:04 – 9:19
VIBRONIC EMISSION SPECTROSCOPY OF JET-COOLED CHLORO-SUBSTITUTED BENZYL-TYPE RADICALS PRODUCED BY CORONA DISCHARGE, Sang Lee

WD04   9:21 – 9:36
HIGH RESOLUTION SPECTRA OF THE SIMPLEST CRIEGEE INTERMEDIATE CH_2OO BETWEEN 880 AND 932 cm^{-1}, Pei-Ling Luo, Yasuki Endo, Yuan-Pern Lee

WD05   9:38 – 9:53
MILLIMETER-WAVE SPECTROSCOPY OF KO: ESTABLISHING THE ELECTRONIC GROUND STATE, Mark Burton, Benjamin Russ, Phillip M. Sheridan, Matthew Bucchino, Lucy M. Ziurys

WD06   9:55 – 10:10
LASER SPECTROSCOPIC DETECTION OF THE JET-COOLED $SnCH_2$ MOLECULE, Tony Smith, Mohammed Gharaibeh, Dennis Clouthier

Intermission

WD07   10:46 – 11:01
HIGH RESOLUTION LASER SPECTROSCOPY OF THE JET-COOLED SiCF FREE RADICAL, Gretchen K Rothschopf, Tony Smith, Dennis Clouthier

WD08   11:03 – 11:18
PROBING SPIN-ORBIT COUPLING OF ORGANOCERIUM RADICALS FORMED IN Ce ATOM REACTIONS WITH ALKYLAMINES., Silver Nyambo, Yuchen Zhang, Dong-Sheng Yang

WD09   11:20 – 11:35
ELECRONIC STRUCTURE OF ALKOXY RADICAL ISOMERS FROM ANION PEI SPECTROSCOPY, Kellyn M. Patros, Jennifer Mann, Caroline Chick Jarrold

WD10   11:37 – 11:52
ANION PHOTOELECTRON IMAGING OF 2-PROPENOL, Marissa Dobulis, Kellyn M. Patros, Jennifer Mann, Caroline Chick Jarrold

WD11   11:54 – 12:09
SUB-DOPPLER INFRARED SPECTROSCOPY OF JET COOLED CH_2I RADICAL: CH_2 STRETCH VIBRATIONS AND "CHARGE-SLOSHING" INTENSITY DYNAMICS, Andrew Kortyna, Daniel Lesko, Preston G. Scrape, David Nesbitt

WE. Clusters/Complexes
Wednesday, June 20, 2018 – 8:30 AM
Room: B102 Chemical and Life Sciences

Chair: Nasser Moazzen-Ahmadi, University of Calgary, Calgary, AB, Canada

WE01   8:30 – 8:45
STRUCTURE OF MICROSOLVATED VERBENONE DETERMINED BY MICROWAVE FOURIER TRANSFORM SPECTROSCOPY AND QUANTUM CHEMICAL CALCULATIONS, Mhamad Chrayteh, Annunziata Savoia, Pascal Dréan, Therese R. Huet

WE02   8:47 – 9:02
MICROSOLVATION COMPLEXES OF ETHYL CARBAMATE STUDIED BY MICROWAVE SPECTROSCOPY., Pablo Pinacho, Juan Carlos Lopez, Zbigniew Kisiel, Susana Blanco

WE03   9:04 – 9:19
STUDYING CO_2 SOLVENT PROPERTIES BY MICROWAVE SPECTROSCOPIC INVESTIGATION OF FLUOROETHYLENE...CO_2...CO_2 TRIMERS, Prashansa Kannangara, Rebecca A. Peebles, Sean A. Peebles, Brooks Pate

WE04   9:21 – 9:36
π-π STACKING IN COMPETITION WITH HYDROGEN BONDING IN THE 1-NAPHTOL DIMER: A CP-FTMW SPECTROSCOPY STUDY, Nathan A. Seifert, Arsh Hazrah, Wolfgang Jäger

Intermission

WE05   10:12 – 10:27
CHARACTERIZATION OF MICROSOLVATED 15C5 CROWN ETHER FROM BROADBAND ROTATIONAL SPECTROSCOPY, Juan Carlos Lopez, Susana Blanco, Cristobal Perez, Melanie Schnell

WE06   10:29 – 10:44
CHARACTERIZATION OF SO_3-SO_2 BY MICROWAVE SPECTROSCOPY AND COMPUTATIONAL CHEMISTRY, Becca Mackenzie, Anna Huff, Ken Leopold

WE07   10:46 – 11:01
FORMAMIDE, WATER, AND THEIR COMPLEXES: A MICROWAVE SPECTROSCOPY STUDY, Susana Blanco, Juan Carlos Lopez, Channing West, Martin S. Holdren, Brooks Pate

WE08   11:03 – 11:18
DOES THE STRUCTURE OF THE POLYCYCLIC AROMATIC HYDROCARBON IMPACT THE AGGREGATION OF WATER ON ITS SURFACE? FLUORENE VS ACENAPHTHENE, Amanda Steber, Sébastien Gruet, Cristobal Perez, Berhane Temelso, Jana Meiser, George C Shields, Melanie Schnell

WE09   11:20 – 11:35
A ROTATIONAL STUDY OF 2-METHOXYBENZOIC ACID AND ITS WATER COMPLEXES, Alberto Macario, Pablo Pinacho, Susana Blanco, Juan Carlos Lopez

WE10   11:37 – 11:52
MICROWAVE SPECTROSCOPY OF 2-METHOXYETHYLAMINE-WATER: STRUCTURAL CHANGES DUE TO HYDROGEN BONDING NETWORKS, Nathan Harper, Brittany Basenback, Ranil Gurusinghe, Michael Tubergen

WF. Metal containing
Wednesday, June 20, 2018 – 8:30 AM
Room: 2079 Natural History

Chair: Lindsay N. Zack, Austin College, Sherman, TX, USA

WF01   8:30 – 8:45
SPECTROSCOPIC INVESTIGATION OF A SERIES OF CERIUM-DOPED BORON CLUSTERS, Jarrett Mason, Josey E Topolski, Caroline Chick Jarrold

WF02   8:47 – 9:02
ROTATIONAL AND ISOTOPIC STUDY OF THE ZnBr RADICAL ($^2\Sigma^+$), Mark Burton, Lucy M. Ziurys

WF03   9:04 – 9:19
INFRARED SPECTRA OF THE Pd$_n$CO (n=2-5) MOLECULES ISOLATED IN SOLID ARGON AND NEON BETWEEN 100 AND 4000 cm^{-1}, Benoît Tremblay, Sidi M.O. Souvi, Esmaïl Alikhani

WF04   9:21 – 9:36
AN ELECTRONIC SPECTROSCOPIC STUDY OF A MOLECULAR BEAM SAMPLE OF YbOH, Timothy Steimle, Nickolas Pilgram, Nicholas R Hutzler

Intermission

WF05   10:12 – 10:27
ALKALINE EARTH MONOALKOXIDE FREE RADICALS AS CANDIDATES FOR LASER COOLING OF POLYATOMIC MOLECULES, Anam C. Paul, Md Asmaul Reza, Ketan Sharma, Terry A. Miller, Jinjun Liu

WF06   10:29 – 10:44
LASER-INDUCED FLUORESCENCE AND DISPERSED-FLUORESCENCE SPECTROSCOPY OF JET-COOLED CALCIUM MONOALKOXIDE RADICALS, Anam C. Paul, Md Asmaul Reza, Pranoy Deb Shuvra, Jinjun Liu

WF07   10:46 – 11:01
COUPLED-CLUSTER CALCULATIONS FOR LOW-LYING ELECTRONIC STATES OF HEAVY-METAL CONTAINING MOLECULES, Lan Cheng

WF08   11:03 – 11:18
HIGH RESOLUTION SPECTROSCOPY OF THE $[18.0]^2\Pi_{3/2}$ - $X^2\Sigma^+$ TRANSITION OF THORIUM NITRIDE, ThN, Anh T. Le, Duc-Trung Nguyen, Timothy Steimle, Lan Cheng

WF09   11:20 – 11:35
HIGH RESOLUTION SPECTROSCOPY OF THE $[18.2]1.5$ - $X^2\Delta_{3/2}$ TRANSITION OF THORIUM MONOCHLORIDE, ThCl., Colan Linton, Duc-Trung Nguyen, Timothy Steimle

WG. Mini-symposium: Frequency-Comb Spectroscopy
Wednesday, June 20, 2018 – 1:45 PM
Room: 116 Roger Adams Lab

Chair: Adam J. Fleisher, National Institute of Standards & Technology, Gaithersburg, MD, USA

WG01   1:45 – 2:00
MID-INFRARED OPTICAL FREQUENCY COMB VIA COHERENT SUPERCONTINUUM PROCESSES IN NANO-PHOTONIC WAVEGUIDES, Hairun Guo, Wenle Weng, Tobias J. Kippenberg

WG02   2:02 – 2:17
HARMONIC FREQUENCY COMB COVERING THE MID-INFRARED MOLECULAR FINGERPRINT REGION, Christian Gaida, Martin Gebhardt, Tobias Heuermann, Thomas Butler, Daniel Gerz, Christina Hofer, Lenard Vamos, Ferenc Krausz, Jens Limpert, Ioachim Pupeza

WG03   2:19 – 2:34
HIGH-POWER MID-IR COMB GENERATION FOR CAVITY-ENHANCED 2DIR SPECTROSCOPY, Myles C Silfies, Yuning Chen, Henry Timmers, Abijith S Kowligy, Alex Lind, Scott Diddams, Thomas K Allison

WG04   2:36 – 2:51
COMB-REFERENCED MOLECULAR BEAM SPECTROSCOPY OF POLYCYCLIC HYDROCARBONS, Masatoshi Misono, Akiko Nishiyama, Masaaki Baba

WG05   2:53 – 3:08
PRIMARY THERMOMETRY FROM A CO_2 OVERTONE LINE VIA COMB-ASSISTED CAVITY-RING-DOWN SPECTROSCOPY, Riccardo Gotti, Luigi Moretti, Davide Gatti, Antonio Castrillo, Gianluca Galzerano, Paolo Laporta, Livio Gianfrani, Marco Marangoni

Intermission

WG06   3:44 – 3:59
TOWARD QUANTUM STATE RESOLVED INFRARED FREQUENCY COMB SPECTROSCOPY OF THE C_{60} FULLERENE, Bryan Changala, Marissa L. Weichman, Kevin Lee, Martin Fermann, Jun Ye

WG07   4:01 – 4:16
CO_2 LINE PARAMETER RETRIEVAL BEYOND THE VOIGT PROFILE USING COMB-BASED FOURIER TRANSFORM SPECTROSCOPY, Alexandra C Johanssson, Anna Filipsson, Lucile Rutkowski, Piotr Maslowski, Aleksandra Foltynowicz

WG08   4:18 – 4:33
PREDICTING *PARA-ORTHO* CONVERSION IN AMMONIA, Guang Yang, Christoph Heyl, Ingmar Hartl, Andrey Yachmenev, Jochen Küpper

WG09   4:35 – 4:50
SEARCH FOR INVERSION SPLITTING OF PHOSPHINE, Shoko Okuda, Hiroyuki Sasada

WG10   4:52 – 5:07
ONLINE GAS MONITORING USING A MID-INFRARED OPO BASED DUAL COMB SPECTROMETER, Frans Harren

WH. Small molecules
Wednesday, June 20, 2018 – 1:45 PM
Room: 100 Noyes Laboratory

Chair: Stephen T Gibson, Australian National University, Canberra, ACT, Australia

WH01   1:45 – 2:00
THE DICARBON BONDING PUZZLE, Benjamin A Laws, Stephen T Gibson

WH02   2:02 – 2:17
ANOMALOUS Q BRANCH INTENSITY IN THE 2+1 REMPI SPECTRUM OF THE $^1\Pi$-$^1\Sigma^+$ TRANSITION IN HIGHLY ROTATIONALLY EXCITED CO PHOTOFRAGMENTS FROM OCS PHOTODISSOCIATION AT 215 NM, Carolyn E. Gunthardt, Colin J. Wallace, Gregory Hall, Simon North

WH03   2:19 – 2:34
BRIDGING THE GAP - NEWLY OBSERVED VIBRATIONAL LEVELS OF A AND B STATES OF CaH, Kyohei Watanabe, Iori Tani, Takumi Namekata, Kaori Kobayashi, Fusakazu Matsushima, Yoshiki Moriwaki, Stephen Cary Ross

WH04   2:36 – 2:51
FLUOROCARBONS IN SATELLITE PLUMES: THE PHOTOSYNTHESIS AND FLUORESCENCE FROM TRIFLUOROMETHYL RADICAL., Justin W. Young, Christopher Annesley

WH05   2:53 – 3:08
COMPUTING SPECTRA OF OPEN-SHELL DIATOMIC MOLECULES WITH DUO, Sergei N. Yurchenko, Jonathan Tennyson, James R. Ashford, Heng Ying Li, Elizaveta Pyatenko, Maire N. Gorman

WH06   3:10 – 3:25
MICROWAVE SPECTRUM AND THEORETICAL INVESTIGATION OF TRIFLUOROACETIC SULFURIC ANHYDRIDE, Anna Huff, Becca Mackenzie, CJ Smith, Ken Leopold

Intermission

WH07   4:01 – 4:16
VERY DIFFERENT CH_3 INTERNAL ROTATION BARRIERS IN THE SYN- AND ANTI- FORMS OF THIOACETIC ACID: MICROWAVE MEASUREMENTS AND ENERGY DECOMPOSITION ANALYSIS, CJ Smith, Anna Huff, Ken Leopold, Huaiyu Zhang, Yirong Mo

WH08   4:18 – 4:33
LINE INTENSITY MEASUREMENTS AND ANALYSIS IN THE ν_3 BAND OF RUTHENIUM TETROXIDE, Jean Vander Auwera, Sébastien Reymond-Laruinaz, Vincent Boudon, Denis Doizi, Laurent Manceron

WH09   4:35 – 4:50
MOLECULAR LINE INTENSITIES OF CARBON DIOXIDE IN THE 1.6 μm REGION DETERMINED BY CAVITY RINGDOWN SPECTROSCOPY, Zachary Reed, David A. Long, Joseph T. Hodges

WH10   4:52 – 5:07
FIRST HIGH RESOLUTION IR SPECTRA OF 2-D_1-PROPANE. THE ν_9 (A_1) B-TYPE BAND NEAR 367.2389 cm^{-1}., Stephen J. Daunt, Robert Grzywacz, Walter Lafferty, Jean-Marie Flaud, Brant E. Billinghurst

WH11   5:09 – 5:24
FIRST FAR-IR SPECTRA OF 2,2-D_2-PROPANE: THE ν_9 (A_1) B-TYPE BAND NEAR 365.3508 cm^{-1}. THE DETERMINATION OF GROUND AND UPPER STATE CONSTANTS., Daniel Gjuraj, Stephen J. Daunt, Robert Grzywacz, Walter Lafferty, Jean-Marie Flaud, Brant E. Billinghurst

WI. Mini-symposium: Far-Infrared Spectroscopy
Wednesday, June 20, 2018 – 1:45 PM
Room: 1024 Chemistry Annex

Chair: Brian Drouin, California Institute of Technology, Pasadena, CA, USA

WI01   *INVITED TALK*   1:45 – 2:15
PHOTONICS-BASED TERAHERTZ SOURCES FOR MOLECULAR SPECTROSCOPY, Jean-François Lampin

WI02   2:19 – 2:34
THZ HETERODYNE SPECTROSCOPY ON THE AILES BEAMLINE OF SOLEIL FACILITY USING THE SYNCHROTRON RADIATION EMITTED BY THE MULTIBUNCH OPERATION MODE, Olivier Pirali, Gaël Mouret, Francis Hindle, Arnaud Cuisset, Jean-François Lampin, Sophie Eliet, Joan Turut, P. Roy

WI03   2:36 – 2:51
GAS-PHASE INFRARED SPECTROSCOPY OF METAL-LIGAND REDOX PAIRS , Musleh Uddin Munshi, Giel Berden, Jonathan K Martens, Jos Oomens

WI04   2:53 – 3:08
HIGH-RESOLUTION TERAHERTZ GAIN SPECTRA OF MID-INFRARED PUMPED NH_3, Martin Micica, Sophie Eliet, A. Pienkina, R. A. Motiyenko, L. Margulès, Mathias Vanwolleghem, Kamil Postava, Jean-François Lampin

Intermission

WI05   *INVITED TALK*   3:44 – 4:14
EXPLORING THE SOLID STATE PHASE TRANSITION IN DL-NORVALINE WITH TERAHERTZ SPECTROSCOPY, Jens Neu, Coleen T. Nemes, Kevin P. Regan, Michael R. C. Williams, Charles A. Schmuttenmaer

WI06   4:18 – 4:33
PULSE-ECHO MILLIMETER WAVE *IN SITU* SENSOR WITH 65 nm CMOS TRANSMITTER AND HETERODYNE RECEIVER ELECTRONICS, Deacon J Nemchick, Brian Drouin, Adrian Tang, Yanghyo Kim, Gabriel Virbila, M.-C. Frank Chang

WI07   4:35 – 4:50
FOURIER TRANSFORM MILLIMETER-WAVE SPECTROMETER WITH ORIGINAL DESIGN, R. A. Motiyenko, L. Margulès, E. A. Alekseev

WI08   4:52 – 5:07
THE ROTATIONAL SPECTRUM OF THE METHANETHIOL ISOTOPOLOG $CH_3^{34}SH$, Olena Zakharenko, Frank Lewen, Stephan Schlemmer, Holger S. P. Müller, V. Ilyushin, E. A. Alekseev, Igor Krapivin, Li-Hong Xu, Ronald M. Lees, Robin T. Garrod, Arnaud Belloche, Karl M. Menten

WJ. Conformers and isomers
Wednesday, June 20, 2018 – 1:45 PM
Room: 217 Noyes Laboratory

Chair: Timothy S. Zwier, Purdue University, West Lafayette, IN, USA

WJ01   1:45 – 2:00
THE ROTATIONAL STUDY OF DOPAC, A NEURAL METABOLITE, Elena R. Alonso, Iker León, Lucie Kolesniková, Carlos Cabezas, José L. Alonso

WJ02   2:02 – 2:17
MM-WAVE AND AB INITIO STUDIES OF THE CONFORMATIONAL LANDSCAPE OF METHOXYPHENOLS IDENTIFIED AS SOA PRECURSORS, Atef Jabri, Anthony Roucou, Daniele Fontanari, Cédric Bray, Francis Hindle, Gaël Mouret, Robin Bocquet, Arnaud Cuisset

WJ03   2:19 – 2:34
CONFORMATIONAL STUDY OF SECONDARY ORGANIC AEROSOL PRECURSORS CONTAINING INTERNAL ROTATIONS: CASES OF METHYL ANISOLE ISOMERS, Atef Jabri, Daniele Fontanari, Anthony Roucou, Guillaume Dhont, Gaël Mouret, Arnaud Cuisset, Wolfgang Stahl, Ha Vinh Lam Nguyen, Isabelle Kleiner

WJ04   2:36 – 2:51
HIGH RESOLUTION SPECTROSCOPY OF 3-METHYLBUTYRONITRILE BETWEEN 2 AND 400 GHZ, Nadine Wehres, Marius Hermanns, Olivia H. Wilkins, Kirill Borisov, Frank Lewen, Jens-Uwe Grabow, Stephan Schlemmer, Holger S. P. Müller

WJ05   2:53 – 3:08
HIGH-RESOLUTION SPECTROSCOPY OF TWO CONFORMERS OF 2-CYANOBUTANE BETWEEN 10 AND 400 GHZ, Marius Hermanns, Nadine Wehres, Frank Lewen, Holger S. P. Müller, Stephan Schlemmer

WJ06   3:10 – 3:25
CONFORMATIONAL ISOMERISM OF 1-IODOPENTANE, Susanna L. Stephens, Joshua A. Signore, Daniel A. Obenchain, Robert Karl Bohn, Stewart E. Novick, S. A. Cooke

Intermission

WJ07   4:01 – 4:16
ISOMER-SPECIFIC SPECTROSCOPY OF ETHYL NAPHTHALENE DERIVATIVES: SPECTROSCOPIC FOUNDATION FOR UNDERSTANDING ETHYL-BRIDGED DINAPHTHYLS, Victoria M. Boulos, Daniel M. Hewett, Timothy S. Zwier

WJ08   4:18 – 4:33
THE ORIGINATION OF SOOT FORMATION: A STUDY ON ETHYL-LINKED NAPHTHALENE DIMERS, Daniel M. Hewett, Victoria M. Boulos, Timothy S. Zwier

WJ09   4:35 – 4:50
STUDYING THE FOLDING PROPENSITY OF ASPARAGINE-CONTAINING PEPTIDES IN A MOLECULAR BEAM, Karl N. Blodgett, Joshua L. Fischer, Dewei Sun, Timothy S. Zwier

WJ10   4:52 – 5:07
INSIGHTS INTO PROTON TRANSFER MECHANISMS IN HALOGEN-SUBSTITUTED MALDI MATRICES, Chelsea N Bridgmohan, Kristopher M Kirmess, Lichang Wang

WJ11   5:09 – 5:24
CONFORMERS OF L-GLUTAMIC ACID: MATRIX ISOLATION FTIR AND *AB-INITIO* STUDIES., Pankaj Dubey, K S Viswanathan

WK. Structure determination
Wednesday, June 20, 2018 – 1:45 PM
Room: B102 Chemical and Life Sciences

Chair: Helen O. Leung, Amherst College, Amherst, MA, USA

WK01   1:45 – 2:00
DESIGN AND APPLICATIONS OF A MASS-CORRELATED BROADBAND MICROWAVE SPECTROMETER, Sean Fritz, Alicia O. Hernandez-Castillo, Brian M Hays, Timothy S. Zwier

WK02   2:02 – 2:17
STRUCTURAL CHARACTERIZATION OF PHENOXY RADICAL USING A MASS-CORRELATED BROADBAND MICROWAVE SPECTROMETER, Alicia O. Hernandez-Castillo, Chamara Abeysekera, John F. Stanton, Timothy S. Zwier

WK03   2:19 – 2:34
AN ARGON-OXYGEN COVALENT BOND IN THE ArOH$^+$ MOLECULAR ION, J. Philipp Wagner, David C McDonald II, Michael A Duncan

WK04   2:36 – 2:51
PROTON IN A DOUBLE-WELL POTENTIAL AS SEEN FROM MICROWAVE AND CORE LEVEL PHOTOEMISSION SPECTROSCOPY, Luca Evangelisti, Weixing Li, Assimo Maris, Sonia Melandri

WK05   2:53 – 3:08
HETERO-OLIGOMERS OF DIFLUOROMETHANE AND 1,1-DIFLUOROETHANE: CONFORMATIONAL EQUILIBRIA, MOLECULAR STRUCTURE AND WEAK HYDROGEN BONDS, Tao Lu, Junha Chen, Qian Gou, Gang Feng

WK06   3:10 – 3:25
STRUCTURE DETERMINATION, CONFORMATIONAL EQUILIBRIA AND WEAK HYDROGEN BONDS IN HETERODIMERS OF FREONS: THE ROTATIONAL STUDY OF CH_2F_2-CF_3CH_2F, Tao Lu, Junha Chen, Qian Gou, Gang Feng

WK07   3:27 – 3:42
THE CONFORMATIONAL PREFERENCE OF A STRONG O-H•••O HYDROGEN BONDED COMPLEX IS DETERMINED BY WEAK C-H•••π INTERACTION: A LIF STUDY OF BINARY COMPLEXES OF P-FLUOROPHENOL WITH 2,5-DIHYDROFURAN AND TETRAHYDROFURAN, Deb Pratim Mukhopadhyay, Souvick Biswas, Tapas Chakraborty

Intermission

WK08   4:18 – 4:33
TUNING OF NON-COVALENT INTERACTIONS IN MOLECULAR COMPLEXES OF FLUORINATED AROMATIC COMPOUNDS, Sonia Melandri, Luca Evangelisti, Assimo Maris, Imanol Usabiaga Gutierrez, Weixing Li, Camilla Calabrese, Laura B. Favero

WK09   4:35 – 4:50
THE CYCLOHEXANOL DIMER AND THE MONOHYDRATE: INTERNAL ROTATION AND CONVERSION FROM TRANSIENT TO PERMANENT CHIRALITY, Marcos Juanes, Iker León, Ruth Pinacho, José Emiliano Rubio, Weixing Li, Luca Evangelisti, Walther Caminati, Alberto Lesarri

WK10   4:52 – 5:07
ROTATIONAL SPECTRA AND GEOMETRIES OF FOUR CONFORMERS OF ISOLATED UROCANIC ACID; AND OF A COMPLEX OF UROCANIC ACID WITH WATER, Nick Walker, Graham A. Cooper, Chris Medcraft, Eva Gougoula

WK11   5:09 – 5:24
ROTATIONAL SPECTROSCOPIC STUDIES ON THE CH_3CN-CO_2 COMPLEX, Sharon Priya Gnanasekar, Elangannan Arunan

WK12   5:26 – 5:41
CONFORMER SPECIFIC METHYL INTERNAL ROTATION AND OBSERVATION OF PHOTODISSOCIATION DYNAMICS: IS METHYL INTERNAL ROTATION COUPLED WITH TORSIONAL VIBRATION?, Heesung Lee, So-Yeon Kim, Jean Sun Lim, Junggil Kim, Sang Kyu Kim

WL. Astronomy
Wednesday, June 20, 2018 – 1:45 PM
Room: 2079 Natural History

Chair: Cristina Puzzarini, University of Bologna, Bologna, Italy

WL01   1:45 – 2:00
EXPLOITING TUNABLE VACUUM ULTRAVIOLET PHOTOIONIZATION COMBINED WITH REFLECTRON TIME-OF-FLIGHT MASS SPECTROMETRY TO UNRAVEL THE NITROGEN CHEMISTRY OF COMPLEX ORGANICS IN THE INTERSTELLAR MEDIUM, Robert Frigge, Andrew Martin Turner, Matthew James Abplanalp, Ralf Ingo Kaiser

WL02   2:02 – 2:17
RADIO ASTRONOMY RECEIVERS AND A GAS REACTION CHAMBER FOR LABORATORY ASTROCHEMICAL SIMULATIONS., Jose Cernicharo, Juan R. Pardo, Juan Daniel Gallego, Pablo de Vicente, Isabel Tanarro, Victor Jose Herrero, José Luis Doménech, Ramón J. Peláez

WL03   2:19 – 2:34
O(^{1}D) INSERTION REACTIONS FOR THE PRODUCTION AND SPECTRAL ANALYSIS OF INTERSTELLAR ORGANIC MOLECULES, Hayley Bunn, Samuel Zinga, Carson Reed Powers, Brian Savino, Morgan N McCabe, Brian M Hays, Susanna L. Widicus Weaver

WL04   2:36 – 2:51
COSMIC RAY-DRIVEN RADIATION CHEMISTRY IN COLD INTERSTELLAR ENVIRONMENTS, Christopher N Shingledecker, Jessica D. Tennis, Romane Le Gal, Eric Herbst

WL05   2:53 – 3:08
PROBING THE PHOTOPRODUCTS OF INTERSTELLAR ICE ANALOGUES VIA LABORATORY SUBMILLIMETER SPECTROSCOPY, Katarina Yocum, Houston H Smith, Stefanie N Milam, Susanna L. Widicus Weaver

WL06   3:10 – 3:25
SUB-DOPPLER INFRARED SPECTROSCOPY OF JET COOLED NDH_3^+: N–H STRETCH VIBRATIONS IN A KEY ASTROCHEMICAL ION, Preston G. Scrape, Andrew Kortyna, Daniel Lesko, David Nesbitt

WL07   3:27 – 3:42
ULTRAVIOLET AND INFRARED OSCILLATOR STRENGTHS FOR OH^+, James Neil Hodges, Dror M. Bittner, Peter F. Bernath

Intermission

WL08   4:18 – 4:33
INFRARED ABSORPTION CROSS SECTIONS OF HYDROCARBONS, Peter F. Bernath, Andy Wong, Dominique Appadoo, Brant E. Billinghurst

WL09   4:35 – 4:50
INFRARED SPECTROSCOPY ON SMALL METAL-BEARING OXIDES, Daniel Witsch, Alexander A. Breier, Guido W Fuchs, Thomas Giesen

WL10   4:52 – 5:07
HIGH ACCURACY THERMOCHEMISTRY AND KINETICS OF THE HCN/HNC SYSTEM, Kelvin Lee, Michael C McCarthy

WL11   5:09 – 5:24
DETECTION OF INTERSTELLAR BENZONITRILE (c-C_6H_5CN), Brett A. McGuire, Andrew M Burkhardt, Sergei Kalenskii, Christopher N Shingledecker, Anthony Remijan, Eric Herbst, Michael C McCarthy

WL12   5:26 – 5:41
SYNTHESIS OF INTERSTELLAR BENZONITRILE (c-C_6H_5CN): A MICROWAVE SPECTROSCOPIC STUDY, Brett A. McGuire, Kelvin Lee, Michael C McCarthy

RA. Plenary
Thursday, June 21, 2018 – 8:30 AM
Room: Foellinger Auditorium

Chair: Leslie Looney, University of Illinois at Urbana-Champaign, Urbana, IL, USA

RA01   8:30 – 9:10
VIBRATIONAL AND ROTATIONAL SPECTROSCOPY IN CRYOGENIC ION TRAPS, Sandra Brünken, Britta Redlich, Pavol Jusko, Oskar Asvany, Stephan Schlemmer

RA02   9:15 – 9:55
ULTRAFAST TRANSIENT ABSORPTION SPECTROSCOPY OF PHOTOCHEMICAL DYNAMICS IN SOLUTION, Daisuke Koyama, Ravi Kumar Venkatraman, Harvey J A Dale, Andrew Orr-Ewing

Intermission

RAO AWARDS   10:35
Presentation of Awards by Gary Douberly, University of Georgia

2017 Rao Award Winners
Bryce Bjork, University of Colorado at Boulder
Anna Huff, University of Minnesota
Christopher Shingledecker, University of Virginia

MILLER PRIZE   10:45
Introduction by Michael Heaven, Emory University

RA03   *Miller Prize Lecture*   10:50 – 11:05
BOND INSERTION IN METAL–CARBON DIOXIDE ANIONIC CLUSTERS STUDIED BY INFRARED PHOTODISSOCIATION SPECTROSCOPY, Leah G Dodson, Michael C Thompson, J. Mathias Weber

COBLENTZ AWARD   11:10
Presentation of Award by Linda Kidder, Coblentz Society

RA04   *Coblentz Society Award Lecture*   11:15 – 11:55
ATTOSECOND TIME-RESOLVED MOLECULAR SPECTROSCOPY, Hans Jakob Wörner

RG. Cold and ultracold molecules
Thursday, June 21, 2018 – 1:45 PM
Room: 116 Roger Adams Lab

Chair: Hideto Kanamori, Tokyo Institute of Technology, Tokyo, Japan

RG01   1:45 – 2:00
ELECTRONIC PHOTODISSOCIATION SPECTROSCOPY OF COLD NITROPHENOLATE IONS. PART I. ORTHO- AND PARA-NITROPHENOLATE, Wyatt Zagorec-Marks, Leah G Dodson, J. Mathias Weber

RG02   2:02 – 2:17
ELECTRONIC PHOTODISSOCIATION SPECTROSCOPY OF COLD NITROPHENOLATE IONS. PART II. META-NITROPHENOLATE, Wyatt Zagorec-Marks, Leah G Dodson, J. Mathias Weber

RG03   2:19 – 2:34
ROTAMERS OF ISOPRENE: INFRARED SPECTROSCOPY IN HELIUM DROPLETS AND AB INITIO THERMOCHEMISTRY, Peter R. Franke, Gary E. Douberly

RG04   2:36 – 2:51
INFRARED SPECTRA OF PROPENE IN HELIUM NANODROPLETS AND SOLID *PARA*-HYDROGEN, Gregory T. Pullen, Peter R. Franke, Gary E. Douberly, Yuan-Pern Lee

RG05   2:53 – 3:08
DETECTION AND SPECTROSCOPY OF POLYATOMIC MOLECULES INSIDE A CRYOGENIC BUFFER GAS CELL, Thomas Wall, Julia Bieniewska, B. E. Sauer, Michael Tarbutt, Benoit Darquie, Trevor Sears

RG06   3:10 – 3:25
CHARACTERIZING MOLECULAR IONS FOR LASER CONTROL, Sruthi Venkataramanababu, Patrick R Stollenwerk, Ivan Antonov, Brian C. Odom

Intermission

RG07   4:01 – 4:16
DETERMINATION OF THE SPIN-ROTATION FINE STRUCTURE OF He_2^+, Paul Jansen, Luca Semeria, Frederic Merkt

RG08   4:18 – 4:33
FINE STRUCTURE OF METASTABLE $^4He_2$ USING ZEEMAN-DECELERATED MOLECULAR-BEAM RESONANCE SPECTROSCOPY, Luca Semeria, Paul Jansen, Josef A. Agner, Hansjürg Schmutz, Frederic Merkt

RG09   4:35 – 4:50
SPECTOSCOPY OF SiO AND SiO^+ IN SUPPORT OF ULTACOLD MOLECULE STUDIES, Ivan Antonov, Patrick R Stollenwerk, Sruthi Venkataramanababu, Brian C. Odom

RG10   4:52 – 5:07
TOWARDS STATE-RESOLVED ULTRACOLD CHEMICAL REACTIONS OF KRb MOLECULES, David Grimes, Ming-Guang Hu, Yu Liu, Andrei Gheorghe, Kang-Kuen Ni

RH. Spectroscopy as an analytical tool
Thursday, June 21, 2018 – 1:45 PM
Room: 100 Noyes Laboratory

Chair: Kyle N. Crabtree, University of California, Davis, CA, USA

RH01   1:45 – 2:00
MEASUREMENTS OF $N_2(A^3\Sigma_u^+, v)$ POPULATIONS IN A NANOSECOND PULSE DISCHARGE BY CAVITY RING-DOWN SPECTROSCOPY, Elijah R Jans, Kraig Frederickson, Igor V. Adamovich

RH02   2:02 – 2:17
ABSOLUTE NUMBER DENSITY MEASUREMENTS OF HYDROPEROXYL RADICAL IN A NANOSECOND PULSE DISCHARGE USING CAVITY RING-DOWN SPECTROSCOPY, Kraig Frederickson, Terry A. Miller, Igor V. Adamovich

RH03   2:19 – 2:34
ACETONE AND METHANE DETECTION WITH WAVELENGTH MODULATION SPECTROSCOPY IN THE NEAR- AND MID-IR, Jinbao Xia, Feng Zhu, James R Bounds, Sasa Zhang, Alexander Kolomenskii, Hans A Schuessler

RH04   2:36 – 2:51
SPECTROSCOPIC CHARACTERIZATION OF SMALL POLAR IMPURITIES IN GASOLINE, Sylvestre Twagirayezu, Alex Mikhonin, Matt Muckle, Justin L. Neill

RH05   2:53 – 3:08
OPTICAL SENSING OF ENVIRONMENTALLY HAZARDOUS HEAVY METALS (Cr^{3+}, Pb^{2+}, Zn^{2+}) AND CANCER CELLS BY FUNCTIONALIZED CORE/SHELL QUANTUM DOTS, Papia Chowdhury

RH06   3:10 – 3:25
ANALYSIS OF PEAR ESTER FLAVORING SAMPLES USING BROADBAND ROTATIONAL SPECTROSCOPY, Channing West, Rachel Bocwinski, Aisling Foley, Sasha Hoyt, Sarah Johnson, Alexander Khlopenkov, Julia Marks, Rachel Schelling, Xuayne Zhu, Jinbum Dupont, Liam Fineman, Justin L. Neill, Brooks Pate

Intermission

RH07   4:01 – 4:16
MOLECULAR COMPOSITION OF GALLBLADDER STONE USING PHOTOACOUSTIC SPECTROSCOPY, Zainab Gazali, Surya Narayan Thakur, Awadhesh Kumar Rai

RH08   4:18 – 4:33
IN SITU CHEMICAL CHARACTERIZATION OF THE MOTILE TO SESSILE TRANSITION OF *PSEDOMONAS AERUGINOSA* COMMUNITIES, Tianyuan Cao, Nydia Morales-Soto, Kristen M. Kramer, Nameera F. Baig, Joshua D. Shrout, Paul W. Bohn

RH09   4:35 – 4:50
UTILISING DIFFUSE REFLECTANCE INFRA-RED SPECTROSCOPY TO MONITOR THE OXIDATION OF BITUMEN AND ASPHALT AS A RESULT OF ARTIFICIAL AND NATURAL AGEING, Hannah Bowden, Matthew Almond, Wayne Hayes, Stuart McRobbie

RI. Mini-symposium: Far-Infrared Spectroscopy
Thursday, June 21, 2018 – 1:45 PM
Room: 1024 Chemistry Annex

Chair: Sandra Brünken, Radboud University, Nijmegen, The Netherlands

RI01   *INVITED TALK*   1:45 – 2:15
POLAR RADIANT ENERGY IN THE FAR-INFRARED EXPERIMENT (PREFIRE), Brian Drouin, Tristan S L'Ecuyer

RI02   2:19 – 2:34
PROGRESS ON THE FT-IR MEASUREMENTS OF WATER CONTINUUM IN THE FAR-INFRARED REGION AT 252 – 296 K, Keeyoon Sung, Brian Drouin, Timothy J. Crawford, Edward H Wishnow

RI03   2:36 – 2:51
MILLIMETER-WAVE CHIRALITY SPECTROMETER (CHIRALSPEC), Shanshan Yu, Theodore J Reck, John Pearson, Michael Malaska, Robert Hodyss, Brooks Pate

RI04   2:53 – 3:08
MICROWAVE SPECTRUM OF 1-ADAMANTANOL $C_{10}H_{15}$–OH, Olivier Pirali, Marie-Aline Martin-Drumel, L. H. Coudert, Manuel Goubet, Sébastien Gruet, Melanie Schnell

Intermission

RI05   *INVITED TALK*   3:44 – 4:14
FAR-INFRARED SPECTROSCOPY OF SHORT-LIVED SPECIES, Hiroyuki Ozeki

RI06   4:18 – 4:33
THZ SPECTROSCOPY OF SULFUR DERIVATIVES OF ASTROPHYSICAL INTEREST, L. Margulès, S. Bailleux, R. A. Motiyenko, J.-C. Guillemin, Jose Cernicharo, Arnaud Belloche, Brett A. McGuire, Anthony Remijan, Olga Dorovskaya, V. Ilyushin

RI07   4:35 – 4:50
WATER VAPOUR AND AMMONIA IN CIRCUMSTELLAR ENVELOPES OF C-RICH EVOLVED STARS, Miroslaw R. Schmidt

RI08   4:52 – 5:07
EXTENDED MEASUREMENTS AND AN EXPERIMENTAL ACCURACY EFFECTIVE HAMILTONIAN MODEL FOR THE $3\nu_2$ AND $\nu_2 + \nu_4$ STATES OF AMMONIA, Jeniveve Pearson, Shanshan Yu, John Pearson, Keeyoon Sung, Brian Drouin, Olivier Pirali

RJ. Electronic structure, potential energy surfaces
Thursday, June 21, 2018 – 1:45 PM
Room: 217 Noyes Laboratory

Chair: Steve Alexandre Ndengue, Missouri University of Science & Technology, Rolla, MO, USA

RJ01 1:45 – 2:00
ROTATION-TUNNELING ANALYSIS OF PROTON-TRANSFER DYNAMICS IN ELECTRONICALLY EXCITED 6-HYDROXY-2-FORMYLFULVENE USING DEGENERATE FOUR-WAVE MIXING, Zachary Vealey, Lidor Foguel, Patrick Vaccaro

RJ02 2:02 – 2:17
NEW ELECTRONIC STATES OF YO IN THE UV REGION, Allan S.C. Cheung, Na Wang, Yuk Wai Ng, Andrew Clark, Wenli Zou

RJ03 2:19 – 2:34
STARK AND ZEEMAN EFFECT IN THE $[18.5]^2\Delta_{3/2} - X^2\Delta_{3/2}$ TRANSITION OF THORIUM MONOFLUORIDE, Duc-Trung Nguyen, Timothy Steimle

RJ04 2:36 – 2:51
LASER INDUCED FLUORESCENCE (LIF) SPECTROSCOPY OF JET COOLED ThO, Joel R Schmitz, Michael Heaven

RJ05 2:53 – 3:08
INELASTIC COLLISIONS OF Ar AND O_3, Sangeeta Sur, Ernesto Quintas Sánchez, Steve Alexandre Ndengue, Richard Dawes

RJ06 3:10 – 3:25
METAL-AMMONIA COMPLEXES DISCLOSE A SECRET PERIODIC TABLE OF SOLVATED ELECTRON PRECURSORS, Evangelos Miliordos

RJ07 3:27 – 3:42
EXTRA HIGH ACCURACY FITTING OF THE PES FOR SUB-PERCENT CALCULATION OF INTENSITIES, Oleg L. Polyansky, Jonathan Tennyson, Vladimir Yu. Makhnev, Aleksandra A. Kyuberis, Nikolay F. Zobov

Intermission

RJ08 4:18 – 4:33
AUTOSURF: A CODE FOR AUTOMATED CONSTRUCTION OF POTENTIAL ENERGY SURFACES, Ernesto Quintas Sánchez, Richard Dawes

RJ09 4:35 – 4:50
APPROXIMATIONS FOR HIGH-ACCURACY THEORETICAL THERMOCHEMISTRY, Bradley Welch, Richard Dawes

RJ10 4:52 – 5:07
VACUUM UV LABORATORY STUDY OF THE PHOTODISSOCIATION OF CS, Zhongxing Xu, Yih-Chung Chang, Kyle N. Crabtree, William M. Jackson, Cheuk-Yiu Ng

RJ11 5:09 – 5:24
HIGH-SPIN ELECTRONIC STATES OF MOLECULAR OXYGEN, Gabriel J. Vázquez, H. P. Liebermann, H. Lefebvre-Brion

RK. Clusters/Complexes
Thursday, June 21, 2018 – 1:45 PM
Room: B102 Chemical and Life Sciences

Chair: Haruki Ishikawa, Kitasato University, Sagamihara, Japan

RK01   1:45 – 2:00
A CONFORMATIONAL STUDY OF *META*-ANISIC ACID AND ITS COMPLEXES WITH FORMIC ACID BY MICROWAVE SPECTROSCOPY, Alberto Macario, Susana Blanco, Javix Thomas, Yunjie Xu, Juan Carlos Lopez

RK02   2:02 – 2:17
CP-FTMW SPECTROSCOPY OF 2-CYANOACETIC ACID, Erika Johnson, Steven Shipman, Iker León, Lucie Kolesniková, Santiago Mata, José L. Alonso

RK03   2:19 – 2:34
HIGH RESOLUTION MICROWAVE SPECTROSCOPY AND STRUCTURE OF THE WEAKLY BOUND Xe$\cdots$OCS COMPLEX, Daniel A. Obenchain, Sven Herbers, Peter Kraus, Dennis Wachsmuth, Jens-Uwe Grabow

RK04   2:36 – 2:51
MICROWAVE SPECTRUM OF THE A INTERNAL ROTOR STATE OF Ar-CH$_3$I, Anna Huff, CJ Smith, Ken Leopold

RK05   2:53 – 3:08
MICROWAVE STUDY OF 2-PHENYLPYRIDINE AND THEIR WATER COMPLEXES, Susana Blanco, Alberto Macario, Juan Carlos Lopez

Intermission

RK06   3:44 – 3:59
INTRA AND INTERMOLECULAR DYNAMICS AND STRUCTURE IN THE FORMANILIDE-(H$_2$O)$_n$ (n=1,2) CLUSTERS, Pablo Pinacho, Susana Blanco, Juan Carlos Lopez

RK07   4:01 – 4:16
OH-π HYDROGEN BOND IN THE COMPLEX OF STYRENE-WATER: A ROTATIONAL STUDY, Yang Zheng, Juncheng Lei, Gang Feng, Qian Gou

RK08   4:18 – 4:33
ISOTOPIC SUBSTITUTIONS UNVEILED THE IDENTIFICATION OF THE MORE STABLE CONFORMER OF FENCHOL AND OF ITS WATER COMPLEX, Elias M. Neeman, Therese R. Huet

RK09   4:35 – 4:50
A ROTATIONAL STUDY OF THE METHYL CARBAMATE-(H$_2$O)$_n$ n=1,2 COMPLEXES: MICROWAVE SPECTRUM, INTERNAL ROTATION AND HYPERFINE STRUCTURE., Pablo Pinacho, Juan Carlos Lopez, Zbigniew Kisiel, Susana Blanco

RK10   4:52 – 5:07
ROTATIONAL SPECTRUM OF THE ISOPRENE-WATER COMPLEX, Brandon Carroll, Michael C McCarthy

RL. Astronomy
Thursday, June 21, 2018 – 1:45 PM
Room: 2079 Natural History

Chair: Brett A. McGuire, National Radio Astronomy Observatory, Charlottesville, VA, USA

RL01   1:45 – 2:00
MOLECULAR COMPLEXITY IN PRESTELLAR CORES, Valerio Lattanzi, Paola Caselli

RL02   *Post-Deadline Abstract*   2:02 – 2:17
MILLIMETER/SUBMILLIMETER-WAVE SPECTROSCOPY OF THE CrP RADICAL $^4\Sigma^-$, Mark Burton, DeWayne T Halfen, Lucy M. Ziurys

RL03   2:36 – 2:51
NEW CARBON-CHAIN MOLECULAR DETECTIONS IN TMC-1 WITH THE GREEN BANK TELESCOPE, Andrew M Burkhardt, Christopher N Shingledecker, Eric Herbst, Sergei Kalenskii, Michael C McCarthy, Anthony Remijan, Brett A. McGuire

RL04   2:53 – 3:08
INVESTIGATING THE DISTRIBUTION OF COMPLEX MOLECULES AT LOW FREQUENCY USING THE KARL G. JANSKY VERY LARGE ARRAY IN SEARCH OF THE EXCITATION OF HNCNH, Anthony Remijan, Brett A. McGuire, Andrew M Burkhardt, Joanna F. Corby

RL05   3:10 – 3:25
DETECTION OF CH_3CN IN DIFFUSE CLOUD TOWARD GALACTIC CENTER SGRB2(M), Mitsunori Araki, Shuro Takano, Yoshiaki Minami, Takahiro Oyama, Nobuhiko Kuze, Kazuhisa Kamegai, Koichi Tsukiyama

RL06   3:27 – 3:42
CONSTRAINING SULFUR ISOTOPE ABUNDANCES IN MOLECULAR CLOUDS: A METEORITIC PERSPECTIVE, Jacob Bernal, Maitrayee Bose, Lucy M. Ziurys

RL07   3:44 – 3:59
FIFI-LS FIR VIEW OF ORION: FINE STRUCTURE AND CO LINES, Frankie Encalada, Leslie Looney, Randolf Klein, Christian Fischer, Sebastian Colditz, Dario Fadda, Norbert Geis, Rainer Hönle, Christof Iserlohe, Alfred Krabbe, Albrecht Poglitsch, Walfried Raab, William Vacca

Intermission

RL08   4:35 – 4:50
THE TRANSITION FROM DIFFUSE ATOMIC CLOUDS TO DENSE MOLECULAR CLOUDS, Johnathan S Rice, Steven Federman

RL09   4:52 – 5:07
DUST POLARIZATION IN THREE PROTOSTELLAR DISKS, Rachel E. Harrison, Leslie Looney, Robert J Harris, Zhi-Yun Li, Ian Stephens, Woojin Kwon

RL10   5:09 – 5:24
IDENTITY OF THE CARRIER OF λ5797 DIFFUSE INTERSTELLAR BAND and λ5800 RED-RECTANGLE EMISSION BAND, Keir Adams, Takeshi Oka

RL11   5:26 – 5:41
CENTRAL 300 PC OF THE GALAXY PROBED BY THE INFRARED SPECTRA OF H_3^+ AND CO PART II. MORPHOLOGY AND DYNAMICS OF THE GAS, Takeshi Oka, Thomas R. Geballe, Miwa Goto, Tomonori Usuda, Benjamin J. McCall, Nick Indriolo

FA. Vibrational structure/frequencies
Friday, June 22, 2018 – 8:30 AM
Room: 116 Roger Adams Lab

Chair: G. S. Grubbs II, Missouri University of Science and Technology, Rolla, MO, USA

FA01   8:30 – 8:45
THE 103 - 360 GHZ ROTATIONAL SPECTRUM OF BENZONITRILE, THE FIRST INTERSTELLAR BENZENE DERIVATIVE DETECTED BY RADIOASTRONOMY, Maria Zdanovskaia, Brian J. Esselman, Hunter Singh Lau, Desiree M. Bates, R. Claude Woods, Robert J. McMahon, Zbigniew Kisiel

FA02   8:47 – 9:02
INFRARED SPECTRUM OF CHLOROMETHYL HYDROPEROXIDE CH_2ClOOH PRODUCED FROM REACTION OF THE CRIEGEE INTERMEDIATE CH_2OO WITH HCl, Wei-Che Liang, Yuan-Pern Lee

FA03   9:04 – 9:19
INFRARED SPECTRA OF C_2H_4BR AND C_2H_4I IN SOLID *PARA*-HYDROGEN: BRIDGED OR OPEN STRUCTURE?, Yu-Hsuan Chen, Yuan-Pern Lee

FA04   9:21 – 9:36
A MULTIMODE-LIKE SELECTION OF CENTERS OF GAUSSIAN BASIS FUNCTIONS WHEN COMPUTING VIBRATIONAL SPECTRA USING COLLOCATION , Sergei Manzhos, Xiao-Gang Wang, Tucker Carrington

FA05   9:38 – 9:53
TRANSIENT RAMAN SPECTRA, STRUCTURE AND THERMOCHEMISTRY OF THE SELENOCYANATE DIMER RADICAL ANION IN WATER, Ireneusz Janik, G. N. R. Tripathi

FA06   9:55 – 10:10
MOLECULAR-SCALE INTERROGATION OF CATALYTIC INTERACTIONS BETWEEN OXYGEN AND COBALT PHTHALOCYANINE USING ULTRAHIGH VACUUM TIP-ENHANCED RAMAN SPECTROSCOPY, Duc Nguyen, Gyeongwon Kang, George C. Schatz, Richard P. Van Duyne

FB. Metal containing
Friday, June 22, 2018 – 8:30 AM
Room: 100 Noyes Laboratory

Chair: Leah C O'Brien, Southern Illinois University, Edwardsville, IL, USA

FB01   8:30 – 8:45
PROBING SELECTIVE BOND ACTIVATION IN ALKYLAMINES: LANTHANUM-MEDIATED C-H AND N-H BOND ACTIVATION STUDIED BY MATI SPECTROSCOPY., Silver Nyambo, Yuchen Zhang, Dong-Sheng Yang

FB02   8:47 – 9:02
FTIR STUDY OF THE REACTIVITY OF HETERONUCLEAR SMALL TRANSITION METAL CLUSTER WITH CARBON MONOXIDE, Mohamad Ibrahim, pascale soulard, Esmaïl Alikhani, Benoît Tremblay

FB03   9:04 – 9:19
MAPPING THE INTRINSIC PHOTOCHEMISTRY OF PhotoCORMS VIA GAS-PHASE LASER SPECTROSCOPY, Rosaria Cercola, Jason M. Lynam, Caroline H. E. Dessent

FB04   9:21 – 9:36
TIME-RESOLVED RELAXATION DYNAMICS OF NEAR-INFRARED EXCITED ELECTRONIC STATES IN TRANSITION METAL COMPLEXES., Darya S. Budkina, Sergey M. Matveev, Christopher M. Hicks, Veniamin A. Borin, Andrey S. Mereshchenko, Alexander N Tarnovsky

FB05   9:38 – 9:53
DISCOVERY OF DATIVE BONDING OF BERYLLIUM FLUORIDE ANION BY PHOTOELECTRON VELOCITY MAP IMAGING SPECTROSCOPY, Mallory Theis, Pearl Jean, Michael Heaven

Intermission

FB06   10:29 – 10:44
SPECTROSCOPY OF TiO SINGLET STATES, Dror M. Bittner, Peter F. Bernath

FB07   10:46 – 11:01
ROTATIONAL ANALYSIS OF SEVERAL VIBRATIONAL BANDS OF THE [7.7] $Y\ ^2\Sigma^+$ - $X\ ^2\Pi_i$ TRANSITION OF ^{63}CuO, Jack C Harms, James J O'Brien, Leah C O'Brien

FB08   11:03 – 11:18
ROTATIONAL ANALYSIS OF AN ELECTRONIC TRANSITION OF CuOH OBSERVED WITH INTRACAVITY LASER SPECTROSCOPY, Jack C Harms, Leah C O'Brien, James J O'Brien

FC. Comparing theory and experiment
Friday, June 22, 2018 – 8:30 AM
Room: 1024 Chemistry Annex

Chair: Jennifer van Wijngaarden, University of Manitoba, Winnipeg, MB, Canada

FC01   8:30–8:45
THE ROTATIONAL SPECTRUM AND POTENTIAL ENERGY SURFACE OF AR-SIO: AN EXPERIMENTAL INVESTIGATION, Michael C McCarthy, Richard Dawes

FC02   8:47–9:02
THE ROTATIONAL SPECTRUM AND POTENTIAL ENERGY SURFACE OF AR-SIO: A THEORETICAL INVESTIGATION, Richard Dawes, Michael C McCarthy

FC03   9:04–9:19
HIGH RESOLUTION ROTATIONAL SPECTROSCOPY OF CH_3^+-He, Matthias Töpfer, Thomas Salomon, Otto Dopfer, Koichi MT Yamada, Hiroshi Kohguchi, Stephan Schlemmer, Oskar Asvany

FC04   9:21–9:36
SLOW PHOTOELECTRON VELOCITY-MAP IMAGING (SEVI) SPECTROSCOPY OF CRYO-COOLED ANIONS, Marissa L. Weichman, Jongjin B. Kim, Jessalyn A. DeVine, Daniel Neumark

Intermission

FC05   10:12–10:27
DFT CALCULATION OF TORSIONAL LEVELS OF ETHANOL MOLECULE INCLUDING NONCOVALENT INTERACTIONS., Uladzimir Sapeshka, George Pitsevich, Dorozhkin Nikolay, Vytautas Balevicius

FC06   10:29–10:44
IMPROVEMENT OF THE DISSOCIATION ENERGY OF THE HYDROGEN MOLECULE (PART ONE), Joël Hussels, Cunfeng Cheng, Ming Li Niu, Hendrick Bethlem, K.S.E. Eikema, Edcel John Salumbides, Wim Ubachs, Maximilian Beyer, Nicolas Hoelsch, Josef A. Agner, Frederic Merkt, Lei-Gang Tao, Shui-Ming Hu, Christian Jungen

FC07   10:46–11:01
IMPROVEMENT OF THE DISSOCIATION ENERGY OF THE HYDROGEN MOLECULE (PART TWO), Maximilian Beyer, Nicolas Hoelsch, Josef A. Agner, Frederic Merkt, Cunfeng Cheng, Joël Hussels, Ming Li Niu, Hendrick Bethlem, K.S.E. Eikema, Edcel John Salumbides, Wim Ubachs, Christian Jungen

FC08   11:03–11:18
BENCHMARK CALCULATION OF K-EDGE IONIZATION ENERGIES USING EXACT-TWO-COMPONENT COUPLED-CLUSTER METHODS, Junzi Liu, Devin A. Matthews, Lan Cheng

FD. Fundamental interest
Friday, June 22, 2018 – 8:30 AM
Room: 217 Noyes Laboratory

Chair: Shui-Ming Hu, University of Science and Technology of China, Hefei, China

FD01   8:30 – 8:45
HAVING A BALL! MICROWAVE SPECTRUM OF THE (NEARLY) SPHERICAL TOP TEFLIC ACID, Sven Herbers, Daniel A. Obenchain, Peter Kraus, Jens-Uwe Grabow

FD02   8:47 – 9:02
HIGH PRECISION SPECTRUM OF THE SECOND OVERTONE OF $^{12}C^{16}O$, An-Wen Liu, Jin Wang, Yu Robert Sun, Shui-Ming Hu

FD03   9:04 – 9:19
IMPLEMENTING THE NEW KELVIN BY MOLECULAR PRECISION SPECTROSCOPY, Elias Moufarej, Olga Kozlova, Catherine Martin, Stephan Briaudeau, Benoit Darquie, Christophe Daussy

FD04   9:21 – 9:36
TESTING THE PARITY SYMMETRY IN COLD CHIRAL MOLECULES USING VIBRATIONAL SPECTROSCOPY, Matthieu Pierens, Louis Lecordier, Anne Cournol, Mathieu Manceau, Sean Tokunaga, Alexander Shelkovnikov, Olivier Lopez, Christophe Daussy, Anne Amy-Klein, Christian Chardonnet, Pierre Asselin, Yann Berger, Therese R. Huet, L. Margulès, R. A. Motiyenko, Richard J. Hendricks, Thomas Wall, Michael Tarbutt, Benoit Darquie

FD05   9:38 – 9:53
VIBRATIONAL MODE MIXING FERMI RESONANCE AND VIBRATIONAL RELAXATION IN THE EXCITED STATE OF HYDROGEN BONDED COMPLEXES OF A PHENOLIC CHROMOPHORE, Deb Pratim Mukhopadhyay, Souvick Biswas, Tapas Chakraborty

FD06   9:55 – 10:10
2-METHYl-1-HEXEN-3-YNE AND 3-HEXYN-2-ONE ADVENTURES IN METHYL GROUP INTERNAL ROTATION, Susanna L. Stephens, Zain Khanna, Robert Karl Bohn, Stewart E. Novick, S. A. Cooke

Intermission

FD07   10:46 – 11:01
ROTATIONAL-PREDISSOCIATION DOUBLE RESONANCE SPECTROSCOPY OF THE He-HCO$^+$ COMPLEX, Thomas Salomon, Matthias Töpfer, Phillip Schreier, Hiroshi Kohguchi, Leonid Surin, Stephan Schlemmer, Oskar Asvany

FD08   11:03 – 11:18
AMMONIA AT 10^6 V/CM IN AN 8K ARGON MATRIX: POLARIZATION, ORIENTATION, AND PENDULARIZATION, Youngwook Park, Robert W Field, Heon Kang

FD09   11:20 – 11:35
QUANTUM CASCADE LASER SPECTROSCOPY OF CARBONYL SULFIDE AND METHANOL ISOTOPOLOGUES IN HELIUM NANODROPLETS, Isaac James Miller, Ty Faulkner, Paul Raston

FE. Ions
Friday, June 22, 2018 – 8:30 AM
Room: B102 Chemical and Life Sciences

Chair: Caroline Chick Jarrold, Indiana University, Bloomington, IN, USA

FE01   8:30 – 8:45
CHARACTERIZING PEPTIDE ALPHA HELICES VIA COLD ION SPECTROSCOPY OF MODEL COMPOUNDS, John T Lawler, Christopher P Harrilal, Timothy Hill, David Fairlie, Scott A McLuckey, Timothy S. Zwier

FE02   8:47 – 9:02
INVESTIGATING ELECTRONIC AND STRUCTURAL CHANGES IMPOSED BY ZWITTERIONIC PARING IN MODEL PEPTIDE SYSTEMS USING IR-UV-IR TRIPLE RESONANCE SPECTROSCOPY, Christopher P Harrilal, Anthony Pitts-McCoy, Scott A McLuckey, Timothy S. Zwier

FE03   9:04 – 9:19
MASS-ANALYZED THRESHOLD IONIZATION SPECTROSCOPY OF P-CHLOROANISOLE, Shen-Yuan Tzeng, Wen-Bih Tzeng

FE04   9:21 – 9:36
PHOTODETACHMENT AND RESONANT PHOTOELECTRON SPECTROSCOPY OF CRYOGENICALLY-COOLED PHENOXIDE ANIONS VIA DIPOLE-BOUND EXCITED STATES, Chen-hui Qian, Guo-Zhu Zhu, Lai-Sheng Wang

FE05   9:38 – 9:53
PHOTOINDUCED CHARGE TRANSFER IN CATION-π COMPLEXES STUDIED WITH VMI, Brandon M. Rittgers, Daniel Leicht, Michael A Duncan

FE06   9:55 – 10:10
INFRARED SPECTROSCOPY OF $Zn(ACETYLENE)_{1-5}^+$: EVIDENCE OF ACETYLENE ACTIVATION BY A METAL RADICAL, Joshua H Marks, Timothy B Ward, Michael A Duncan

Intermission

FE07   10:46 – 11:01
SINGLE ATOM CATALYTIC CYCLOTRIMERIZATION OF $V(ACETYLENE)_3^+$ STUDIED WITH INFRARED SPECTROSCOPY, Joshua H Marks, Timothy B Ward, Michael A Duncan

FE08   11:03 – 11:18
THRESHOLD IONIZATION SPECTROSCOPY AND SPIN-ORBIT COUPLING OF LnNH (Ln = La and Ce) FORMED BY Ln REACTIONS WITH AMMONIA, Yuchen Zhang, Silver Nyambo, Dong-Sheng Yang

FE09   11:20 – 11:35
TWO-PHOTON IONIZATION STUDY OF THE LOW LYING STATES OF UN^+, Robert A. VanGundy, Thomas D Persinger, Michael Heaven

FE10   11:37 – 11:52
CESIUM IONIZATION AND RECOMBINATION, Sean Michael Bresler, Michael Heaven

FF. Atmospheric science
Friday, June 22, 2018 – 8:30 AM
Room: 2079 Natural History

Chair: Jacob Stewart, Connecticut College, New London, CT, USA

FF01   8:30 – 8:45
POSITIONS, INTENSITIES AND AIR-BROADENED LINE SHAPE PARAMETERS FOR THE 1←0 BANDS OF CO ISOTOPOLOGUES, V. Malathy Devi, D. Chris Benner, Keeyoon Sung, Timothy J. Crawford, Gang Li, Robert R. Gamache, Mary Ann H. Smith, Iouli E Gordon, Arlan Mantz

FF02   8:47 – 9:02
SPECTROSCOPIC STUDY OF SELF- AND AIR-BROADENED METHANE IN THE 4100-4300 cm^{-1} REGION, Adriana Predoi-Cross, V. Malathy Devi, Keeyoon Sung, Andrei V. Nikitin, Mary Ann H. Smith

FF03   9:04 – 9:19
A SPECTROSCOPIC PERTURBATION ORIGIN FOR SULFUR MASS INDEPENDENT FRACTIONATION VIA THE B-X SYSTEM OF S_2, Alexander W Hull, Shuhei Ono, Robert W Field

FF04   9:21 – 9:36
A POSSIBLE MECHANISM FOR SULFUR MASS INDEPENDENT FRACTIONATION IN THE B-X SYSTEM OF S_2, Alexander W Hull, Shuhei Ono, Robert W Field

FF05   9:38 – 9:53
NEAR-GLOBAL ATMOSPHERIC DISTRIBUTIONs OF CARBONYL SULFIDE (OCS) ISOTOPOLOGUES, Mahdi Yousefi, Peter F. Bernath, Chris Boone

Intermission

FF06   10:29 – 10:44
INFRARED SPECTROSCOPIC CHARACTERIZATION OF THE STRUCTURES OF SULFURIC ACID/AMINE/WATER CLUSTERS, Yi Yang, Sarah Waller, Eleanor Castracane, Emily E. Racow, John J. Kreinbihl, Kathleen A. Nickson, Christopher J Johnson

FF07   10:46 – 11:01
RATE CONSTANTS AND MECHANISM FOR THE REACTION OF ALKANES WITH ELECTRONICALLY EXCITED SO_2, Jay A Kroll, Veronica Vaida

FF08   11:03 – 11:18
HIGH RESOLUTION MICROWAVE SPECTROSCOPY IN A CRYOGENIC BUFFER GAS CELL: BRANCHING RATIOS AND REACTIVE INTERMEDIATES IN THE OZONOLYSIS OF ISOPRENE, Jessie P Porterfield, Sandra Eibenberger, David Patterson, Michael C McCarthy

MA. Plenary
Monday, June 18, 2018 – 8:30 AM
Room: Foellinger Auditorium

Chair: Martin Gruebele, University of Illinois at Urbana-Champaign, Urbana, IL, USA

Welcome 8:30
Susan Martinis, Vice Chancellor for Research
University of Illinois at Urbana-Champaign

MA01 8:40 – 9:20
UNDERSTANDING MOLECULES WITH NEW TOOLS

<u>JUN YE</u>, *JILA, National Institute of Standards and Technology and Univ. of Colorado Department of Physics, University of Colorado, Boulder, Boulder, CO, USA.*

Broad advances in the capabilities of controlling and spectroscopic investigation of molecules have enabled new scientific discoveries in molecular structure and interaction dynamics. We will present examples for some of the latest work.

MA02 9:25 – 10:05
ULTRAFAST VIBRONIC DYNAMICS OF FUNCTIONAL ORGANIC POLYMER MATERIALS: COHERENCE, CONFINEMENT, AND DISORDER

<u>IRENE BURGHARDT</u>, *Institute of Physical and Theoretical Chemistry, Goethe University Frankfurt, Frankfurt, Germany.*

This talk addresses quantum dynamical studies of ultrafast photo-induced energy and charge transfer in functional organic materials, complementing time-resolved spectroscopic observations [1] which underscore that the elementary transfer events in these molecular aggregate systems can be guided by quantum coherence, despite the presence of static and dynamic disorder. The intricate interplay of electronic delocalization, coherent vibronic dynamics, and trapping phenomena requires a quantum dynamical treatment that goes beyond conventional mixed quantum-classical simulations. Our approach combines first-principles parametrized Hamiltonians, based on TDDFT and/or high-level electronic structure calculations, with accurate quantum dynamics simulations using the Multi-Configuration Time-Dependent Hartree (MCTDH) method [2]. The talk will specifically focus on (i) exciton dissociation and free carrier generation in regioregular donor-acceptor assemblies [3-5], (ii) exciton multiplication in acene materials [6] and (iii) the elementary mechanism of exciton migration and creation of charge-transfer excitons in polythiophene and poly-(p-phenylene vinylene) type materials [7]. Special emphasis is placed on the influence of structural (dis)order and molecular packing, which can act as a determining factor in transfer efficiencies. Against this background, we will comment on the role of temporal and spatial coherence along with a consistent description of the transition to a classical-statistical regime.

[1] A. De Sio and C. Lienau, Phys. Chem. Chem. Phys. 19, 18813 (2017).
[2] G. A. Worth, H.-D. Meyer, H. Köppel, L. S. Cederbaum, and I. Burghardt, Int. Rev. Phys. Chem. 27, 569 (2008).
[3] M. Polkehn, H. Tamura, P. Eisenbrandt, S. Haacke, S. Méry, and I. Burghardt, J. Phys. Chem. Lett. 7, 1327 (2016).
[4] M. Polkehn, P. Eisenbrandt, H. Tamura, and I. Burghardt, Int. J. Quantum Chem. 118:e25502. (2018).
[5] M. Polkehn, H. Tamura, and I. Burghardt, J. Phys. B: At. Mol. Opt. Phys. 51, 014003 (2018).
[6] H. Tamura, M. Huix-Rotllant, I. Burghardt, Y. Olivier, and D. Beljonne, Phys. Rev. Lett. 115, 107401 (2015).
[7] R. Binder, M. Polkehn, T. Ma, and I. Burghardt, Chem. Phys. 482, 16 (2017).

Intermission

MA03
ELECTRONIC STRUCTURES OF MIXED METAL SUB-OXIDE CLUSTERS

10:40 – 11:20

CAROLINE CHICK JARROLD, *Department of Chemistry, Indiana University, Bloomington, IN, USA.*

Metal oxide clusters possess electronic, chemical, and physical properties that reflect the complex properties of defective bulk metal oxide materials, the importance of which is difficult to overstate when considering their ubiquity in applications ranging from catalysis to spintronics. Choice of binary metal combinations adds an important dimension in efforts to enhance and tune the properties of oxides, which are further affected by manipulating the oxidation state.

We have explored the intrinsically local phenomena arising from mixed metal oxides, particularly in lower-than-traditional oxidation states (sub-oxides), by applying anion photoelectron spectroscopy and density functional theory calculations to the study of small mixed metal sub-oxide clusters. Anion photoelectron spectroscopy is a mass-selective method that probes the energies of the manifold of low-lying electronic states inherent in neutral sub-oxide species. By coupling experimental with computational results, a detailed picture of how mixed metal composition impacts molecular and electronic structure emerges. For example, profoundly asymmetric metal-oxygen bond formation in near-neighbor mixed transition or lanthanide metal oxides can be reconciled with relative oxophilicities, and the prevalence of antiferromagnetic spin states can be reconciled with localization of metal atomic orbitals in mixed systems. Striking charge separation is observed in trans-periodic mixed metal oxides that combine transition and post-transition metals, or transition and lanthanide metals, combinations that are evocative of strongly-interacting catalyst-support systems. Finally, we consider whether more can be gleaned from anion photoelectron spectroscopy of these exceptionally complex systems by exploiting the electron-kinetic-energy-dependent neutral-electron interactions.

MA04
EXPLORATIONS OF INFRARED SPECTRA OF CRIEGEE INTERMEDIATES AND THEIR REACTIONS

11:25 – 12:05

YUAN-PERN LEE, *Applied Chemistry, National Chiao Tung University, Hsinchu, Taiwan, Institute of Atomic and Molecular Sciences, Academia Sinica, Taipei, Taiwan.*

Criegee intermediates, carbonyl oxides produced in ozonolysis of unsaturated hydrocarbons, play important roles in atmospheric chemistry. A new production scheme using photolysis of $R_2CI_2 + O_2$ facilitated the production and direct detection of Criegee intermediates with various spectral techniques and has stimulated rapidly expanding research.[a,b] Our understanding of important atmospheric reactions involving Criegee intermediates is becoming clarified because of the direct probing of Criegee intermediates in kinetic experiments. The infrared spectra of CH_2OO,[c,d] CH_3CHOO,[e] and $(CH_3)_2COO$[f] have been recorded with a step-scan FTIR with resolution 0.25 to 1 cm^{-1}; rotational contours with unresolved rotational lines were reported. On employing a quantum cascade laser coupled with a Herriot cell, we recorded spectra of the O-O stretching bands of CH_2OO and CH_3CHOO in the region 880-932 cm^{-1} at resolution 0.002 cm^{-1}. In addition to improved rotational parameters, perturbation was observed at high-J levels of K_a = 3, 6, and 11 of CH_2OO. Distinct lines of *syn*- and *anti*-CH_3CHOO were also observed. Kinetic investigations based on this new experimental scheme will be presented. Taking advantage of the wide spectral coverage of an FTIR, we investigated the mechanism of the reactions of CH_2OO with SO_2, HNO_3, HCl, and $HCOOH$. For example, in the reaction of $CH_2OO + HCOOH$, eight observed bands are assigned to hydroperoxymethyl formate HPMF (P5). In the later reaction period, three bands are assigned to an isomer HPMF (P6) and three bands to the final product, *anti*-FAN. According to our kinetic analysis, only P5, not P6, decomposes to form FAN.

---

[a] Y.-P. Lee, *J. Chem. Phys.* **143**, 020901 (2015).
[b] D. L. Osborn, C. A. Taatjes, *Int. Rev. Phys. Chem.* **34**, 309 (2015).
[c] Y. T. Su, Y.-H. Huang, H. A. Witek, Y.-P. Lee, *Science* **340**, 174 (2013).
[d] Y.-H. Huang, J. Li, H. Guo, Y.-P. Lee, *J. Chem. Phys.* **142**, 214301 (2015).
[e] H.-Y. Lin, Y.-H. Huang, X. Wang, J. M. Bowman, Y. Nishimura, H. A. Witek, Y.-P. Lee, *Nature Comm.* **6**, 7012 (2015).
[f] Y.-Y. Wang, C.-Y. Chung, Y.-P. Lee, *J. Chem. Phys.* **145**, 154303 (2016).

MG. Mini-symposium: Frequency-Comb Spectroscopy

Monday, June 18, 2018 – 1:45 PM

Room: 116 Roger Adams Lab

Chair: Marissa L. Weichman, JILA, Boulder, CO, USA

MG01   **INVITED TALK**   1:45 – 2:15

PRECISION RAMSEY-COMB SPECTROSCOPY OF MOLECULAR HYDROGEN IN THE DEEP-UV

L.S. DREISSEN, R.K. ALTMANN, C. ROTH, M.G.J. FAVIER, J. KRAUTH, EDCEL JOHN SALUMBIDES, WIM UBACHS, K.S.E. EIKEMA, *Department of Physics and Astronomy, VU University, Amsterdam, Netherlands.*

High-precision spectroscopy experiments with simple atomic and molecular systems provide important benchmarks for tests of bound-state Quantum Electrodynamics (QED). In recent years there has been significant progress in both experiment [1,2] and theory [3] for QED tests in H_2. Even investigations of the proton radius puzzle [4] might become feasible in this molecule if the dissociation energy can be determined at a level of 10 kHz [3]. For this latter target, we developed an excitation method, Ramsey-comb spectroscopy, that enables kHz-level precision spectroscopy in the deep-UV. The method is based on excitation with amplified and upconverted frequency comb laser pulses [5]. It has allowed us to measure the $EF^1\Sigma_g^+ - X^1\Sigma_g^+(0,0)$ Q1 two-photon transition at 202 nm with an accuracy of 73 kHz [6]. This result is two orders of magnitude better than obtained with previous experiments, and combined with future improved measurements of the EF ionization energy, this could lead to a dissociation energy with an uncertainty below 100 kHz. New measurements from V=1 and N=0 are now in preparation. Moreover, a new setup has been constructed to extend Ramsey-comb spectroscopy to the vacuum- and extreme-ultraviolet spectral region through high-harmonic generation, and an experiment is in preparation to demonstrate this.

[1] J. Liu et al., J. Chem. Phys. 130, 174306 (2009)
[2] W. Ubachs et al., J. Mol. Spectr. 320, 1-12 (2016)
[3] M. Puchalski et al., Phys. Rev. A 95, 052506 (2017)
[4] R. Pohl et al., Science 353, 669 (2016)
[5] J. Morgenweg et al., Nature Physics 10, 30-33 (2014)
[6] R.K. Altmann et al., Phys. Rev. Lett. 120, 043204 (2018)

MG02   2:19 – 2:34

PRECISION MEASUREMENT OF THE IONIZATION ENERGY OF METASTABLE He_2

PAUL JANSEN, LUCA SEMERIA, FREDERIC MERKT, *Laboratorium für Physikalische Chemie, ETH Zurich, Zurich, Switzerland.*

Predicting the energy-level structure of molecules from first principles is one of the major goals and tasks of theoretical molecular physics and chemistry. Few-electron molecules are particularly attractive systems for comparison with experimental results because numerically "exact" predictions of molecular properties can in principle be obtained, i.e., predictions that are only limited in accuracy by the uncertainties of fundamental constants.

We present absolute-frequency measurements of transitions from rotational levels of the a $^3\Sigma_u^+$ ($v''=0$) metastable state of He_2 to np Rydberg states. The transition frequencies are determined by one-photon UV spectroscopy in slow molecular beams using a narrow-band laser system referenced to a frequency comb. The ionization energy of metastable He_2 and the rotational structure of the X^+ $^2\Sigma_u^+$ ($v^+=0$) ground state of He_2^+ have been determined with unprecedented precision and accuracy using Rydberg-series extrapolation.

MG03   2:36–2:51

PRECISION SPECTRA OF $A\,^2\Sigma^+, v' = 0 \leftarrow X\,^2\Pi_{3/2}, v'' = 0, J'' = 3/2$ TRANSITIONS IN ^{16}OH AND ^{16}OD

ARTHUR FAST, *Precision Infrared Spectroscopy on Small Molecules, Max Planck Institute for Biophysical Chemistry, Göttingen, Germany*; JOHN FURNEAUX, *Homer L Dodge Department of Physics and Astronomy, University of Oklahoma, Norman, OK, USA*; SAMUEL MEEK, *Precision Infrared Spectroscopy on Small Molecules, Max Planck Institute for Biophysical Chemistry, Göttingen, Germany.*

The hydroxyl radical, OH, is a prototypical open-shell diatomic molecule that is important in a variety of fields, including atmospheric chemistry, interstellar chemistry, crossed beam molecular collision studies, and Stark deceleration. In laboratory studies, OH is commonly detected with rotational state selectivity by measuring laser-induced fluorescence from ultraviolet $A\,^2\Sigma^+ - X\,^2\Pi$ transitions. Previous studies have determined the absolute frequencies of these transitions to within approximately 0.003 cm^{-1} (100 MHz) [a]. This level of accuracy is quite sufficient for excitation with commonly-used frequency-doubled pulsed dye lasers, which typically have a bandwidth on the order of 0.1 cm^{-1}, but for driving the transitions with a continuous-wave (CW) laser with a linewidth on the order of 1 MHz or less, the transition frequencies must be known much more exactly.

In this talk, I would like to present our recent high-precision measurements of the $A\,^2\Sigma^+, v' = 0 \leftarrow X\,^2\Pi_{3/2}, v'' = 0, J'' = 3/2$ transitions in ^{16}OH and ^{16}OD. Using a frequency-doubled CW dye laser which is stabilized and monitored with the help of an optical frequency comb, we have measured transitions to the 12 lowest levels of the $A\,^2\Sigma^+, v' = 0$ vibronic state of ^{16}OH with an uncertainty of less than 100 kHz (10^{-10} relative uncertainty) and are currently completing measurements of transitions to the 16 lowest A levels in ^{16}OD with an expected uncertainty of 100–200 kHz. These measurements have enabled us to determine the ^{16}OH $A\,^2\Sigma^+, v' = 0$ band origin with three orders of magnitude higher precision and the rotational constant with two orders of magnitude higher precision than previously possible. Similar improvements are expected in the corresponding constants of ^{16}OD, as well as in its spin-rotation constant γ, which has not been measured in microwave double-resonance experiments as in ^{16}OH [b].

[a] G. Stark et al. *J. Opt. Soc. Am. B*, 11:3–32, 1994
[b] J. J. ter Meulen et al. *Chem. Phys. Lett.*, 129:533–537, 1986

MG04   2:53–3:08

DETERMINATION OF ROVIBRATIONAL INTERVALS IN H_2^+ WITH SUB-MHZ ACCURACY

MAXIMILIAN BEYER, NICOLAS HOELSCH, FREDERIC MERKT, *Laboratorium für Physikalische Chemie, ETH Zurich, Zurich, Switzerland*; CHRISTIAN JUNGEN, *Laboratoire Aimé Cotton, CNRS, Orsay, France.*

H_2^+ is the simplest of all molecules and as such an important system for the development of molecular quantum mechanics. The rovibrational energy-level structure of this one-electron system can be calculated extremely precisely by quantum-chemical methods[a]. By comparison with the results of precise spectroscopic measurements of rovibrational intervals, fundamental constants or particle properties, such as the proton-to-electron mass ratio or the proton size, can be determined[b]. Because the rotational and vibrational transitions of H_2^+ are electric-dipole forbidden, the experimental data on its energy-level structure are limited.

We present the determination of spin-rovibrational intervals in H_2^+ from high-resolution measurements of the Rydberg spectrum of H_2 and Rydberg-series extrapolation using multichannel quantum defect theory[c]. Choosing suitable double-well valence states of H_2, characterized by long lifetimes and favorable Franck-Condon factors to different vibrational states in the ion, allows us to excite Rydberg states that converge on selected rovibrational levels of H_2^+.

For the excitation of Rydberg states, a resonant three-photon excitation scheme was employed, using pulsed VUV and VIS laser sources to reach the intermediate valence state and a continuous-wave (cw) near-infrared laser source for the excitation to the Rydberg states. The valence state - Rydberg state intervals could be measured with a relative accuracy of 3E-10 using an optical frequency comb for the frequency calibration of the cw laser and minimizing systematic uncertainties[d].

[a] V. I. Korobov, L. Hilico, and J.-Ph. Karr, Phys. Rev. A 89, 032511 (2014)
[b] J.-Ph. Karr, L. Hilico, J. C. J. Koelemeij, and V. I. Korobov, Phys. Rev. A 94, 050501(R) (2016)
[c] D. Sprecher, Ch. Jungen and F. Merkt, J. Chem. Phys. 140, 104303:1-18 (2014)
[d] M. Beyer, N. Hölsch, J. A. Agner, J. Deiglmayr, H. Schmutz, and F. Merkt, Phys. Rev. A 97, 012501 (2018)

MG05 3:10 – 3:25

DIRECT FREQUENCY-COMB-DRIVEN RAMAN TRANSITIONS IN THE TERAHERTZ RANGE

<u>CYRILLE SOLARO</u>, STEFFEN MEYER, KARIN FISHER, MICHAEL DePALATIS, MICHAEL DREWSEN, *Department of Physics and Astronomy, University of Aarhus, Aarhus, Denmark.*

I will present our recent results on the use of a femtosecond frequency comb to coherently drive stimulated Raman transitions between terahertz-spaced atomic energy levels [1]. More specifically, we address the $3d\,^2D_{3/2}$ and $3d\,^2D_{5/2}$ fine structure levels of a single trapped ^{40}Ca$^+$ ion and spectroscopically resolve the transition frequency with a relative accuracy of 5.5×10^{-12} ! The achieved accuracy is nearly a factor of five better than the previous best Raman spectroscopy [2], and is currently limited by the inaccuracy of our atomic clock reference. Using direct frequency comb Raman spectroscopy on four other isotopes ^{42,44,46,48}Ca$^+$, in combination with precise measurements of the $4s\,^2S_{1/2} - 3d\,^2D_{5/2}$ transition, we were also able to improve bounds on new physics beyond the standard model [3,4].

Furthermore, I will discuss the population dynamics of frequency-comb-driven Raman transitions which can be fully predicted from the spectral properties of the femtosecond frequency comb. We achieved Rabi oscillations with a contrast of 99.3(6)% and milliseconds coherence time (see figure)! The technique can be easily generalized to transitions in the sub-kHz [5] to tens of THz range and should be applicable for driving, e.g., spin-resolved rovibrational transitions in molecules and hyperfine transitions in highly charged ions.

[1] C. Solaro et al. arXiv:1712.07429 (2017)
[2] R. Yamazaki et al. PRA 77,012508 (2008)
[3] S. Meyer et al. in preparation
[4] J.C. Berengut et al. arXiv:1704.05068 (2017)
[5] D. Hayes et al. PRL 104,140501 (2010)

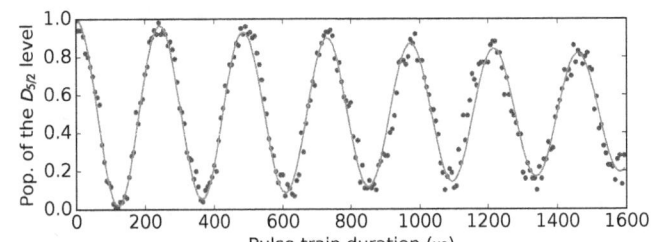

Intermission

MG06 4:01 – 4:16

RELATIVE INTENSITY OF A CROSSOVER RESONANCE TO LAMB DIPS OBSERVED IN STARK SPECTROSCOPY OF METHANE II

<u>SHOKO OKUDA</u>, HIROYUKI SASADA, *Department of Physics, Faculty of Science and Technology, Keio University, Yokohama, Japan.*

We carried out Stark-modulation spectroscopy of the ν_3 band of methane [1]. Figure shows observed spectra of the $P(4)E$, $Q(4)E$, and $R(4)E$ transitions with the selection rule of $\Delta M = \pm 1$, where M is the angular momentum quantum number along the Stark filed. Each triplet includes two Lamb-dips from $|M''| = 2$ to $|M'| = 1$ and from $|M''| = 0$ to $|M'| = 1$ and a crossover resonance (COR) at the middle. The COR is the largest in intensity in the triplet for the Q- and R-branch transitions, and middle for the P-branch transition. The COR of the Λ-type three-level system overlaps in frequency with that of the V-type three-level system of $|M''| = 1$ and $|M'| = 0$ and 2, and the relative intensity of the COR to the Lamb dips is analyzed using a steady-state solution of rate equations. The model fairly agrees with the observed relative intensity, and detailed analysis is in progress. [1] S. Okuda, H. Sasada, J. Opt. Soc. Am. B, **34**, 2558-2568 (2017).

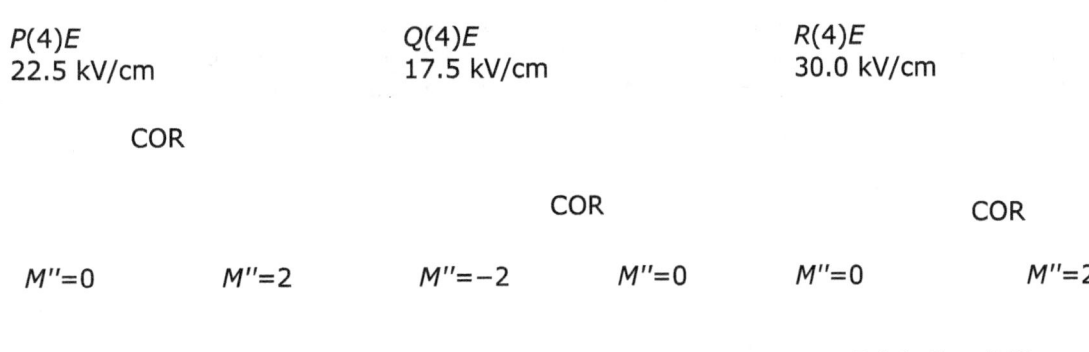

MG07

ACCURATE COMB-ASSISTED CAVITY RING DOWN SPECTROSCOPY OF MOLECULAR HYDROGEN

<u>SAMIR KASSI</u>, TIM STOLTMANN, MAGDALENA KONEFAL, ALAIN CAMPARGUE, *UMR5588 LIPhy, Université Grenoble Alpes/CNRS, Saint Martin d'Hères, France.*

Because molecular hydrogen is the simplest molecule, it is considered as the best candidate for a direct comparison of experiment against high level ab initio calculations, both in terms of transition frequencies and line strength. Unfortunately, this apparent simplicity is not only spoiled by the weakness of the transition, but also by its surprisingly complex line profile that hampers accurate parameters to be straightforwardly derived. To address that problem, we have recorded with unprecedented sensitivity pure H_2 Q(1) 2-0 and D_2 S(2) 2-0 transitions around 1.24 and 1.59 μm, respectively, down to a pressure of 100 Pa. A limit of detection of about 2×10^{-12} cm^{-1} was achieved with the two accurate comb-referenced cavity ring down spectrometers used. Effective parameters were determined for different line profiles (NGP, SDNGP, HTP), allowing line reproduction down to the noise level. The zero pressure parameters will be presented and discussed.

MG08

COMB-REFERENCED COHERENT RAMAN SPECTROSCOPY ON PURE H_2

<u>MARCO LAMPERTI</u>, LUCILE RUTKOWSKI, DAVIDE GATTI, RICCARDO GOTTI, GIULIO CERULLO, DARIO POLLI, MARCO MARANGONI[a], *Dipartimento di Fisica, Politecnico di Milano, Milano, Italy*; FRANCK THIBAULT, *Institut de Physique de Rennes, Université de Rennes 1, Rennes, France*; PIOTR MASLOWSKI, SZYMON WOJTEWICZ, PIOTR WCISLO, *Institute of Physics, Faculty of Physics, Astronomy and Informatics, Nicolaus Copernicus University, Torun, Poland.*

H_2 is a benchmark system for testing quantum electrodynamics and physics beyond the standard model via highly accurate measurements of transition frequencies, which has been the subject of many works during the past decades[b,c,d]. However, retrieving the unperturbed transition frequencies requires to measure spectra at very low pressure, where the low density combined with weak quadrupole transition moments makes it challenging to achieve high signal-to-noise ratios. An alternative approach is to model very precisely the transition profiles at higher pressure in order to correct for the strong Dicke narrowing and speed-dependent collisional effects which distort the absorption profiles.

We present a new approach to measure H_2 transition frequencies in the fundamental rovibrational band with high accuracy and signal-to-noise ratio. The approach uses an optical frequency comb to calibrate the frequency spacing between a cw pump and a cw Stokes beams that interact with H_2 in a multipass cell by stimulated Raman scattering. Specifically, we focus on the Q(1) transition of the 1-0 band of pure H_2 at 4155.25cm^{-1}. The pump laser emits at 737.8 nm and is kept fixed while the Stokes laser is swept over 0.3cm^{-1} around 1064 nm. The wavelength of the Stokes laser is referenced to a local oscillator, which in turn is locked to an optical frequency comb along with the pump laser. The frequency comb is obtained using an Er:fiber amplified femtosecond laser with stabilized repetition rate and offset frequency to a reference Rb standard. The profiles measured at various pressures, spanning from 0.1 to 5 atm, are fitted using Hartmann-Tran profiles. The retrieved parameters are compared to *ab-initio* values based on H_2-H_2 quantum scattering calculations.

---

[a]The authors acknowledge the CH2ROME project, R164WYYR8N, Italian Ministry of Education and Research.
[b]L. A. Rahn, R. L. Farrow, and G. J. Rosasco, Physical Review A, 43(11), 6075 (1991).
[c]G. D. Dickenson, M. L. Niu, et al., Physical review letters, 110(19), 193601 (2013).
[d]S. Kassi and A. Campargue, Journal of Molecular Spectroscopy, 300, 55-59 (2014).

MG09 4:52 – 5:07

SUB-DOPPLER SPECTROSCOPY OF THE ν_2 FUNDAMENTAL BAND AND FIRST HOT BAND OF THE H_3^+ CATION

CHARLES R. MARKUS, PHILIP A. KOCHERIL, ANNE MARIE ESPOSITO, ALEX W SCHRADER, *Department of Chemistry, University of Illinois at Urbana-Champaign, Urbana, IL, USA*; BENJAMIN J. McCALL, *Departments of Chemistry and Astronomy, University of Illinois at Urbana-Champaign, Urbana, IL, USA.*

The simplest polyatomic molecule, H_3^+, serves as an important benchmark for *ab initio* theory and is an important constituent of the interstellar medium (ISM). In the ISM, H_3^+ initiates a chain of ion-neutral reactions which leads to more complex chemistry, and observations of H_3^+ can be used to measure interstellar conditions such as the cosmic ray ionization rate.[a] For *ab initio* theorists, accurate calculations of the rovibrational structure of H_3^+ require going beyond the Born-Oppenheimer approximation, and for its low-lying rovibrational states, agreement between theory and experiment has reached 0.001 cm^{-1}.[b] As these calculations begin to rival experimental measurements, new data are needed to benchmark future *ab initio* approaches.

Using the technique Noise-Immune Cavity-Enhanced Optical Heterodyne Velocity Modulation Spectroscopy (NICE-OHVMS)[c] to perform sub-Doppler spectroscopy and an optical frequency comb to accurately calibrate the frequency, we have expanded our survey of H_3^+ to include transitions from higher rotational levels in the fundamental band and transitions in the $2\nu_2^2 \leftarrow \nu_2^1$ hot band. Using combination differences, we have determined a number of energy level spacings in the ground state with an accuracy of ~ 5 MHz, which are directly compared with state of the art *ab initio* calculations. We also discuss our progress towards calculating "forbidden" rotational transitions, including a possible astrophysical maser,[d] which requires our newly measured hot band transitions.

---

[a] N. Indriolo, *Phil. Trans. R. Soc. A*, **370**, 5142 (2012).
[b] L. G. Diniz, J. R. Mohallem, A. Alijah, M. Pavanello, L. Adamowicz, O. L. Polyansky, and J. Tennyson, *Phys. Rev. A.*, **88**, 032506 (2013).
[c] J. N. Hodges, A. J. Perry, P. A. Jenkins II, B. M. Siller, and B. J. McCall, *J. Chem. Phys*, **139**, 164201 (2013).
[d] J. H. Black, *Phil. Trans. R. Soc. A*, **358**, 2515 (2000)

MG10 5:09 – 5:24

SUB-DOPPLER FREQUENCY METROLOGY IN HD FOR TEST OF FUNDAMENTAL PHYSICS

F.M.J. COZIJN, EDCEL JOHN SALUMBIDES, K.S.E. EIKEMA, WIM UBACHS, *Department of Physics and Astronomy, VU University, Amsterdam, Netherlands*; PATRICK DUPRÉ, *Laboratoire de Physico-Chimie de l'Atmosphère, Université du Littoral Côte d'Opale, Dunkerque, France.*

Molecular hydrogen has evolved into a benchmark quantum test system for fundamental physics. Accurate results on the vibrational splitting in hydrogen isotopologues can be exploited to provide a test of QED in the smallest neutral molecule, and open up an avenue to resolve the proton radius puzzle, as well as constrain putative fifth forces and extra dimensions.

We will present the first sub-Doppler determination of weak dipole transitions in the (2,0) overtone band of HD at $\lambda \sim 1.38\,\mu$m. To saturate and detect the weak absorption we have implemented a technique called Noise-Immune Cavity-Enhanced Optical Heterodyne Molecular Spectroscopy (NICE-OHMS). To obtain an absolute frequency during the measurements, the spectroscopy laser is simultaniously locked onto a Cs-clock referenced optical frequency comb. The obtained Doppler-free linewidth of ~ 300kHz (FWHM) could give access and insight into the underlying hyperfine structure. Our current determination of the obtained transition frequencies is around 30 kHz; a 1000-fold improvement on the previous Doppler-broadened determination.

MH. Mini-symposium: New Ways of Understanding Molecular Spectra

Monday, June 18, 2018 – 1:45 PM

Room: 100 Noyes Laboratory

Chair: Stephan Schlemmer, I. Physikalisches Institut, Köln, Germany

---

MH01 **INVITED TALK** 1:45 – 2:15

ALGEBRAIC APPROACHES AND THEIR CONNECTION WITH PHASE SPACE METHODS: APPLICATIONS TO SPECTROSCOPY

RENATO LEMUS, *Estructura de la Materia, Instituto de ciencias Nucleares, Mexico City, Mexico.*

First the salient features of the $U(\nu+1)$ algebraic approach associated to ν equivalent oscillators are presented. Then we introduce the 1D case through the connection of the $U(2)$ algebra with the Morse/Pöshl-Teller potentials with the goal of describing the vibrational degrees of freedom of non linear polyatomic molecules. The coordinates and momenta are then identified and generalized to any potential, providing the possibility to solve the 1D Schrödinger equation for general potentials by purely algebraic means using the concept of transformation brackets. A new procedure to calculate of Franck-Condon factors is presented. Because of their importance in linear molecules the $U(3)$ model is introduced, emphasizing its connection with configuration space. It is shown the application of the $U(2) \times U(3) \times U(2)$ algebraic approach to describe the Raman spectroscopy of the CO_2 molecule. The $U(3)$ model is applied to consider general potentials to describe linear-to-bend transition in triatomic molecules. Finally the $U(4)$ model is introduced to describe 3D systems for general potentials. The Hydrogen atom as well as the 3D Morse systems are analyzed by purely algebraic means as a benchmark to show how to apply the algebraic method for potentials with spectroscopic interest.

MH02 2:19 – 2:34

VIBRATIONAL QUANTUM GRAPHS AND THEIR APPLICATION TO THE QUANTUM DYNAMICS OF CH_5^+

CSABA FÁBRI, *Laboratory of Molecular Structure and Dynamics, Eötvös University, Budapest, Hungary;* ATTILA CSÁSZÁR, *Research Group on Complex Chemical Systems, MTA-ELTE, Budapest, Hungary.*

The first application of the quantum graph model to vibrational quantum dynamics of molecules is reported. The usefulness of the approach is demonstrated for the astructural molecular ion CH_5^+, an enigmatic system of high-resolution molecular spectroscopy and molecular physics, challenging our traditional understanding of chemical structure and rovibrational quantum dynamics. The vertices of the quantum graph correspond to different versions of the molecule (120 in total for CH_5^+), while the differently colored edges represent different collective nuclear motions transforming the distinct versions into one or another. These definitions allow the mapping of the complex low-energy vibrational quantum dynamics of CH_5^+ onto the motion of a one-dimensional particle confined in a quantum graph. The quantum graph model provides a simple and intuitive qualitative understanding of the intriguing low-energy vibrational dynamics of CH_5^+ and is able to reproduce, with just two adjustable parameters related to the two different motions (indicated by the red and blue lines in the figure), the lowest vibrational energy levels of CH_5^+ (and CD_5^+) with remarkable accuracy.

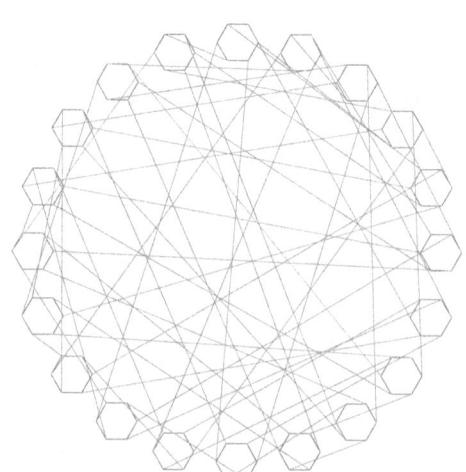

MH03   2:36 – 2:51

A VARIATIONAL METHOD FOR COMPUTING VIBRATIONAL SPECTRA OF MOLECULES WITH UP TO 18 ATOMS

PHILLIP THOMAS, TUCKER CARRINGTON, *Department of Chemistry, Queen's University, Kingston, ON, Canada.*

I shall present an improvement and applications of the Hierarchical Intertwined Reduced-Rank Block Power Method (J. Chem. Phys, 146, 204110 (2017)) for solving the vibrational Schroedinger equation. The improvement decreases the memory required to compute a spectrum. Variational calculations for molecules with a dozen atoms are now possible on a desktop computer. The memory cost scales linearly with the number of atoms in the molecule. We apply the HI-RRBPM to compute vibrational spectra of uracil and naphthalene, with 12 and 18 atoms, respectively. The HI-RRBPM uses a direct product basis but: 1) it is not necessary to store a direct-product-basis matrix representation of the Hamiltonian matrix (for naphthalene the size of the matrix would be $\sim 10^{48}$); 2) it is not necessary to store vectors whose length is equal to the size of the direct-product basis. This is accomplished by using sum-of-product (SOP) basis functions stored in a canonical polyadic tensor format and generated by evaluating matrix-vector products. The number of terms in the SOP basis functions is minimized by optimising the factors. Representing vibrational wavefunctions as optimised SOPs reveals the essential entangledness and provides new understanding. The method only works if the Hamiltonian is itself a SOP.

MH04   2:53 – 3:08

A NONDIRECT PRODUCT DISCRETE VARIABLE REPRESENTATION-LIKE METHOD FOR CALCULATING VIBRATIONAL SPECTRA OF POLYATOMIC MOLECULES

EMIL J ZAK, TUCKER CARRINGTON, *Department of Chemistry, Queen's University, Kingston, ON, Canada.*

We present a new method for solving the vibrational Schroedinger equation for polyatomic molecules. It has the following advantages: 1) the size of the matrix eigenvalue problem is the size of the required pruned (nondirect product) polynomial-type basis; 2) it requires solving a regular, and not a generalized, symmetric matrix eigenvalue problem; 3) accurate results are obtained even if quadrature points and weights are not good enough to yield a nearly exact overlap matrix; 4) the potential matrix is diagonal; 5) the matrix-vector products required to compute eigenvalues and eigenvectors can be evaluated by doing sums sequentially, despite the fact that the basis is pruned. To achieve these advantages we use sets of nested Leja points and appropriate Leja quadrature weights and special hierarchical basis functions. Matrix-vector products are inexpensive because transformation matrices between the basis and the grid, and their inverses, are lower triangular. Vibrational energy levels of CH_2NH are calculated with the new method. For this purpose a simple harmonic oscillator kinetic energy operator and a quartic force field are used.

MH05   3:10 – 3:25

THE $\tilde{A}^2 E''$ STATE OF NO_3: NEW VIBRONIC CALCULATIONS

BRYAN CHANGALA, *JILA, National Institute of Standards and Technology and Univ. of Colorado Department of Physics, University of Colorado, Boulder, CO, USA*; JOHN F. STANTON, *Physical Chemistry, University of Florida, Gainesville, FL, USA.*

The first excited electronic state of the NO_3 radical is a dark state just less than 1 eV above the ground state, and its vibronic features have been observed optically by the Hirota, Okumura and Miller groups. In addition, it has been directly accessed by the Neumark group via photodetachment of the nitrate anion. This $^2E''$ state represents a case where the Jahn-Teller effect is neither weak enough to be treated with low-order (linear or quadratic) models nor is it profoundly distorted to the degree that it can be accurately represented by a quasistatic lower-symmetry geometry. While efforts to treat this state with polynomial Jahn-Teller model Hamiltonians have been carried out, none of these can really be regarded as being quantitatively successful. In this work, we use newly available methodology to treat this electronic state with a semi-global potential based on permutationally invariant polynomials and report results of the corresponding energy levels and simulated photodetachment spectrum.

Intermission

MH06

MULTI-CHANNEL QUANTUM DEFECT THEORY CALCULATION OF VIBRATIONAL AUTOIONIZATION RESONANCE WIDTH OF $v = 1$, $n^* \approx 14$ CaF RYDBERG STATE

JUN JIANG, TIMOTHY J BARNUM, ROBERT W FIELD, *Department of Chemistry, MIT, Cambridge, MA, USA.*

Vibrational auto-ionization resonance widths (γ) of $v = 1$, $n^* \approx 14$ Rydberg states of CaF are calculated in this work, based on results of a global multi-channel quantum defect fit. The calculation indicates that the $n.36\ p\Pi$ eigen-channel has the shortest vibrational auto-ionization lifetime, ~ 10 ps, which is at least $4\times$ shorter than the lifetime of all other CaF eigen-channels, in agreement with experimental observations. In addition, the calculation successfully reproduces the experimental observations that γ of the $14.36\ p\Pi^-$ rotational sequence (where the superscript '-' indicates negative Kronig symmetry) are nearly N-independent, while those of the $14.36\ p\Pi^+$ rotational sequence (where the superscript '+' indicates positive Kronig symmetry) decrease quickly as a function of N, i.e. $\gamma(N = 10) \approx \frac{1}{2}\gamma(N = 1)$. By examining the eigen-channel composition of the two rotational sequences of state of opposite Kronig symmetry, we are able to show that the significantly faster decrease of γ for the $14.36\ p\Pi^+$ rotational sequence is caused by the stronger l-uncoupling interaction in the positive Kronig symmetry manifold. Based on a valence-precursor model (first suggested by Mulliken), the significantly faster vibrational auto-ionization rate of the $n.36 p\Pi$ eigen-channel is explained based on the electronic properties of its valance-precursor state, the $C^2\Pi$ state, for which the electron density is polarized toward the fluorine atom.

MH07

ENERGETICS AND COUPLINGS IN OLIGOACENE-BASED SINGLET FISSION: EFFICIENT AND ACCURATE DENSITY FUNCTIONAL THEORY TELLS THE STORY

ZHOU LIN, HIKARI IWASAKI, HONG-ZHOU YE, TROY VAN VOORHIS, *Department of Chemistry, Massachusetts Institute of Technology, Cambridge, MA, USA.*

In the sub-picosecond singlet fission (SF) process, two coherent triplet excitons ($^1(T_1T_1)$) are generated from one singlet exciton (S_1) through a mysterious "multi-exciton" (^{1}ME) intermediate: $S_1 \longleftrightarrow {}^1\text{ME} \longleftrightarrow {}^1(T_1T_1) \longrightarrow 2T_1$.[a] Organic semiconducting materials that undergo exothermic SF double the efficiency associated with high-energy incident photons and allow the photovoltaics to exceed the Shockley–Queisser limit of 33.7%.[b] Understanding the character of ^{1}ME serves as the key to deciphering the ultrafast SF mechanism. Based on a popular hypothesis, the complete SF procedure involves two steps of charge transfer (CT) and ^{1}ME is a superposition of localized and charge-separated excited states.[c] However, the straightforward experimental and theoretical evidence to support this hypothesis is still in darkness due to the challenges to measure ultrafast photochemistry and to describe charge-separated and multiply-excited energy levels. Herein we modeled the local SF reactions occurring in hexacene, pentacene and bis(6,13-bis(triisopropylsilylethynyl)pentacene)benzene in the crystalline or solution phase. The efficient and accurate density functional theory (DFT) based approaches developed in the earlier[d] and present study were utilized to evaluate energies of S_1, ^{1}ME and $^1(T_1T_1)$, as well as the non-adiabatic coupling between each pair of them. Our results shed light on the roles of localized and charge-separated excited states in the complete SF mechanism and propose the strategy for the rational design of oligoacene-based SF materials.

---

[a]M. B. Smith and J. Michl, *Chem. Rev.* **110**, 6891 (2010).
[b]W. Shockley and H. J. Queisser, *J. Appl. Phys.* **32**, 510 (1961).
[c]T. C. Berkelbach, M. S. Hybertsen, and D. R. Reichman, *J. Chem. Phys.* **141**, 074705 (2014).
[d]S. R. Yost, J. Lee, M. W. B. Wilson, T. Wu, D. P. McMahon, R. R. Parkhurst, N. J. Thompson, D. N. Congreve, A. Rao, K. Johnson, M. Y. Sfeir, M. G. Bawendi, T. M. Swager, R. H. Friend, M. A. Baldo and T. Van Voorhis, *Nat. Chem.*, **6**, 492 (2014).

MH08   4:35 – 4:50

COMPUTATIONAL SPECTROSCOPY OF NCS IN THE RENNER-DEGENERATE ELECTRONIC STATE $\tilde{X}\,^2\Pi$

JENS FREUND, SARAH CAROLL GALLEGUILLOS KEMPF, PER JENSEN, *Faculty of Mathematics and Natural Sciences, University of Wuppertal, Wuppertal, Germany*; UMPEI NAGASHIMA, *Foundation for Computational Science, Kobe, Japan*; TSUNEO HIRANO, *Department of Chemistry, Ochanomizu University, Tokyo, Japan*.

$\tilde{X}\,^2\Pi$ NCS is a Renner-degenerate linear molecule whose rovibronic spectrum is greatly complicated by the Renner effect and all-pervading resonances. As an alternative avenue to understanding this spectrum, we have calculated values of the ro-vibronic energies, intensities, and rotational constants by direct numerical solution of the rovibronic Schrödinger equation with the RENNER program.[a] All values obtained are in good agreement with the available experimental data. Rovibronic spectra are also simulated. The Renner calculations are based on three-dimensional potential energy surfaces and dipole moment surfaces computed *ab initio* for NCS in the $\tilde{X}\,^2\Pi$ electronic ground state at the core-valence, full-valence MR-SDCI+Q/[aug-cc-pCVQZ(N, C, S)] level of theory.

[a] J. Freund, S. C. Galleguillos Kempf, P. Jensen, U. Nagashima, T. Hirano, *J. Mol. Spectrosc.* **345**, 31–38 (2018). DOI: 10.1016/j.jms.2017.11.010; T. Hirano, U. Nagashima, P. Jensen, J. Mol. Spectrosc., (2018), https://doi.org/10.1016/j.jms.2017.12.011.

MH09   4:52 – 5:07

RO-VIBRATIONALLY AVERAGED STRUCTURE OF $^2\Pi$ NCS: RE-INTERPRETATION OF THE B_0 VALUES

TSUNEO HIRANO, *Department of Chemistry, Ochanomizu University, Tokyo, Japan*; UMPEI NAGASHIMA, *Foundation for Computational Science, Kobe, Japan*; PER JENSEN, *Faculty of Mathematics and Natural Sciences, University of Wuppertal, Wuppertal, Germany*.

We have constructed *ab initio* 3D potential energy surfaces (PESs) for $\tilde{X}\,^2\Pi$ NCS in core-valence SDCI+Q/[aCVQZ(N,C,S)] calculations. The B_0 value predicted from these PESs deviates only 0.05% from the corresponding experimental values for NC^{32}S and NC^{34}S. Since we have quite accurate 3D PESs, we can determine both the equilibrium structure and the r_0 structure accurately: r_e(N–C) = 1.1778 Å, r_e(C–S) = 1.6335 Å, and $\angle_e$(N–C–S) = 180°. The ro-vibrationally averaged structure, determined as expectation values over DVR3D wavefunctions, has $\langle r(\text{N–C})\rangle_0$ = 1.1836 Å, $\langle r(\text{C–S})\rangle_0$ = 1.6356 Å, and $\langle \angle(\text{N–C–S})\rangle_0$ = 172.5°. The 3D PESs show that the $\tilde{X}\,^2\Pi$ NCS has its potential energy minimum at a linear configuration, and hence it is a "linear molecule." Experimentally, B_0 values are reported for two isotopologues only.[a] Using the expectation values given above as the initial guess, a bent r_0 structure having an $\langle\angle(\text{N–C–S})\rangle_0$ of 172.2° is deduced from the experimentally reported B_0 values for NC^{32}S and NC^{34}S. It shows that the linear molecule NCS has a "bent" ro-vibrationally averaged structure, confirming our previous predictions:[b] any linear molecule is observed as being bent on ro-vibrational average. See Ref. c [c] for further discussion of this molecule.

$^2\Pi$ NCS is a typical Renner molecule. The Renner spectroscopy of this molecule will be presented in a separate talk.[d]

[a] A. Maeda, H. Habara, T. Amano, *Mol. Phys.*, **105**, 477–495 (2007).
[b] T. Hirano, U. Nagashima, *J. Mol. Spectrosc.*, **314**, 35–47 (2015); T. Hirano, U. Nagashima, P. Jensen, *J. Mol. Spectrosc.* **343**, 54–61 (2018).
[c] T. Hirano, U. Nagashima, P. Jensen, *J. Mol. Spectrosc.* (2018), https://doi.org/10.1016/j.jms.2017.12.011.
[d] J. Freund et al, "Computational spectroscopy of NCS in the Renner-degenerate Electronic state $\tilde{X}\,^2\Pi$."

MH10

A COLLOCATION-BASED MULTI-CONFIGURATION TIME-DEPENDENT HARTREE METHOD FOR COMPUTING VIBRATIONAL SPECTRA

ROBERT WODRASZKA, TUCKER CARRINGTON, *Department of Chemistry, Queen's University, Kingston, ON, Canada.*

It is possible to compute vibrational spectra using the Heidelberg implementation of the multi-configuration time-dependent Hartree method. However, the Heidelberg program can only be used if the potential energy surface is a sum of products (SOP). I shall present a new collocation-based MCTDH approach that can be used with general potential energy surfaces. This is imperative if one wishes to compute very accurate spectra. Collocation obviates the need for quadrature and facilitates using complicated kinetic energy operators. When the basis is good, the accuracy of collocation solutions to the Schroedinger equation is not sensitive to the choice of the collocation points. We test the collocation MCTDH equations by showing that they can be used to compute accurate vibrational energy levels of CH_3. MCTDH, with or without collocation, uses a direct-product basis. I shall demonstrate that by using so-called hierarchical basis functions, it is possible to both benefit from the advantages of collocation and prune the MCTDH basis. These new computational tools will make it possible to use MCTDH-type methods to compute very accurate spectra.

MI. Linelists, Lineshapes, Collisions
Monday, June 18, 2018 – 1:45 PM
Room: 1024 Chemistry Annex

Chair: Matthew J. Cich, TOPTICA Photonics Inc., Farmington, NY, USA

MI01   1:45 – 2:00

VELOCITY-CHANGING COLLISIONS IN SUB-DOPPLER AND DOPPLER-BROADENED LINES

TREVOR SEARS[a], *Department of Chemistry, Stony Brook University, Stony Brook, NY, USA*; GREGORY HALL[b], *Division of Chemistry, Department of Energy and Photon Sciences, Brookhaven National Laboratory, Upton, NY, USA*; SYLVESTRE TWAGIRAYEZU, *Chemistry and Biochemistry, Lamar University, Beaumont, TX, USA*.

The role of velocity changing collisions (VCCs) in pressure-dependent line shapes is revisited, highlighting their contributions to pressure broadening in sub-Doppler saturation line shapes and the conditions required for collisional narrowing in isolated Doppler- and pressure-broadened lines. As reported at last year's meeting (Paper WJ06, 72nd ISMS), we have observed the self-broadening of sub-Doppler saturation dip absorption lines in the $v_1 + v_3$ band of acetylene near 1.5μm in frequency comb-referenced measurements. The saturation line shapes are well described by Voigt functions with a fixed, narrow Gaussian component and a Lorentzian component that increases linearly with pressure up to 0.04 mbar. This sub-Doppler pressure broadening exceeds the normal pressure broadening of a full Doppler line observed at higher pressures. Velocity changes following large cross-section, elastic, collisions are dominated by a sharply spiked exponential cusp in the laboratory-frame collision kernel. The VCCs will contribute to the total broadening when the typical change in Doppler detuning associated with small angle elastic collisions exceeds the pressure-dependent homogeneous line width associated with inelastic damping. At higher pressures, the homogeneous width becomes larger than this collisional frequency shift, and the additional damping effect of VCCs becomes negligible. The pressure at which the change in slope of the line width vs. pressure will occur depends on details of the elastic collision kernel. A Monte Carlo sampling model of elastic and inelastic collision rates and cusp-like elastic collision kernels has been developed to generate electric field time correlation functions whose real Fourier transforms depict the pressure dependent line shapes. Useful physical insights follow. In order to produce collisional (Dicke) narrowing, multiple velocity changing collisions must generate large changes in the Doppler shift of a given absorbing molecule prior to its first inelastic collision.

[a] Also: Division of Chemistry, Brookhaven National Laboratory, Upton, NY 11973
[b] Work at Brookhaven National Laboratory was carried out under Contract No. DE-SC0012704 with the U.S. Department of Energy, Office of Science, and supported by its Division of Chemical Sciences, Geosciences and Biosciences within the Office of Basic Energy Sciences.

MI02   2:02 – 2:17

SUB-DOPPLER SPECTROSCOPY OF THE ν_3 BAND OF METHANE

PHILIP A. KOCHERIL, CHARLES R. MARKUS, ANNE MARIE ESPOSITO, ALEX W SCHRADER, *Department of Chemistry, University of Illinois at Urbana-Champaign, Urbana, IL, USA*; THOMAS S DIETER, *Department of Physics, University of Illinois at Urbana-Champaign, Urbana, IL, USA*; BENJAMIN J. McCALL, *Departments of Chemistry and Astronomy, University of Illinois at Urbana-Champaign, Urbana, IL, USA*.

Methane has been observed in brown dwarfs[a] and planetary atmospheres, including planets in our solar system[b] and extrasolar planets.[c] Methane is also a potent greenhouse gas[d] and relevant to ozone formation and depletion in Earth's atmosphere.[e] As the simplest stable hydrocarbon, methane is also a benchmark for state-of-the-art *ab initio* calculations.[f] While methane is a strong absorber due to its characteristically large transition dipole moments, transition frequencies were historically limited by Doppler broadening, and many frequencies are still known only to Doppler-limited precision. We have constructed a double-pass saturation experiment to perform sub-Doppler spectroscopy of rovibrational transitions of methane. With the accuracy provided by optical frequency combs, we have measured 22 methane transitions from the ν_3 band in the 3 μm region to MHz-level uncertainty, improving the accuracy of the rest frequencies by at least an order of magnitude. This data can be used for higher-precision models of methane as an *ab initio* benchmark.

[a] B. Oppenheimer, S. Kulkarni, K. Matthews *et al.*, *Astrophys. J.* (1998), **502(2)**, 932-943.
[b] T. Owen, R. Cess, *Astrophys. J.* (1975), **197**, L37-L40.
[c] M. Swain, G. Vasisht, G Tinetti *et al.*, *Nature* (2008), **452**, 329-331.
[d] M. Khalil, *Annu. Rev. Energy Environ.* (1999), **24**, 645-661.
[e] O. Boucher, P. Friedlingstein, B. Collins *et al.*, *Environ. Res. Lett.* (2009), **4(4)**, 044007.
[f] A. Nikitin, M. Rey, V. Tyuterev, *J. Quant. Spectrosc. Radiat. Transf.* (2017), **200**, 90-99.

MI03   2:19 – 2:34

H_2 BROADENING IN THE ν_3 AND ν_4 BANDS OF CH_4 AT ROOM TEMPERATURE

<u>EHSAN GHARIB-NEZHAD</u>, *School of Molecular Sciences, Arizona State University, Tempe, AZ, USA*; ALAN N HEAYS, JAMES R LYONS, MICHAEL R LINE, *School of Earth and Space Exploration, Arizona State University, Tempe, AZ, USA*; GLENN STARK, *Department of Physics, Wellesley College, Wellesley, MA, USA*; HANS A BECHTEL, *Advanced Light Source, Lawrence Berkeley National Laboratory, Berkeley, CA, 94720, USA.*

Methane (CH_4) is the dominant carbon-bearing molecule in terrestrial and exoplanetary atmospheres where the temperature is below 1000 K. Therefore, knowing its pressure-induced H_2-broadened absorption cross section is fundamental for exoplanetary atmospheric modeling. In this study, the pressure-induced H_2-broadening coefficients of CH_4 are determined in the spectral regions 2800-3200 (ν_3) and 1200-1400 cm^{-1} (ν_4). The laboratory transmission spectra in this study were recorded at high resolution (i.e., 0.005 and 0.01 cm^{-1}) at room temperature with an FTIR 125HR Bruker spectrometer at Lawrence Berkeley National Laboratory. The CH_4 pressure was constant during the entire experiment (29 mtorr), and elevated H_2 pressures were used in the range 100-700 torr. The Lorentzian coefficients are determined by a nonlinear regression approach in order to model the relationship between the linewidth and its corresponding pressure. Our preliminary results show that the Lorentzian coefficients of different lines in these two bands fall in the range 0.06-0.09 cm^{-1}/atm, consistent with the previous available measurements. Atmospheric modeling will be employed using exoplanet forward transmission modeling to highlight the importance of H_2-broadening of CH_4 to exoplanetary observations.

MI04   2:36 – 2:51

COLLISION INDUCED ABSORPTION OF THE $a^1\Delta_g$-$X^3\Sigma^-_g$ BAND OF OXYGEN NEAR 1.27 μM BY CAVITY RING DOWN SPECTROSCOPY

<u>DIDIER MONDELAIN</u>, ALAIN CAMPARGUE, SAMIR KASSI, *UMR5588 LIPhy, Université Grenoble Alpes/CNRS, Saint Martin d'Hères, France.*

Collision induced absorption (CIA) coefficients of the $a^1\Delta_g$-$X^3\Sigma^-_g$(v=0-0) band of oxygen have been measured using cavity ring down spectroscopy (CRDS) technique at room temperature. More precisely, the B_{O2-O2}, B_{O2-N2} and B_{O2-Air} coefficients have been determined with a reduced uncertainty from series of low density spectra (from 0.36 to 0.85 amagat) of pure oxygen and N_2+O_2 mixture with O_2=20.95%. For that 12 distributed feed-back laser diodes were used below 7920 cm^{-1} together with an external cavity diode laser above this wavenumber. We particularly paid attention to the base line stability (2×10^{-10} cm^{-1}) during the entire measurements. CIA was obtained from the difference between the absorbing samples spectra and argon spectra recorded for the same densities after removal of the local contribution of the absorption lines. The low densities at which the spectra were recorded were very useful to reliably remove this local contribution. The retrieved coefficients were compared to the CIA reported in HITRAN2016. A good overall agreement is found but differences between 5 and 8% for B_{O2-Air} coefficients are observed below 7850 cm^{-1}.

MI05   2:53 – 3:08

LOW-TEMPERATURE HIGH PRECISION MEASUREMENTS OF LINE MIXING and COLLISIONAL INDUCED ABSORPTION IN THE OXYGEN A-BAND

<u>ERIN M. ADKINS</u>, MÉLANIE GHYSELS, DAVID A. LONG, JOSEPH T. HODGES, *Chemical Sciences Division, National Institute of Standards and Technology, Gaithersburg, MD, USA.*

Because of the constant mixing ratio of molecular oxygen (O_2) within the Earth's atmosphere, the O_2 A-band is commonly used in satellite and remote sensing measurements as a measure of the airmass. A recent collaborative effort has produced a self-consistent integrated spectroscopic model for the O_2 A-Band that simultaneously accounts for high-order line-shapes, line mixing (LM), and collisional induced absorption (CIA).[a] This model has improved OCO-2 mission retrievals of dry air CO_2, however, limitations in existing spectroscopic models still lead to airmass dependent biases. Currently, model development is limited by a lack of high resolution experimental data at low temperatures and in the R-branch. To address this, measurements of the entire O_2 A-band were recently made with a variable-temperature cavity ring-down spectrometer (CRDS) over a range of temperatures, pressures, and molar fractions. Because of the limited dynamic range of the CRDS system, at high molar fractions of O_2 saturation can occur at the line cores of strong transitions. Therefore, a range of molar fraction O_2 samples were employed. Low mole fraction data, which was unaffected by saturation provided information on the temperature dependence of high-order line-shape parameters. Conversely, high molar fraction data provided information on LM and CIA effects that dominate absorption in the troughs between saturated transitions. By combining this high-resolution experimental data, that covers both the entire O_2 A-Band as well as a range of temperatures, with existing datasets, these results aim to improve on LM and CIA models for the next iteration of the global O_2 A-Band model.

---

[a]B. J. Drouin, D. C. Benner, L. R. Brown, et al., Multispectrum analysis of the oxygen A-band, J. Quant. Spectrosc. Radiat. Transfer, 2017, 186: p. 118-138.

Intermission

MI06   3:44 – 3:59

HIGH PRECISION LINE PARAMETERS OF N_2O NEAR 1.5μm BY CAVITY RING-DOWN SPECTROSCOPY

<u>GU-LIANG LIU</u>, PENG KANG, TIAN-PENG HUA, SHUI-MING HU, AN-WEN LIU, *Hefei National Laboratory for Physical Science at Microscale, University of Science and Technology of China, Hefei, China.*

Accurate parameters of the N_2O transitions in the 1.5μm region are needed for monitoring global N_2O concentration in the atmosphere. The strongest band in this region is the 0003-0000 band. In HITRAN database, some parameters of this band are given by calculation, others are given by experiments but they are obtained by the Voigt profile, which is now well known can lead to significant deviations. The ro-vibrational transitions of the 0003-0000 band with line intensities in the order of 10^{-24} to 10^{-23}cm^{-1}/(molecule·cm^{-2}) have been recorded using a laser-locked cavity ring-down spectrometer with high sensitivity as well as high precision. The positions were determined with an uncertainty of sub-MHz. The line intensities and Nitrogen induced pressure broadening coefficients were also derived with accuracies better than 0.8% and 1%, respectively. Comparisons of the line parameters determined in this work with literature experimental values and those from HITRAN2016 database are given.

MI07 4:01 – 4:16

UPDATES AND CURRENT STATUS OF THE HITRAN APPLICATION PROGRAMMING INTERFACE (HAPI)

ROMAN V KOCHANOV[a], IOULI E GORDON, LAURENCE S. ROTHMAN, YAN TAN, *Atomic and Molecular Physics, Harvard-Smithsonian Center for Astrophysics, Cambridge, MA, USA*; JOSHUA KARNS, WYATT MATT, *Computer Science, State University of New York at Oswego, Oswego, NY, USA*; CHRISTIAN HILL, *Atomic and Molecular Data Unit, International Atomic Energy Agency, Vienna, Austria.*

The HITRAN Application Programming Interface (HAPI)[b] is a powerful tool for working with spectroscopic data in the gas phase. HAPI provides access to the capabilities of the HITRAN*online* (http://hitran.org) web information system with the recent edition of the HITRAN2016 spectroscopic database[c]. Besides an access to HITRAN*online*, HAPI allows working with user-supplied data. Among the capabilities are data filtering and analysis, as well as modeling of gas absorption with the fine tuning of many parameters (gas mixture, path length, instrumental function, temperature, and pressure). In this talk we present the update for HAPI (v.2.0) which has the following features: 1) access to line-by-line spectroscopic transitions and experimental cross-sections from HITRAN2016; 2) access to the metadata for molecules from the line-by-line part, and more than 300 molecules from the cross-section part, as well as for the database bibliography; 3) seamless use of the foreign broadening and shifting parameters, and non-Voigt line profiles, relevant for atmospheric and planetary applications; 4) use of the custom CPF implementations; 5) updated partition sums from the recent TIPS software[d] covering wider temperature ranges; 6) line mixing support. The new version features HAPIEST (HAPI and Efficient Spectroscopic Tools) – a portable graphical user interface providing access to HAPI features. HAPI v.2.0 is available at the official HITRAN*online* site as well as through the Github repository (https://github.com/hitranonline/hapi). The HAPIEST open source package with binary installers will be available at HITRAN*online* upon release. This effort is supported through the NASA AURA (NNX 17AI78G) and NASA PDART grants (NNX16AG51G).

---

[a]QUAMER, Tomsk State University, Tomsk 634050, Russia
[b]Kochanov RV, Gordon IE, Rothman LS et al. JQSRT 2016;177:15–30. doi:10.1016/j.jqsrt.2016.03.005.
[c]Gordon IE, Rothman LS, Hill C, Kochanov RV, Tan Y et al. JQSRT 2017;203:3-69. doi:10.1016/j.jqsrt.2017.06.038.
[d]Gamache RR, Roller C, Lopes E, Gordon IE, Rothman LS et al. JQSRT 2017;203:70–87. doi:10.1016/j.jqsrt.2017.03.045.

MI08 4:18 – 4:33

NEW MEASUREMENTS OF THE WATER VAPOR ABSORPTION CROSS SECTION IN THE BLUE-VIOLET RANGE BY CAVITY-ENHANCED DIFFERENTIAL OPTICAL ABSORPTION SPECTROSCOPY

RANDALL CHIU, *Department of Chemistry and Biochemistry, University of Colorado Boulder, Boulder, CO, United States*; OLEG L. POLYANSKY, *Department of Physics and Astronomy, University College London, Gower Street, London WC1E 6BT, United Kingdom*; RAINER VOLKAMER, *Department of Chemistry and Biochemistry, University of Colorado, Boulder, CO, USA.*

The absorption cross section of water vapor in the blue-violet range (415-460 nm) is currently not well known, and many weak spectral lines are not included in either the HIgh resolution TRANsmission molecular absorption (HITRAN) database or its HIgh TEMPerature companion, HITEMP. Direct measurements of the absorption cross section of water vapor in this region have been limited by the slant column density (SCD) of gaseous water molecules achievable in a laboratory setting. We use cavity-enhanced differential optical absorption spectroscopy (CE-DOAS) to generate water vapor SCDs comparable to those in field measurements. Our cavity consists of high-reflectivity ($R > 0.99995$) mirrors separated by 80 cm to realize effective path lengths up to 16 km; water vapor is generated from deionized water in a double-bubbler system. Broadband light sources (LEDs) with peak intensities at 420 and 455 nm allow us to measure multiple lines at moderately high spectral resolution (0.15 nm). The first spectra were measured at room temperature (298 K), but our setup allows us to explore temperature variations.

Our goals are to refine available line lists for gas-phase water by combining laboratory measurements with quantum chemical calculations, and to reevaluate field measurements. In particular, we will revisit field data from University of Colorado Airborne Multi-AXis Differential Optical Absorption Spectroscopy (CU AMAX-DOAS) instrument during the Tropical Ocean tRoposphere Exchange of Reactive halogen species and Oxygenated VOC (TORERO) campaign. As part of TORERO, comparisons of in situ and remote-sensing measurements of water vapor were performed, and AMAX-DOAS fits exhibited cosmetic residual structures when using HITRAN and HITEMP reference spectra. Our work has the potential to improve trace gas retrievals from many current and planned satellites, e.g. Ozone Monitoring Instrument (OMI), TROPOspheric Monitoring Instrument (TROPOMI), and Tropospheric Emissions: Monitoring of Pollution (TEMPO); and from aircraft-based remote-sensing instruments such as the CU AMAX-DOAS.

MI09 4:35–4:50

COMPLETE PHOTOABSORPTION LINELIST FOR CO AND ITS ISOTOPOLOGUES BETWEEN 101 AND 115 NM

<u>ALAN HEAYS</u>, *LERMA2, CNRS UMR8812, Observatoire de Paris, MEUDON, France*; JEAN LOUIS LEMAIRE, *CNRS, Institut des Sciences Moleculaires d'Orsay, Orsay, France*; MICHELE EIDELSBERG, *Meudon, Observatoire de Paris, Paris, France*; LISSETH GAVILAN, *CNRS/INSU, UPMC Univ Paris 06, Paris, France*; GLENN STARK, *Department of Physics, Wellesley College, Wellesley, MA, USA*; STEVEN FEDERMAN, *Physics and Astronomy, University of Toledo, Toledo, OH, USA*; JAMES R LYONS, *School of Earth and Space Exploration, Arizona State University, Tempe, AZ, USA*; WIM UBACHS, *Department of Physics and Astronomy, VU University, Amsterdam, Netherlands*; NELSON DE OLIVEIRA, *DESIRS Beamline, Synchrotron SOLEIL, Saint Aubin, France*.

The photoabsorbing bands of CO and its isotopologues appearing between 101 and 115 nm provide more than half of its photodissociative potential in the interstellar medium and planetary atmospheres, and are responsible for the well-known fractionation of C and O isotopes due to self-shielding.

An experimental study of this region over several years using the undulator radiation source and vacuum-ultraviolet Fourier-transform spectroscopy facilities at the SOLEIL synchrotron [1] is complete. Line frequencies [2] and oscillator strengths [3], and widths [in prep.] are deduced, and in some cases extrapolated, to provide updated and reliable cross sections over a range of temperatures, including for the rare ^{17}O isotopologues.

1 N. de Oliveira et al. (2016). The high-resolution absorption spectroscopy branch on the VUV beamline DESIRS at SOLEIL. J. Synchrotron Radiat. 23:887.

2 J.L. Lemaire et al. (2018). Atlas of new and revised high-resolution spectroscopy of six CO isotopologues in the 101-115 nm range. Astron. Astrophys. (accepted)

3 G. Stark et al. (2014). High-resolution oscillator strength measurements of the v=0,1 bands of the B-X, C-X, and E-X systems in five isotopologues of carbon monoxide. Astrophys. J. 788:68

MI10 4:52–5:07

SHIFTS AND BROADENING IN THE CN A $^2\Pi - X\,^2\Sigma^+$ (2-0) BAND INDUCED BY ARGON COLLISIONS

<u>JAMES LOCKHART</u>, *Division of Chemistry, Brookhaven National Laboratory, Long Island, NY, USA*; TREVOR SEARS[a], GREGORY HALL, *Division of Chemistry, Department of Energy and Photon Sciences, Brookhaven National Laboratory, Upton, NY, USA*.

Selected P-and R-branch transitions of the CN A $^2\Pi - X\,^2\Sigma^+$ (2-0) band have been recorded at room temperature as a function of argon pressure, using frequency modulation spectroscopy. The experimental line shapes have been successfully fit using a Quadratic Speed-Dependent Voigt (QSDV) model at total pressures ranging from 1 - 160 Torr. The pressure broadening coefficients derived from the QSDV analysis are nearly independent of the rotational quantum state and rotational branch. The pressure-dependence of the line shifts, in contrast, displays a distinctive variation with rotational state and branch, with larger and strongly J-dependent shifts for P-branch lines, compared to smaller and more J-independent shifts for R-branch lines. The pressure-induced shifts provide a challenge to first principles scattering calculations on validated potential energy surfaces. Previously puzzling measurements on fewer lines in the (1-0) band are confirmed and extended by the present measurements.

Work at Brookhaven National Laboratory was carried out under Contract No. DE-SC0012704 with the U.S. Department of Energy, Office of Science, and supported by its Division of Chemical Sciences, Geosciences and Biosciences within the Office of Basic Energy Sciences.

---

[a] Also at: Department of Chemistry, Stony Brook University, Stony Brook, NY, U.S.A.

MI11 5:09 – 5:24

UNUSUAL POWER DEPENDENT PEAK SPLITTING IN REMPI SPECTRUM

<u>JUNGGIL KIM</u>, JEAN SUN LIM, SANG KYU KIM, *Chemistry, Korea Advanced Institute of Science and Technology, Daejeon, Republic of Korea.*

Dynamic Stark effects, such as Autler-Townes splitting (ATS) and Electromagnetically induced transparency (EIT), has been known to happen in simple atoms or quantum confined systems so far. We have been observed similar power-dependent peak splitting of resonant two-photon ionization (R2PI) spectrum in a polyatomic molecule. In its R2PI spectrum, doublet structures start to appear even at a very weak nanosecond-laser field and show vibronic mode-specificity. Prominent isotope substitution effect indicates that this phenomenon comes from the excited state dynamics.

MJ. Radicals

Monday, June 18, 2018 – 1:45 PM

Room: 217 Noyes Laboratory

Chair: Melanie A.R. Reber, University of Georgia, Athens, GA, USA

MJ01   1:45 – 2:00

AN UPDATED LOOK AT THE INFRARED SPECTRUM OF FULVENALLENE AND FULVENALLENYL

<u>ALAINA R. BROWN</u>, JOSEPH T. BRICE, PETER R. FRANKE, GARY E. DOUBERLY, *Department of Chemistry, University of Georgia, Athens, GA, USA.*

The closed shell species fulvenallene (C_7H_6) and the fulvenallenyl radical (C_7H_5) are produced via thermal decomposition of phthalide ($C_8H_6O_2$) in a continuous-wave SiC pyrolysis furnace. Prompt pick-up and solvation of these species in helium droplets allows for the measurement of well-resolved infrared spectra in the CH stretching region. VPT2+K simulations based on a hybrid CCSD(T) force field with quadratic (cubic and quartic) force constants computed using the ANO1 (ANO0) basis set are used to predict anharmonic frequencies for both species. The 3300 cm^{-1} region of the spectrum contains the acetylenic stretch of fulvenallenyl which serves as a sensitive marker for the extent of delocalization between the conjugated propargyl and cyclopentadienyl subunits of the radical. This delocalization is explored with spin density calculations at the B3LYP/aug-cc-pVTZ and ROHF-CCSD(T)/ANO1 levels of theory.

MJ02   2:02 – 2:17

INFRARED SPECTRA OF THE 1,1-DIMETHYLALLYL AND 1,2-DIMETHYLALLYL RADICALS ISOLATED IN SOLID *PARA*-HYDROGEN

<u>JAY C. AMICANGELO</u>, *School of Science (Chemistry), Penn State Erie, Erie, PA, USA*; YUAN-PERN LEE, *Applied Chemistry, National Chiao Tung University, Hsinchu, Taiwan, Institute of Atomic and Molecular Sciences, Academia Sinica, Taipei, Taiwan.*

The reaction of hydrogen atoms (H) with isoprene (C_5H_8) in solid *para*-hydrogen (p-H_2) matrices at 3.2 K has been studied using infrared spectroscopy. The production of H atoms for reaction with C_5H_8 was essentially a three step process. First, mixtures of C_5H_8 and Cl_2 were co-deposited in p-H_2 at 3.2 K for several hours, then the matrix was irradiated with ultraviolet light at 365 nm to produce Cl atoms from the Cl_2, and finally the matrix was irradiated with infrared light to induce the reaction of the Cl atoms with p-H_2 to produce HCl and H atoms. Upon infrared irradiation, a series of new lines appeared in the infrared spectrum, with the strongest lines appearing at 776.0 and 766.7 cm^{-1}. To determine the grouping of lines to distinct chemical species, secondary photolysis was performed using a 365-nm light-emitting diode and a low-pressure mercury lamp in combination with filters. Based on the secondary photolysis, it was determined that the majority of the new lines belong to two distinct chemical species, designated as set X (3030.6, 1573.2, 1452.0, 1435.6, 1123.2, 1051.4, 982.7, 922.5, 792.5, 776.0, 699.2, 524.7, 469.0 cm^{-1}) and set Y (3110.1, 2972.0, 1564.4, 1471.1, 1430.2, 1379.7, 1376.2, 1335.4, 1233.0, 1205.4, 1050.1, 766.7, 570.0 cm^{-1}). The most likely reactions to occur under the low temperature conditions in solid p-H_2 are the addition of the H atom to the four alkene carbon atoms to produce the corresponding hydrogen atom addition radicals (HC_5H_8). Quantum-chemical calculations were performed at the B3PW91/6-311++G(2d,2p) level for the four possible HC_5H_8 radicals in order to determine the relative energetics and the predicted vibrational spectra for each radical. The addition of H to each of the four carbons is exothermic, with relative energies of 0.0, 93.3, 77.0, and 8.4 kJ/mol for the addition to carbons 1 – 4, respectively. When the lines in set X and Y are compared to the scaled harmonic and anharmonic vibrational spectra, the best agreement for set X is with the radical produced by the addition to carbon 4 (1,2-dimethylallyl radical) and the best agreement for set Y is with the radical produced by addition to carbon 1 (1,1-dimethylallyl radical).

MJ03
MATRIX-ISOLATION FTIR SPECTROSCOPY OF THE 1-BUTYN-3-YL RADICAL

GLENNA J. BROWN, MARTHA ELLIS, LAURA R. McCUNN, *Department of Chemistry, Marshall University, Huntington, WV, USA.*

The 1-butyn-3-yl radical (C_4H_5) is thought to play a role in the formation of hydrocarbons in the interstellar medium and planetary atmospheres, but it is not well characterized. In this study, the 1-butyn-3-yl radical was formed by the pyrolysis of gas-phase 3-bromo-1-butyne at temperatures of 800-1200 K. Nascent radicals were isolated in an argon marix, followed by FTIR spectroscopy. Vibrational bands in the experimental spectra were matched to frequencies predicted by Gaussian 09. Pyrolysis of 3-methyl-1-butyne was also investigated as a possible pyrolytic precursor to the 1-butyn-3-yl radical under similar conditions. Evidence of 1-butyn-3-yl formation was observed, but other radicals may have formed as well.

MJ04
SUB-DOPPLER INFRARED SPECTROSCOPY OF JET COOLED BENZYl $C_6H_5CH_2$: A CLASSIC RESONANTLY STABILIZED ORGANIC HYDROCARBON RADICAL

ANDREW KORTYNA, *JILA, National Institute of Standards and Technology and Univ. of Colorado, Boulder, CO, USA*; DANIEL LESKO, *Department of Chemistry and Biochemistry, University of Colorado, Boulder, CO, USA*; PRESTON G. SCRAPE, DAVID NESBITT, *JILA, National Institute of Standards and Technology and Univ. of Colorado, Boulder, CO, USA.*

The benzyl radical ($C_6H_5CH_2$) is a classic example of a relatively long lived, resonantly stabilized transient molecule. Its relative stability makes it a likely intermediate in the formation of polycyclic aromatic hydrocarbons and ultimately soot during the combustion of fossil fuels. Benzyl radical is also thought to be candidate for detection in interstellar molecular clouds. Benzyl radical is generated by seeding benzyl chloride in a rare gas He/Ne mixture through a pulsed slit discharge, with the radical formation process likely dominated by electron dissociative attachment. The radicals are subsequently cooled in a slit jet supersonic expansion to a 15K rotational temperature. Narrow band infrared radiation is produced through difference frequency generation of two single-mode visible lasers. High frequency stability (± 11 MHz) is achieved through servo-locking techniques, and meticulous suppression of noise permits detection sensitivities approaching the quantum shot-noise limit. The slit jet has an inherent sub-Doppler resolution of 60 MHz. The present work reports the first ro-vibrationally resolved infrared spectra of antisymmetric (ν_3, B_2) and symmetric (ν_4, A_1) CH ring stretch modes in benzyl radical, with band origins (3073.2350 ± 0.0005 cm^{-1} and 3067.0576 ± 0.0006 cm^{-1}, respectively) and rotational constants determined by least-squares fits to an asymmetric top Hamiltonian. Surprisingly, the benzyl spectrum shows little evidence of dark-state perturbations despite the relatively large number of atoms (N = 14) in this molecule. This is most likely due to the highly delocalized resonance structure of the benzyl radical, which generates a large barrier ($\Delta E = 11.5$ kcal/mol) for internal rotation of the methylene group, suppresses the rovibrational density of states, and makes the tunneling splittings too small to detect with the present sub-Doppler resolution. Particular effort is directed toward accurate determination of the ground state rotational constants, with a goal of assisting microwave search for benzyl radical in the interstellar medium.

MJ05

2:53 – 3:08

ANALYSIS OF THE $\tilde{A} - \tilde{X}$ BANDS OF THE ETHYNYL RADICAL NEAR 1.48 μm AND RE-EVALUATION OF $\tilde{X}$ STATE ENERGIES

EISEN C. GROSS, *Department of Chemistry, Stony Brook University, Stony Brook, NY, USA*; ANH T. LE[a], GREGORY HALL, *Division of Chemistry, Department of Energy and Photon Sciences, Brookhaven National Laboratory, Upton, NY, USA*; TREVOR SEARS[b], *Department of Chemistry, Stony Brook University, Stony Brook, NY, USA.*

We report the observation and analysis of spectra in part of the near-infrared spectrum of C_2H, originating in rotational levels in the ground and lowest two excited bending vibrational levels of the ground $\tilde{X}\,^2\Sigma^+$ state. In the analysis, we have combined present and previously reported high resolution spectroscopic data for the lower levels involved in the transitions to determine significantly improved molecular constants to describe the fine and hyperfine split rotational levels of the radical in the zero point, $v_2 = 1$ and the $^2\Sigma^+$ component of $v_2 = 2$. Two of the upper state vibronic levels involved, a $^2\Pi$ symmetry level at 6819.3 cm^{-1} and a $^2\Sigma^+$ one at 7527.1 cm^{-1}, had not been previously observed. The data and analysis indicate the electronic wavefunction character changes with bending vibrational excitation in the ground state and provide avenues for future measurements of reactivity of the radical as a function of vibrational excitation.

Work at Brookhaven National Laboratory was carried out under Contract No. DE-SC0012704 with the U.S. Department of Energy, Office of Science, and supported by its Division of Chemical Sciences, Geosciences and Biosciences within the Office of Basic Energy Sciences.

---

[a] Now at: School of Molecular Sciences, Arizona State University, Tempe, AZ 85287, USA
[b] Also: Chemistry Division, Brookhaven National Laboratory, Upton, NY 11973-5000, USA

MJ06

Post-Deadline Abstract

3:10 – 3:25

DECOMPOSITION OF VIBRONIC AND RENNER-TELLER STRUCTURE IN C_2H AND C_2D FROM ANION HIGH-RESOLUTION PHOTOELECTRON IMAGING

STEPHEN T GIBSON, BENJAMIN A LAWS, *Research School of Physics and Engineering, Australian National University, Canberra, ACT, Australia.*

The ethynyl radial, C_2H, has a complex spectral structure due to vibronic coupling between the ground $\tilde{X}\,^2\Sigma^+$ and low-lying $\tilde{A}\,^2\Pi$ electronic states, and a Renner-Teller interaction within the Π state.

A good understanding of the low-lying rovibrational structure has come from measurements, including slow electron velocity-map imaging of anion photoelectron spectra[a], and *ab initio* calculations[b], that give wavefunction character.

In this work, high-resolution photoelectron velocity-map imaging of C_2H^- and C_2D^- photodetachment (the 355 nm wavelength illustrated), provide a quantitative comparison over an extended energy range, to reveal unassigned structure, anomalous intensities, and illustrate the dramatic difference between isotopologues in the region of the A-state. These measurements, together with the measured photoelectron angular distributions, provide new insight into the non-adiabatic couplings of ethynyl.

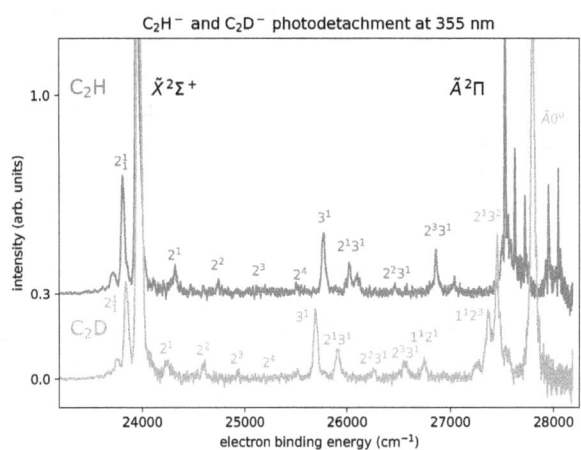

---

[a] J. Zhou *et al. J. Chem. Phys.* **127**, 114313 (2007).
[b] R. Tarroni and S. Carter, *J. Chem. Phys.* **119**, 12878 (2003).

Research supported by the Australian Research Council Discovery Project Grant DP160102585.

MJ07 3:27 – 3:42
OPTICAL DETECTION OF TWO NEW ISOMERS OF C_7H_7 IN A TOLUENE DISCHARGE

MEREDITH WARD, JONATHAN FLORES, SEDERRA D ROSS, *Chemistry, University of Massachusetts Boston, Boston, Massachusetts, United States*; DAMIAN L KOKKIN, *Department of Chemistry, Marquette University, Milwaukee, WI, USA*; NEIL J REILLY, *Chemistry, University of Massachusetts Boston, Boston, Massachusetts, United States*.

In an effort to optically detect reactive intermediates implicated in mechanisms of C_7H_7 decomposition and formation, we have been interrogating radical products formed in supersonically cooled discharges of various C_7H_8 precursors. So far, two C_7H_7 isomers (neither of which is benzyl or tropyl) have been observed in the 470 – 455 nm region in resonant two-color two-photon ionization spectra. Both isomers were first observed in our laboratory in a toluene discharge, but they can be more efficiently produced from other precursors: one of them, for which we have measured an adiabatic ionization energy (AIE) of 6.92 eV, is highly conspicuous in a discharge of 1,6-heptadiyne; the other, with an AIE of 7.16 eV, is most efficiently generated from cycloheptatriene. Both species possess low frequency vibrational modes suggestive of acyclic structures, and because they absorb at similar wavelengths to the 1-vinylpropargyl radical, may also incorporate substituted propargyl chromophores. Quantum chemical calculations, additional chemical tests, and measurements of ground state vibrational frequencies by dispersed fluorescence are on-going to conclusively assign each spectrum to a particular carrier. The implications of our surprisingly facile discovery of two putatively unknown isomers will be discussed in the context of recent investigations of combustion and pyrolysis processes that begin or terminate with C_7H_7.

Intermission

MJ08 4:18 – 4:33
SUB-DOPPLER INFRARED SPECTROSCOPY OF JET COOLED CH_2Br RADICAL: CH_2 STRETCH VIBRATIONS

ANDREW KORTYNA, PRESTON G. SCRAPE, *JILA, National Institute of Standards and Technology and Univ. of Colorado, Boulder, CO, USA*; DANIEL LESKO, *Department of Chemistry and Biochemistry, University of Colorado, Boulder, CO, USA*; DAVID NESBITT, *JILA, National Institute of Standards and Technology and Univ. of Colorado, Boulder, CO, USA*.

Bromomethyl radical (CH_2Br) has recently been used as a novel precursor for producing the simplest Criegee intermediates (CH_2OO). With the goal of spectroscopically investigating a Criegee intermediate, we have pursued high resolution characterization of the CH_2Br radical in our slit jet discharge spectrometer. The bromomethyl radical is generated by seeding CH_2Br_2 into a Ne/He/H_2 mixture in a pulsed slit discharge. The radical is produced through either electron dissociative attachment to form bromine anions or hydrogen abstraction of bromine, with subsequent cooling in a supersonic expansion to about 15 K. Infrared absorption in the CH_2 symmetric stretch vibrational band is fully resolved at high single-to-noise ratios for both the ^{79}Br and ^{81}Br isotopologues . The sub-Doppler rotational structure is fitted to a rigid-rotor Hamiltonian with spin-rotation coupling, generating principal rotational constants and the spin-orbit coupling tensor for the vibrationally excited state. The results are consistent with a vibrationally averaged planar π-radical with unpaired electron spin density in a partially filled p_π-orbital on the central C atom. Relative band intensities in the symmetric and antisymmetric CH_2 stretch manifolds provide further elucidation of the "charge-sloshing" mechanism noted in CH_2F, CH_2Cl, and CH_2I radical species due to vibrationally mediated shifts in electron density along the carbon-halogen bond axis.

MJ09 4:35 – 4:50

LIF SPECTROSCOPY OF A $^1\Sigma$ SPECIES CONTAINING Si: LINEAR SiOSi ?

<u>MASARU FUKUSHIMA</u>, TAKASHI ISHIWATA, *Information Sciences, Hiroshima City University, Hiroshima, Japan.*

In our past SiCN investigation[a], we found unknown bands with $^1\Sigma - ^1\Sigma$ rotational structure in the laser induced fluorescence (LIF) excitation spectrum of SiCN. From the rotational constants, the spectral species may possibly be attributed to SiOSi. Although the most stable geometry of the ground electronic state is reported to be cyclic structure[b], our CCSD(T) calculation with arg-cc-pCVTZ indicates the linear geometry, $^1\Sigma_g^+$, lying $\sim$2,000 cm^{-1} above it. The potential energy surface calculated is very strange, and it indicates a barrier between the two geometries, $\sim$10,000 cm^{-1} from the bottom. The dispersed fluorescence (DF) spectra from the single vibronic levels have fairly long progressions with very harmonic structure, but no hot-band structure. More precise computational works are underway, and we will discuss the assignment of the spectral species in this talk.

---

[a]M. Fukushima and T. Ishiwata, J. Chem. Phys. 145, 124304 (2016).
[b]S. J. Paukstis, et al., J. Chem. Phys. A 106, 8435 (2002).

MJ10 4:52 – 5:07

THERMAL DECOMPOSITION OF THE LIGNIN MODEL COMPOUNDS: SALICYLALDEHYDE AND CATECHOL

THOMAS ORMOND, *Department of Chemistry and Biochemistry, University of Colorado, Boulder, CO, USA*; JOSHUA H BARABAN, *Chemistry, Ben-Gurion University of the Negev, Beer-Sheva, Israel*; <u>JESSIE P PORTERFIELD</u>, *Department of Chemistry and Biochemistry, University of Colorado, Boulder, CO, USA*; ADAM M SCHEER, *Renewable Energy, Pacific Gas and Electric, San Francisco, CO, USA*; PATRICK HEMBERGER, *General Energy, Paul Scherrer Institute, Villigen, Switzerland*; TYLER TROY, *Chemical Science Division, Lawrence Berkeley National Laboratory, Berkeley, CA, USA*; MUSAHID AHMED, *UXSL, Chemical Sciences Division, Lawrence Berkeley National Laboratory, Berkeley, CA, USA*; MARK R NIMLOS, DAVID ROBICHAUD, *Biomass Molecular Science , National Renewable Energy Laboratory , Golden, CO, USA*; JOHN W DAILY, *Department of Mechanical Engineering, University of Colorado Boulder, Boulder, CO, USA*; BARNEY ELLISON, *Department of Chemistry and Biochemistry, University of Colorado, Boulder, CO, USA.*

The nascent steps in the pyrolysis of the lignin components, salicylaldehyde (o-HOC$_6$H$_4$CHO) and catechol (o-HOC$_6$H$_4$OH), have been studied in a set of heated micro-reactors. The micro-reactors are small (roughly 1 mm ID x 3 cm long); transit times through the reactors are about 100 μsec. Temperatures in the micro-reactors can be as high as 1600 K and pressures are typically a few hundred Torr. The products of pyrolysis are identified by a combination of photoionization mass spectrometry and matrix isolation infrared spectroscopy. The main pathway by which salicylaldehyde decomposes is a concerted fragmentation: o-HOC$_6$H$_4$CHO (+ M) $\rightarrow$ H$_2$ + CO + C$_5$H$_4$=C=O. At temperatures above 1300 K, fulveneketene loses CO to yield a mixture of HC$\equiv$C–C$\equiv$C–CH$_3$, HC$\equiv$C–CH$_2$–C$\equiv$CH, and HC$\equiv$C–CH=C=CH$_2$. These alkynes decompose to a mixture of radicals (HC$\equiv$C–C$\equiv$C–CH$_2$ and HC$\equiv$C–CH–C$\equiv$CH and H atoms. H-atom chain reactions convert salicylaldehyde to phenol: o-HOC$_6$H$_4$CHO + H $\rightarrow$ C$_6$H$_5$OH + CO + H. Catechol has similar chemistry to salicylaldehyde. Electrocyclic fragmentation produces water and fulveneketene: o-HOC$_6$H$_4$OH (+ M) $\rightarrow$ H$_2$O + C$_5$H$_4$=C=O. These findings have implications for the pyrolysis of lignin itself.

MJ11

5:09 – 5:24

CAVITY RING-DOWN SPECTROSCOPY OF 1-, 2- AND 3-METHYL ALLYL PEROXY RADICALS

<u>MD ASMAUL REZA</u>, HAMZEH TELFAH, ANAM C. PAUL, JAHANGIR ALAM, JINJUN LIU, *Department of Chemistry, University of Louisville, Louisville, KY, USA.*

Peroxy radicals are key reaction intermediates formed during the oxidation of hydrocarbons in the atmosphere and in low-temperature combustion. Allyl group-containing peroxy radicals are particularly important because they are generated in large quantities by the OH-initiated oxidation of isoprene, the most abundant non-methane biogenic hydrocarbon. In this talk, room-temperature cavity ring-down (CRD) spectra of the $\tilde{A} \leftarrow \tilde{X}$ electronic transition of 1-, 2- and 3-methyl allyl peroxy radicals will be reported. Peroxy radicals were produced in 193 nm photolysis of selected methyl-substituted allyl chlorides, e.g., 1-chloro-2-butene, 3-chloro-2-methyl-1 propene, and 3-chloro-1-butene, in the presence of O_2. Vibronic structure of the experimentally observed spectra are simulated using calculated electronic transition frequencies, vibrational frequencies, and Franck-Condon factors. Spectroscopic detection and characterization of isoprene peroxy radicals[a] are underway.

[a] A. P. Teng, J. D. Crounse, and P. O. Wennberg, J. Am. Chem. Soc. 139, 5367 (2017).

MJ12

5:26 – 5:41

OBSERVATION OF THE $\tilde{A} \leftarrow \tilde{X}$ ELECTRONIC TRANSITIONS OF TETRAHYDROPYRANYL AND TETRAHYDROFURANYL PEROXY RADICALS BY ROOM-TEMPERATURE CAVITY RING-DOWN SPECTROSCOPY

<u>HAMZEH TELFAH</u>, MD ASMAUL REZA, ANAM C. PAUL, JINJUN LIU, *Department of Chemistry, University of Louisville, Louisville, KY, USA.*

Peroxy radicals are important chemical reaction intermediates in low-temperature combustion systems as well as in the Earth's troposphere; hence spectroscopic detection of peroxies and studies of their kinetics are essential to improving the efficiency of internal combustion engines and reducing air pollution. In this talk, we report the room-temperature cavity ring-down (CRD) spectra of the $\tilde{A} \leftarrow \tilde{X}$ electronic transition of the tetrahydropyranyl peroxy (THPOO) and tetrahydrofuranyl peroxy (THFOO) radicals. Both THP and THF are building blocks of lignocellulose-derived biofuels. THPOO and THFOO therefore play critical roles in the oxidation of biofuels.[a, b, c] In the present experiment, they are produced via hydrogen abstraction of THP and THF by chlorine atoms followed by oxygen addition. Chlorine atoms are produced in 193 nm photolysis of oxalyl chloride $(COCl)_2$. The presence of oxygen in the ring defines 3 distinct positions on THP (α,β and γ) and 2 for THF (α and β), which leads to different regioisomers. Moreover, compared to chain alkyl peroxy radicals, cyclic ones possess significantly different conformational landscapes. Quantum chemical calculations have been performed and provide electronic transition frequencies, vibrational frequencies, Franck-Condon factors, as well as relative energies of isomers and conformers. Spectral simulation using these calculated result suggests that all isomers and most of the possible conformers contribute to the experimentally observed spectra. Also determined in the spectral simulation is the branching ratios of reactions that produce different regioisomers. The CRD technique has been used for lifetime measurements and investigation of the ring-opening mechanism. Comparison between the target molecules and the controls, namely, homocyclic peroxy radicals, will be briefly discussed.

[a] Chen, M. W. et al., Phys. Chem. Chem. Phys., 2018, DOI: 10.1039/c7cp08164b.
[b] Rotavera, B. et al., Proc. Combust. Inst., 2017, 36 (1), 597–606.
[c] Antonov, I. O. et al., J. Phys. Chem. A, 2016, 120 (33), 9823-9840.

MK. Clusters/Complexes

Monday, June 18, 2018 – 1:45 PM

Room: B102 Chemical and Life Sciences

Chair: Susana Blanco, Universidad de Valladolid, Valladolid, Spain

---

MK01   1:45 – 2:00

TEMPERATURE DEPENDENCE OF HYDROGEN-BONDED STRUCTURES OF PHENOL CATION INVESTIGATED BY UV PHOTODISSOCIATION SPECTROSCOPY

ITARU KURUSU, REONA YAGI, RYOTA KATO, HIKARU SATO, MASATAKA ORITO, YASUTOSHI KASAHARA, HARUKI ISHIKAWA, *Department of Chemistry, School of Science, Kitasato University, Sagamihara, Japan.*

To investigate the temperature effect on microscopic hydration structures in clusters, we have recorded ultraviolet photodissociation spectra of hydrated phenol cation, $[PhOH(H_2O)_5]^+$, under the temperature-controlled condition[a]. The temperature dependence in the spectra clearly exhibits that there are two isomers in the present experimental condition and that the relative populations between them changes with an elevation of the temperature. Among many optimized structures obtained by the DFT calculations, two distinct hydration motifs, ring-with-tail and chain type motifs, are assigned for the isomers observed in our experiment. The change in the relative populations based on our observation is quantitatively interpreted by statistical mechanical estimation based on the DFT calculations. A ring with tail type hydration motif is dominant in cold condition, whereas a chain-like motif is dominant in hot condition. Moreover, possible cooling paths from the chain-like to ring-with-tail type motifs are discussed. In the present paper, temperature effects on the structures of the other hydrogen-bonded phenol cation clusters than $[PhOH(H_2O)_5]^+$ are also introduced.

---

[a]H. Ishikawa, I. Kurusu, R. Yagi, R. Kato, Y. Kasahara, *J. Phys. Chem. Lett.* **8**, 2641 (2017).

MK02   2:02 – 2:17

OBSERVATION OF THE IR-INDUCED ISOMERIZATION OF HYDRATED PHENOL CATIONS TRAPPED IN THE COLD ION TRAP

HIKARU SATO, RYOTA KATO, YASUTOSHI KASAHARA, HARUKI ISHIKAWA, *Department of Chemistry, School of Science, Kitasato University, Sagamihara, Japan.*

Gas-phase hydrated clusters are treated as a microscopic model of hydration networks. Recently, we have revealed the temperature-dependence of hydration structures of hydrated phenol cation, $[PhOH(H_2O)_5]^+$ [a]. In the cold condition (30 K), only an isomer having a ring-with-tail type hydration motif (*Rt* isomer) exists, whereas chain-like (*C*) isomers are dominant in the hot condition (150 K). Since isomerizations among the isomers having distinct hydration motifs can be related to structural fluctuations in the bulk systems, we have been investigating the isomerization between these two isomers induced by the IR vibrational excitation for the further understanding of the microscopic hydration. At first, we observed an IR spectrum of the *Rt* isomer at 30 K, and found a OH stretch band that is specific for the *Rt* isomer at 3330 cm^{-1}. Next, the IR laser light at 3330 cm^{-1} was irradiated to the *Rt* isomers in the cold trap. After 3 μs from the IR excitation, we observed UV photodissociation spectra. As a result, an increase of the intensity at the band of the *C* isomer (25400 cm^{-1}) was clearly observed. This change indicates the IR induced isomerization of the *Rt* isomer to the *C* isomer. Moreover, we observed a cooling of the *C* isomer produced by the IR excitation by the collisions with He buffer gas in the trap.

---

[a]H. Ishikawa, I. Kurusu, R. Yagi, R. Kato, Y. Kasahara, *J. Phys. Chem. Lett.* **8**, 2641 (2017).

MK03

FINE AND HYPERFINE STRUCTURE IN THE MID-INFRARED ABSORPTION SPECTRUM OF THE AR-NO COMPLEX

CHUANXI DUAN, *College of Physical Science and Technology, Central China Normal University, Wuhan, China.*

The absorption spectrum of the open-shell complex Ar-NO was recorded in the region of NO fundamental band using distributed feed-back quantum cascade lasers in a supersonic slit-jet expansion. Five sub-bands with P' up to 7/2 and J' up to 12.5 was observed. Three different Hamiltonians[a,b,c] were used to analyze the observed fine structure. The hyperfine splittings due to the N(I=1) nuclei could also be resolved in the subbands P = 1/2-1/2 and P = 1/2-3/2. The progress on the analysis of the fine and hyperfine structure will be presented.

---

[a] Y. Kim, H. Meyer, Int. Rev. Phys. Chem. 20, 219 (2001).
[b] W. M. Fawzy, J. T. Hougen, J. Mol. Spectrosc. 137, 154 (1989).
[c] J. Liu, T.A. Miller, J. Phys. Chem. A 118,11871 (2014).

MK04

SPECTRA OF C_6H_6-Rg_n (n=1,2) IN THE 3 MIRCON INFRARED BAND SYSTEM OF BENZENE

A. J. BARCLAY, *Department of Physics and Astronomy, University of Calgary, Calgary, AB, Canada*; BOB McKELLAR, *Steacie Laboratory, National Research Council of Canada, Ottawa, ON, Canada*; NASSER MOAZZEN-AHMADI, *Physics and Astronomy/Institute for Quantum Science and Technology, University of Calgary, Calgary, AB, Canada.*

Benzene-noble gas complexes were one of the earliest topics of interest in spectroscopic investigation of van der Waals (vdW) complexes. Smalley et al.[a] observed C_6H_6-$(He)_{1,2}$ vdW complexes in the late 1970s by means of electronic spectroscopy. A recent study on the same species was done by Hayashi and Oshima[b] at higher resolution (250 MHz). Here, we present an extensive infrared observation of C_6H_6-Rg_n (n=1,2) with the rare gas being He, Ne, or Ar, in the 3 micron region. The spectra were observed using a tunable optical parametric oscillator to probe a pulsed supersonic-jet expansion from a slit nozzle.

Benzene monomer is known to have a complex band system in this region.[c] The strongest band, centered around 3047.91 cm^{-1}, belongs mainly to the C-H stretching fundamental ν_{12} of symmetry E_{1u}. Other strong perpendicular bands occurring just above the main band as a result of intensity borrowing via anharmonic resonances between the fundamental ν_{12} and the combinations are $\nu_2 + \nu_{13} + \nu_{18}$, occurring near 3079 cm^{-1}, and $\nu_{13} + \nu_{16}$ and $\nu_3 + \nu_{10} + \nu_{18}$, both occurring near 3100 cm^{-1}. The latter two bands are separated by merely 1.45 cm^{-1}. Although data analysis and observation are presently ongoing, we observe analogous bands for C_6H_6-Rg_n (n=1,2). Spectra were assigned to a symmetric top with C_{6v} symmetry with the rare gas atom being located on the C_6 symmetry axis. Spectra of the C_6H_6-Rg_2 trimers are in agreement with a D_{6h} symmetry structure, where the rare gas atoms are positioned above and below the plane of the Benzene monomer. Although jet conditions have resulted in excellent signal to noise for the dimer and trimer spectra, we have not been able to identify any lines which might be due to tetramers or larger clusters. We intend to pursue the search for large clusters using a cooled nozzle.

---

[a] S. M. Beck, M. G. Liverman, D. L. Monts and R. E. Smalley, J. Chem. Phys. 70, 232 (1979).
[b] M. Hayashi and Y. Ohshima, Chem. Phys. 419, 131 (2013).
[c] J. Pliva and A.S. Pine, J. Mol. Spectrosc. 126, 82 (1987).

MK05

CHARACTERIZATION OF OCS-HCCCH AND N2O-HCCCH DIMERS: THEORY AND EXPERIMENT

A. J. BARCLAY, *Department of Physics and Astronomy, University of Calgary, Calgary, AB, Canada*; ANDREA PIETROPOLLI CHARMET, *Dipartimento di Scienze Molecolari e Nanosistemi, Università Ca' Foscari, Venezia, Italy*; K. H. MICHAELIAN, *CanmetENERGY, Natural Resources Canada, Edmonton, Alberta, Canada*; NASSER MOAZZEN-AHMADI, *Physics and Astronomy/Institute for Quantum Science and Technology, University of Calgary, Calgary, AB, Canada*.

The infrared spectra of the weakly-bound dimers OCS-HCCCCH, in the region of the ν_1 fundamental band of OCS (2050 cm^{-1}), and N$_2$O-HCCCCH, in the region of the ν_1 fundamental band of N$_2$O (2200 cm^{-1}), are observed in a pulsed supersonic slit jet expansion probed with tunable diode/QCL lasers. Both OCS-HCCCCH and N$_2$O-HCCCCH were found to have planar structure with side-by-side monomer units having nearly parallel axes. These bands have hybrid rotational structure which allow for estimates of the orientation of OCS and N$_2$O in the plane of their respective dimers. Analogous bands for OCS-DCCCCD and N$_2$O-DCCCCD were also observed and found to be consistent with the normal isotopologues. Various levels of ab initio calculations were performed to find stationary points on the potential energy surface, optimized structures and interaction energies. Three stable geometries were found for OCS-HCCCCH and two for N$_2$O-HCCCCH. The rotational parameters at CCSD(T*)-F12c level of theory give results in very good agreement with those obtained from the observed spectra. In both dimers, the experimental structure corresponds to the lowest energy isomer.

Intermission

MK06

LASER SPECTROSCOPIC STUDY ON PHENOL-ETHYLDIMETHYLSILANE DIHYDROGEN-BONDED CLUSTER

MASAAKI UCHIDA, TAKUTOSHI SHIMIZU, YASUTOSHI KASAHARA, HARUKI ISHIKAWA, *Department of Chemistry, School of Science, Kitasato University, Sagamihara, Japan*; YOSHITERU MATSUMOTO, *Department of Chemistry, Faculty of Science, Shizuoka University, Shizuoka, Japan*.

Dihydrogen bond is a hydrogen bond which acts between two H atoms having opposite partial charges. Among various kinds of dihydrogen bond systems, we have been investigating the Si-H$\cdots$H-O type dihydrogen bond[a][b]. On the course of our study, we found that the competition between the dihydrogen bond and dispersion interactions determines the structures of phenol-alkylsilane 1:1 dihydrogen-bonded clusters. However, since there are many isomers due to intermolecular orientation as well as conformation of alkyl groups, we have not yet determined their structures completely. In the present study, we have carried out a laser spectroscopic study on the phenol-ethyldimetylsilane (PhOH-EDMS) dihydrogen bonded clusters. Since EDMS has a simple structure, the number of the isomers is expected to be small. We recorded laser-induced fluorescence (LIF), UV-UV hole-burning, and IR spectra of jet-cooled PhOH-EDMS clusters. As a result, we identified two isomers, A and B, based on the UV-UV hole-burning spectra. The 0-0 band of the isomer A is redshifted by -83.3 cm^{-1} compared with that of the PhOH monomer and exhibits a simple and long progression of 16.6 cm^{-1} interval of the intermolecular vibration. On the other hand, the redshift of the 0-0 band of the isomer B is much smaller (-20.3 cm^{-1}) and exhibits rather congested band patterns. The redshifts of the OH stretching band of these isomers are -27 and -20 cm^{-1} for the isomers A and B, respectively. Based on the comparison of spectral features observed with those predicted by the DFT calculations, we determined the structure of the isomers A and B. Details will be presented in the paper.

---

[a] H. Ishikawa, A. Saito, M. Sugiyama, N. Mikami, *J. Chem. Phys.* **123**, 224309 (2005).
[b] H. Ishikawa, T. Kawasaki, R. Inomata, *J. Phys. Chem. A* **119**, 601 (2015).

MK07

IR SPECTROSCOPIC STUDY ON PHENOL-TRIETHYLSILANE DIHYDROGEN-BONDED CLUSTER IN THE ELECTRONIC EXCITED STATE

TAKUTOSHI SHIMIZU, MASAAKI UCHIDA, YASUTOSHI KASAHARA, HARUKI ISHIKAWA, *Department of Chemistry, School of Science, Kitasato University, Sagamihara, Japan*; YOSHITERU MATSUMOTO, *Department of Chemistry, Faculty of Science, Shizuoka University, Shizuoka, Japan.*

To reveal detailed characters of the dihydrogen bond at molecular level, we have been carrying out IR spectroscopic study on the Si-H$\cdots$H-O type dihydrogen-bonded clusters[a,b]. It was found that the structures of the phenol-alkylsilane 1:1 clusters are determined by the competition between the dihydrogen bond and the dispersion interaction in the case of the S_0 state of the neutral clusters. On the other hand, the dihydrogen bond exhibit a dominant contribution in the cationic states. Based on these results, it is expected the balance between the dihydrogen bond and the dispersion interaction is expected to change in the S_1 state compared with the S_0 state. Thus, we have carried out an IR spectroscopic study on the phenol-alkylsilane clusters, in the present study.

In the present paper, we will report mainly on the results of the phenol-triethylsilane (PhOH-TES) clusters. It is already reported that three isomers appear in the fluorescence excitation spectrum of PhOH-TES. Using the vibronic bands of these isomers as excitation transitions, IR spectra in the S_1 state were observed by the UV-IR double resonance technique. All of the isomers exhibit much larger redshifts of the OH stretching band compared with those in the S_0 state. It indicates the strengthening of the dihydrogen bond in the S_1 state. In addition, all the isomer exhibit Franck-Condon-like patterns. The patterns change by the intermediate vibrational levels selected by the UV transitions. Similar Franck-Condon-like pattern in the IR transition is reported in the literature[c]. This result indicates a strong coupling between the OH stretch and the intermolecular vibrational mode. This coupling is considered to be a characteristic feature in the S_1 state.

[a] H. Ishikawa, A. Saito, M. Sugiyama, N. Mikami, *J. Chem. Phys.* **123**, 224309 (2005).
[b] H. Ishikawa, T. Kawasaki, R. Inomata, *J. Phys. Chem. A* **119**, 601 (2015).
[c] A. V. Zabuga, M. Z. Kamrath, T. R. Rizzo, *J. Phys. Chem. A* **119**, 10494 (2015).

MK08

COMPOUND-MODEL MORPHED POTENTIAL FOR THE HYDROGEN BOND HCN–HF

LUIS A. RIVERA-RIVERA, *Department of Physical Sciences, Ferris State University, Big Rapids, MI, USA*; ROBERT R. LUCCHESE, JOHN W. BEVAN, *Department of Chemistry, Texas A & M University, College Station, TX, USA.*

A five-dimensional compound-model morphed potential has been generated for the prototype hydrogen-bonded dimer HCN-HF. The potential includes the intermolecular degree of freedom and the HF stretching vibration. Five morphing parameters only are optimized correcting for inadequacies in the underlying *ab initio* potentials. The morphing transformation utilized a rotationally resolved spectroscopic database composed of microwave and infrared spectroscopic information. Band origin fundamental vibrational frequencies in HCN-HF are fitted to an average absolute error of 0.006 cm^{-1}. The calculated value of the ground state dissociation energy, D_0 = 1969 cm^{-1} is in excellent agreement with the experimental value of 1970(10) cm^{-1} [Oudejans and Miller, Chem. Phys. 239 (1998) 345]. Limitations of the morphing methodology and its potential applications will be discussed.

MK09   4:35 – 4:50

SPECTRAL SHIFTS AND STRUCTURES OF O-H•••π HYDROGEN BONDED COMPLEXES BETWEEN CARBOXYLIC ACIDS AND BENZENE: A MATRIX ISOLATION INFRARED SPECTROSCOPIC STUDY

PUJARINI BANERJEE, INDRANI BHATTACHARYA, <u>DEB PRATIM MUKHOPADHYAY</u>, TAPAS CHAKRABORTY, *Physical Chemistry, Indian Association for the Cultivation of Science, Kolkata, India.*

A series of binary O-H•••π hydrogen bonded complexes between benzene and different carboxylic acids, formic, acetic and trifluoroacetic acid, have been synthesized under matrix isolated conditions. The ν_{O-H} stretching frequencies of homologous acids of different aqueous phase acidities (pK_as) appear at very similar frequencies as long as they are dispersed within the matrix as monomers. The effect of chemical substitution is manifested only upon complex formation, and the magnitude of ν_{O-H} shifting ($\Delta\nu_{O-H}$) is found to relate with corresponding pK_a values of the acids. Local charge transfer effects are found to be good descriptors of the observed spectral shifts. The shift in the case of formic acid has been used to test the accuracy of electronic structure calculations in determining correct geometries of such weakly H-bonded complexes. It has been determined that popular DFT and DFT-D functionals with modestly sized basis sets are sufficient for correct predictions of the ν_{O-H} spectral shifts observed in our study. On the other hand, larger sized basis sets with diffuse functions do not always predict structures that are consistent with the observed shifts. It is suggested that energetic contribution of the carbonyl sub-group is not balanced properly when diffuse functions are used in geometry optimizations, and apparently some of the repulsive components are overestimated. This underscores the need for experimental benchmarking in determining correct geometries of H-bonded complexes.

MK10   4:52 – 5:07

INFRARED SPECTROSCOPY OF ALANINE AND ITS WATER CLUSTER ISOLATED IN SOLID PARAHYDROGEN

<u>BRENDAN MOORE</u>, SHIN YI TOH, YING-TUNG ANGEL WONG, PAVLE DJURICANIN, TAKAMASA MOMOSE, *Department of Chemistry, University of British Columbia, Vancouver, BC, Canada.*

High-resolution infrared spectra of β-alanine and its water clusters have been studied using solid *para*-H_2 FT-IR matrix-isolation spectroscopy. It is known that zwitterion forms of amino acids are more stable than neutral forms in water solutions and biological environments, but it is still under debate whether zwitterions are stable in small alanine-water clusters. We have investigated the stabilization effect of water molecules on the zwitterion form of β-alanine by codepositing H_2O and β-alanine in solid *para*-H_2. Through a comparison with theoretical calculations, as well as with crystalline β-alanine FT-IR spectra, the characteristic NH_3 N-H bending vibrational frequency for the zwitterionic form was identified. Analysis of the spectral peak temporal behavior shows that other proposed zwitterion peaks behave similarly to the characteristic NH_3 spectral peak. It has been shown that water can stabilize the zwitterionic form of gas phase amino acids, causing the zwitterion to form preferentially over the neutral form under certain conditions. The β-alanine zwitterion formation rate may be attributed to aggregation of small water clusters in the solid *para*-H_2 matrix. These findings provide insight into the behavior of amino acid zwitterion formation.

ML. Structure determination
Monday, June 18, 2018 – 1:45 PM
Room: 2079 Natural History

Chair: S. A. Cooke, Purchase College SUNY, Purchase, NY, USA

ML01   1:45 – 2:00

PUSHING THE LA-MB-FTMW TO THE LIMIT: CONTROLLED DOUBLE PULSE LASER ABLATION FOURIER TRANSFORM MICROWAVE SPECTROSCOPY

<u>IKER LEÓN</u>, ELENA R. ALONSO, SANTIAGO MATA, JOSÉ L. ALONSO, *Grupo de Espectroscopia Molecular, Lab. de Espectroscopia y Bioespectroscopia, Unidad Asociada CSIC, Universidad de Valladolid, Valladolid, Spain.*

In a quest for improving current laser-ablation devices in combination with molecular-beam Fourier-transform microwave spectroscopy (LA-MB-FTMW)[a,b] we present a double pulse laser ablation system scheme to improve the experimental conditions during the ablation of any organic molecule. As a proof of concept, we determine the substitution structure (rs) of aspirin, with molecular beam Fourier transform microwave spectroscopy. Furthermore, to prove the universality of this technique we extended it to the study of metal clusters. As it will be shown, the main advantages of this set up are a considerable generation of neutral molecules/clusters with an excellent S/N ratio, a reduced integration time and a considerable damage reduction caused to the sample. This is possible due to a softer ablation occurrence. The benefits of using these two lasers scheme would be also beneficial for chirped pulse Fourier transform microwave (CP-FTMW) spectroscopy.

[a] Bermúdez, C.; Mata, S.; Cabezas, C.; Alonso, J. L., Tautomerism in Neutral Histidine. Angew. Chem. Int. Ed. 53, 11015–11018 (2014)
[b] Sanz, M. E.; Cabezas, C.; Mata, S.; Alonso, J. L. Rotational Spectrum of Tryptophan. J. Chem. Phys., 140 (20), 204308 (2014).

ML02   2:02 – 2:17

A PICTURE OF THE GLY-PRO SEQUENCE OF COLLAGEN IS CAUGHT IN GAS PHASE

<u>IKER LEÓN</u>, ELENA R. ALONSO, CARLOS CABEZAS, SANTIAGO MATA, JOSÉ L. ALONSO, *Grupo de Espectroscopia Molecular, Lab. de Espectroscopia y Bioespectroscopia, Unidad Asociada CSIC, Universidad de Valladolid, Valladolid, Spain.*

The structure of the Gly-Pro and Pro-Gly dipeptides have been studied by laser ablation chirped pulse and molecular beam Fourier transform microwave (LA-CP-FTMW and LA-MB-FTMW) spectroscopy.[a,b] While three conformers of Gly-Pro have been characterized only one conformer has been found in Pro-Gly. Interestingly, the resemblance of the structure determined in Gly-Pro and that observed in the crystal of a collagen-like peptide is striking.

[a] Bermúdez, C.; Mata, S.; Cabezas, C.; Alonso, J. L., Tautomerism in Neutral Histidine. Angew. Chem. Int. Ed. 53, 11015–11018 (2014).
[b] J. L. Alonso and J. C. López, in Gas-Phase IR Spectroscopy and Structure of Biological Molecules, eds. A. M. Rijs and J. Oomens, Topics in., 2015, vol. 364, pp. 335–402.

ML03   2:19 – 2:34

USING HYPERFINE STRUCTURE TO QUANTIFY THE EFFECTS OF SUBSTITUTION IN 2-, 3-, AND 4-PICOLYLAMINE

LINDSEY M McDIVITT, TIMOTHY J McMAHON, JOSIAH R BAILEY, CALEB B SHIERY, <u>RYAN G BIRD</u>, *Chemistry, University of Pittsburgh Johnstown, Johnstown, PA, USA.*

Last year we presented preliminary results on three methylamine substituted pyridines, 2-, 3-, and 4-picolylamine. After helpful feedback, the microwave spectra of all three molecules were recollected over the frequency range of 7-18 GHz using by zero-fitting the free induction decay. Each molecule showed a distinctive quadrupole splitting, which is representative of the local electronic environment around the two different ^{14}N nuclei, with the pyridine nitrogen being particularly sensitive to the pi-electron distribution within the ring. An extended Townes and Dailey analysis was used to determine the lone pair density around each nitrogen and compared to that of benzylamine and pyridine. Results of this analysis and how it explains the configuration of the methylamine group with respect to the pyridine ring in each of the picolylamines will be discussed.

ML04 2:36 – 2:51

ELECTRON-WITHDRAWING EFFECTS ON THE MOLECULAR STRUCTURE OF 2- AND 3-NITROBENZONITRILE REVEALED BY BROADBAND MICROWAVE SPECTROSCOPY

JACK B. GRANEEK, MELANIE SCHNELL, *FS-SMP, Deutsches Elektronen-Synchrotron (DESY), Hamburg, Germany.*

Nitrobenzonitrile consists of two electron-withdrawing groups, which have negative inductive and mesomeric effects on the phenyl ring resulting in interesting physical properties. The rotational spectra of 2- and 3-nitrobenzonitrile were recorded via chirped-pulse Fourier transform microwave spectroscopy in the frequency range of 2–8 GHz. For both molecules, the main isotopologues and all isotopologues of the respective ^{13}C-, ^{15}N-, ^{18}O-monosubstituted species in their natural abundance were assigned. These assignments allowed for the structural determination of 2- and 3-nitrobenzonitrile via Kraitchman's equations as well as a mass-dependent least-squares fitting approach. Structural changes caused by steric interaction and competition for the electron density of the phenyl ring highlight how these strong electron-withdrawing substituents affect one another according to their respective positions on the phenyl ring.

ML05 2:53 – 3:08

ROTATIONAL SPECTRUM OF METHYL PIVALATE

NOBUHIKO KUZE, *Faculty of Science and Technology, Sophia University, Tokyo, Japan*; YOSHIYUKI KAWASHIMA, *Applied Chemistry, Kanagawa Institute of Technology, Atsugi, Japan.*

The rotational spectrum of methyl pivalate (t-BuC(O)OCH$_3$) in the ground vibrational state was observed by molecular beam-Fourier transform microwave spectroscopy. Thirty b-type rotational transitions were assigned. Some high-K_a lines were found to be split and we have interpreted these splittings in terms of the internal rotation of the methyl group. Some forbidden transitions were also observed in case where $K_a = 2$ levels were involved in the internal rotation with E state. The observed spectra including the forbidden transitions were analyzed using XIAM program. Spectral assignment confirmed that the observed spectral splitting was due to the methyl rotation of the methoxy group. In comparison with the theoretical rotational constants, observed rotational constants indicated the molecule took a trans-zigzag conformation for C-C-C-O-C skeleton with C_s symmetry.

ML06 3:10 – 3:25

FTMW SPECTRA OF CH$_3$OC(O)NCO AND CH$_3$OC(O)N$_3$

SHINICHIRO WATANABE, *Faculty of Science and Technology, Sophia University, Tokyo, Japan*; YOSHIYUKI KAWASHIMA, *Applied Chemistry, Kanagawa Institute of Technology, Atsugi, Japan*; NOBUHIKO KUZE, *Faculty of Science and Technology, Sophia University, Tokyo, Japan.*

The rotational spectra of methoxycarbonyl isocyanate (CH$_3$OC(O)NCO) and methyl azidoformate (CH$_3$OC(O)N$_3$) in the ground vibrational state were observed by molecular beam-Fourier transform microwave spectroscopy. Observed spectral lines for a-type transitions were assigned. Splittings of the spectral lines were observed by the internal rotation of the CH$_3$ group and the hyperfine structure of the ^{14}N atom. Comparison of the observed spectroscopic constants with the calculated ones led to the conclusion that the assigned spectrum was due to the *syn-syn* form that NCO or N$_3$ group. Determined parameters by the spectral analysis of methoxycarbonyl isocyanate were rotational, centrifugal distortion and nuclear electric quadrupole coupling constants including the potential barrier V_3 to internal rotation of the methyl group. At this stage, we could not analyze the complicate nuclear electric quadrupole coupling splittings of the observed spectrum of methyl azidoformate due to the hyperfine structure.

ML07　　　　　　　　　　　　　　　　　　　　　　　　　　　　　　　　　　　　　3:27 – 3:42

CARBON-13 STUDIES OF SULFUR-TERMINATED CARBON CHAINS: CHEMICAL BONDING, MOLECULAR STRUCTURES AND FORMATION PATHWAYS

<u>MICHAEL C McCARTHY</u>, *Atomic and Molecular Physics, Harvard-Smithsonian Center for Astrophysics, Cambridge, MA, USA*; KELVIN LEE, *Radio and Geoastronomy Division, Harvard-Smithsonian Center for Astrophysics, Cambridge, MA, USA.*

The rotational spectra of the singly-substituted ^{13}C isotopic species of a number of sulfur-terminated carbon chains have been detected between 5 and 40 GHz using a supersonic jet in combination with a cavity Fourier transform microwave spectrometer. The chains include both closed-shell molecules (e.g., H_2C_3S) and radicals (e.g., HCCS, HC_3S, and C_4S). The experiments were carried out with precursors enriched in ^{13}C, either $H^{13}C^{13}CH$ or $^{13}CS_2$. From the ^{13}C hyperfine coupling constants, the unpaired electronic density along the chain can be quantified for the radical species, while precise experimental structures (r_0) can be derived for each molecule by a least-squares fit to the rotational constants. The use of $^{13}CS_2$ in particular provides clues as to the dominant formation pathway for each chain in our discharge nozzle. Somewhat surprisingly, the ^{13}C from this precursor appears to be substituted in one of three distinct ways: random, a specific C site, or not at all. This propensity appears to be molecule specific, implying that both neutral-radical and radical-radical reactions are important.

Intermission

ML08　　　　　　　　　　　　　　　*Post-Deadline Abstract*　　　　　　　　　　　　　　　4:18 – 4:33

THE CURIOUS(ER) CASE OF MASS 63: A THRILLING AND CONFUSING SEQUEL

<u>CHRISTOPHER C BLACKSTONE</u>, ANDREI SANOV, *Chemistry and Biochemistry, University of Arizona, Tucson, AZ, USA.*

Produced in a reaction or interaction either between methanol and methoxide or methoxide and oxygen, an unexpected and difficult to identify species has appeared in our mass spectrum. We seek to narrow its molecular formula to one of CH_3O_3 or $C_2H_7O_2$ using fully deuterated methanol as a precursor to separate the two formulas by mass, and to further use the photoelectron spectrum and a fragmentation spectrum of the fully deuterated species to most accurately determine the molecular identity and completely assign its photoelectron spectrum.

ML09　　　　　　　　　　　　　　　　　　　　　　　　　　　　　　　　　　　　　4:35 – 4:50

CONFORMATION-SPECIFIC IR AND UV SPECTROSCOPY OF A MODEL SYNTHETIC FOLDAMER: β^3-ALA TRIPEPTIDE

<u>DEWEI SUN</u>, *Department of Chemistry, Purdue University, Valparaiso, IN, USA*; JOSHUA L. FISCHER, KARL N. BLODGETT, TIMOTHY S. ZWIER, *Department of Chemistry, Purdue University, West Lafayette, IN, USA.*

With the development of designed foldamers that that display biological activity and aid in drug delivery processes, there is an increasing interest in exploring the inherent conformational properties of model foldamers to understand better the subtle counter-balance of forces at play. β-amino acids have one additional backbone carbon atom that extends the spacing between amide groups, thereby providing a flexibility of construction, leading to secondary structures that either mimic or are complementary to those found in nature. Here we explore the secondary structures of foldamers in which the position of substitution of the methyl side chains of the Ala residues are switched from the β^2 to the β^3 position. In particular, we present data on a β^3-Ala tripeptide, Ac-β^3-Ala-β^3-Ala-β^3-Ala-NHBn under jet-cooled, isolated conditions. This talk will present conformation-specific IR and UV spectra using the techniques of resonant ion-dip infrared (RIDIR) spectroscopy and IR-UV holeburning, respectively. Following an exhaustive computational search of the conformational potential energy surface, low-lying minima are optimized and IR spectra calculated for comparison with the experimental RIDIR spectra. A single conformer is observed experimentally and assigned to a conformer with a C12/C8 hydrogen-bonded architecture that incorporates a 12-membered turn that is the β-peptide analog of a β-turn.

ML10   4:52 – 5:07
THE EFFECTS OF C-TERMINAL FLUOROPHORE CAPS ON THE STRUCTURE OF AMINOISOBUTYRIC ACID DIPEPTIDES

JOSHUA L. FISCHER, *Department of Chemistry, Purdue University, West Lafayette, IN, USA*; BRAYAN R. ELVIR, *Department of Chemistry and Biochemistry, Fairfield University, Fairfield, CT, USA*; KARL N. BLODGETT, *Department of Chemistry, Purdue University, West Lafayette, IN, USA*; MATTHEW A. KUBASIK, *Department of Chemistry and Biochemistry, Fairfield University, Fairfield, CT, USA*; TIMOTHY S. ZWIER, *Department of Chemistry, Purdue University, West Lafayette, IN, USA*.

Aminoisobutyric acid (Aib) is an achiral synthetic amino acid with a high propensity to form 3_{10} helices, the second most biologically abundant helix. Aib is also known to induce helix formation in other peptides, and can persuade helix formation from a consecutive sequence of as little as three residues. Studying polyAib in the gas-phase provides a spectroscopic opportunity to examine the inherent hydrogen bonding that directs this helix formation. Since Aib is achiral, there is no infrared (IR) distinction between left- and right-handed helices, preventing powerful IR-based structural differentiation of left- and right-handed structures. However, the introduction of chirality through a cap offers a means by which the balance between the two can be tipped. However, it could also induce alterations to the structure. This talk describes conformation-specific IR and ultraviolet (UV) double-resonance spectroscopy of a series of three Aib dipeptides with three UV-active C-terminal caps: NH-benzyl (NHBn) as a reference structure due to its common use as a fluorophore in similar studies, NH-p-fluorobenzyl (NHBnF), and alpha-methylbenzylamine (AMBA), a chiral cap. These molecules are brought into the gas phase via laser desorption and cooled in a supersonic expansion, enabling us to probe the inherent conformational preferences of the isolated molecules. For both the NHBn and NHBnF cap, a single conformer is observed, with infrared spectra assignable to nearly identical type II β-turn structure. Additionally, the higher oscillator strength of the NHBnF cap enabled UV-UV holeburning, not readily accomplished with the NHBn cap. For AMBA, two unique conformers were found, one of which was a nearly identical type II β-turn, while the minor conformer possessed two sequential γ-turns formed between adjacent amide groups, stabilized by an NH-π interaction.

ML11   5:09 – 5:24
GAS PHASE INFRARED SPECTROSCOPY OF ISOMERIC BENZYL AND TROPYLIUM CATIONS

J. PHILIPP WAGNER, *Department of Chemistry, University of Georgia, Athens, GA, USA*; DAVID C McDONALD, *Chemistry, University of Georgia, Athens, GA, USA*; MICHAEL A DUNCAN, *Department of Chemistry, University of Georgia, Athens, GA, USA*.

The isomeric benzyl and tropylium $C_7H_7^+$ cations have been of great interest to physical organic and gaseous ion chemists for many decades. Still, infrared spectroscopic characterization of these ions in the gas phase could so far only be achieved for their methylated derivatives but not for the $C_7H_7^+$ ions themselves. Thus, we set out to produce both relevant isomers of this elusive ion in a cold molecular beam experiment through ionization of different precursor molecules comprising the preformed 6- and 7-membered rings. We measured their IR spectra via photodissociation with a tunable OPO/OPA laser system in combination with the argon messenger atom technique. The obtained spectra were assigned with the aid of second order vibrational perturbation theory utilizing dispersion-corrected density functional theory.

ML12

NMR, RAMAN AND DFT STUDY OF TAUTOMERIC EQUILIBRIUM AND SLOW SELF-ASSEMBLING IN LYOTROPIC CHROMONIC LIQUID CRYSTAL SSY AQUEOUS SOLUTIONS

NOMEDA RIMA VALEVICIENE, *Faculty of Medicine, Vilnius University, Vilnius, Lithuania*; VYTAUTAS BALEVICIUS, KESTUTIS AIDAS, LAURYNAS DAGYS, ARUNAS MARSALKA, KRISTINA KRISTINAITYTE, *Faculty of Physics, Vilnius University, Vilnius, Lithuania*; VALERIY POGORELOV, YELYZAVETA CHERNOLEVSKA, YEVHENII VASKIVSKYI, IRYNA DOROSHENKO, *Faculty of Physics, Taras Shevchenko National University of Kyiv, Kyiv, Ukraine*; GEORGE PITSEVICH, *Faculty of Physics, Belarusian State University, Minsk, Belarus*; ULADZIMIR SAPESHKA, *Physics, University of Illinois at Chicago, Chicago, IL, USA*.

Temperature and composition effects on tautomeric equilibrium and slow self-assembling in various phases of Sunset Yellow FCF (SSY) aqueous solutions were studied by ^{1}H, ^{15}N NMR and Raman spectroscopy. The solutions were prepared according to the phase diagram in order to pass through all phase transitions between isotropic phase (I) and principal chromonic phases - nematic (N) and columnar (M) ones. It was shown that the tautomeric equilibrium in SSY is strongly shifted towards the hydrazone form. The corresponding equilibrium constant $pK_T = -2.5$ was calculated using the DFT SMD model. The dominance of hydrazone form was confirmed experimentally using the long-range ^{1}H–^{15}HN correlation, widely known as HMBC. The peak at 14.2 - 14.7 ppm found in ^{1}H NMR spectra in all phases can be attributed to the proton in the N–H...O bridge. It evidences that: i) the growing of SSY aggregates is accompanied by the segregation of water in the intercolumnar areas with no exchange with the N–H protons in the internal layers of the columnar stacks; ii) the life time of those aggregates is $\geq 10^{-7} - 10^{-8}$ s or even longer. The water confined in intercolumnar areas can be considered as the neat. Its molecular motion is changing at heating and crossing the border from M to N phases in similar manner as at the melting. The equilibration time for N+M→M is very long, most probably because of supramolecular restructuring, i.e. the growing of stack aggregates in M phase. If the sample is cooled down relatively fast to the temperature below N→M transition, the structural changes are behind, and the system falls to the supercooled state. Then the system evolves via slow self-assembling from this state to the equilibrium. This process can last over hours or even tens of days.

TA. Mini-symposium: Frequency-Comb Spectroscopy

Tuesday, June 19, 2018 – 8:30 AM

Room: 116 Roger Adams Lab

Chair: Frans Harren, Radboud University, Nijmegen, Netherlands

TA01   INVITED TALK   8:30 – 9:00

DUAL THZ COMB SPECTROSCOPY

<u>TAKESHI YASUI</u>, *Graduate School of Technology, Industrial and Social Sciences, Tokushima University, Tokushima, Japan.*

Optical frequency combs are innovative tools for broadband spectroscopy because a series of comb modes can serve as frequency markers that are traceable to a microwave frequency standard. However, a mode distribution that is too discrete limits the spectral sampling interval to the mode frequency spacing even though individual mode linewidth is sufficiently narrow. Here, using a combination of a spectral interleaving and dual-comb spectroscopy in the terahertz (THz) region, we achieved a spectral sampling interval equal to the mode linewidth rather than the mode spacing. The spectrally interleaved THz comb was realized by sweeping the laser repetition frequency and interleaving additional frequency marks. In low-pressure gas spectroscopy, we achieved an improved spectral sampling density of 2.5MHz and enhanced spectral accuracy of 8.39×10^{-7} in the THz region. The proposed method is a powerful tool for simultaneously achieving high resolution, high accuracy, and broad spectral coverage in THz spectroscopy.

TA02   9:04 – 9:19

DOPPLER-LIMITED BROADBAND DUAL-COMB SPECTROSCOPY AT 3 μm

<u>ZAIJUN CHEN</u>, THEODOR W. HÄNSCH, NATHALIE PICQUÉ, *Laser Spectroscopy Division, Max Planck Institute of Quantum Optics, Garching, Germany.*

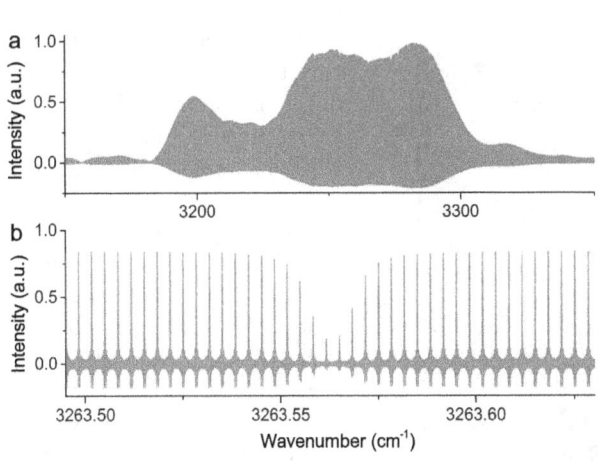

We present a new scheme of mid-infrared dual-comb Fourier transform spectroscopy. It provides broad-spectral bandwidth absorption and dispersion spectra at a resolution limited by the Doppler width of the molecular profiles, with a negligible instrumental lineshape and self-calibration of the wavenumber scale within the accuracy of an atomic clock. The averaging times can be arbitrarily long and the spectra do not involve any phase corrections or other types of processing that could generate systematic effects or artifacts. Such results build on our concept of feed-forward dual-comb spectroscopy[a], first demonstrated in the near-infrared 1.5 μm region. Here, we illustrate our latest results around 3000 cm^{-1} measured at a signal-to-noise ratio that exceeds 1000 and a resolution of 3 10^{-3} cm^{-1}. A spectrum with resolved comb lines of acetylene in the region of the ν_3 band, measured within 35 minutes, is shown over its entire span in Fig.a. In an expanded view (Fig.b), the $P(13)$ line of the ν_3 band of $^{12}C_2H_2$ is sampled by the individual comb lines of perfect cardinal sine instrumental line shape. Our technique opens up novel opportunities for broadband and precise investigation of molecular profiles.

---

[a] Z. Chen, M. Yan, T. W. Hänsch, and N. Picqué, A phase-stable dual-comb interferometer, preprint at arXiv:1705.04214 (2017).

TA03   9:21 – 9:36

MASSIVELY PARALLEL DETECTION OF TRACE MOLECULES AND ISOTOPOLOGUES WITH A SUBHARMONIC MID-IR DUAL COMB SYSTEM

ANDREY MURAVIEV, *CREOL, The College of Optics & Photonics, University of Central Florida, Orlando, Fl, USA*; VIKTOR O SMOLSKI, *Mid-IR lasers, IPG Photonics, Birmingham, AL, USA*; ZACHARY E LOPARO, *Mechanical and Aerospace Engineering, University of Central Florida, Orlando, Fl, USA*; KONSTANTIN L VODOPYANOV, *CREOL, The College of Optics & Photonics, University of Central Florida, Orlando, Fl, USA*.

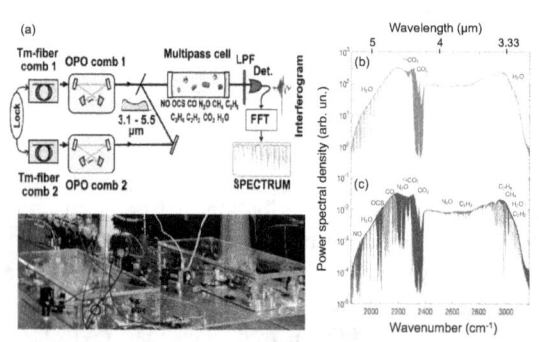

We use a pair of highly-coherent subharmonic GaAs optical parametric oscillators with an instantaneous span 3.1-5.5 μm to demonstrate fast acquisition of 350,000 mode-resolved spectral data points and perform parallel detection in a mixture of 22 molecular species including N_2O, NO, CO, OCS, CH_4, C_2H_6, C_2H_4, C_2H_2, CO_2, H_2O, and their isotopologues containing ^{33}S, ^{34}S, ^{13}C, ^{15}N, ^{18}O, ^{17}O, and ^2H (deuterium) isotopes. We demonstrate all the benefits of the mid-IR dual-comb spectroscopy including broadband coverage, fast acquisition of massive spectral data, ppb-level sensitivity, comb-tooth resolved spectra (with finesse 4000) and absolute optical frequency referencing to atomic clock. We sampled molecular spectra with the comb-tooth spacing (115 MHz), however, thanks to the narrow comb teeth (3-kHz absolute and 25-mHz relative linewidth between the two combs), much higher spectral resolution can be obtained in the scanning comb-tooth resolved mode. The Figure shows: (a) schematic of the dual-comb setup, (b) log-scale optical spectrum retrieved from a single coherently-averaged interferogram with an evacuated multipass gas cell, and (c) when the cell was filled with a mixture of gases. The two spectra are vertically offset for clarity.

TA04   9:38 – 9:53

DUAL-COMB SPECTROSCOPY USING QUANTUM AND INTERBAND CASCADE LASERS[a]

JONAS WESTBERG, LUKASZ A. STERCZEWSKI[b], GERARD WYSOCKI, *Department of Electrical Engineering, Princeton University, Princeton, NJ, USA*.

Dual-comb spectroscopy (DCS) using quantum cascade laser (QCL) or interband cascade laser (ICL) frequency combs presents an opportunity for miniaturized and fully electronically controlled broadband spectrometers with no moving-parts and all-electrical control that can serve as alternatives to systems based on broadly tunable lasers or external cavity lasers. In contrast to systems based on conventional tunable laser sources, a DCS system gives instantaneous access to the optical information across the entire spectral bandwidth through a multi-parallel heterodyning process, which enables acquisition times in the μs-range. The main drawback of quantum and interband cascade laser frequency combs is the excess phase noise observed, which affects the averaging capabilities of the DCS systems. In principle, coherent averaging can be implemented using active feedback control, but this adds additional complexity to the systems, which negates the intrinsic advantage of the monolithic comb emitters. Here, we report on DCS using free-running lasers, where a coherent averaging algorithm is implemented to correct for phase noise via purely computational means. The correction algorithm leverages the temporal mode coherence masked by the noise and is generally applicable to all DCS systems affected by excessive phase noise. Spectroscopic detection of molecular species with broadband spectral signatures in the mid-infrared will be discussed. Further details on the spectroscopic systems, as well as a discussion of current limitations and future directions of this DCS technique will be given.

[a]The authors acknowledge support from the DARPA SCOUT program (W31P4Q161001) and Thorlabs Inc.
[b]Faculty of Electronics, Wroclaw University of Science and Technology, Wroclaw 50370, Poland.
L. A. Sterczewski acknowledge support from the Kosciuszko Foundation Research Grant.

Intermission

TA05 10:29 – 10:44

FREQUENCY-AGILE COMBS: APPLICATIONS IN SPECTROSCOPY AND PHYSICAL METROLOGY

DAVID A. LONG, ADAM J. FLEISHER, *Material Measurement Laboratory, National Institute of Standards and Technology, Gaithersburg, MD, USA*; FENG ZHOU, YILIANG BAO, JASON J GORMAN, THOMAS W LEBRUN, *Physical Measurement Lab, National Institute of Standards and Technology, Gaithersburg, MD, USA*; DAVID F. PLUSQUELLIC, *Physical Measurement Laboratory, National Institute of Standards and Technology, Boulder, CO, USA*; JOSEPH T. HODGES, *Material Measurement Laboratory, National Institute of Standards and Technology, Gaithersburg, MD, USA*.

I will be discussing our recent work on the development and application of frequency-agile optical frequency combs generated with electro-optic modulators. Through the use of an arbitrary waveform generator we are able to digitally control the resulting frequency comb and then employ it in a self-heterodyne configuration in which we can exploit the common-mode nature of the probe and local oscillator beams for facile coherent averaging. Importantly, this approach allows for complete control over the comb tooth spacing, down to the 100's of Hz level for ultrahigh resolution measurements. We have recently applied these approaches to atomic spectroscopy as well as for physical metrology using optical microcavities, areas in which the agility and resolution of these frequency combs are ideally suited.

TA06 10:46 – 11:01

DUAL-COMB SPECTROSCOPY USING A DUAL-COMB FIBER LASER BASED ON HYBRID PULSE FORMATION MECHANISMS

TING LI, XIN ZHAO, ZIJUN YAO, YUEHAN WU, JIE CHEN, ZHENG ZHENG, *School of Electronic Information Engineering, Beihang University, Beijing, China*.

Single-cavity dual-comb fiber laser can generate two ultra-short pulse trains from the same cavity, which could have relatively low common-mode noises and good mutual coherence. They had been demonstrated to be applicable to various dual-comb metrology applications including optical spectroscopy[a], absolute distance measurement[b] and so on. Here, broadband dual-comb spectroscopy measurement using a broadband mode-locked dual-comb fiber laser based on hybrid pulse formation mechanisms is demonstrated. Through the dispersion management in the cavity, the bandwidths of both pulses are greatly increased. Dual-comb pulse with broader spectra can be generated when the pump power is above the threshold power. After jointly being amplified by an Erbium-doped fiber amplifier, the 3-dB spectral bandwidths, at the center wavelength of $1550nm$ 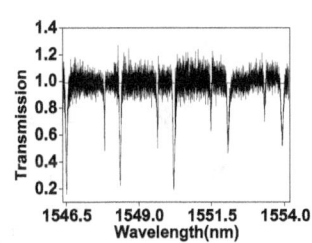 and $1555nm$, have been broadened from $4.4nm$ and $1.3nm$[c] to $58.8nm$ and $8.5nm$, respectively. Also, the 10-dB width of the overlapped spectra is over $16nm$, covering $1546nm$ to $1562nm$. The repetition rate difference of two pulse trains is $37Hz$, because of the small difference in their center wavelengths and the low cavity dispersion. Then, typical asynchronous experimental setup is used to measure transmission spectrum of an on-chip microring resonator by averaging over 30 interferograms. All the resonance spectral features are clearly resolved. This kind of dual-comb fiber lasers could offer extra options for low-complexity, dual-comb metrology applications.

[a]X. Zhao, et al., Optics Express, vol. 24, 21833, 2016.
[b]B. Lin, et al., IEEE Photonics Journal, 99, 2017.
[c]Y. Liu, et al., Optics Express, vol. 24, 21392, 2016.

TA07 — Post-Deadline Abstract — 11:03–11:18

DUAL COMB GENERATION FROM A SINGLE FIBER LASER CAVITY VIA SPECTRAL SUBDIVISION

GEORG WINKLER, JAKOB FELLINGER, OLIVER H HECKL, *Christian Doppler Laboratory for Mid-IR Spectroscopy and Semiconductor Optics, University of Vienna, Vienna, Austria.*

Dual comb spectroscopy holds the promise of bringing the bandwidth, resolution and sensitivity advantages of direct frequency comb spectroscopy to the mainstream. Replacing complex Fourier transform interferometer (FTIR) or virtually imaged phased array (VIPA) detectors with a simple photo diode potentially leads to compact, robust, field-usable measurement setups. In exchange, some of that complexity is then shifted to the laser source, usually leading to the requirement of two identical mode-locked lasers, actively stabilized to each other. However, such a dual comb source can be significantly simplified by generating two pulse trains using a single laser cavity, with the advantage of passive mutual coherence due to common-mode noise cancellation in the down converted radio frequency (RF) comb.

We introduce a new, particularly flexible, method to generate a single-cavity dual comb, exploiting independent mode-locking within two isolated spectral regions of the gain profile, created by intra-cavity spectral filtering. By setting a non-zero cavity dispersion we generate stable pulse trains with different repetition rates. These are made to overlap spectrally via successive spectral broadening in a non-linear fiber, yielding a compact, fully useable dual comb source. The underlying concept is generally applicable to nearly all kinds of passively mode-locked lasers, including state-of-the-art all-fiber types.

We demonstrate its feasibility in a nonlinear polarization evolution (NPE) mode-locked Yb:fiber laser. Spectral filtering is introduced in the free-space section of the grating compressor by mechanically blocking a central frequency band using a needle-shaped object. Fine control over its position and thickness allows for easy tuneability of the spectral separation and dual-laser operation. The laser features two independently mode-locked pulse trains centered around 1015 nm and 1040 nm, respectively, with a spectral width of about 10 nm each. Their pump-power limited repetition rates are around 20 MHz with a difference tuneable from 10 kHz down to 750 Hz. In a proof-of-principle experiment the laser output was spectrally broadened and further amplified, producing a spectral overlap around 1030 nm with an average power of 40 mW over a bandwidth of 10 nm. Once properly stabilized and brought to higher repetition rates we expect to produce a stable downconverted RF comb. Even more robust implementations might for example involve the integration of spectral filtering in form of fiber Bragg gratings or dielectric filters into and all-fiber figure-9 laser. Nonlinear conversion schemes will further extend the spetral range into the Mid-IR or XUV regions.

TA08 — 11:20–11:35

PHASE-CONTROLLED DUAL-COMB COHERENT ANTI-STOKES RAMAN SPECTROSCOPIC IMAGING

HAOYUN WEI, *Department of Precision Instrument, Tsinghua University, Beijing, China*; KUN CHEN, *Department of Chemistry, University of California at Berkeley, Berkeley, CA, USA*; TAO WU, YAN LI, *Department of Precision Instrument, Tsinghua University, Beijing, China.*

Coherent Raman microscopy can intrinsically enable label-free imaging by measuring vibrational spectra of biomolecules. Nevertheless, trade-off between high chemical-specificity and high imaging-speed currently exists in transition from spectroscopy to spectroscopic imaging when capturing dynamics in complex living systems. Here, we present a dual-comb scheme to substantially beat this trade-off and facilitate high-resolution broadband coherent anti-Stokes Raman spectroscopic imaging by combining dual-comb asynchronous optical scanning and phase-controlled spectral focusing excitation. A rapid measurement of vibrational micro-spectroscopy on sub-microsecond scale over a spectral span 700 cm^{-1} with solid signal-to-noise ratio provides access to well-resolved molecular signatures within the fingerprint region. We demonstrate this high-performance spectroscopic imaging for spatially inhomogeneous distributions of chemical substances. This technique offers an unprecedented route for broadband CARS imaging with hundreds of kHz spectroscopic refresh rate using available 1-GHz oscillators.

TA09

ATTENUATED TOTAL REFLECTANCE SPECTROSCOPY OF LIQUIDS USING A MID-IR DUAL FREQUENCY COMB SPECTROMETER

DANIEL I. HERMAN, GABRIEL YCAS, ELEANOR WAXMAN, FABRIZIO GIORGETTA, ESTHER BAUMANN, NATHAN R. NEWBURY, IAN CODDINGTON, *Applied Physics Division, NIST, Boulder, CO, USA*.

Since its inception, dual comb spectroscopy (DCS) has enabled a variety of new spectroscopic applications[a]. Here, we report on the extension of a broad bandwidth, high coherence DCS system to mid-infrared (MIR) spectroscopy of liquid-phase samples in the C-H stretch region ($\sim$2800 to 3100 cm^{-1}). Each comb originates from a self-referenced Er:fiber oscillator with a repetition rate of 200 MHz. Using a two-branch difference frequency generation (DFG) configuration, each comb is downconverted to the MIR, which yields output tunable from 2.6 to 5.2 μm with up to 100 mW in a single instantaneous bandwidth. Using attenuated total reflectance (ATR) coupling techniques, spectra of the pure isopropanol C-H stretch are recorded in 1.5 second integration periods consisting of $\sim$100 phase-corrected, averaged interferograms and show excellent agreement with reference spectra. A fitted average of 50 integration periods is shown in the abstract image. Initial estimates show that baseline fluctuations on the order of 2% will dominate sensitivity models. Further, we demonstrate progress towards the use of our DCS system as an analytical tool for organic reaction monitoring. We use an aldol condensation reaction as a test case where we detect C$_{sp^3}$-H stretches in the reactant (acetone) and C$_{sp^2}$-H stretches in the product (mesityl oxide). Methods for quantitative analysis of the aldol condensation will be discussed. Finally, this talk will explore the compatibility of DCS and newly-developing micro-reactor waveguide technology.

---

[a]Ian Coddington, Nathan Newbury, and William Swann, "Dual-comb spectroscopy," Optica 3, 414-426 (2016)

TB. Mini-symposium: New Ways of Understanding Molecular Spectra
Tuesday, June 19, 2018 – 8:30 AM
Room: 100 Noyes Laboratory

Chair: Renato Lemus, Universidad Nacional Autonoma de Mexico, Mexico City, CDMX, Mexico

---

TB01 8:30–8:45

NUMERICAL ANALYSIS OF VIBRONIC STRUCTURE OF THE SiCN $\tilde{X}\ ^2\Pi$ SYSTEM[a]

<u>MASARU FUKUSHIMA</u>, TAKASHI ISHIWATA, *Information Sciences, Hiroshima City University, Hiroshima, Japan*.

The laser induced fluorescence (LIF) spectrum of the $\tilde{A}\ ^2\Delta - \tilde{X}\ ^2\Pi$ transition was obtained for SiCN generated by laser ablation under supersonic free jet expansion. The vibrational structure, particularly that associated with the bending mode, of the dispersed fluorescence (DF) spectra from single vibronic levels (SVL's) is too complicated to analyze by the usual formulation derived from perturbational approach. Successful analysis requires us to numerically diagonalize the vibronic Hamiltonian, in which Renner-Teller (R-T), anharmonicity, spin-orbit (SO), Herzberg-Teller (H-T), Fermi, and Sears interactions have been considered, where the Sears resonance is a second-order interaction combined from SO and H-T interactions with $\Delta K = \pm 1$, $\Delta \Sigma = \mp 1$, and $\Delta P = 0$. Accurate results were obtained from this procedure reproducing experimental observations within the deviations of our instrumental resolution, ~ 5 cm^{-1}. The mixing coefficients of the two vibronic levels are comparable to those obtained from computational studies[b].

---

[a] M. Fukushima and T. Ishiwata, J. Chem. Phys. 145, 214304 (2016).
[b] V. Brites, A. O. Mitrushchenkov, and C. Léonard, J. Chem. Phys. 138, 104311 (2013); C. Léonard, Private communication.

TB02 8:47–9:02

ARE LINEAR MOLECULES REALLY LINEAR?   I. THEORETICAL PREDICTIONS

<u>TSUNEO HIRANO</u>, *Department of Chemistry, Ochanomizu University, Tokyo, Japan*; UMPEI NAGASHIMA, *Foundation for Computational Science, Kobe, Japan*; PER JENSEN, *Faculty of Mathematics and Natural Sciences, University of Wuppertal, Wuppertal, Germany*.

In spectroscopic parlance, a linear triatomic molecule is one whose potential energy minimum occurs at a linear geometry. We have recently discussed[a,b,c] that any linear triatomic molecule will be observed as being "bent" on ro-vibronic average in any ro-vibronic state. As quantum mechanics asserts, we have to characterize Nature through "observation." Theoretically we make observations of molecular structures by calculating the expectation values of the structural parameters over the relevant ro-vibronic wavefunctions.

In computational molecular spectroscopy studies, we have shown that for many linear triatomic molecules such as $^6\Delta$ FeNC, $^6\Delta$ FeCN, $^2\Pi$ BrCN$^+$, $^3\Phi$ CoCN, $^2\Delta$ NiCN, $^1\Sigma^+$ CsOH, $^3\Sigma^-$ FeCO, and $^2\Pi$ NCS, the ro-vibrationally averaged structure (zero-point structure, for example) is slightly bent with a bond angle supplement $180° - \angle$(A-B-C) [where $\angle$(A-B-C) is the bond angle] in the range from 7.5° (NCS) to 22.5° (C$_3$). We have also described the theoretical background[b] for this fact using a Laguerre-Gauss type wavefunction for the doubly degenerate bending oscillator; the average "bentness" is basically caused by the inseparability of the bending motion from the free rotation about the molecular axis.

Our finding is in contradiction to the well-established paradigm in spectroscopy that the ro-vibrationally averaged structure of a linear molecule is linear. In particular, it throws doubt on the so-called r_0 structures routinely determined for linear triatomic molecules under the *a priori* assumption that ro-vibrationally averaged bond-angle of a linear molecule should be 180°.

In the following talk, we discuss how experimentally derived rotational-constant values are to be interpreted.

---

[a] T. Hirano, U. Nagashima, *J. Mol. Spectrosc.*, **314**, 35–47 (2015).
[b] T. Hirano, U. Nagashima, P. Jensen, *J. Mol. Spectrosc.* **343**, 54–61 (2018).
[c] T. Hirano, U. Nagashima, P. Jensen, *J. Mol. Spectrosc.* (2018), https://doi.org/10.1016/j.jms.2017.12.011; and references therein.

TB03 9:04 – 9:19

ARE LINEAR MOLECULES REALLY LINEAR?   II. RE-INTERPRETATION OF EXPERIMENTAL B_0-VALUES.

TSUNEO HIRANO, *Department of Chemistry, Ochanomizu University, Tokyo, Japan*; UMPEI NAGASHIMA, *Foundation for Computational Science, Kobe, Japan*; PER JENSEN, *Faculty of Mathematics and Natural Sciences, University of Wuppertal, Wuppertal, Germany*.

As discussed in the preceding talk, any linear triatomic molecule will be observed as being "bent" on ro-vibronic average in any ro-vibronic state.[a,b,c] Experimentally derived B_0 constants are the results of the "observation" of Nature. This suggests that the observed B_0 values are in fact those for the ro-vibrationally averaged bent structures. The easiest way to check this proposition is to interpret the set of B_0 values of isotopologues taking the bond-angle as a "variable," discarding the preconceived, conventional notion that the ro-vibrationally averaged bond angle of a linear molecule is 180°.

We have shown in previous publications[a] that bond length values derived from a set of experimental B_0 values under the assumption of a linear r_0 structure, is not the ro-vibrationally averaged bond lengths, but their projections onto the molecular axis. Therefore, when the projection angle is not accounted for, the bond length values obtained from the B_0 values may differ significantly from the averaged bond lengths.

We will show how we can derive physically sound ro-vibrational structures from the experimentally reported B_0 values, taking the FeCO, NCS, HCO$^+$, HCN, and C$_3$ molecules as examples. The averaged bond-angle deviations from the linearity, derived from experimentally reported B_0 values of multiple isotopologues, are 7.8°, 9.5°, 12.5°, 14.3°, and 23.4°, respectively, for NCS, FeCO, HCO$^+$, HCN, and C$_3$ in their respective vibrational ground states.

Thus, we can conclude that both theoretically (as described in the preceding talk) and experimentally (as shown here), the ro-vibrationally averaged structure of a linear molecule is observed as being bent.

[a] T. Hirano, U. Nagashima, *J. Mol. Spectrosc.*, **314**, 35–47 (2015)
[b] T. Hirano, U. Nagashima, P. Jensen, *J. Mol. Spectrosc.* **343**, 54–61 (2018).
[c] T. Hirano, U. Nagashima, P. Jensen, *J. Mol. Spectrosc.* (2018), https://doi.org/10.1016/j.jms.2017.12.011; and references therein.

TB04 9:21 – 9:36

DIPOLE MOMENTS OF LINEAR MOLECULES: A COMPUTATIONAL MOLECULAR SPECTROSCOPY STUDY

HUI LI, *Institute of Theoretical Chemistry, Jilin University, Changchun, China*; PER JENSEN, *Faculty of Mathematics and Natural Sciences, University of Wuppertal, Wuppertal, Germany*; TSUNEO HIRANO, *Department of Chemistry, Ochanomizu University, Tokyo, Japan*.

Computation of the dipole moment in a ro-vibrationally averaged molecular state of a triatomic molecule requires the 3D potential energy surface and the associated ro-vibrational wavefunctions. Consequently, in most cases, the experimentally derived value of the dipole moment in the ro-vibronic ground state is compared with the theoretical dipole moment value for the "equilibrium geometry." We have proposed in recent publications[a,b,c] that any linear molecule, whose potential energy minimum occurs at a linear configuration, is observed as being bent on ro-vibrational average. That gives rise to the question recently asked by a reviewer of our paper:[c] Why does the averaged dipole moment vanish for a symmetrical triatomic molecule of type ABA, such as CO_2?

In the present talk we show that there is no contradiction, for linear triatomic molecules, between our proposition of a bent averaged geometry and the experimentally derived, vibrationally averaged dipole moment value. We must consider two facts: 1) for a linear molecule, the rotation about the a axis, which approximately coincides with the molecular axis, cannot be separated from the bending motion described by variation of the bond angle in the instantaneous molecular plane, and 2) the dipole moment function is an odd function of the angle describing this rotation. Therefore, only the a axis component of the dipole moment can be observed (and it, too, vanishes by symmetry for an ABA molecule).

Taking CO_2 and HCO$^+$ as examples, we show, from the theoretical view point, how the dipole moment is observed in the experimental study, typically in Stark spectroscopy. Our theoretically predicted dipole moment value for the ro-vibronic ground state of HCO$^+$, 3.933 D, is in good agreement with that determined from Stark experiments,[d] 3.921(31) D (quoted uncertainty in parentheses, in units of the last digit).

[a] T. Hirano, U. Nagashima, *J. Mol. Spectrosc.*, **314**, 35–47 (2015)
[b] T. Hirano, U. Nagashima, P. Jensen, *J. Mol. Spectrosc.* **343**, 54–61 (2018).
[c] T. Hirano, U. Nagashima, P. Jensen, *J. Mol. Spectrosc.* (2018), https://doi.org/10.1016/j.jms.2017.12.011.
[d] B. J. Mount, M. Redshaw, E. G. Myers, *Phys. Rev. A*, **85**, 012519 (2012).

Intermission

TB05 10:12 – 10:27

THE JAHN-TELLER EFFECT AS A TREATMENT OF MOLECULAR ANHARMONICITY

<u>DAVID S. PERRY</u>, BISHNU P. THAPALIYA[a], MAHESH B. DAWADI[b], *Department of Chemistry, The University of Akron, Akron, OH, USA.*

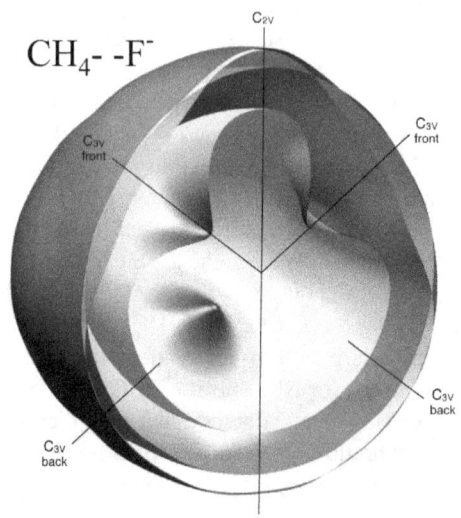

An important aspect of vibrational anharmonicity results from the substantial changes in molecular geometry and bonding that occur in the coordinate space of large-amplitude nuclear motion. Examples of such large-amplitude motion include torsional motion, inversion, the intermolecular motions within clusters, and reaction coordinates. Here we show that the Jahn-Teller formalism, when suitably extended, provides a precise description of the variation of the small-amplitude vibrational frequencies in a large-amplitude coordinate space. The locations where the small-amplitude frequencies cross are vibrational conical intersections (CIs) and multiple CIs may occur in one molecular system. In this work, we expand the motion of one molecular fragment relative to the other in spherical harmonics to allow an even-handed treatment of large-amplitude motion in 4π steradians. The molecular systems treated include CH_3OH, CH_3SH, and the complexes of CH_4 with F^- and Na^+ ions. The Jahn-Teller formalism provides a general treatment of near-resonant interactions including their explicit dependence on large-amplitude nuclear coordinates. It also includes a crude adiabatic basis, which allows for convenient computation of the fully coupled quantum nuclear dynamics. The opportunities and limitations of this approach will be discussed.

[a]Present Address: Department of Chemistry, University of Tennessee, 1420 Circle Dr., Knoxville, TN 37996.
[b]Present address: Department of Chemistry, University of Texas Rio Grande Valley, Edinburg, TX, 78539

TB06 10:29 – 10:44

COMPARING EXPERIMENTAL AND CALCULATED SPECTRAL PARAMETERS FOR JAHN-TELLER ACTIVE MOLECULES. PART I

<u>KETAN SHARMA</u>, SCOTT M. GARNER, TERRY A. MILLER, *Department of Chemistry and Biochemistry, The Ohio State University, Columbus, OH, USA*; JOHN F. STANTON, *Physical Chemistry, University of Florida, Gainesville, FL, USA.*

Great advances in quantum chemistry calculations for molecules over the last decade has left a cogent need to make theory testable by experiments. One area of fundamental interest is the computation of both spectroscopic and dynamical properties of molecules near conical intersections between different electronic states. Experiments on naturally occurring and spectroscopically accessible conical intersections in Jahn-Teller molecules provide excellent benchmarks for computations. This talk discusses an approach which allows direct computation of experimental parameters determined from rotationally resolved experiments. An Effective Rotational Hamiltonian (ERH) for Jahn-Teller active system with a C_3 or C_5 symmetry axis is presented. Methods for calculating the parameters of the ERH using electronic structure and vibrational mode data from coupled cluster calculations are reported. Furthermore theory is advanced to develop a prescription for obtaining observable rovibronic parameters like the rotational distortion parameter, h_1, and transition intensities for systems with a C_3 and C_5 symmetry axis. An important goal of these calculations is to compare experimentally observable parameters to those from highly developed theoretical tools presently available.

TB07   10:46 – 11:01
COMPARING EXPERIMENTAL AND CALCULATED SPECTRAL PARAMETERS FOR JAHN-TELLER ACTIVE MOLECULES: PART II

SCOTT M. GARNER, KETAN SHARMA, TERRY A. MILLER, *Department of Chemistry and Biochemistry, The Ohio State University, Columbus, OH, USA*; JOHN F. STANTON, *Physical Chemistry, University of Florida, Gainesville, FL, USA.*

In Part I, the theoretical motivations for studying spectroscopic and dynamic properties of molecules near conical intersections were outlined. Jahn-Teller active molecules were identified as excellent candidates for these studies, as vibronic eigenfunctions are calculatable utilizing either electronic structure methods or by fitting experimental results from vibronic spectra. Here, we discuss the utility of these eigenfunctions in the calculation of matrix elements necesary toward understanding rovibronic experimental spectra of Jahn-Teller molecules. As an example, the determination of the Watson distortion term, h_1, of the Jahn-Teller rotational Hamiltonian will be presented. In the cyclopentadienyl radical, h_1 in the vibrationless level of the ground electronic state, $\tilde{X}^2 E_1''$, is well determined experimentally and reproducable via our theoretical methods. Tabulated h_1 values for vibrationally excited states provide insights toward rotational structure, and thus identification and assignment, of observable vibrational transitions. For the nitrate radical, comparison of computational and experimental results for h_1 in the $\tilde{A}^2 E''$ state provide a background for evaluating the potential energy surface of NO_3.

TB08   11:03 – 11:18
UNDERSTANDING QUANTUM YIELDS IN NAPHTHALENES AND BORON-DIPYRROMETHENES: TOWARDS A PREDICTION OF NON-RADIATIVE DECAY PATHWAYS IN ORGANIC OPTOELECTRONIC MATERIALS

ZHOU LIN, ALEXANDER W. KOHN, TROY VAN VOORHIS, *Department of Chemistry, Massachusetts Institute of Technology, Cambridge, MA, USA.*

In recent years, organic optoelectronic materials have attracted considerable attention due to their ease of production and fabrication and great potentials in industrial applications. However, their efficiencies can be strongly limited by non-radiative decay pathways in which the excited energies are lost in the form of thermal energy. In the present study, we focused on two families of organic optoelectronic materials, the derivatives of naphthalene and those of boron-dipyrromethene (BODIPY), and characterized their efficiencies using the fluorescence quantum yield (ϕ_{fl}) – the probability that a photoexcited molecule fluoresces a photon rather than undergoing a non-radiative decay. To predict these ϕ_{fl}'s, we developed inexpensive, accurate and transferable semi-empirical methods based on time-dependent density functional theory (TDDFT) and its multiple variants, and examined two non-radiative decay mechanisms – internal conversion (IC) and intersystem crossing (ISC). We managed to predict radiative rates (k_{fl}), IC rates (k_{IC}), and ISC rates (k_{ISC}), and controlled the mean absolute error (MAE) within 0.4 orders of magnitude. These results, in combination, allowed us to reproduce ϕ_{fl}'s with an MAE of 0.2 for organic optoelectronic materials in question. During this process, we discovered two novel IC pathways through low-energy distorted transition states and intermediates, and indicated that the energy gap law is inadequate to estimate the activation energy (E_a) and account for k_{IC} and k_{ISC}. The present study paves the way for the rational design of organic optoelectronic devices by high-throughput computations.

TC. Chirality and stereochemistry

Tuesday, June 19, 2018 – 8:30 AM

Room: 1024 Chemistry Annex

Chair: Sonia Melandri, University of Bologna, Bologna, Italy

TC01   8:30–8:45

REACTION FLASK ANALYSIS OF THE ASYMMETRIC HYDROGENATION OF ARTEMISINIC ACID

REILLY E. SONSTROM, *Department of Chemistry, The University of Virginia, Charlottesville, VA, USA*; JUSTIN L. NEILL, *BrightSpec Labs, BrightSpec, Inc., Charlottesville, VA, USA*; B FRANK GUPTON, YUAN YANG, *Chemical and Life Science Engineering, Virginia Commonwealth University, Richmond, VA, USA*; LUCA EVANGELISTI, *Dipartimento di Chimica G. Ciamician, Università di Bologna, Bologna, Italy*; BROOKS PATE, *Department of Chemistry, The University of Virginia, Charlottesville, VA, USA*.

There is currently a search for a reliable, low cost, synthetic or semi-synthetic method of production for artemisinin – a potent antimalarial drug with limited natural supply. Synthesis of artemisinin from artemisinic acid can be broken down into two key steps: the asymmetric hydrogenation of AA to dihydroartemisinic acid (DHAA) and the oxidation and complex rearrangement of DHAA to form artemisinin. This work reports the reaction flask analysis of the stereospecific conversion of AA to DHAA using chirped-pulse Fourier transform microwave spectroscopy (CP-FTMW). Successful monitoring of this reaction requires resolution of multiple species: artemisinic acid (AA), (R,R)-dihydroartemisinic acid (DHAA), (R,S)-dihydroartemisinic acid (epiDHAA), and the over-reduced form tetrahydroartemisinic acid (THAA). The rotational spectra of these compounds have been obtained through measurements on purified samples with quantities in the 20-100 mg level. For two species (AA and (R,R)-DHAA) the broadband rotational spectrum had 13C-level sensitivity permitting a carbon framework structure determination. For the analysis of the reaction mixture a 70 mg sample was provided. We were able to identify all species in the reaction mixture without further purification. Using dipole moments from quantum chemistry, the relative abundance of each species in the reaction mixture was determined: 14.85% AA, 57.28% DHAA, 9.19% epiDHAA, and 18.67% THAA.

TC02   8:47–9:02

FAST CHIRAL MONITORING IN A CONTINUOUS PHARMACEUTICAL SYNTHESIS BY MOLECULAR ROTATIONAL RESONANCE SPECTROSCOPY

JUSTIN L. NEILL, MATT MUCKLE, *BrightSpec Labs, BrightSpec, Inc., Charlottesville, VA, USA*; BROOKS PATE, *Department of Chemistry, The University of Virginia, Charlottesville, VA, USA*; YUAN YANG, B FRANK GUPTON, *Chemical and Life Science Engineering, Virginia Commonwealth University, Richmond, VA, USA*.

We present the successful application of MRR spectroscopy in the microwave region to monitor the output of a continuous pharmaceutical synthesis. Microwave spectroscopy has an excellent capability to distinguish isomers and other structurally similar compounds, and techniques have been developed recently that are also sensitive to enantiomeric excess. A Balle-Flygare-style Fourier transform microwave spectrometer was employed as the detector in this study, along with a new solutions sampling interface that injects crude product solution directly from the reactor, bakes off the solvent, and volatilizes the analyte mixture for analysis. The reaction under study was the catalytic asymmetric hydrogenation of artemisinic acid to produce a stable intermediate in the synthesis of artemisinin, an important antimalarial.

The instrument is fully automated and consumes less than 1 mg of analyte in order to analyze the composition of 4 species with a detection limit of approximately 75 ppmw in the solution: the starting material, desired product, epimer of the product, and an overreduction byproduct that is not readily detectable by HPLC or NMR. This talk will describe the results of this study and prospects for future application in pharmaceutical process development.

TC03 9:04 – 9:19

MICROWAVE SPECTRUM AND MOLECULAR STRUCTURE OF THE CHIRAL TAGGING CANDIDATE, 3,3-DIFLUORO-1,2-EPOXYPROPANE, AND ITS COMPLEX WITH THE ARGON ATOM

<u>HELEN O. LEUNG</u>, MARK D. MARSHALL, *Chemistry Department, Amherst College, Amherst, MA, USA.*

Continuing our efforts in characterizing small molecules for use as potential chiral tags for the conversion of enantiomeric molecules into spectroscopically distinct diasteromeric complexes for chiral analysis, we examine the microwave spectrum and molecular structure of 3,3-difluoro-1,2-epoxypropane. This compound is available as a high vapor pressure liquid, both in enantiomerically pure form and as a racemic mixture, and it is easily incorporated into a free jet expansion for complex formation and spectroscopic analysis. Like the structurally similar 3,3,3-trifluoro-1,2-epoxypropane, it has a simple, hyperfine-free rotational spectrum. This spectrum has been obtained for the most abundant and four singly-substituted isotopologues, all in natural abundance, and the structure of the molecule determined. In addition, the spectrum and structure of the 3,3-difluoro-1,2-epoxypropane-argon complex are obtained.

TC04 9:21 – 9:36

EFFECTS OF CHIRALITY IN HOMODIMERS OF 3,3,3-TRIFLUORO-1,2-EPOXYPROPANE

<u>MARK D. MARSHALL</u>, HELEN O. LEUNG, *Chemistry Department, Amherst College, Amherst, MA, USA*; NATHAN A. SEIFERT, YUNJIE XU, WOLFGANG JÄGER, *Department of Chemistry, University of Alberta, Edmonton, AB, Canada.*

We are investigating the suitability of 3,3,3-trifluoro-1,2-epoxypropane [2-(trifluoromethyl)-oxirane, or TFO] as a tag for chiral analysis through conversion of enantiomers into structurally distinct diastereomers via the formation of non-covalently bound heterodimers. This method can determine the absolute stereochemistry and enantiomeric composition of an analyte and shows promise for great impact in analytical chemistry. Using density functional theory, we examine the possible conformations of both homochiral (RR or SS) and heterochiral (RS or SR) homodimers of TFO to guide the search for the microwave spectra of these species. Several conformers are found for each, but the lowest energy heterochiral TFO dimer is a microwave silent one with an inversion center. However, a spectrum is observed that can be assigned to the lowest energy geometry of the homochiral TFO dimer.

TC05 9:38 – 9:53

CHIRAL ANALYSIS OF BIOLOGICALLY RELEVANT SAMPLES USING BROADBAND ROTATIONAL SPECTROSCOPY

<u>MARÍA MAR QUESADA-MORENO</u>, ANNA KRIN, MELANIE SCHNELL, *FS-SMP, Deutsches Elektronen-Synchrotron (DESY), Hamburg, Germany.*

Terpenes are the main constituents of essential oils and are responsible for their chemical and biological activities. Additionally, terpenes found in essential oils are often structurally similar, like thymol, carvacrol, p-cymene and terpinen-4-ol in the case of thyme essential oil. These oils are likely to be enantio-enriched because of their natural origin [1]. Broadband rotational spectroscopy offers unparalleled features that make it a unique tool to analyze complex molecular mixtures as those present in essential oils. Even structurally similar molecules as diastereoisomers can be easily detected as they would show a different rotational spectrum. Additionally, the combination of the microwave three-wave mixing (M3WM) technique with the broadband capabilities allow one to distinguish enantiomers within a mixture of chiral molecules and to determine their enantiomeric excess in the gas phase.

Here, we present recent results on the analysis of two thyme essential oils from Spain obtained from the leaves of *Thymus vulgaris* and using the above-mentioned techniques. It is important to bear in mind that the chemical composition of the essential oils coming from the same plant species shows variations according to the environment, growth region and cultivation practices. In our case, terpene compositions of the two studied thyme oils change, even coming from the same country. Linalool is mainly present in one of the oils, whereas thymol is present in the other. The analyses of essential oils with these techniques could extent the use of rotational spectroscopy as a chemical analytical tool (a new application still to be explored in more detail).

[1] V. A. Shubert, D. Schmitz, C. Pérez, C. Medcraft, A. Krin, S. R. Domingos, D. Patterson, M. Schnell, *J. Phys. Chem. Lett.* 7 (2016) 341–350.

TC06 9:55 – 10:10

ENANTIOMERIC EXCESS MEASUREMENTS OF ISOPULEGOL USING CHIRAL TAG SPECTROSCOPY

KEVIN J MAYER, CAITLIN EMBLY, BROOKS PATE, *Department of Chemistry, The University of Virginia, Charlottesville, VA, USA*; LUCA EVANGELISTI, *Dipartimento di Chimica G. Ciamician, Università di Bologna, Bologna, Italy*.

Chiral analysis was performed on samples of isopulegol and its isomers using chiral tag rotational spectroscopy. Isopulegol, with three chiral centers, has 8 stereoisomers. There are four diastereomers with distinct geometries and the diastereomer ratio can be determined using traditional rotational spectroscopy. To determine the enantiomeric ratio for each diastereomer the chiral tagging method was used to convert these enantiomers into distinguishable diastereomer complexes. Isopulegol was placed into the nozzles of a chirped-pulsed Fourier transform microwave spectrometer and was heated to 323K. The isopulegol was complexed with a 0.1% mixture of propylene oxide in neon as the carrier gas. The measurement methodology for EE determinations is: 1) a 400K average spectrum is measured using the enantiopure S-propylene oxide, 2) the tag is purged by flowing pure neon over the sample and heating, and 3) a 400K average spectrum using racemic propylene oxide is measured. Enantiopure samples of (-)-isopulegol and (+)-isopuelgol were purchased from Sigma Aldrich and used to create standards of 0, 5, 10, 30, 55, 80, and 90 enantiomeric excess of (-)-isopulegol. The calibration curve was fit using a linear expression with zero offset giving a slope of 1.005 ± 0.007 ($R^2 = 0.99935$). These results demonstrate that the method has linear performance over the full EE scale. The reference solution with EE=80 was measured in six separate runs to assess reproducibility. The average of the measurements was 80.595% with a standard deviation of 0.274. A sample of isopulegol provided as a mixture of isomers (Alfa Aesar) was analyzed using the chiral tag method. The enantiomeric excess for the two most abundant diastereomers were determined: isopulegol: EE=4.1(4) and neoisopulegol: EE=4.8(4). The similar enantiomeric excess values for these isomers is consistent with the usual production method for isopulegol where the EE of the reagent (citronellal) sets the EE for all four diastereomer products.

Intermission

TC07 10:46 – 11:01

DETERMINATION OF ABSOLUTE CONFIGURATION IN CEDROL USING CHIRAL TAG ROTATIONAL SPECTROSCOPY

LUCA EVANGELISTI, *Dipartimento di Chimica G. Ciamician, Università di Bologna, Bologna, Italy*; BROOKS PATE, TAYLOR SMART, CHANNING WEST, MARTIN S. HOLDREN, KEVIN J MAYER, *Department of Chemistry, The University of Virginia, Charlottesville, VA, USA*.

Rotational spectroscopy has been extended for use in chiral analysis by both the chiral tagging and three-wave mixing methods. The chiral tagging method uses complex formation between two species in a pulsed jet: the first, a small chiral tag of known absolute configuration and the second, an unknown analyte. The complexing of the tag with the unknown analyte converts the enantiomers to diastereomers. Cedrol ((1S,2R,5S,7R,8R)-2,6,6,8-tetramethyltricyclo[$5.3.1.0^{1.5}$]undecan-8-ol, $C_{16}H_{26}O$, MW 222.37) was chosen as a test molecule for the chiral tagging method to assess the potential for extending the approach to larger molecules. In addition to its large size, cedrol has conformational flexibility from both the internal rotation of the hydroxyl group and a ring pucker coordinate. As the analyte molecules become larger, there is the potential that the difference in the rotational constants of the diastereomer complexes differ by amounts too small to reliably predict by quantum chemistry making difficult to produce a high-confidence assignment of the absolute configuration. The chiral tag measurement of cedrol used propylene oxide as the tag. The estimates of the rotational constants and dipole moments of the complexes were obtained using the B3LYP-D3BJ method and def2TZVP basis set. Six isomers for the cedrol / propylene oxide complex were observed using an enantiopure tag. The challenges in identifying these isomers in the quantum chemistry calculations will be discussed. The agreement between the experimental and theoretical rotational constants was at the $\sim$1% level and this makes it possible to assign the absolute configuration. The quality of the theoretical structure of the cedrol / propylene oxide complex was assessed by determining the experimental carbon atom framework geometry through 13C isotopic substitution using natural abundance.

TC08 11:03 – 11:18

ENANTIOMERIC EXCESS MEASUREMENTS USING MICROWAVE THREE-WAVE MIXING

MARTIN S. HOLDREN, BROOKS PATE, CAITLIN EMBLY, ARTHUR WU, KEVIN J MAYER, JAMES DITTMAN, PATRICK BUONICONTI, GOLARA HAGHTALAB, BRIANNA MITCHELL, *Department of Chemistry, The University of Virginia, Charlottesville, VA, USA*; LUCA EVANGELISTI, *Dipartimento di Chimica G. Ciamician, Università di Bologna, Bologna, Italy.*

Microwave three-wave mixing was demonstrated in 2013 by Patterson, Schnell, and Doyle (D. Patterson, M. Schnell, and J.M Doyle, Nature 497, 475- 478 (2013)) to distinguish a pair of enantiomers by rotational spectroscopy. This is possible because the product of the three electric dipole components in the principal axis system has opposite sign for enantiomers. The three-wave measurement produces a coherent emission signal proportional to the product of the dipole moment components making it possible to distinguish enantiomers through the phase of the signal. The enantiomeric excess (EE) is then proportional to the signal amplitude. Microwave three-wave mixing experiments were performed to quantify the EE and assess the limits of low EE measurements of the molecules propylene oxide, 1,2-propanediol, and isopulegol. Challenges in these measurements will be discussed including the need for low-frequency coherence transfer pulses due to phase matching requirements in generation of the chiral signal. Other measurement issues like the possibility of off-resonance direct excitation of the chiral transition that can limit instrument performance will be described. Lastly, a test of the linearity of the three-wave mixing signals as a function of EE using isopulegol reference mixtures will be presented.

TC09 11:20 – 11:35

CP-FTMW SPECTROSCOPY OF THE LOW ENERGY CONFORMERS OF TWO CHIRAL ALCOHOLS: MYRTENOL AND NOPOL

GALEN SEDO, *Department of Natural Sciences, University of Virginia's College at Wise, Wise, VA, USA*; FRANK E MARSHALL, G. S. GRUBBS II, *Department of Chemistry, Missouri University of Science and Technology, Rolla, MO, USA.*

New microwave spectra of two bicyclic monoterpenols have been observed using CP-FTMW spectroscopy in the 6-18 GHz region of the electromagnetic spectrum. These spectra have given insight into the low energy conformers of the two molecules, with at least one conformer of nopol and multiple conformers of myrtenol having been observed. Rotational constants of the assigned structures will be reported and discussed in comparison to theory. The study of these molecules adds to a growing body of work on the spectroscopy of chiral monoterpenes.[a,b]

---

[a] J. Mol. Spectrosc. 342 (2017) 109-115
[b] J. Mol. Spectrosc. 336 (2017) 22-28

TC10 11:37 – 11:52

A COMPARATIVE STUDY OF CHIRAL ANALYSIS OF FENCHYL ALCOHOL USING NUCLEAR MAGNETIC RESONANCE, INFRARED, AND ROTATIONAL SPECTROSCOPY

KEVIN J MAYER, SUPRAJA CHITTARI, ALYSA MODI, ERIC ODERMATT, CHARLES SPIVEY, JULIAN STASHOWER, BROOKS PATE, *Department of Chemistry, The University of Virginia, Charlottesville, VA, USA.*

The analysis of chiral molecules with multiple chiral centers is a challenging problem in analytical chemistry. The goal of the analysis is to determine the fractional composition for each unique stereoisomer. In the most general case, a molecule with N chiral centers will have 2^N distinct stereoisomers. Half of these, 2^{N-1}, will be molecules with distinct molecular structures (the diastereomers). The diasteormer composition can be analyzed by normal spectroscopy methods because they have distinct spectra. For each diastereomer, there are the two non-superimposable mirror images (the enantiomers) and additional measurement methodology is required to determine the enantiomeric ratio using spectroscopy. Furthermore, in many applications the "unwanted" isomers (diastereomers and/or enantiomers) will be present as low-abundance impurities placing strong demands on the dynamic range of the spectroscopic technique. A commercial sample of (1R)-endo-(+)-Fenchyl alcohol (C10H18O, four stereoisomers) has been analyzed using nuclear magnetic resonance (NMR), infrared (IR), and rotational spectroscopy. The commercial sample has a small amount (3%) of the diastereomer as an impurity. The ability to quantitatively identify the diastereomer impurity using quantum chemistry estimates of the NMR, IR, and rotational spectrum parameters will be discussed. The enantiomer analysis uses chiral resolving agents for NMR spectroscopy, vibrational circular dichroism (VCD) for IR spectroscopy, and chiral tag rotational spectroscopy. The ability of these techniques to verify the stereochemistry of the dominant (1R)-endo-(+)-Fenchyl alcohol will be discussed. The ability to identify the enantiomeric excess of fenchyl alcohol and the possibility of performing enantiomer analysis on the low abundance diastereomer using the direct sample without purification will also be presented.

TC11 11:54 – 12:09

DESIGNING CHIRAL TAGS TO IMPROVE ABSOLUTE CONFIGURATION DETERMINATION BY ROTATIONAL SPECTROSCOPY

LUCA EVANGELISTI, *Dipartimento di Chimica G. Ciamician, Università di Bologna, Bologna, Italy*; CHANNING WEST, *Department of Chemistry, The University of Virginia, Charlottesville, VA, USA*; MARK D. MARSHALL, HELEN O. LEUNG, *Chemistry Department, Amherst College, Amherst, MA, USA*; BROOKS PATE, *Department of Chemistry, The University of Virginia, Charlottesville, VA, USA.*

Determining the absolute configuration (AC) of a chiral analyte is a challenging analytical problem. In 2013, Patterson, Schnell, and Doyle showed that molecular rotational spectroscopy can be used to determine AC through the phase of a coherent emission signal generated by special three-wave mixing cycles. An alternate approach to determine AC is to convert enantiomers into diastereomers. Diastereomers possess distinct geometries, and, therefore, produce distinct rotational spectra. The enantiomer-to-diastereomer approach is accomplished in rotational spectroscopy by creating weakly bound complexes of the analyte with an enantiopure chiral tag molecule in a pulsed jet expansion. The AC of cedrol ($C_{15}H_{26}O$) has previously been determined by rotational spectroscopy using propylene oxide (C_3H_6O) as the tag. The size difference of cedrol relative to propylene oxide results in diastereomer complexes possessing similar rotational constants. The difference in rotational constants between the diastereomer complexes is small enough that it approaches the confidence limit of the computational methods used (B3LYP D3BJ 6-311G++(d,p)). Therefore, full carbon structures of the complexes were required to confidently assign the AC. The current study exploits a structural motif of the complexes to increase confidence in the AC assignment without needing to experimentally determine the carbon structure. To a good approximation, the change in inertia between the diastereomer complexes arises from changing the position of the propylene oxide methyl group. By modifying propylene oxide, the AC of cedrol is determined without observing ^{13}C spectra in natural abundance. Two approaches are tested. The first method involves using a propylene oxide derivative with a -CF$_3$ group instead of a methyl group, which provides a larger inertia difference between the diastereomer complexes. The second approach involves using an isotopically labeled tag. Quantum calculations can predict isotopic shifts in rotational constants to a higher degree of accuracy than the constants themselves, allowing for higher confidence in the AC assignment.

TD. Dynamics and kinetics
Tuesday, June 19, 2018 – 8:30 AM
Room: 217 Noyes Laboratory

Chair: James Lockhart, Brookhaven National Laboratory, Long Island, NY, USA

TD01   8:30 – 8:45

STATE-TO-STATE ROTATIONAL RATE COEFFICIENTS FOR AMMONIA SELF COLLISIONS FROM PUMP-PROBE CHIRPED-PULSE EXPERIMENTS

CHRISTIAN ENDRES, PAOLA CASELLI, *The Center for Astrochemical Studies, Max-Planck-Institut für extraterrestrische Physik, Garching, Germany*; STEPHAN SCHLEMMER, *I. Physikalisches Institut, Universität zu Köln, Köln, Germany.*

Rotational state populations of ammonia are inferred from chirped pulse spectra of its tunneling doublets in a room temperature K-band waveguide experiment where many tunneling doublets can be addressed by a single chirped pulse excitation. The thermal distribution of states is altered by a pump pulse where the population of the tunneling doublet of a single rotational state is inverted by a π-pulse within roughly 100 ns. The resulting deviation from equilibrium is then propagating to other states due to collisions and interrogated by a probe pulse from which the state populations of many rotational states are inferred at once. From the free induction decays (FID) of the individual states the relaxation time of the radiation-induced superposition state of the two level tunneling system (T2) is inferred. Also the collisional relaxation time (T1) for the difference in the population of the two-level system is determined. These values exhibit a linear pressure dependence, the slope of which agrees very well with previous measurements [a].

Analysis of probe FID signals from these pump-probe experiments reveals the well known hierachy of collisional relaxation in ammonia which was first found by Oka fifty years ago through steady state intensity measurements [b]. Collision-induced transitions within the tunneling doublet ($\Delta J = 0$) determined from T1 measurements are faster than $\Delta J = \pm 1$ transitions. Of those the $\Delta K = 0$ transitions are much faster than those with $\Delta K \neq 0$. Due to this hierachy of inelastic processes and thanks to the fast optical pumping experiments state-to-state rates can be measured. As a result, from the pressure dependence of the measured rates state-to-state rate coefficients are determined. Those rate coefficients agree very well with results of simulations of all coupled states which fit with the temporal behavior of the complete pump probe experiments where many individual (J,K) rotational states can be addressed step by step by separate probe-pump-probe pulse sequences.

[a] P. E. Wagner et al 1981 J. Phys. B: At. Mol. Phys., vol. 14, 4763.
[b] T. Oka, 1968, J.Chem.Phys., vol 48, 4919

TD02   8:47 – 9:02

WAVEGUIDE CP-FTMW SPECTROSCOPY OF ETHYL CYANOFORMATE: EXPLORING THE POSSIBILITY OF SEEING EXCHANGE-AVERAGED STATES

STEVEN SHIPMAN, J. H. WESTERFIELD, *Department of Chemistry, New College of Florida, Sarasota, FL, USA.*

The microwave spectrum of ethyl cyanoformate was recorded at 261 K from 8.7 - 26.5 GHz with waveguide chirped-pulse Fourier transform microwave spectroscopy (CP-FTMW). Ethyl cyanoformate has been studied previously by Suenram, True, and Bohn in a Stark modulation cell.[a] In that work, the data showed contributions from two stable ethyl cyanoformate conformers as well as an additional series of broad features ascribed to an exchange-averaged state, in analogy to similar features routinely seen in NMR chemical exchange measurements. In recent years, True has provided estimates for the lifetimes of these exchange species.[b]

Our high resolution spectra of a thermalized sample of ethyl cyanoformate at 261 K show contributions from *syn-anti* and *syn-gauche* conformers in agreement with the previous work. Using our standard data collection methods, however, we do not see evidence for the presence of exchange averaged bands. In this talk, we will highlight the similarities and differences of our work to previous work on this molecule and discuss the possibilities of observing these types of species with our spectrometer.

[a] Suenram, R.D., True, N.S., and Bohn, R.K., Journal of Molecular Spectroscopy, 69, 435-444 (1978).
[b] True, N.S., Journal of Physical Chemistry A, 113, 6936-6946 (2009).

TD03   9:04 – 9:19

INFRARED SPECTROSCOPIC STUDIES OF ORTHO-PARA CONVERSION IN SOLID HYDROGEN CATALYZED BY HYDROGEN ATOMS

<u>DAVID T. ANDERSON</u>, MORGAN E. BALABANOFF, AARON I. STROM, *Department of Chemistry, University of Wyoming, Laramie, WY, USA.*

Our group has been studying the reactions of hydrogen atoms (H atoms) with various molecules (NO, N_2O, CH_3OH) in solid hydrogen for the last several years.[a,b,c] One interesting puzzle that we have been unable to solve is how to detect the concentration of H atoms using FTIR spectroscopy. One possibility to estimate the H atom concentration is to measure the conversion of ortho-H_2 to para-H_2 within the solid that is catalyzed by the presence of H atoms. The H atom is a good ortho-para catalyst because it is paramagnetic and mobile within the solid even at extremely low temperatures. We have recently conducted a number of studies where we purposely synthesize solid para-H_2 samples with approximately 3% ortho-H_2 concentrations (slightly elevated). In the absence of H atoms, the ortho-H_2 concentration in the solid is stable on the order of days due to slow self-conversion. We can quantitatively detect the ortho-H_2 fraction using the overlapping $Q_1(0)+S_0(1)$ and $Q_1(1)+S_0(1)$ double transitions of solid molecular hydrogen. By rapidly generating H atoms via in situ photolysis of various H atom precursor molecules (NO and N_2O), we can initiate ortho-para conversion and follow the ortho-H_2 fraction in real time. This H atom catalyzed ortho-para conversion data therefore has the time dependent H atom concentration encoded in the signal; the challenge is to extract it. We observe qualitative differences in the shape of the ortho-H_2 fraction decay curve depending on the specific precursor used, the specific photolysis conditions, and the temperature of the sample. We will present the latest results and analysis at the meeting.

[a] F.M. Mutunga, S.E. Follett, D.T. Anderson, *J. Chem. Phys.* **139**, 151104 (2013).
[b] M. Ruzi, D.T. Anderson, *J. Phys. Chem. A* **119**, 12270 (2015).
[c] M.E. Balabanoff, M. Ruzi, D.T. Anderson, *Phys. Chem. Chem. Phys.* **20**, 422 (2018).

TD04   9:21 – 9:36

INFRARED EMISSION FROM UV-IRRADIATED MIXTURES OF CH_2I_2 AND O_2 PROBED WITH A STEP-SCAN FTIR SPECTROMETER

<u>TING-YU CHEN</u>, *Department of Applied Chemistry, National Chiao Tung University, Hsinchu, Taiwan;* YUAN-PERN LEE, *Applied Chemistry, National Chiao Tung University, Hsinchu, Taiwan, Institute of Atomic and Molecular Sciences, Academia Sinica, Taipei, Taiwan.*

The Criegee intermediates, carbonyl oxides proposed by Criegee in 1949 as key intermediates in the ozonolysis of alkenes, play important roles in organic chemistry and atmospheric chemistry. The simplest Criegee intermediate is CH_2OO. In the reaction of O_3 with C_2H_4, some CH_2OO thus produced are internally excited so that they decompose to form OH, CO, CO_2, and other compounds. Recently a new scheme for production of CH_2OO in laboratories, ultraviolet (UV) irradiation of diiodomethane (CH_2I_2) in O_2, has enabled direct detection of CH_2OO with various methods and stimulated active research on Criegee intermediates. Even though $\sim 25\%$ of CH_2OO was reported to decompose at pressure smaller than 60 Torr,[a] and infrared absorption of internally excited CO and CO_2 was reported,[b] no investigation on the dynamics of the decomposition products in the reaction $CH_2I + O_2$ has been reported.

We employed a step-scan Fourier-transform infrared (FTIR) spectrometer to record temporally resolved emission upon irradiation of mixtures of CH_2I_2, O_2, and Ar at 248 and 308 nm. IR emission of CO, CO_2, OH, CH_2I, and H_2CO in the region $1860-4900$ cm^{-1} was recorded. At total pressure 8 Torr and irradiation wavelength 248 nm, rotationally resolved lines of CO ($v \leq 11$, $J \leq 19$) in region $1860-2300$ cm^{-1} were observed; the rotational distribution is Boltzmann with temperature near 300 K, but the vibrational distribution is bimodal, with two components having averaged vibrational energies of 99 and 18 kJ mol^{-1}. Emission of OH ($v \leq 3$, $J \leq 5.5$) in region $2980-3600$ cm^{-1} was observed with ambient rotational distribution and average vibrational energy of 41 kJ mol^{-1}. The branching ratio of CO : OH is 60:40. Emission of highly internally excited CO_2 was also observed; its average internal energy was estimated. The effects of pressure and irradiation wavelength on the emission of these species will be discussed.

[a] W.-L. Ting, C.-H. Chang, Y.-F. Lee, H. Matsui, Y.-P. Lee, and J. J.-M. Lin, *J. Chem. Phys.* 2014, **141**, 104308.
[b] Y. T. Su, H.-Y. Lin, R. Putikam, H. Matsui, M. C. Lin, and Y. P. Lee, *Nat. Chem.* 2014, **6**, 477.

TD05 9:38 – 9:53

ULTRAFAST DYNAMICS OF DIBROMOCYCLOALKANES

DARYA S. BUDKINA, KANYKEY E. KARABAEVA, R. MARSHALL WILSON, ALEXANDER N TARNOVSKY, *Department of Chemistry and Center for Photochemical Sciences, Bowling Green State University, Bowling Green, OH, USA.*

Carbocyclic geminal dibromides are important intermediates in synthesis of complex heterocyclic molecules and natural products.[a] Ultrafast time-resolved techniques can help to obtain information about the electronic structure of the intermediates which appears in the chemistry and photochemistry of these molecules. Also, relaxation dynamics and reactivity of these intermediates in different solvents can be characterized. This information can be used for the design of different compounds with desired properties. In the current work, 1,1-dibromocycloalkanes with 3-, 4-, and 5-member rings were synthesized using previously described procedures.[b][c][d] For all three samples, ultrafast time-resolved absorption experiments with UV-excitation were performed in acetonitrile and methylcyclohexane solvents. It was shown that excitation of dibromocycloalkane solutions with 250 nm short (40 fs) pulses forms excited state absorption (ESA) of parent molecules in the spectral range from 360 to 760 nm. Within 800 fs, ESA decays with beginning of formation product bands. Next within 50 ps, these broad bands reach their maximum intensity and form well-defined broad peaks centered at 550, 600 and 400 nm for 3-, 4-, and 5-member rings respectively. These product species are long-lived and begin to decay at 1 ns. Obtained results suggest the formation of isomer products $(CH_2)n - C - Br - Br$ (n=1-3), similar to the isomeric species (HBrCBr-Br) observed in isomerization of bromoform.[e][f]

---

[a] Ganesh, N. V.; Jayaraman, N. J. Org. Chem. 2007, 72, 5500-5504.
[b] Napolitano, E.; Fiachi, R.; Mastrorilli, E. Synt. Comm. 1986, 122-125.
[c] ingh, R. K.; Danishefsky, S. J.Org.Chem. 1975, 40, 2969-2970.
[d] Blankenship, C.; Paquette, L. A. Synt. Comm. 1984, 14, 983-987.
[e] Pal, S. K.; Mereshchenko, A. S.; Butaeva, E. V.; El-Khoury, P. Z.; Tarnovsky, A. N. J. Chem. Phys. 2013, 138, 124501.
[f] Mereshchenko, A. S.; Butaeva, E. V.; Borin, V. A.; Eyzips, A.; Tarnovsky, A. N. Nat. Chem. 2015, 7, 562-568.

Intermission

TD06 10:29 – 10:44

BEYOND VPTn-SCTST FOR CHEMICAL KINETICS: COMPLEX SCALING AND CURVILINEAR VIBRATIONAL SELF-CONSISTENT FIELD THEORY

BRYAN CHANGALA, *JILA, National Institute of Standards and Technology and Univ. of Colorado Department of Physics, University of Colorado, Boulder, CO, USA*; JOHN F. STANTON, *Quantum Theory Project, University of Florida, Gainesville, Florida, USA.*

Semi-classical transition state theory (SCTST) is a powerful tool for calculating quantitative chemical reaction rates based on local properties of the transition state (TS) itself without detailed knowledge of the entire reaction path. The widely successful variant of SCTST based on second-order vibrational perturbation theory (VPT2) routinely provides accurate kinetic rate predictions. However, the fundamental formulation and approximations of VPT2 (and higher order extensions) often cause it to fail in the presence of significant anharmonicity near the TS. Curvilinear vibrational self consistent field theory (VSCF) together with its second order perturbative extension (VMP2) are an effective alternative to standard VPT2 for the bound state vibrational problem of highly anharmonic PES minima, suggesting similar improvements are possible at the TS. This talk will present some ideas and exploratory results towards applying complex scaled VSCF to the TS problem.

TD07

MCTDH ROVIBRATIONAL STATES AND STATE-TO-STATE INELASTIC SCATTERING CALCULATIONS ON THE H_2O-H_2 SYSTEM

STEVE ALEXANDRE NDENGUE, RICHARD DAWES, *Department of Chemistry, Missouri University of Science and Technology, Rolla, MO, USA*; YOHANN SCRIBANO, *Laboratoire Univers et Particules, Universite de Montpellier, Montpellier, France*; FABIEN GATTI, *CNRS, Institut des Sciences Moleculaires d'Orsay, Orsay, France*.

Water, an essential ingredient of life, is prevalent in space and various media. H_2O in the gas phase is the major polyatomic species in the interstellar medium (ISM) and a primary target of current studies of collisional dynamics. In recent years a number of theoretical and experimental studies have been devoted to H_2O-X (with X=He, H_2, D_2, Ar, ...) elastic and inelastic collisions in an effort to understand rotational distributions of H_2O in molecular clouds. In this work we are following those studies and will present benchmark calculations of rovibrational states and resonances of the H_2O-H_2 cluster in the rigid rotor approximation using the MultiConfiguration Time Dependent Hartree (MCTDH) approach. We will also present the first state-to-state inelastic scattering results of the H_2O+H_2 process in the rigid rotor approximation using the MCTDH approach. These calculations will serve as a foundation for similar triatomic – linear molecule interactions which are usually computationally expensive using standard calculations methods.

TD08

RELAXATION OF VIBRATIONAL, ROTATIONAL, AND CORIOLIS ENERGIES OF AN EXCITED NITROMETHANE MOLECULE IN ARGON BATH

LUIS A. RIVERA-RIVERA, *Department of Physical Sciences, Ferris State University, Big Rapids, MI, USA*; ALBERT F. WAGNER, *Chemical Sciences and Engineering Division, Argonne National Laboratory, Argonne, IL, USA*.

Our previous work [Rivera-Rivera *et al.* *J. Chem. Phys.* 142, 014303 (2015)] used classical molecular dynamics simulations to study the pressure effects on the relaxation of a nitromethane (CH_3NO_2) molecule in an argon bath at 300 K and pressure ranging from 10 to 400 atm. The molecule was instantaneously excited by statistically distributing 50 kcal/mol among all its internal degrees of freedom. The saved CH_3NO_2 positions and momenta are then used to separate the vibrational and rotational energy of the molecule following the methodology developed by Rhee and Kim [*J. Chem. Phys.* 107, 1394 (1997)]. The vibrational, rotational, and Coriolis energies exhibited multi-exponential decay. It is also found, that at later times the three energies decay approximately exponentially with similar decay rates. The mode-specific decomposition of these three energies produces, for each of the eight studied pressures, approximately 30 separate decay curves whose signal rises above statistical noise. Which vibrational and rotational modes these decay curves represent, and how their pressure dependence varies, gives insight into how excess energy equilibrates in CH_3NO_2.

TD09 11:20 – 11:35

CONFORMATIONAL DYNAMICS OF THE CYTOCHROME P450CAM-PUTIDAREDOXIN COMPLEX PROBED VIA 2D IR SPECTROSCOPY

SASHARY RAMOS, EDWARD BASOM, MEGAN THIELGES, *Department of Chemistry, Indiana University, Bloomington, IN, USA.*

Protein conformational dynamics are at the root of many biological processes but are difficult to characterize experimentally because they occur at timescales that can range from picoseconds to milliseconds and longer. Sophisticated techniques must be used to measure dynamics occurring on fast timescales (fs or ps), such as dynamics of protein side chains or solvent in protein microenvironments. Two-dimensional infrared (2D IR) spectroscopy has emerged as a powerful tool for the measurement of protein dynamics and conformational heterogeneity at the picosecond timescale due to its high temporal and spatial resolution. However, the IR spectrum of a protein is typically severely congested due to the large number of similarly bonded atoms. For this reason, protein 2D IR is paired with site-specific incorporation of spectrally resolved IR probes that are active in the transparent frequency region (1800-2500 cm^{-1}) and thus act as vibrational reporters. Putidaredoxin is known to play an effector role on cytochrome P450cam, however the conformation of the cytochrome P450cam-putidaredoxin (P450-Pdx) complex is currently debated. The conformational dynamics of the P450-Pdx complex were measured using heme-bound CO as a vibrational probe of local environment. To further examine the proposed conformational state of the P450-Pdx complex, the dynamics of a P450cam mutant (L358P) thought to behave similarly to the putidaredoxin complex were also measured. The information gathered from 2D IR experiments has provided new insight into the conformational states exhibited by the P450-Pdx complex.

TD10 11:37 – 11:52

RESOLVING ULTRAFAST PHOTOCHEMISTRY OF COORDINATION COMPLEXES USING HIGH HARMONIC GENERATION XANES SPECTROSCOPY

ELIZABETH S RYLAND, JOSH VURA-WEIS, KAILI ZHANG, *Department of Chemistry, University of Illinois at Urbana-Champaign, Urbana, IL, USA*; MUFFADDAL BURHANI, *Chemistry, University of Illinois at Urbana-Champaign, Urbana, IL, United States*; MAX A VERKAMP, *Department of Chemistry, University of Illinois at Urbana-Champaign, Urbana, IL, USA*; MING-FU LIN, *Stanford Synchrotron Radiation Lightsource, SLAC National Accelerator Laboratory, Menlo Park, CA, USA*; KRISTIN BENKE, MICHAELA CARLSON, *Department of Chemistry, University of Illinois at Urbana-Champaign, Urbana, IL, USA.*

Extreme ultraviolet (XUV) spectroscopy is an inner shell technique that probes the $M_{2,3}$-edge excitation of atoms. Absorption of the XUV photon causes a $3p \rightarrow 3d$ transition, the energy and multiplet of which is directly related to the element and ligand environment. This in-lab technique is thus element-, oxidation state-, spin state-, and ligand field specific and is a useful tool for the study of electron and energy transfer processes in materials and chemical biology.

With the use of this technique and semi-empirical simulations, I have collected ultrafast transient $M_{2,3}$-edge absorption data of four different metalloporphyrinates (M = Fe, Co, Ni, Mn) in order to resolve the early time relaxation mechanism of these catalytically-relevant coordination complexes with femtosecond time resolution. This is the first instance of using tabletop transient XUV/VUV spectroscopy on coordination complexes and furthermore highlights the importance of directly probing of the metal center in these systems. I will additionally present ongoing work on applying this technique to the study of heterobimetallic systems with directly-interacting dual metal centers within a non-innocent ligand scaffold. The relation of function to metal-specific photodynamics will help lay essential groundwork for the development of multimetallic catalysts with efficiencies comparable to those found in nature.

TD11

FEMTOSECOND EXTREME ULTRAVIOLET SPECTROSCOPY OF SEMICONDUCTOR CARRIER DYNAMICS

<u>JOSH VURA-WEIS</u>, MAX A VERKAMP, *Department of Chemistry, University of Illinois at Urbana-Champaign, Urbana, IL, USA.*

Extreme ultraviolet (XUV) transient absorption spectroscopy is emerging as a powerful, element-specific tool for measuring femtosecond to attosecond dynamics in molecular and solid-state systems. Tabletop XUV transient absorption spectra retain the element specificity of hard x-ray absorption while providing a straightforward mapping of the unoccupied valence and conduction band density of states. The presence of distinct signals for holes and electrons in the XUV region is especially powerful, as the dynamics of these carriers are often convolved in transient UV/visible measurements. In this work we measure the rate of charge transfer across $TiO_2/CH_3NH_3PbI_3$ and $CH_3NH_3PbI_3/NiO$ interfaces, and highlight the competition between carrier cooling in the perovskite absorber and charge injection into the electron/hole collection layers

TE. Clusters/Complexes
Tuesday, June 19, 2018 – 8:30 AM
Room: B102 Chemical and Life Sciences

Chair: Juan Carlos Lopez, Universidad de Valladolid, Valladolid, Spain

TE01 8:30 – 8:45

INFRARED PHOTODISSOCIATION SPECTROSCOPIC AND THEORETICAL STUDY OF BORON CARBONYL COMPLEXES

JIAYE JIN, GUANJUN WANG, MINGFEI ZHOU, *Fudan University, Department of Chemistry, Shanghai, China.*

The structures and bonding of boron compounds remain an outstanding question in cluster science.[a] Here, we will highlight our recent studies on the formation and infrared spectroscopic characterization of several boron carbonyl cation complexes. The $B_3(CO)_3^+$ is characterized to have a planar D_{3h} symmetry, which features the smallest π aromatic system B_3^+.[b] The $B_3(CO)_4^+$ complex is determined to have a D_{2d} $(OC)_2B=B=B(CO)_2$ structure and $^1A_1$ electronic ground state with a linear boron skeleton.[c] The $B_3(CO)_5^+$ is predicted to have a chain boron framework with C_{2v} symmetry. Bonding analysis reveals that $B_3(CO)_{3-5}^+$ complexes have chemical bonding patterns similar to $C_3H_3^+$, C_3H_4, and $C_3H_5^-$, indicating that isolobal relationship of CO/H^- and B^-/C is useful in bridging the boron carbonyl complexes and the hydrocarbon molecules.

[a] A. P. Sergeeva, L. S. Wang, A. I. Boldyrev, et al. *Acc. Chem. Res.* **47**, 1349 (2014).
[b] J. Y. Jin, M. F. Zhou, G. Frenking, et al. *Angew. Chem. Int .Edit.* **55**, 2078 (2016).
[c] J. Y. Jin, G. J. Wang, M. F. Zhou, *J. Phys. Chem. A.* (2018, in press).

TE02 8:47 – 9:02

WITHDRAWN

TE03 9:04 – 9:19

SPIN-ORBIT STATE-SELECTIVE AUTODETACHMENT OF VIBRATIONALLY EXCITED CCP$^-$

G. STEPHEN KOCHERIL, JOSEPH CZEKNER, LING FUNG CHEUNG, LAI-SHENG WANG, *Department of Chemistry, Brown University, Providence, RI, USA.*

The linear dicarbon phosphide molecule (CCP) has a $^2\Pi$ ground electronic state with a small spin-orbit splitting into $^2\Pi_{1/2}$ and $^2\Pi_{3/2}$ states. It has a reasonably large dipole moment and has been observed in interstellar space. We have studied CCP$^-$ ion using high-resolution photoelectron imaging and observed dipole-bound excited states for CCP$^-$ right below the detachment threshold. Resonant photoelectron spectra have been obtained by exciting the anion to specific vibrational levels of the dipole-bound states. We have observed a dipole-bound state for each spin-orbit state and the vibrational autodetachment is state-selective, providing the first spectroscopic evidence that the dipole-bound electron does not couple to the neutral core.

TE04 9:21 – 9:36

INFRARED SPECTRA OF C_2H_4 DIMER AND TRIMER

A. J. BARCLAY, KOOROSH ESTEKI, *Department of Physics and Astronomy, University of Calgary, Calgary, AB, Canada*; BOB McKELLAR, *Steacie Laboratory, National Research Council of Canada, Ottawa, ON, Canada*; NASSER MOAZZEN-AHMADI, *Physics and Astronomy/Institute for Quantum Science and Technology, University of Calgary, Calgary, AB, Canada.*

Spectra of ethylene dimers and trimers are studied in the ν_{11} and (for the dimer) ν_9 fundamental band regions of C_2H_4 ($\sim$2990 and 3100 cm^{-1}) using a tunable optical parametric oscillator source to probe a pulsed supersonic slit jet expansion. The deuterated trimer has been observed previously, but this represents the first rotationally resolved spectrum of $(C_2H_4)_3$. The results support the previously determined cross-shaped (D_{2d}) dimer and barrel-shaped (C_{3h} or C_3) trimer structures. However, the dimer spectrum in the ν_9 fundamental region of C_2H_4 is apparently very perturbed and a previous rotational analysis is not well verified.

TE05 9:38 – 9:53

GAS PHASE SPECTROSCOPY ON HYDRATED CLUSTERS OF OXYTOCIN BY ELECTROSPRAY IONIZATION / COLD ION TRAP TECHNIQUE

MIZUKI TABATA, MASAAKI FUJII, *Chemical Resources Laboratory, Tokyo Institute of Technology, Yokohama, Japan.*

Oxytocin (OT) is the first structure-identified peptide hormone, which has a cyclic part formed by disulfide-bridge and a C-terminal α-amidated tail. Since some studies show that oxytocin alleviates mental disorders and postpartum hemorrhage, it has been developed as pharmaceuticals, recently more focused on oral drugs. However, low membrane permeability of oxytocin impedes its oral administration. One of the important factors which affect the membrane permeability is how dehydrated structure in the membrane changes into the hydrated structure. In this work, to elucidate the effect of the hydration on oxytocin structure, infrared spectroscopy was employed on isolated hydrated oxytocin clusters (OTH$^+\cdot$(D$_2$O)$_n$, n=1-4), combining with theoretical calculations. The results of this study suggest that the hydration of only a single water molecule interrupts the intramolecular hydrogen bond between the cyclic moiety and the tail part. The structural deformation by the hydration will be discussed.

TE06 9:55 – 10:10

BLUE SHIFTED HYDROGEN BOND IN CH/D$_3$CN...HCCL$_3$ COMPLEXES

BEDABYAS BEHERA, PUSPENDU KUMAR DAS, *Department of Inorganic and Physical Chemistry, Indian Institute of Science, Bangalore, India.*

H-bonded complexes between CHCl$_3$ and CH/D$_3$CN have been identified by FTIR spectroscopy in the gas phase at room temperature. With increasing partial pressure of the components, The C-H stretching fundamental shifts to the blue which has been identified as due to C-H...N interaction. The C-H stretching frequency of CHCl$_3$ with CH$_3$CN and CD$_3$CN are shifted by +8.7 and +8.6 cm^{-1}, respectively. By using quantum chemical calculations at the MP2/6-311++G** level, we predict the geometry, electronic structural parameters, binding energy, and spectral shift in the H-bonded complexes. The potential energy scans of the above complexes as a function of C...N distance shows that the H-bonding interaction is predominantly due to contribution of two opposing forces i.e., electrostatic attraction between H and N which leads to the C-H bond elongation with consequent red-shift and the electronic and nuclear repulsion between the C and N which results in C-H bond contraction and blue-shift of the C-H stretching frequency. The net effect of these two opposing forces at the equilibrium complex geometry dictates the nature of the shift although the influence of the surrounding atoms bonded to the atoms that are directly involved in the H-bonding cannot be fully underestimated. The total interaction energy (-14.23 kJ/mol) is characterized by Morokuma energy decomposition analysis where the binding in CH/D$_3$CN...CHCl$_3$ is dominated by electrostatic attraction (-25.86 kJ/mol). The attraction, however, is considerably suppressed by exchange repulsion (+19.54 kJ/mol). Other components like polarization (-5.44 kJ/mol) and charge transfer (-5.06 kJ/mol) make significant contribution to the interaction energy.

Intermission

TE07   10:46 – 11:01
BORAZINE AND BENZENE: CHALK AND CHEESE.
MATRIX ISOLATION INFRARED AND AB INITIO STUDIES.

KANUPRIYA VERMA, K S VISWANATHAN, *Chemical Science, Indian Institute of Science Education and Research, MOHALI, PUNJAB, India.*

Borazine, also referred to as inorganic benzene, displays an interesting case study of non-covalent interactions involving the N-H group as well as the π-system, unlike the benzene system, whose interaction is dominated by only the π-system. Borazine is a multifunctional molecule with the N-H serving as a proton donor and the partially delocalized π electron cloud as a proton acceptor. The interaction of borazine with various π systems, such as acetylene, benzene and phenylacetylene was studied experimentally, using matrix isolation IR spectroscopy and computationally, using ab initio calculations. Computations were carried out at the M06-2X and MP2 levels of theory using both 6-311++G** and aug-cc-pVDZ basis sets. In the case of all the π systems studied in our work, the N-H group of borazine was found to serve as the proton donor. In the case of both borazine-acetylene and borazine-phenylacetylene, a bent NH...C structure, where the N-H of borazine was the proton donor to the carbon of the acetylenic group was found to be the global minimum. However, in the case of borazine-benzene, a parallel displaced structure was found to be the global minimum at both levels of theory. Borazine was also different from benzene in that it displayed some unconventional bonding scenarios such as the dihydrogen bond in the borazine dimer and a boron bond in the borazine-water systems. Details of the experimental data and computational results will be presented.

TE08   11:03 – 11:18
INTERROGATION OF $MoO_yC_nH_n^-$ CHEMIFRAGMENTS ILLUMINATES $Mo-(\eta^2-$ACETYLENE) INTERACTIONS WITHIN $Mo_xO_y^-$ AND ETHYLENE REACTIONS

JOSEY E TOPOLSKI, RICHARD N SCHAUGAARD, MANISHA RAY, KRISHNAN RAGHAVACHARI, CAROLINE CHICK JARROLD, *Department of Chemistry, Indiana University, Bloomington, IN, USA.*

In an effort to determine the feasibility of producing hydrogen gas via $H_2O + C_2H_4 \rightarrow H_2 + CH_3CHO$, cluster reactivity studies were completed between Mo_xO_y cluster anions and H_2O, C_2H_4, and mixtures of both H_2O and C_2H_4. These studies unveiled the evolution of several chemifragmentation products. To better understand the molecular-scale interactions along these chemifragmentation pathways, the photoelectron spectra and supporting theoretical calculations were analyzed for each cluster. In this talk, spectra and computational results for a series of monometallic chemifragment cluster anions formed from reactions with C_2H_4 will be presented. The analysis indicates that experimental spectra are most consistent with η^2-acetylene complexes, however vinylidene complexes cannot be definitively ruled out for all clusters. The results of this work deepen our understanding of the side reactions which occur in this multi-reactant system.

TE09   11:20 – 11:35
PROPYLBENZENE-$(H_2O)_n$ CLUSTERS: EFFECT OF THE ALKYL CHAIN ON THE π H-BOND

PIYUSH MISHRA, JOSHUA L. FISCHER, *Department of Chemistry, Purdue University, West Lafayette, IN, USA*; EDWIN SIBERT, *Department of Chemistry, University of Wisconsin–Madison, Madison, WI, USA*; TIMOTHY S. ZWIER, *Department of Chemistry, Purdue University, West Lafayette, IN, USA.*

This talk focuses on the mass resolved- resonant 2-photon ionization (R2PI), resonant ion-dip infrared spectroscopy (RIDIR) and IR-UV holeburning (IR-UV HB) spectroscopy of propylbenzene(pBz)-$(H_2O)_n$ clusters and the comparison with their Benzene(Bz)-$(H_2O)_n$ cluster counterparts, which are a well-studied prototype system for the π H-bond. Since the pBz monomer exists in *gauche* and *trans* conformers, one anticipates the presence of pBz-H_2O complexes with H_2O on the same or opposite sides of the ring as the *gauche* or *trans* propyl chain. Indeed, local minima associated with these four complexes were identified by dispersion-corrected DFT calculations. R2PI and IR-UV HB spectra of pBz-H_2O show long Franck-Condon progressions associated with the set of conformers of the complex. The OH stretch RIDIR spectra consist of a single transition in the symmetric stretch region, and a doublet with varying spacing in the antisymmetric stretch region, indicating coupling to a large-amplitude motion (LAM). The changes in the OH stretch region indicate that the water molecule bound to propylbenzene undergoes more restricted motion on the π cloud than its Bz-H_2O counterpart. The potential energy surface for H_2O tumbling on the pBz π cloud was mapped out, and used as the basis for calculating from first principles the OH stretch infrared spectrum. Comparison with the spectrum for Bz-H_2O further illustrates the source and restrictions of the LAM of H_2O in pBz compared to Bz. OH stretch IR spectra of the higher water clusters pBz-$(H_2O)_n$ with n=3, 4 are very similar to their Bz-$(H_2O)_n$ counterparts, existing as H-bonded cycles, with no evidence of LAM on the aromatic π cloud.

TE10 11:37 – 11:52

WEAKLY-BOUND COMPLEXES OF FURAN AND WATER AS INVESTIGATED BY MATRIX ISOLATION FTIR

TYLER G. FULLER, SCHUYLER P LOCKWOOD[a], <u>JOSH NEWBY</u>, *Department of Chemistry, Hobart and William Smith Colleges, Geneva, NY, USA.*

Weakly bound complexes containing aromatic species have been the subject of study for many years. Here, a study of the 1:1 complexes of furan (C_4H_4O) with water will be presented. In this work, matrix isolation FTIR and computational methods were used to examine stable geometries of this dimer system. Density functional theory and MP2 methods were used to identify four minimum energy geometries. Three interaction motifs are recognized in these structures: C−H···O, O−H···O, and O−H···π. The four structures were found to be within 610 cm^{-1}(7.3 kJ/mol) of each other across all computational methods. Matrix isolation FTIR spectroscopy was used to explore mixtures of furan with H_2O in a nitrogen matrix at 15 K. Spectra acquired show several peaks that were not associated with water or furan monomers and have been assigned to FUW. Additionally, mixtures of furan with D_2O and HDO were deposited and their spectra recorded. Characteristic shifts of each isotopologue were identified and used to characterize the geometry of the FUW complex. Both computational and spectroscopic results point to the formation of a single complex geometry that interacts through a standard O−H···O hydrogen bond.

[a]Current address: Department of Chemistry, University of Massachusetts Amherst, Amherst, MA 01003

TE11 11:54 – 12:09

CHARACTERIZATION OF A HYDROGEN PEROXIDE-BENZENE COMPLEX USING MATRIX ISOLATION INFRARED SPECTROSCOPY

<u>JAY C. AMICANGELO</u>, YUDHISHTARA PAYAGALA, DYLAN JOHNSON, CATHERINE KAISER, *School of Science (Chemistry), Penn State Erie, Erie, PA, USA.*

Matrix isolation infrared spectroscopy was used to characterize a 1:1 complex of hydrogen peroxide (H_2O_2) with benzene (C_6H_6). Co-deposition experiments with H_2O_2 and C_6H_6 were performed at 20 K using argon as the matrix gas. New infrared peaks attributable to the H_2O_2-C_6H_6 complex were observed near the O-H stretching vibrations and the OH bending vibrations of the H_2O_2 monomer and near the hydrogen out-of-plane bending vibration of the C_6H_6 monomer. The initial identification of the newly observed infrared peaks to those of a H_2O_2-C_6H_6 complex was established by performing several concentration studies in which the sample-to-matrix ratios of the monomers were varied between 1:100 to 1:1600, by comparing the resulting co-deposition spectra with the spectra of the individual monomers, and by matrix annealing experiments (30 – 35 K). Co-deposition experiments were also performed using isotopically labeled hydrogen peroxide (D_2O_2 and HDO_2) and benzene (C_6D_6) and the analogous peaks for the isotopically labelled complexes were observed. Quantum chemical calculations were performed for the H_2O_2-C_6H_6 complex at the MP2/aug-cc-pVDZ level of theory in order to explore the intermolecular potential energy surface of the complex and to obtain optimized complex geometries and predicted vibrational frequencies of the complex, which were compared to the experimental infrared spectra.

TF. Atmospheric science

Tuesday, June 19, 2018 – 8:30 AM

Room: 2079 Natural History

Chair: Peter F. Bernath, Old Dominion University, Norfolk, VA, USA

TF01   8:30 – 8:45

SPECTROSCOPIC DATABASES FOR THE VAMDC PORTAL: NEW TOOLS AND IMPROVEMENTS

CYRIL RICHARD, VINCENT BOUDON, *Laboratoire ICB, CNRS/Université de Bourgogne, DIJON, France*; NICOLAS MOREAU, MARIE-LISE DUBERNET, *LERMA2, CNRS UMR8812, Observatoire de Paris, MEUDON, France*.

Dijon spectroscopic databases include calculated line lists, in positions and intensities, that are obtained from experimental spectroscopic analyses. They contain 6 molecules: CH_4, C_2H_4, CF_4, SF_6, GeH_4 and RuO_4 and are all compatibles with the XSAMS (XML Schema for Atoms, Molecules, and Solids) format adopted with the Virtual Atomic and Molecular Data Centre (VAMDC) Project. VAMDC, the worldwide consortium which federates atomic and molecular databases through an e-science infrastructure, aims to provide a unique access point for scientists seeking the best atomic and molecular data for their studies. So far, development of new tools allows to easily download and compare data issued from different databases in a single XML document or into the HITRAN2004 format. Making the comparison that easy will help data users in the choice of data that best match their needs. It will also help data producers by checking the consistency of their data.

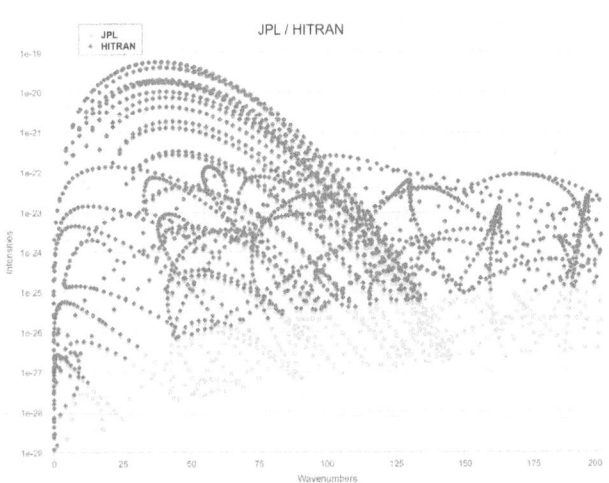

TF02   8:47 – 9:02

GLOBAL ANALYSES OF SF_6 HIGH-RESOLUTION SPECTRA FOR THE VAMDC/SHeCaSDa DATABASE

VINCENT BOUDON, HANZHANG KE, CYRIL RICHARD, *Laboratoire ICB, CNRS/Université de Bourgogne, DIJON, France*; MBAYE FAYE, , *LISA CNRS et Universités Paris Est et Paris Diderot , Creteil, France*; LAURENT MANCERON, *Synchrotron SOLEIL, CNRS-MONARIS UMR 8233 and Beamline AILES, Saint Aubin, France*.

Sulfur hexafluoride (SF_6) is a very strong greenhouse gas of anthropogenic origin. Its relatively high molecular mass leads to a very dense molecular spectrum which is quite difficult to analyze. Our groups performed recently strong experimental and theoretical efforts in order to assign and model many cold and hot bands of SF_6, allowing to strongly improve the modeling of its atmospheric absorption[a]. We present here a synthesis of the latest available data and global analyses performed thanks to the formalism and programs developed in Dijon. In particular, new progress has been realized towards the modeling of hot bands in the strongly absorbing ν_3 region around 948 cm^{-1}. The results have been used tu fully update the SHeCaSDa database (http://vamdc.icb.cnrs.fr/PHP/shecasda.php) which is also accessible from the VAMDC portal (http://portal.vamdc.org) and will be included in future HITRAN and GEISA updates.

---

[a]M. Faye *et al.*, J. Quant. Spectrosc. Radiat Transfer **190**, 38,47 (2017) ; M. Faye *et al.*, J. Mol. Spectrosc. in press (2018).

TF03　　9:04–9:19

NEW DATA AND ANALYSIS FOR SF$_6$ ABSORPTION MODELLING IN THE 10 MICRON ATMOSPHERIC WINDOW

MBAYE FAYE, , *LISA CNRS et Universités Paris Est et Paris Diderot , Creteil, France*; VINCENT BOUDON, MICHEL LOETE, CYRIL RICHARD, *Laboratoire ICB, CNRS/Université de Bourgogne, DIJON, France*; P. ROY, *AILES beamline, Synchrotron SOLEIL, Saint Aubin, France*; LAURENT MANCERON, *AILES Beamline, Synchrotron SOLEIL, Saint-Aubin, France*.

Modelling correctly the SF$_6$ atmospheric absorption requires the knowledge of the spectroscopic parameters of all states involved in the many hot bands in the 10 μm atmospheric window used for remote sensing. Since a direct analysis of the hot bands near the ν_3 absorption of SF$_6$ in this atmospheric window is not possible, due to their overlapping, we use another strategy, gathering information in the far and mid infrared regions on initial and final states to compute the relevant total absorption. In this talk, we present new results of an analysis of spectra recorded at the AILES beam line at the SOLEIL Synchrotron facility. For these measurements, we used an IFS125HR interferometer in the 100 to 3200 cm^{-1} range, coupled to a cryogenic multiple pass cell. The optical path length was varied from 45 to 141 m with 223 and 153 K temperatures. New information has been obtained on the $\nu_3 + \nu_5$ band which, combined with improved parameters for ν_5, is used to model the important $\nu_3+\nu_5$ - ν_5 hot band contribution [1]. Also, data have been obtained on the ν_3 band of the ^{36}SF$_6$ isotopic species present in very low abundance (0.0002) [2]. These new parameters will be included in the XTDS model [3] and VAMDC/SheCaSDa database [4], thus improving the previous SF$_6$ parameters.

[1] M. Faye, L. Manceron, P. Roy, V. Boudon, M. Loete, "First analysis of the $\nu_3+\nu_5$ combination band of SF$_6$ observed at Doppler-limited resolution and effective model for the $\nu_3+\nu_5$-ν_5 hot band" J. Mol. Spectrosc., in press.

[2] M. Faye, L. Manceron, P. Roy, V. Boudon, M. Loete, "First high resolution analysis of the ν_3 band of the ^{36}SF$_6$ isotopologue", J. Mol. Spectrosc., in press.

[3] C. Wenger, V. Boudon, M. Rotger, M. Sanzharov, and J.-P. Champion,"XTDS and SPVIEW: Graphical tools for Analysis and Simlation of High Resolution Molecular Spectra", J. Mol. Spectrosc. 251, 102 (2008).

[4] http://vamdc.icb.cnrs.fr/PHP/shecasda.php

TF04　　9:21–9:36

TRENDS IN ATMOSPHERIC HCL, HFC − 134A AND CHF$_3$ FROM THE ACE SATELLITE MISSION

ANTON MADUSHANKA FERNANDO, *Department of Physics, Old Dominion University, Norfolk, VA, USA*; PETER F. BERNATH, *Department of Chemistry and Biochemistry, Old Dominion University, Norfolk, VA, USA*; CHRIS BOONE, *Department of Chemistry, University of Waterloo, Waterloo, ON, Canada*.

The Montreal Protocol has banned the production of major ozone depleting substances such as chlorofluorocarbons (CFCs) to protect the Earth's ozone layer. These chlorinated compounds eventually form a large reservoir of HCl in the upper atmosphere. Therefore, the success of the Montreal Protocol can be determined by measuring the stratospheric HCl abundance. The Atmospheric Chemistry Experiment Fourier Transform Spectrometer (ACE-FTS) measures infrared solar occultation spectra of the Earth's atmosphere from which altitude profiles of HCl volume mixing ratios (VMRs) are determined. The stratospheric HCl VMR time series has a linear trend of -4.9$\pm$0.4% per decade for 2004-2017, highlighting the continuing success of the Montreal Protocol. The emission of hydrofluorocarbons (HFCs) has increased in order to replace the CFCs since they are not restricted by the Montreal Protocol. As a result, the atmospheric abundance of HFCs are rapidly increasing. Even though these HFCs do not contribute to the depletion of the ozone layer, they are powerful greenhouse gases with large global warming potentials. In 2016, the Kigali Amendment to the Montreal Protocol was introduced in order to phaseout long-lived HFCs. The VMRs of the two most abundant HFCs in the atmosphere, HFC − 134a and CHF$_3$, are also measured by the ACE-FTS. These measurements will be useful for monitoring the Kigali Amendment to the Montreal Protocol. The ACE observations and trends for atmospheric HCl, HFC − 134a and CHF$_3$ will be presented.

Intermission

TF05 10:12 – 10:27

QUANTITATIVE INFRARED SPECTROSCOPY OF HALOGENATED SPECIES FOR ATMOSPHERIC REMOTE SENSING

JEREMY J. HARRISON, *Department of Physics and Astronomy, University of Leicester, Leicester, United Kingdom.*

Fluorine- and chlorine-containing molecules in the atmosphere are very strong greenhouse gases, meaning that even small amounts of these gases contribute significantly to the radiative forcing of climate. In addition, a number of these molecules, such as chlorofluorocarbons (CFCs) and hydrochlorofluorocarbons (HCFCs), are harmful to the Earth's ozone layer and for this reason their use is regulated by the 1987 Montreal Protocol. The recent Kigali Amendment has added hydrofluorocarbons (HFCs) to the list of controlled substances, coming into effect on 1 January 2019. HFCs, which do not deplete stratospheric ozone, were introduced as refrigerant replacements for CFCs and HCFCs. They are potent greenhouse gases, with global-warming potentials many times greater than carbon dioxide, and are increasing in the atmosphere at a very fast rate.

A number of satellite instruments, in particular the ACE-FTS (Atmospheric Chemistry Experiment – Fourier Transform Spectrometer), can monitor many of these species by detecting infrared radiation that has passed through the Earth's atmosphere. However, the quantification of their atmospheric abundances crucially requires accurate quantitative infrared spectroscopy. This talk will focus on new and improved laboratory spectroscopic measurements for a number of important halogenated species.

TF06 10:29 – 10:44

ACCURATE LABORATORY DETERMINATION OF THE MID AND SHORT WAVE INFRARED WATER VAPOR SELF-CONTINUUM. NEW MEASUREMENTS AND TEST OF THE MT_CKD MODEL

DIDIER MONDELAIN, ALAIN CAMPARGUE, *UMR5588 LIPhy, Université Grenoble Alpes/CNRS, Saint Martin d'Hères, France*; ROBERTO GRILLI, *Institut des Géosciences de l'Environnement, Université Grenoble Alpes, Saint Martin d'Hères, France*; SAMIR KASSI, *UMR5588 LIPhy, Université Grenoble Alpes/CNRS, Saint Martin d'Hères, France*; LOÏC LECHEVALLIER, *Institut des Géosciences de l'Environnement, Université Grenoble Alpes, Saint Martin d'Hères, France*; LUCILE RICHARD, DANIELE ROMANINI, SEMYON VASILCHENKO, IRENE VENTRILLARD, *UMR5588 LIPhy, Université Grenoble Alpes/CNRS, Saint Martin d'Hères, France.*

The semi empirical MT_CKD model of the absorption continuum of water vapor is widely used in atmospheric radiative transfer codes of the atmosphere of Earth and, recently, of exoplanets, but lacks of experimental validation in the atmospheric windows. We report on accurate water vapor absorption continuum measurements by Cavity Ring Down Spectroscopy (CRDS) and Optical-Feedback-Cavity Enhanced Laser Spectroscopy (OF-CEAS) at selected spectral points of the transparency windows centered around 4.0, 2.1, 1.6 and 1.25 μm. Temperature dependence of the absorption continuum is also measured in the 23-50 °C range for a few spectral points. The self-continuum water vapor absorption is derived either from the baseline variation of water vapor spectra recorded for a series of pressure values over a small spectral interval or from baseline monitoring at fixed laser frequency during pressure ramps. After subtraction of the local water monomer lines contribution, self-continuum cross-sections, C_S, are accurately determined from the pressure squared dependence of the continuum absorption measured up to about 15 Torr. The derived water vapor self-continuum provides a unique set of water vapor self-continuum cross-sections for a test of the MT_CKD model in four transparency windows.

TF07 10:46 – 11:01

A HIGHLY ACCURATE *AB INITIO* DIPOLE MOMENT SURFACE FOR WATER: TRANSITIONS EXTENDING INTO THE ULTRAVIOLET

EAMON K CONWAY[a], *Atomic and Molecular Physics, Harvard-Smithsonian Center for Astrophysics, Cambridge, MA, USA*; ALEKSANDRA A. KYUBERIS, *Microwave Spectroscopy, Institute of Applied Physics, Nizhny Novgorod, Russia*; OLEG L. POLYANSKY, JONATHAN TENNYSON, *Department of Physics and Astronomy, University College London, Gower Street, London WC1E 6BT, United Kingdom.*

We present a new *ab initio* dipole moment surface (DMS) for the water molecule valid for transitions which stretch into the near ultraviolet. Intensities computed using this surface agree very well with precise laboratory measurements designed to aid atmospheric observations. This work is based on a data set encompassing 17 628 multi-reference configuration interaction configurations that were calculated with the aug-cc-pCV6Z basis set with the Douglass-Kroll-Hess Hamiltonian to second order and required approximately 116 years of CPU running time to complete.

Compared to recent experimental measurements in the far infrared region[b], this new DMS significantly improves agreement with theory for transitions in the previously problematic bands (121), (300) and (102). For highly energetic overtones located in both the visible and ultraviolet regimes, we successfully predict the intensity of all measured bands to within 10% of the latest atmospheric observations[c]. These include bands at 487 nm (303), 471 nm (511), and 363 nm (900), for which previous models underestimated the intensity by up to 139%. Absorption features are also predicted in the 290 nm to 355 nm window and the theoretical shape demonstrates reasonably good behaviour with previously measured cross sections. The 10 ν_1 band is identified as the strongest absorber in this region and the maximum intensity is approximately 6.3×10^{-27} cm per molecule, which should be observable in atmospheric spectra.

---

[a]Department of Physics and Astronomy, University College London, Gower Street, London WC1E 6BT, United Kingdom
[b]M. Birk et al., J. Quant. Spectrosc. Rad. Transf, 203, 88-102 (2017)
[c]J. Lampel et al., Atmos. Chem. Phys, 17, 1271-1295, (2017)

TF08 11:03 – 11:18

HIGH-RESOLUTION INFRARED SPECTROSCOPY OF ISOPRENE AND METHYL VINYL KETONE IN THE 10 μm REGION

MICHAEL CYRUS IRANPOUR, MINH NHAT TRAN, MARCUS VINICUS PEREIRA, TYLER HOADLEY, JACOB STEWART, *Department of Chemistry, Connecticut College, New London, CT, USA.*

Isoprene (C_5H_8) is the most abundant biogenic volatile organic compound (BVOC) emitted by plants into Earth's atmosphere and plays a key role in the chemistry of the troposphere. Isoprene (and other BVOCs) play a role in the formation of secondary organic aerosols and production of tropospheric ozone, a pollutant that is a major part of photochemical smog. One attractive means of measuring isoprene levels in the atmosphere is the use of infrared spectroscopy to monitor strong absorption bands that lie in the "atmospheric window" between 8-14 μm. We have measured the high-resolution infrared spectrum of isoprene in the region of the strong ν_{26} vibrational band near 10 μm using a quantum cascade laser-based spectrometer. This work will support future efforts to use the ν_{26} band for sensing applications. We will discuss the assignment of several weaker bands we have observed near the main ν_{26} band and a rotational analysis based on the strong Q-branches. In addition, we have also observed the high-resolution spectrum of methyl vinyl ketone (C_4H_6O), an oxidation product of isoprene, in this same frequency region, and will present a preliminary analysis of our spectrum.

TG. Mini-symposium: Frequency-Comb Spectroscopy

Tuesday, June 19, 2018 – 1:45 PM

Room: 116 Roger Adams Lab

Chair: Hairun Guo, École polytechnique fédérale de Lausanne, Lausanne, Switzerland

TG01   *INVITED TALK*   1:45 – 2:15

EXTENDING FREQUENCY COMB SPECTROSCOPY TO THE MID AND FAR INFRARED RANGE

PAOLO DE NATALE, *Istituto Nazionale di Ottica, CNR, Firenze, Italy.*

Visible and near-IR optical frequency combs have become key metrological tools for atomic and molecular spectroscopy, in the last 20 years. Their extension to the mid-IR and THz spectral ranges required appropriate and non-trivial nonlinear techniques for down-conversion. Such developments significantly improved the accuracy of frequency measurements across vibrational and rotational spectra of molecules, thus enormously enlarging the quality and application range of spectroscopic measurements. Very recently, Quantum Cascade Lasers have shown a great potential for comb generation in spectral intervals chosen by design with all-in-one miniature devices. I will discuss progress in mid-IR and THZ combs for application to high resolution molecular spectroscopy and I will show very recent results for QCLs comb generation and control.

TG02   2:19 – 2:34

HIGH-PRECISION MID-IR MOLECULAR SPECTROSCOPY WITH TRACEABILITY TO PRIMARY FREQUENCY STANDARDS USING SUB-Hz FREQUENCY COMB-STABILIZED QCLS

DANG BAO AN TRAN, ROSA SANTAGATA, OLIVIER LOPEZ, SEAN TOKUNAGA, *Laboratoire de Physique des Lasers, CNRS, Université Paris 13, Sorbonne Paris Cité, Villetaneuse, France*; MICHEL ABGRALL, YANN LE COQ, RODOLPHE LE TARGAT, WON-KYU LEE, DAN XU, PAUL-ERIC POTTIE, *LNE-SYRTE, Observatoire de Paris, PSL Research University, CNRS, Sorbonne Universités, UPMC Univ. Paris 06, Paris, France*; ANNE AMY-KLEIN, BENOIT DARQUIE, *Laboratoire de Physique des Lasers, CNRS, Université Paris 13, Sorbonne Paris Cité, Villetaneuse, France.*

High-precision measurements with molecules may refine our knowledge of various fields of physics, from atmospheric and interstellar physics to the standard model or physics beyond it. Many of them can be cast as absorption frequency measurements, particularly in the mid-infrared molecular fingerprint region, creating the need for narrow-linewidth lasers of well-controlled frequency. We present a new technology for precision vibrational spectroscopy using quantum cascade lasers (QCLs)[a]. Via an optical frequency comb, a QCL is stabilized at the sub-Hz level on an ultra-stable near infrared reference signal provided by the French metrology institute (SYRTE) to our laboratory and transferred through a 43-km long fiber cable[b]. The stability of the reference is transfered to the QCL, which therefore exhibits a relative frequency stability lower than 2×10^{-15} between 1 and 100 s. Moreover its absolute frequency is known with an uncertainty below 10^{-14} thanks to the traceability to the primary standards of SYRTE. The setup allows the QCL to be widely scanned over ~ 1 GHz while maintaining the highest stabilities and accuracies. We report saturated absorption spectroscopy investigations conducted around 10 μm on osmium tetroxyde in a Fabry-Perot cavity and methanol in a multipass cell at low pressures ranging from 0.01 to 10 Pa, allowing central frequencies, pressure shifts and broadenings to be determined with record uncertainties. By combining wide tuneability, high sensitivity and resolution, this setup constitutes a key technology for precise spectroscopic measurements with molecules.

---

[a] Argence et al., *Quantum cascade laser frequency stabilization at the sub-Hz level*, Nature Photon. **9**, 456 (2015).
[b] W.-K. Lee et al., *Hybrid fiber links for accurate optical frequency comparisons*, Appl. Phys. B **123**, 161 (2017).

TG03 2:36 – 2:51

SINGLE-SHOT SUB-MICROSECOND SPECTROSCOPY OF THE BACTERIORHODOPSIN PHOTOCYCLE WITH QUANTUM CASCADE LASER FREQUENCY COMBS

MARKUS MANGOLD, *IRsweep AG, IRsweep AG, Stäfa, Switzerland*; TILMAN KOTTKE, JESSICA KLOCKE, *Physical and Biophysical Chemistry, Uni Bielefeld, Bielefeld, Germany*; ANDREAS HUGI, PITT ALLMENDINGER, *IRsweep AG, IRsweep AG, Stäfa, Switzerland*; JEROME FAIST, *Institute for quantum electronics, ETH Zurich, Zurich, Switzerland*.

Time-resolved vibrational spectroscopy is an important tool for understanding biological processes and chemical reaction pathways [1]. Today, all available methods to our knowledge require many repetitions of an experiment to acquire a microsecond time-resolved mid-infrared spectrum.

We present the IRspectrometer, a quantum cascade laser dual frequency comb spectrometer [2-3]. It allows for parallel acquisition of hundreds of mid-infrared wavelengths with microsecond time resolution. The formation of the light-activated L, M and N-states in bacteriorhodopsin – which only have μs to ms lifetimes – has been recorded by analyzing the infrared response of bacteriorhodopsin to 10 ns visible light pulses with sub-microsecond time-resolution. The different wavelengths were all measured in parallel thanks to the dual-comb approach. The spectra as well as the kinetics show good agreement with those from step-scan FT-IR measurements. As a benchmark, the spectral signature of several intermediate states of the bacteriorhodopsin photocycle has been recorded in a single shot measurement. This approach greatly reduces the complexity of time-resolved spectroscopy measurements in the mid-infrared which currently require many repetitions.

References

[1] Ritter, E. et al. "Time-Resolved Infrared Spectroscopic Techniques as Applied to Channelrhodopsin." Front. Mol. Biosci. 2:38.

[2] Hugi, A. et al. "Mid-Infrared Frequency Comb Based on a Quantum Cascade Laser." Nature 492: 229–33

[3] Villares, G. et al. "Dual-comb spectroscopy based on quantum-cascade-laser frequency combs." Nat. Commun. 5:5192

TG04 2:53 – 3:08

WIDELY TUNABLE UV/VIS CAVITY-ENHANCED ULTRAFAST SPECTROSCOPY AND EXCITED STATE PROTON TRANSFER IN JET-COOLED MOLECULES AND CLUSTERS

YUNING CHEN, *Department of Chemistry, Stony Brook University, Stony Brook, NY, USA*; MYLES C SILFIES, *Department of Physics, Stony Brook University, Stony Brook, NY, USA*; THOMAS K ALLISON, *Department of Chemistry, Stony Brook University, Stony Brook, NY, USA*.

Ultrafast optical spectroscopy methods, such as transient absorption spectroscopy and 2D spectroscopy, are typically restricted to optically thick samples, such as solids and liquid solutions. We have developed a technique, Cavity-Enhanced Ultrafast Spectroscopy, to study dynamics in a molecular beam with femtosecond temporal resolution. By coupling frequency combs into optical cavities, we previously demonstrated ultrafast transient absorption measurements with a detection limit of $\Delta\text{OD} = 2 \times 10^{-10}(10^{-9}/\sqrt{\text{Hz}})$.[a] In this talk, I will present a widely tunable version of this spectrometer operating at probe wavelengths between 450 and 700 nm (8000 cm^{-1}) using only one set of dispersion managed cavity mirrors. The tunable probe comb is generated using an intracavity doubled optical parametric oscillator. I will discuss the technical details of this spectrometer and its application to the dynamics of excited state intramolecular proton transfer (ESIPT) in jet-cooled molecules and clusters.

---

[a]M. A. R. Reber, Y. Chen, and T. K. Allison, Optica **3**, 311 (2016)

Intermission

TG05 3:44 – 3:59

WIDE-BANDWIDTH COMB-ASSISTED SPECTROSCOPY IN THE FINGERPRINT REGION AND APPLICATION TO THE ν_1 FUNDAMENTAL BAND OF $^{14}N_2^{16}O$

BIDOOR ALSAIF, *Clean combustion research center, King abdullah university for science and technology, Thuwal, Saudi arabia*; MARCO LAMPERTI, DAVIDE GATTI, PAOLO LAPORTA, *Dipartimento di Fisica, Politecnico di Milano, Milano, Italy*; MARTIN FERMANN, *Laser Research, IMRA AMERICA, Inc, Ann Arbor, MI, USA*; AAMIR FAROOQ, *Clean combustion research center, King abdullah university for science and technology, Thuwal, Saudi arabia*; OLEG LYULIN, *Laboratory of Theoretical Spectroscopy, Institute of Atmospheric Optics, Tomsk, Russia*; ALAIN CAMPARGUE, *UMR5588 LIPhy, Université Grenoble Alpes/CNRS, Saint Martin d'Hères, France*; MARCO MARANGONI, *Dipartimento di Fisica, Politecnico di Milano, Milano, Italy*.

Most spectroscopic data available in databases such as HITRAN are retrieved from FTIR measurements and suffer from uncertainties at the MHz level. Much more accurate data, by up to four orders of magnitude, can be achieved using an optical frequency comb to calibrate the frequency axis of a cw laser source. Actually, in the mid-infrared region, at least beyond 5 μm, the only available commercial solution for a widely tunable cw laser is represented by extended cavity quantum cascade lasers (EC-QCLs), whose locking to an optical frequency comb has been so far inhibited by a large amount of frequency noise, leading to linewidths of about 20 MHz [a].

In this work we overcome this limitation and describe a spectrometer that relies on the frequency locking of an EC-QCL tunable in the 7.55-8.2 μm range to a 1.9 μm Tm fiber comb [b]. It is applied to the first comb-calibrated direct characterisation of the ν_1 fundamental band of N_2O, specifically of nearly 70 lines in the 1240 – 1310 cm^{-1} range, from P(40) to R(31). The spectroscopic constants of the upper state are derived from a fit of the line centers with an average rms uncertainty of 4.8×10^{-6} cm^{-1} (144 kHz). The coupling of the spectrometer to a high-finesse optical cavity to the purpose of enhancing its sensitivity and addressing weaker absorbers, is also discussed.

[a] Knabe K., Williams P. A., Giorgetta F., Armacost C. M., Crivello S., Radunsky M., and Newbury N., Opt. Express 20, 12432-12442 (2012)
[b] Lamperti M., Alsaif B., Gatti D., Fermann M., Farooq A., and Marangoni M., Sci. Rep. 8,1292 (2018)

TG06 4:01 – 4:16

OPTICAL FEEDBACK STABILIZED LASER CAVITY RING DOWN SPECTROSCOPY: FROM SATURATED SPECTROSCOPY TO ISOTOPIC RATIO.

SAMIR KASSI, TIM STOLTMANN, *UMR5588 LIPhy, Université Grenoble Alpes/CNRS, Saint Martin d'Hères, France*; MATHIEU DAËRON, MATHIEU CASADO, AMAELLE LANDAIS, *Laboratoire des Sciences du Climat et de l'Environnement, Univ Paris Saclay, CEA, CNRS, UVSQ, Gif sur Yvette, France*; ALAIN CAMPARGUE, *UMR5588 LIPhy, Université Grenoble Alpes/CNRS, Saint Martin d'Hères, France*.

Optical feedback frequency stabilized cavity ring-down spectroscopy is a very high resolution (sub kHz) and sensitive technique (2×10^{-12} cm^{-1} limit of detection). Both aspects will be emphasized through the two followingh applications:

To illustrate the high resolution, Doppler-free saturated-absorption Lamb dips were measured at sub-Pa pressures on rovibrational lines of H_2O near 7180 cm^{-1}. The saturation of the considered lines was so high that at the early stage of the ring down, the cavity loss rate remained unaffected by the absorption. By referencing the laser source to an optical frequency comb, transition frequencies were determined down to 100 Hz precision and kHz accuracy.

To highlight the high precision and stability of the instrument, we will present the first optical absorption measurements of O-17 anomalies in CO_2 with a precision better than 10 ppm, matching the requirements for paleo-environmental applications.

TG07   4:18–4:33
FEED-FORWARD COHERENT LINK FROM A COMB TO A DIODE LASER : APPLICATION TO SATURATED CAVITY RING DOWN SPECTROSCOPY

RICCARDO GOTTI, *Dipartimento di Fisica, Politecnico di Milano, Milano, Italy*; MARCO PREVEDELLI, *Dipartimento di Fisica e Astronomia, Università di Bologna, Bologna, Italy*; SAMIR KASSI, *UMR5588 LIPhy, Université Grenoble Alpes/CNRS, Saint Martin d'Hères, France*; MARCO MARANGONI, *Dipartimento di Fisica, Politecnico di Milano, Milano, Italy*; DANIELE ROMANINI, *UMR5588 LIPhy, Université Grenoble Alpes/CNRS, Saint Martin d'Hères, France.*

We applied a feed-forward frequency control scheme to establish a phase-coherent link from an optical frequency comb to a distributed feedback (DFB) diode laser: This allowed us to exploit the full laser tuning range (up to 1 THz) with the linewidth and frequency accuracy of the comb modes. The approach relies on the combination of an RF single-sideband modulator (SSM) and of an electro-optical SSM, providing a correction bandwidth in excess of 10 MHz and a comb-referenced RF-driven agile tuning over several GHz. As a demonstration, we obtain a 0.3 THz cavity ring-down scan of the low-pressure methane absorption spectrum. The spectral resolution is 100 kHz, limited by the self-referenced comb, starting from a DFB diode linewidth of 3 MHz. To illustrate the spectral resolution, we obtain saturation dips for the $2\nu_3$ R(6) methane multiplet at μbar pressure. Repeated measurements of the Lamb-dip positions provide a statistical uncertainty in the kHz range.

TG08   4:35–4:50
BROADBAND CALIBRATION-FREE COMPLEX REFRACTIVE INDEX SPECTROSCOPY IN A CAVITY USING A COMB-BASED FOURIER TRANSFORM SPECTROMETER

ALEXANDRA C JOHANSSSON, LUCILE RUTKOWSKI, ANNA FILIPSSON, THOMAS HAUSMANINGER, GANG ZHAO, OVE AXNER, ALEKSANDRA FOLTYNOWICZ, *Department of Physics, Umea University, Umea, Sweden.*

Fabry-Perot cavities provide high sensitivity to molecular absorption and dispersion since the mode position, width and amplitude are modified in the vicinity of molecular transitions. Moreover, the mode shift and broadening are directly proportional to the real and imaginary parts of the molecular complex index of refraction, but independent of cavity parameters, such as the cavity length and mirror reflectivity, which reduces the influence of systematic errors. Previous demonstrations of cavity enhanced complex refractive index spectroscopy were based on continuous wave lasers and limited to individual absorption lines[a,b]. Here we use an Er:fiber frequency-comb-based Fourier transform spectrometer with sub-nominal resolution[c,d] to measure a broadband transmission spectrum of a cavity filled with 1% of CO_2 in N_2 at 750 Torr. From Lorentzian fits to each cavity mode we retrieve mode positions and widths, which in turn yield high precision dispersion and absorption spectra of the entire $3\nu_1+\nu_3$ absorption band of CO_2 at 1.6 μm. Fits to these spectra yield line intensities that agree to within 0.6%. Thus comb-based Fourier transform spectroscopy enables broadband cavity mode characterization and calibration-free determination of both the real and imaginary parts of entire molecular absorption bands with high accuracy and precision.

---

[a]Cygan, A., et al., Opt. Express 21, 29744-29754 (2013).
[b]Cygan, A., et al.,Opt. Express 23, 14472-14486 (2015).
[c]Masłowski, P., et al., Phys. Rev. A 93, 021802 (2016).
[d]Rutkowski, L., et al., Opt. Express 25, 21711-21718 (2017).

TG09 4:52 – 5:07

FOURIER-TRANSFORM COMPLEX REFRACTIVE INDEX SPECTROSCOPY AT Hz-LEVEL WITH OPTICAL FREQUENCY COMBS

DOMINIK CHARCZUN, GRZEGORZ KOWZAN, AKIKO NISHIYAMA, AGATA CYGAN, DANIEL LISAK, RYSZARD S. TRAWIŃSKI, PIOTR MASLOWSKI, *Institute of Physics, Faculty of Physics, Astronomy and Informatics, Nicolaus Copernicus University, Torun, Poland*; MICHAEL DEBUS, PHILIPP HUKE, *Institut für Astrophysik, Georg-August-Universität, Göttingen, Germany*; DOROTA TOMASZEWSKA, GRZEGORZ SOBOŃ, *Faculty of Electronics, Wrocław University of Science and Technology, Wrocław, Poland*.

Two relatively new spectroscopic techniques suitable for implementation in a broadband FTS are cavity mode-width spectroscopy and cavity mode dispersion spectroscopy [1,2]. They require scanning the cavity resonances to obtain information about their width and position, yielding information about molecular absorption and dispersion. Previously they used continuous wave lasers, showing signal-to-noise ratio and resolution similar to the well-established cavity ring-down spectroscopy. However in this implementation they shared the same limits of measurement range and relatively slow acquisition. Meanwhile the optical frequency comb-based cavity-enhanced FTS with sub-nominal resolution [3,4] is a perfect match for those methods, allowing for simultaneous measurement of thousands of cavity modes without the limit of cavity dispersion and the requirement of a reference measurement [5]. Here we present the measurements of 10 kHz HWHM cavity resonances, which are some of the narrowest features ever measured by the FTS, from which we derive the absorption and dispersion spectra of the 0-3 band of CO in Ar.

1. A. Cygan *et al.*, Opt. Express 21-24, 29744-29754 (2013).
2. A. Cygan *et al.*, J. Chem. Phys. 144, 214202-11 (2016).
3. P. Maslowski *et al.*, Phys. Rev. A 93, 021802(R) (2016).
4. L. Rutkowski *et al.*, Journal of Quantitative Spectroscopy and Radiative Transfer 204, 63-73 (2018).
5. L. Rutkowski *et al.*, Opt. Express 25-18, 21711-21718 (2017).

TG10 5:09 – 5:24

BROADBAND CAVITY-ENHANCED MOLECULAR ABSORPTION AND DISPERSION SPECTROSCOPY WITH A FREQUENCY COMB-BASED VIPA SPECTROMETER

GRZEGORZ KOWZAN, DOMINIK CHARCZUN, AGATA CYGAN, RYSZARD S. TRAWIŃSKI, DANIEL LISAK, PIOTR MASLOWSKI, *Institute of Physics, Faculty of Physics, Astronomy and Informatics, Nicolaus Copernicus University, Torun, Poland*.

Cavity mode-width spectroscopy (CMWS) [1] and cavity mode-dispersion spectroscopy (CMDS) [2] techniques provide a way to simultaneously determine absorption and dispersion of a sample in an optical cavity. It was shown recently that CMDS can also be efficiently combined with optical frequency combs (OFCs) [3] to perform dispersion measurements in a broad spectral range. In this work, we utilize a near-infrared frequency comb and a VIPA spectrometer to retrieve absorption and dispersion of an atmospheric pressure CO-N_2 sample in a high-finesse cavity, by measuring shapes and positions of 7-kHz-wide cavity modes at Hz-level precision. A Pound-Drever-Hall lock of a CW laser to the cavity and a phase-lock of the OFC to the CW laser allow for arbitrary detuning between comb and cavity modes, while cavity length stabilization to a Rb frequency standard provides absolute frequency scale. The signal-to-noise ratios for CMWS and CMDS spectra were, respectively, 190 and 380.

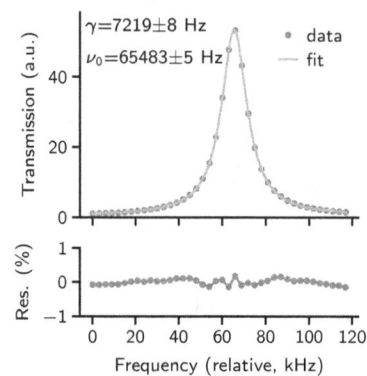

To the best of our knowledge, the 7-kHz-wide cavity resonances shown in this work are the narrowest spectral features measured directly with a frequency comb. The presented technique is capable of fast acquisition and ultrahigh resolution in a broad spectral range, which makes it particularly suitable for spectroscopy of cold molecules or monitoring of chemical kinetics.

[1] A. Cygan *et al*, Opt. Express 21, 29744 (2013).
[2] A. Cygan *et al*, Opt. Express 23, 14472 (2015).
[3] L. Rutkowski *et al*, Opt. Express 25, 21711–21718 (2017).

TH. Mini-symposium: New Ways of Understanding Molecular Spectra

Tuesday, June 19, 2018 – 1:45 PM

Room: 100 Noyes Laboratory

Chair: Per Jensen, University of Wuppertal, Wuppertal, Germany

TH01 1:45 – 2:00
AUTOMATED ASSIGNMENT OF ROTATIONAL SPECTRA USING ARTIFICIAL NEURAL NETWORKS

DANIEL P. ZALESKI, KIRILL PROZUMENT, *Chemical Sciences and Engineering Division, Argonne National Laboratory, Argonne, IL, USA.*

Last year at this conference several approaches to utilize machine learning[a] to train a computer to recognize the patterns inherit in rotational spectra were presented[b]. It was shown that the recognized patterns could be used to identify (or classify) a rotational spectrum by its Hamiltonian type, but at the time, the rotational constants were not recovered. Here, we describe a feed forward artificial neural network that has been trained to identify different types of rotational spectra and determine the parameters of the molecular Hamiltonians. The network requires no user interaction beyond loading a "peak pick", and can return fits within a fraction of a second. The rotational constants are typically deduced with the accuracy of 1–10 MHz. We will describe how the network works and provide benchmarking results.

[a] Bishop, C M. "Neural networks for pattern recognition." Oxford university press, 1995.
[b] Zaleski, D. P.; Prozument, K. Identifying Broadband Rotational Spectra with Neural Networks, International Symposium on Molecular Spectroscopy, June 21; 2017.

TH02 2:02 – 2:17
NEW APPROACHES TO DECODING ROTATIONAL SPECTRA: APPLICATIONS TO FLUOROETHYLENE MICRO-SOLVATION BY CO_2

REBECCA A. PEEBLES, PRASHANSA KANNANGARA, SEAN A. PEEBLES, *Department of Chemistry, Eastern Illinois University, Charleston, IL, USA*; BROOKS PATE, *Department of Chemistry, The University of Virginia, Charlottesville, VA, USA.*

Chirped-pulse Fourier-transform microwave (CP-FTMW) spectrometers such as the instrument at the University of Virginia can acquire spectra with high sensitivity in a short amount of time. This necessitates new approaches to spectroscopic analysis to ensure identification of all species in each recorded spectrum. With an intensity range covering four orders of magnitude after averaging 1 million free induction decays, a recent spectrum of a fluoroethylene (FE)/CO_2 mixture in the 2 – 8 GHz range has over 11,000 lines with signal-to-noise ratio above ∼2.5. These transitions may arise from a combination of monomer, dimer and larger cluster species, including low abundance isotopes and complexes with carrier gas, water or other contaminants.

Our current focus is identifying spectra of FE(CO_2)$_n$ clusters, containing progressively larger numbers of CO_2 molecules. Several methods have been used to facilitate identification of lines for these spectra, which are expected to lose 1-2 orders of magnitude of intensity for each increase in cluster size. These approaches include subtraction of transitions that are observed in the FE-only spectrum from the spectrum of the FE/CO_2 mixture, visual identification of patterns characteristic of asymmetric molecules, and application of extended cross correlation (XCC) techniques.[a] In the XCC approach, several spectra with systematically varied conditions (such as pressure or concentration) are compared, and a combination of graphs and computerized algorithms is used to identify transitions that behave similarly under the changing conditions. In addition to applications for fundamental spectroscopic studies, this approach has potential application to identification of the components of complex mixtures.

[a] N.P. Jacobson, S.L. Coy, R.W. Field, *J. Chem. Phys.*, **107** (1997) 8349.

TH03   2:19 – 2:34

ATTEMPTS TO SOLVE O_2-CONTAINING VAN DER WAALS INTERACTIONS USING SPFIT AND SPCAT WITH MICROWAVE MEASUREMENT PRECISION: PROBLEMS, PITFALLS, AND SUCCESSES.

FRANK E MARSHALL, NICOLE MOON, AMANDA JO DUERDEN, G. S. GRUBBS II, *Department of Chemistry, Missouri University of Science and Technology, Rolla, MO, USA.*

Although there is a vast amount of van der Waals complexes containing small molecular species, there have only been a small number of studies containing an O_2 binding partner. Within this grouping, only five are known to have been attempted using high-resolution microwave techniques with only two- O_2-HF[a] and O_2-H_2O[b]- appearing in the literature. This void is presumably due to the $^3\Sigma$ ground state of O_2 and how this coupling complicates spectral assignment. However, these sorts of couplings determined from high-resolution analysis add rich and important information useful in complex structure determination. Until now, the only analyses of such complexes have been done utilizing Hund's case *a* asymmetric models[c,d] or Hund's case *b* linear Hamiltonians. However, further study either resulted in inaccurate predictions of close transitions (O_2-H_2O) or difficult adjustments to similar systems due to extra complexity (O_2-HCl from O_2-HF). This made a more standardized approach, such as using the ubiquitous program suite of SPFIT/SPCAT by Pickett[e] seem like a more flexible solution. However, SPFIT/SPCAT uses a Hund's case *b* approach which needed to be sorted out and the problems and pitfalls of these analyses will be discussed. This talk will include how to be predictive using O_2-H_2O and O_2-HF as examples as well as discuss work currently being pursued on the molecules O_2-HCl and O_2-OCS to extend the analyses. Structural parameters from these newly determined parameters and their interpretations will also be discussed.

[a] S. Wu, G. Sedo, E. M. Grumstrup, K. R. Leopold, *J. Chem. Phys.*, **127** (2011) 204315-1-204315-11.
[b] Y. Kasai, E. Dupuy, R. Saito, K. Hashimoto, A. Sabu, S. Kondo, Y. Sumiyoshi, Y. Endo. *Atmos. Chem. Phys.*, **11** (2011) 8607–8612.
[c] W. M. Fawzy, *J. Mol. Spec.*, **191** (1998) 68-80.
[d] H.-B. Qian, S. J. Low, D. Seccombe, B. Howard, *J. Chem. Phys.*, **107** (1997) 7651-7657.
[e] H. M. Pickett, *J. Mol. Spec.*, **148** (1991) 371-377.

TH04   2:36 – 2:51

STEPPING ACROSS THE DISSOCIATION THRESHOLD OF THE $I^-\cdot(H_2O)$ COMPLEX: RESONANCE ENHANCED TWO-COLOR IR-IR PHOTODISSOCIATION (R2PD)

NAN YANG, CHINH H. DUONG, PATRICK J KELLEHER, *Department of Chemistry, Yale University, New Haven, CT, USA*; JUSTIN J TALBOT, *Henry Eyring Center for Theoretical Chemistry, University of Utah, Salt Lake City, Utah, USA*; RYAN P STEELE, *Department of Chemistry, University of Utah, Salt Lake City, UT, USA*; MARK JOHNSON, *Department of Chemistry, Yale University, New Haven, CT, USA.*

The $X^-\cdot(H_2O)$ (X=I, Cl, F) clusters ion provide a microscopic window into the intracluster energy relaxation dynamics that ultimately lead to dissociation when excited above threshold (Do). Here we explore the spectra of simple binary complexes of one water molecule with an iodide ion when the system is excited to vibrational levels that span the energy range through Do. This is accomplished by recording the vibrational photodissociation spectra of different vibrational excited states of $X^-\cdot(H_2O)$ using a two-color, IR-IR photodissociation technique. We first quantify the ground state spectra by recording the single photon absorption spectra of cryogenically cooled cluster ions with messenger tag technique. Then we fix the pump laser frequency on a transition known to the ground state cluster ion and scan the probe laser to obtain the excited state photodissociation spectra. Owing to the long lifetime of the vibration excitations in this class of clusters, each pump laser excitation frequency yields a different spectrum (the traces on the right in the figure). This provides an opportunity to obtain a kind of 2DIR spectroscopy (left bottom figure) of cryogenically cooled gas phase ions. The result is a cluster variation of the vibrationally-mediated photodissociation experiments pioneered in the 1990s by Crim and Rizzo on polyatomic molecules. In the cluster regime, we find remarkably long lived (greater than 50 μs!) vibrational levels 300 cm^{-1} above the dissociation threshold and surprisingly localized excitations for the bound OH stretches. This feature allows us to follow many pathways up the vibrational landscape far beyond the dissociation limit.

TH05

DECIPHERING THE EXCITED-STATE VIBRATIONAL SIGNATURES OF THE WATER-IODIDE BINARY COMPLEX THROUGH QUANTUM SIMULATIONS.

<u>JUSTIN J TALBOT</u>, RYAN P STEELE, *Department of Chemistry, University of Utah, Salt Lake City, UT, USA*; NAN YANG, CHINH H. DUONG, PATRICK J KELLEHER, MARK JOHNSON, *Department of Chemistry, Yale University, New Haven, CT, USA*.

The water-iodide monomer (I^-H_2O), nominally the simplest of the halide-water complexes, challenges our current understanding of ion hydration. Most notably, this seemingly simple complex displays multiple resonant vibrational transitions, a low tunneling barrier, and a strong transition dipole moment along the bound OH-I coordinate. These effects combine to yield spectroscopic signatures that deviate strongly from traditional harmonic analyses and are even difficult to qualitatively reproduce with anharmonic methodologies. Among these signatures is a quartet of peaks in the 3300 – 3500 cm-1 range that is unexplained using traditional single-photon spectroscopy. Challenging both experiment and theory alike, this situation required the interpretation of newly developed IR-IR 2-color photodissociation spectroscopy that probes well below the dissociation threshold. In this work, we use both exact eigensolver techniques and a newly developed vibrationally adiabatic model, along with a new potential energy surface[a] to computationally explore the excited-state spectra. The resulting analyses identify the source of the strength of the resonant transitions, directly assign the vibrational and rotational spectroscopic signatures, and discern the electronic origin of these surprising effects in this fundamental model of ion hydration.

[a]Bajaj, P. Gotz, A.W., Paesani, F. Journal of Chemical Theory and Computation, 2016, 12 (6), 2698-2705.

TH06

FROM LINES TO STATES WITHOUT A MODEL

<u>STEFAN BRACKERTZ</u>, STEPHAN SCHLEMMER, OSKAR ASVANY, *I. Physikalisches Institut, University of Cologne, Cologne, Germany*.

The fundamental Ritz combination principle [a] originally found for atoms has also been applied to molecules as a method to reconstruct the energy states from measured lines without relying on any model Hamiltonian. In 2006 Nesbitt and coworkers [b] proposed to apply it to protonated methane, CH_5^+. We used this idea to reconstruct a part of its ground state energies employing spectra of combination differences (CDs) determined from very high resolution ro-vibrational data [c]. Since then the method has been significantly improved [d] as the CD lines essentially represent kernel density estimations, a well-known tool in mathematics. Furthermore, a combinatorial approach has been developed to reconstruct vibrational ground states as well as vibrationally excited states from the CD spectra without relying on measurements at different temperatures. As a result, 1063 of the 2897 measured lines of CH_5^+ being part of four different symmetry species could be assigned. This allowed for a comparison of the measurements with the analytical model of Schmiedt et al. [e] as well as with the *ab initio* calculations of Wang and Carrington [f].

[a]W. Ritz, On a new law of series spectra, Astrophys. J. 28 (1908) 237.

[b]C. Savage, F. Dong, D.J. Nesbitt, Toward a quantum-mechanical understanding of the high-resolution infrared spectrum of CH_5^+, in: Contribution TA05, 61st International Symposium on Molecular Spectroscopy, Columbus, OH, USA, 2006.

[c]Oskar Asvany, Koichi M. T. Yamada, Sandra Brünken, Alexey Potapov, Stephan Schlemmer. Experimental ground-state combination differences of CH_5^+. Science, 347(6228):1346–1349, 2015.

[d]S. Brackertz, S. Schlemmer, O. Asvany, Searching for new symmetry species of CH_5^+ – From lines to states without a model, J. Mol. Spectrosc. 342 (2017) 73–82.

[e]H. Schmiedt, P. Jensen, S. Schlemmer, Rotation-vibration motion of extremely flexible molecules – the molecular superrotor, Chem. Phys. Lett. 672 (2017) 34–46.

[f]X.-G. Wang, T. Carrington, Calculated rotation-bending energy levels of CH_5^+ and a comparison with experiment, J. Chem. Phys. 144 (2016) 204304.

Intermission

TH07   *Journal of Molecular Spectroscopy Review Lecture*   3:58 – 4:31

THE ANALYSIS OF COMPLEX CHEMICAL MIXTURES BY BROADBAND ROTATIONAL SPECTROSCOPY

BROOKS PATE, *Department of Chemistry, The University of Virginia, Charlottesville, VA, USA*; JUSTIN L. NEILL, *BrightSpec Labs, BrightSpec, Inc., Charlottesville, VA, USA*.

Broadband rotational spectroscopy has several experimental advantages as a technique for the analysis of complex chemical mixtures without the need for chromatography to separate the distinct chemical species prior to analysis. The technique has high spectral resolution so that mixtures with a large number of components can be analyzed without spectral overlap, the frequency accuracy is excellent so that library spectra can be transferred between instruments, and the measurement has high dynamic range so that low level impurities can be detected in the presence of dominant species like the solvent in a direct-from-flask reaction mixture. It also has the special feature of a spectroscopic signature that is dependent on the molecular mass distribution so that isomers can be resolved – a problem that can be a challenge for the high-sensitivity analytical chemistry methods based on mass spectrometry. The problem of decomposing a measured spectrum into the individual rotational spectra of each different sample molecule is common to many applications of broadband rotational spectroscopy including reaction product screening in laboratory astrochemistry, the identification of different molecular clusters in the study of weakly bound complexes, and the analysis of chemical samples from a variety of chemistry fields including pharmaceutical science. In this talk we will discuss strategies that have been developed by the spectroscopy community to solve the problem of decomposing a measurement into its constituent spectra. These approaches include traditional analytical chemistry approaches like the creation of large chemical libraries. Instrumental methods that exploit broadband detection to implement efficient double-resonance methods will also be summarized. Two approaches that may deserve additional consideration in the future will also be discussed. One approach measures the spectrum as a function of a continuously variable external parameter and then uses computer algorithms to group transitions with similar parametric dependence as a way to separate the measurement into molecule-specific spectra. The second approach uses the fact that the Hamiltonian for rotational spectroscopy is known and that only certain patterns of transitions are consistent with it. An early example of this idea is the AUTOFIT routine. The possibility of extending this approach into a fully automated computer analysis of the spectrum will be considered.

TH08   4:35 – 4:50

INFRARED PHOTODISSOCIATION SPECTROSCOPY OF THE EXOTIC H_6^+ CATION IN THE GAS PHASE

DAVID C McDONALD, *Chemistry, University of Georgia, Athens, GA, USA*; J. PHILIPP WAGNER, MICHAEL A DUNCAN, *Department of Chemistry, University of Georgia, Athens, GA, USA*.

H_6^+ is generated in a supersonic expansion via pulsed electrical discharge of hydrogen. H_n^+ clusters are extracted into a reflectron time-of-flight mass spectrometer and probed with infrared photodissociation spectroscopy (IRPD) in the 2050 – 4600 cm^{-1} region. H_6^+ was mass selected and found to have three distinct photodissociation channels by loss of one hydrogen atom, one hydrogen molecule or both. Each channel results in different spectra as a result of mode specific dissociation channels. The ground $^2D_{2d}$ state is 4 kcal/mol lower in energy than the ^2c_S state with a 7 kcal/mol barrier. We believe we are probing the $^2D_{2d}$ structure with the three H_m^+ (m=3,4,5) fragment channels as a result of rapid interconversion between the two states after IR photon absorption.

TH09 4:52 – 5:07

SURFACE AND SPECTROSCOPIC PROPERTIES OF 1,8-DIAZAFLUOREN-9-ONE IN TITANIUM DIOXIDE THIN FILMS

ANETA LEWKOWICZ, ANNA SYNAK, MICHAŁ MOŃKA, PIOTR BOJARSKI, KAROL SZCZODROWSKI, *Faculty of Mathematics, Physics and Informatics/Institute of Experimental Physics, University of Gdańsk, Gdańsk, Poland*; ROBERT BOGDANOWICZ, *Faculty of ETI, Gdańsk University of Technology, Gdańsk, Poland, Poland*; JAKUB KARCZEWSKI, *Faculty of Applied Physics, Gdańsk University of Technology, Gdańsk, Poland, Poland*.

Thin films of 1,8-diazafluoren-9-one (DFO) in titanium dioxide were synthesized using the sol-gel method. Particular attention will be given towards preparation of DFO in titanium dioxide as a potential luminescent probe of amino acids. The photophysical properties of DFO in titanium dioxide thin films were identified by a variety of spectroscopic methods including: stationary absorption and emission, the fluorescence intensity decay profiles. Atomic force microscopy (AFM), scanning electron microscopy (SEM), confocal microscopy, Raman spectroscopy, and X-ray diffraction (XRD) techniques were applied to obtain structural characteristic of the prepared films.

This research has been supported by the grant NCN 2017/01/X/ST5/01541, MINIATURA 1, A. Lewkowicz.

TH10 5:09 – 5:24

DETERMINATION OF GLYCOL CONTAMINATION IN ENGINE OIL BY INFRARED AND UV-VIS SPECTROSCOPY

TORREY E. HOLLAND, ROBINSON KARUNANITHY, *Physics, Southern Illinois University Carbondale, Carbondale, IL, USA*; ALI MAZIN ABDUL-MUNAIM, *Plant, Soil, and Agricultural Systems, Southern Illinois University Carbondale, Carbondale, IL, USA*; P SIVAKUMAR, *Physics, Southern Illinois University Carbondale, Carbondale, IL, USA*; DENNIS G. WATSON, *Plant, Soil, and Agricultural Systems, Southern Illinois University Carbondale, Carbondale, IL, USA*.

We investigated the ethylene glycol, which is the crucial ingredient in the automotive antifreeze coolants, the content of engine oil at various levels of contamination using Fourier transform infrared (FT-IR) spectroscopy and ultraviolet-visual spectroscopy (UV-Vis). It is known that glycol in SAE 15W-40 diesel engine lubricating oil has relatively strong signatures in the infrared spectrum, some of which overlap with other molecular bonds that may already be present in engine oil. Therefore, our aim is to correlate this FT-IR data with a UV-Vis spectrograph such that detection of glycol's presence can be improved significantly.

TI. Instrument/Technique Demonstration

Tuesday, June 19, 2018 – 1:45 PM
Room: 1024 Chemistry Annex

Chair: Elangannan Arunan, Indian Institute of Science, Bangalore, India

TI01   1:45 – 2:00

SUB-NANOMETER IMAGING OF ELECTRONICALLY EXCITED QUANTUM DOTS: STARK EFFECT, ORIENTATION DEPENDENCE AND ENERGY TRANSFER

DUC NGUYEN, *Chemistry, University of Illinois at Urbana-Champaign, URBANA-CHAMPAIGN, IL, USA*; MARTIN GRUEBELE, *Department of Chemistry, University of Illinois at Urbana-Champaign, Urbana, IL, USA*; JOSEPH LYDING, *Department of Electrical and Computer Engineering, University of Illinois at Urbana-Champaign, Urbana-Champaign, IL, USA.*

Single-molecule adsorption scanning tunneling microscopy (SMA-STM) is a powerful spectroscopy method capable of imaging absorption at sub-nanometer spatial resolution. Herein, we use SMA-STM to investigate electronically excited quantum dots (QDs). Absorption images of individual QDs vary significantly from dot-to-dot, resulting from heterogeneity and defects. Single QD absorption is strongly dependent on the applied electric field, reflecting different excited states being probed. Details on the three-dimensional geometry of the QD excited states are obtained by using the STM tip to nudge and roll the QDs on the surfaces, then image at different angles. Orientation-dependent imaging, in combination with density functional theory calculations of a model QD, reveals presence of surface localized defects. Finally, the energy transfer in arrays of QDs is imaged and manipulated in real space at individual dot level. This study establishes SMA-STM as a powerful method to study electronically excited nanostructures and energy transfer at sub-nm spatial resolution.

TI02   2:02 – 2:17

LASER ABLATION-RESONANCE ENHANCED PHOTOIONIZATION MASS SPECTROMETRY (LA-REPMS) OF PARTICLE-BASED ASSAYS TO IMPROVE EARLY DETECTION OF CANCER

CHRISTOPHER MANDRELL, JESSICA C JURAK, P SIVAKUMAR, *Physics, Southern Illinois University Carbondale, Carbondale, IL, USA.*

Early detection of cancer has a drastic impact on the successful treatment of the disease. However, detection of early signs of cancer is a challenge especially for a type of cancer such as epithelial ovarian cancer (EOC), with few or no symptoms at the early-stages. Development of a noninvasive method that can improve the detection of biomarkers with sufficient selectivity, sensitivity, and reproducibility is a promising approach to overcome the challenges of early detection. This study aims to develop novel optical and mass spectrographic techniques to detect biomolecules in complex matrices. To accomplish this, Laser Ablation-Resonance Enhanced Photoionization Mass Spectrometry is combined with nano- and micro-particle immunoassay to improve the detectability in a complex media. While there are many commercial mass spectrometry configurations available, none of them meet our specific needs, so a significant portion of the effort in this research to date has been dedicated to designing and building the custom apparatus to meet our needs. We present an overview of the design, testing, and preliminary studies on biomolecules.

TI03   2:19 – 2:34

LED-CAVITY ENHANCED ABSORPTION SPECTROSCOPY FOR SENSING THE ATMOSPHERE

HONGMING YI, *Physical Measurement Lab, National Institute of Standards and Technology, Gaithersburg, MD, USA*; TAO WU, *Physics, Nanchang Hangkong University, Nanchang, China*; EIRC FERTEIN, CÉCILE COEUR, *Laboratoire de Physico-Chimie de l'Atmosphère, Université du Littoral Côte d'Opale, Dunkerque, France*; WEIXIONG ZHAO, GUISHI WANG, XIAOMING GAO, WEIJUN ZHANG, *LAPC, AIOFM, Hefei, China*; WEIDONG CHEN, *Laboratoire de Physico-Chimie de l'Atmosphère, Université du Littoral Côte d'Opale, Dunkerque, France.*

We present our recent progress in the development and applications of cavity enhanced absorption spectroscopy technique based on light emitting diode (LED) for optical monitoring of chemically reactive atmospheric species (HONO, NO3, NO2) in intensive campaigns and in smog chamber studies.

TI04

COMPARISON OF CAVITY ENHANCED FARADAY ROTATION SPECTROSCOPY TECHNIQUES

<u>LINK PATRICK</u>, *Electrical Engineering, Princeton Unviersity, Princeton, NJ, USA*; JONAS WESTBERG, GERARD WYSOCKI, *Department of Electrical Engineering, Princeton University, Princeton, NJ, USA.*

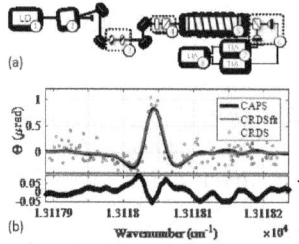

Fig. 1 (a) CAPSFRS setup. (b) FRS measurements of the $^PP_1(1)$ transition of oxygen measured by CAPSFRS (via LIA) and CRDFRS (via ring-down fitting).

Cavity enhanced absorption techniques derive their sensitivity from an increase in the effective light-matter interaction length provided by a high finesse cavity. However, absorption measurements are often affected by spectrally interfering molecular species, which can hinder selectivity. This issue is addressed by Faraday rotation spectroscopy (FRS), which selectively probes the molecular dispersion of paramagnetic gaseous species (e.g. O_2, NO, NO_2, OH, etc.) subjected to an external magnetic field. Immunity to interfering diamagnetic compounds is thereby obtained allowing reliable quantitative concentration assessments of paramagnetic species in the presence of spectrally interfering molecules, such as H_2O and CO_2. Recently, white-noise limited performance over extended averaging times (minutes/hours) was achieved by combining cavity ring-down (CRD) and FRS. While CRD-FRS provides excellent sensitivities down to noise-equivalent rotation angles of 1.3×10^{-9} rad rtHz^{-1}, it requires fast detectors, high bandwidth digitization electronics and high throughput data analysis which significantly increases the system complexity and cost. To address these limitations, cavity attenuated phase shift (CAPS) FRS and integrated cavity output spectroscopy (ICOS) FRS has been developed. Here, CAPS-FRS ICOS-FRS, and CRD-FRS systems are compared by detecting oxygen at the $^PP_1(1)$ transition in the A-electronic band around 762.3 nm. The FRS-based techniques are fully self-referencing and require no additional off-resonance calibration, which provides a powerful, yet simple alternative for cavity-enhanced spectroscopy targeting paramagnetic species. The FRS techniques allow for continuous measurements in a line-locked mode, which further increases the system effective duty-cycle and improves the sensing performance. A comparison of long-term system performance, system modeling, as well as system improvements will be presented in detail.

Acknowledgments: The authors acknowledge funding from the National Science Foundation CBET grant #1507358 and from Thorlabs Inc. LP acknowledges the National Science Foundation Graduate Fellowship support.

TI05

APPLICATION OF COHERENT ANTI-STOKES RAMAN SCATTERING THERMOMETRY IN TURBULENT AND LAMINAR FLAMES

<u>AMAN SATIJA</u>, ZIQIAO CHANG, *Mechanical Engineering, Purdue University, West Lafayette, IN, USA*; DONG HAN, *FM Global, FM Global, Boston, USA*; ALBYN LOWE, *School of Aerospace, Mechanical and Mechatronic Engineering, University of Sydney, Sydney, Australia*; LEVI MICHAEL THOMAS, JAY P GORE, *Mechanical Engineering, Purdue University, West Lafayette, IN, USA*; ASSAAD R MASRI, *School of Aerospace, Mechanical and Mechatronic Engineering, University of Sydney, Sydney, Australia*; ROBERT P. LUCHT, *Mechanical Engineering, Purdue University, West Lafayette, IN, USA*.

Coherent anti-Stokes Raman scattering (CARS) is a non-linear spectroscopic combustion diagnostic technique used for measurement of temperature and species concentration. Broadband CARS spectra can be acquired with a single laser shot with high spatial and temporal resolution. We present two distinct applications of a nanosecond dual-pump vibrational CARS system. The first experiment aimed to study the effect of simulated exhaust-gas-recirculation, via addition of CO_2 to the fuel stream, on the flame structure of lean CH_4-air pilot-assisted turbulent premixed flames. For this experiment over 20,000 single-shot spectra were acquired and spectrally fitted to develop a detailed temperature map of the flame flow-field. In the second experiment, laminar flames with varying soot loading were stabilized over a "Yale burner". This burner, in the combustion community, is a canonical system for the development of soot models. The principal challenge in this experiment was obtaining a CARS signal with an adequate signal to noise ratio in the presence of strong soot-emission background. Our measurements in both experiments will serve as benchmark data for the development of combustion computational models.

TI06 3:10 – 3:25

LASER-BASED MOLECULAR SPECTROSCOPY FOR MONITORING EMISSION IN ANIMAL FARMING

MICHAL NIKODEM, DOROTA STACHOWIAK, *Laser Sensing Laboratory, Wroclaw Research Centre EIT+, Wroclaw, Poland.*

Monitoring gas emission becomes an important issue in the livestock sector. For example, industrial animal farming is responsible for substantial part of total anthropogenic emission of methane and ammonia. Here we present practical aspects of molecular spectroscopy by demonstrating a laser-based system for sensing of methane (near 1651 nm), ammonia (near 1531 nm) and hydrogen sulfide (near 1575 nm) using wavelength modulation spectroscopy (WMS) and a multi-pass cell. This instrument is designed for sequential detection of three species emitted in pig farming facility. Laser-based molecular spectroscopy in the near-infrared region provides unique opportunity for maintenance-free continuous operation at relatively low cost, and with sensitivity and accuracy at single ppmv levels for all three gases. System characterization in laboratory conditions will be presented. We will also demonstrated results of field tests and discuss technical challenges when moving spectroscopic systems from laboratory conditions to real-world environments.

Authors acknowledge support by the National Centre for Research and Development (NCBiR) award LIDER/023/379/L-5/13/NCBR/2014.

TI07 3:27 – 3:42

HIGH-RESOLUTION LINEAR SPECTROSCOPY ON A MICROMETRIC LAYER OF MOLECULAR VAPOR

JUNIOR LUKUSA MUDIAYI, BENOIT DARQUIE, ISABELLE MAURIN, SEAN TOKUNAGA, ALEXANDER SHELKOVNIKOV, *Laboratoire de Physique des Lasers, CNRS, Université Paris 13, Sorbonne Paris Cité, Villetaneuse, France*; JOSE ROBERTO RIOS LEITE, *Departamento de Física UFPE, Universidade Federal de Pernambuco, Recife, Brazil*; DANIEL BLOCH, ATHANASIOS LALIOTIS, *Laboratoire de Physique des Lasers, CNRS, Université Paris 13, Sorbonne Paris Cité, Villetaneuse, France.*

Molecular rovibrational transitions can potentially provide high-resolution frequency references for the visible to the mid infrared including the telecommunications window at 1.5μm (C_2H_2 or HCN). Going towards compact miniaturized molecular spectroscopy is an extreme challenge as molecular transition probabilities are weak and long propagation length is required. Compact systems based on hollow core fibers filled with acetylene gas (C_2H_2) at high pressures have been presented[a] but the propagation length remains macroscopic. Additionally, collisional broadening (high pressures) or the Doppler effect hinder the available resolution.

Here we present selective reflection (SR) spectroscopy measurements on a molecular gas of NH_3 and SF_6 molecules at 10.6μm. Frequency modulated SR is a high-resolution (sub-Doppler) technique, linear in laser power that is essentially only sensitive to a layer of molecules whose depth is defined by the wavelength of optical excitation ($\sim \lambda/2\pi$), even when the cell remains macroscopic. Initial measurements, were performed with a CO_2 laser, subsequently changed for a more user friendly QCL laser rendered compatible with high-resolution spectroscopy. The core of our experiments is performed on the P(1) line of ammonia at 948.23cm^{-1} with best resolution limited by the laser linewidth (~ 0.5MHz). For pressures below 50mTorr (collisional broadening ~ 1.4MHz) the hyperfine structure of ammonia can be clearly resolved.

Our experiments pave the way towards miniaturized molecular frequency references such as nanocells[b,c], and allow us to envisage the first precision measurements of the Casimir-Polder molecule-surface interaction.

---

[a] F. Benabid, *Compact, stable and efficient all-fibre gas cells using hollow-core photonic crystal fibres*, Nature **434**, 488 (2005).
[b] G. Dutier et al., *Collapse and revival of a Dicke-type coherent narrowing in a sub-micron thick vapor cell*, Europhys. Lett. **63**, 35 (2003)
[c] J.-M. Hartmann et al., *Infrared look at the spectral effects of submicron confinements of CO_2 gas*, Phys. Rev. A **93**, 012516 (2016).

Intermission

TI08 4:18–4:33

DEVELOPMENT OF A HYBRID LASER-MASS SPECTROMETER WITH TWO INSTRUMENT ARMS: IRMPD AND HENDI COLD ION SPECTROSCOPIC EXPERIMENTS

MATTHIAS HEGER, JOSEPH CHERAMY, FAN XIE, ZHIHAO CHEN, HAOLU WANG, WOLFGANG JÄGER, YUNJIE XU, *Department of Chemistry, University of Alberta, Edmonton, AB, Canada.*

In this talk, we present a new hybrid laser-mass spectrometer with two instrument arms: one for infrared multiphoton dissociation (IRMPD) and the other for helium nanodroplet isolation (HENDI) cold ion spectroscopic experiments. The spectrometer features an electrospray source for ion production, using either traditional atmospheric nanospray or a novel Subambient Pressure Ionization with Nanoelectrospray (SPIN) design.[1] Ionic species of interest are isolated in a quadrupole mass filter and stored in a Paul trap for OPO IR laser irradiation. The ion trap is coupled to a TOF tube for fast wide-range mass spectrometry at high resolution. The second arm contains in addition an ion deflector and a linear quadruple ion trap which is currently being integrated into an existing HENDI spectrometer. This will enable higher resolution investigations of mass-selected ionic species in an ultra-cold environment.

IRMPD spectroscopy has emerged as a powerful asset in the spectroscopist's toolbox to investigate the isomerization and aggregation behavior of small molecules.[2] For example, this technique has been coupled with mass spectrometry to elucidate the structures of some amino acid complexes of interest, most notably perhaps the "magic number" serine octamer.[3] To demonstrate the current capabilities of the instrument, we extended previous studies on the gas-phase structures of protonated and sodiated asparagine monomers[4] into the IR regime between 3300 and 4000 cm−1. In addition, a theoretical re-assessment of the structural and spectroscopic properties of these species was undertaken with modern quantum-chemical approaches. We present this study as a stepping stone for further research into the structural and energetic preferences of small non-covalently bound amino acid aggregates.

[1] J. S. Page, K. Tang, R. T. Kelly, R. D. Smith, Anal. Chem. 2008, 80, 1800; I. Marginean, et al. Anal. Chem. 2012, 84, 9208. [2] N. C. Polfer, Chem. Soc. Rev., 2011, 40, 2211. [3] F. X. Sunahori, G. Yang, E. N. Kitova, J. S. Klassen, Y. Xu, Phys. Chem. Chem. Phys. 2013, 15, 1873. [4] A. L. Heaton, V. N. Bowman, J. Oomens, J. D. Steill, P. B. Armentrout, J. Phys. Chem. A 2009, 113, 5519.

TI09 4:35–4:50

AC STARK EFFECT OBSERVED IN A MICROWAVE-(SUB)MILLIMETERWAVE DOUBLE RESONANCE EXPERIMENT

KEVIN ROENITZ, BRIAN M HAYS[a], CARSON REED POWERS, MORGAN N McCABE, HOUSTON H SMITH, SUSANNA L. WIDICUS WEAVER, *Department of Chemistry, Emory University, Atlanta, GA, USA;* STEVEN SHIPMAN, *Department of Chemistry, New College of Florida, Sarasota, FL, USA.*

A microwave-(sub)millimeter wave double resonance experiment was developed in order to improve the ease with which one is able to assign spectral lines by displaying the interconnectivity of states. This experiment combines chirped-pulse microwave spectroscopy and fast-sweep (sub)millimeter spectroscopy to increase the speed of data acquisition. During the experiment, the splitting of spectral lines was detected instead of the traditional movement of population seen in other double resonance experiments. The line splitting was determined to be caused by the AC Stark effect. The experimental design along with the resulting double resonance methanol spectra will be presented.

[a]Institut de Physique de Rennes Département Physique Moléculaire UMR 6251 du CNRS - Université de Rennes 1 Bat. 11c, Campus de Beaulieu 263 Avenue du Général Leclerc 35042 RENNES CEDEX, FRANCE tel.: +33 2 23 23 64 83

TI10   4:52 – 5:07

A USB - TO - W-BAND TRANSMITTER: MILLIMETER-WAVE MOLECULAR SPECTROSCOPY WITH CMOS TECHNOLOGY

<u>DEACON J NEMCHICK</u>, BRIAN DROUIN, ADRIAN TANG, YANGHYO KIM, *Jet Propulsion Laboratory, California Institute of Technology, Pasadena, CA, USA*; GABRIEL VIRBILA, M.-C. FRANK CHANG, *Electrical Engineering, University of California - Los Angeles, Los Angeles, CA, USA.*

The distinct rotational signatures of gas-phase molecular species in the millimeter (mm) and sub-millimeter (sub-mm) spectral regions have long assisted remote sensing communities in the interrogation of atmospheric and astrophysical media. *In situ* studies employing highly-mobile instrumentation have not been able to reproduce the success of their remote-based counterparts largely due to the unaccommodating size and power requirements of traditional mm and sub-mm wave hardware. The Laboratory Studies and Atmospheric Observations group at JPL has embraced the marriage of novel custom-designed CMOS source and heterodyne detection electronics, which often leverage advances in the mobile phone industry, and traditional cavity enhanced laboratory techniques to combat the issues that have plagued the deployment of in situ mm wave sensors.

One device emerging from these efforts is a freestanding CMOS-based transmitter tunable to sub-500 Hz resolution over the operational bandwidth of 90 – 105 GHz. For prototyping purposes this transmitter, the output of which can be both frequency and amplitude modulated, has been deployed as the radiation source in a high-resolution sub-Doppler (Lamb-dip) absorption spectrometer. The presented experimental findings have shown that this device, which effectively functions as a USB powered/controlled W-band source, has sufficient output power ($\sim$2 mW peak) to perform spectral-hole burning saturation experiments and a phase-noise floor low enough to determine spectral line positions with a precision of 1 part in 10^9 and accuracy within the error of measurements made with traditional millimeter-wave sources.[a] These findings highlight the promise of exploiting CMOS architectures for use in gas specific, low-power, and potentially low-cost *in situ* sensors.

---
[a]D. J. Nemchick *et al.*, "Sub-Doppler Spectroscopy With a CMOS Transmitter," *IEEE Trans. THz Sci. Technol.*, vol. 8, no. 1, pp. 121-126, 2018.

TI11   5:09 – 5:24

THE CONFORMER SPECIFIC ROOM-TEMPERATURE ROTATIONAL SPECTRUM OF ALLYL CHLORIDE UTILIZING STRONG FIELD COHERENCE BREAKING

<u>ERIKA RIFFE</u>, ERIKA JOHNSON, STEVEN SHIPMAN, *Department of Chemistry, New College of Florida, Sarasota, FL, USA*; SEAN FRITZ, ALICIA O. HERNANDEZ-CASTILLO, TIMOTHY S. ZWIER, *Department of Chemistry, Purdue University, West Lafayette, IN, USA.*

The 8-26.5 GHz conformer specific rotational spectrum of allyl chloride was recorded in a room temperature spectrometer using the strong field coherence breaking (SFCB) technique.[a] Allyl chloride, which has *cis*- and *skew*- conformers as well as ^{35}Cl and ^{37}Cl isotopologues, was chosen as the initial molecule for testing this method in the room temperature chirped pulse waveguide setup.[b] This data was compared to results from the SFCB technique performed at Purdue University using a jet expansion at 1-2K. The application of this and other methods for the simplification of room temperature spectra will be discussed.

---
[a]Hernandez-Castillo, A.O., Abeysekera, C., Hays, B.M., Zwier, T.S. "Broadband Multi-Resonant Strong Field Coherence Breaking as a Tool for Single Isomer Microwave Spectroscopy." J. Chem. Phys. 145, 114203 (2016).
[b]Reinhold, B., Finneran, I.A., Shipman, S.T. "Room temperature chirped-pulse Fourier transform microwave spectroscopy of anisole." J. Mol. Spec. 270, 89 (2011).

A 6–18 GHZ DIRECT DIGITAL SYNTHESIS TUNABLE SEGMENTED CHIRPED PULSE FOURIER TRANSFORM MICROWAVE SPECTROMETER

HALEY N. SCOLATI, SOMMER L. JOHANSEN, ANNA L PISCHER, KYLE N. CRABTREE, *Department of Chemistry, The University of California, Davis, CA, USA.*

Chirped pulse Fourier transform spectroscopy (CP-FTMW) has become a widely used technique for the detection of molecular rotational spectra owing to its broad frequency coverage. Traditional CP-FTMW set ups involve top-quality broadband arbitrary waveform generators (AWG), high-power amplifiers, and digitizers, which are expensive due to their specifications. One method to lower costs with only a mild sacrifice of efficiency is to divide the total bandwidth into smaller sections and step from section to section with a tunable local oscillator; these so-called "segmented" CP-FTMW spectrometers have much lower costs by decreasing the required amplifier power and digitizer bandwidth. Inspired by the work of Finneran et al. (Rev. Sci. Inst. 84, 2013, 083104), our group has designed a 6–18 GHz segmented CP-FTMW broadband spectrometer that also replaces the AWG with a direct digital synthesizer (DDS), further lowering the spectrometer cost. To our knowledge, this is the first instrument in which a DDS has been coupled with the segmented approach to achieve a tunable intermediate frequency (IF). Design, cost analysis, progress, and performance will be discussed in this talk.

TJ. Rotational structure/frequencies
Tuesday, June 19, 2018 – 1:45 PM
Room: 217 Noyes Laboratory

Chair: Ranil Gurusinghe, West-Ward Pharmaceuticals, Bedford, OH, USA

TJ01 1:45 – 2:00

HIGH SENSITIVITY CRDS OF CO_2 IN THE 1.74 μM TRANSPARENCY WINDOW. A VALIDATION TEST FOR THE SPECTROSCOPIC DATABASES

PETER ČERMÁK, *Department of Experimental Physics, Comenius University, Bratislava, Slovakia*; EKATERINA KARLOVETS, *Laboratory of Quantum Molecular Mechanics and Radiation Processes, Tomsk State University, Tomsk, Russia*; DIDIER MONDELAIN, SAMIR KASSI, *UMR5588 LIPhy, Université Grenoble Alpes/CNRS, Saint Martin d'Hères, France*; VALERY PEREVALOV, *Molecular Spectroscopy, V.E. Zuev Institute of Atmospheric Optics, Tomsk, Russia*; ALAIN CAMPARGUE, *UMR5588 LIPhy, Université Grenoble Alpes/CNRS, Saint Martin d'Hères, France*.

The very weak absorption spectrum of natural CO_2 near 1.74 μm (5702 - 5879 cm^{-1}) is studied at high sensitivity. The investigated region corresponds to a transparency window of very weak opacity which is of particular interest for Venus. Very weak lines with intensity value as low as 10^{-30} cm/molecule at 296 K are detected by Cavity Ring Down Spectroscopy. On the basis of the predictions of effective Hamiltonian models, 1135 lines of six carbon dioxide isotopologues - $^{12}C^{16}O_2$, $^{13}C^{16}O_2$, $^{16}O^{12}C^{18}O$, $^{16}O^{12}C^{17}O$, $^{16}O^{13}C^{18}O$ and $^{16}O^{13}C^{17}O$ - were rovibrationnally assigned to 26 bands. The accurate spectroscopic parameters of 16 bands are determined from standard band-by-band analysis (typical rms deviations of the line positions are 8×10^{-4} cm^{-1}). These newly observed bands include perturbed bands, weak hot bands and bands of minor isotopologues (in particular $^{16}O^{12}C^{18}O$ in natural abundance) and provide critical validation tests for the most recent spectroscopic databases. The comparison to the Carbon Dioxide Spectroscopic Databank (CDSD), HITRAN2016 database and recent ab initio line lists will be presented. Deficiencies are evidenced for some weak perpendicular bands of the HITRAN2016 list and identified as due to inaccurate CDSD intensities which were preferred to *ab initio* intensities. New results based on ^{18}O enriched CO_2 spectra will also be detailed.

TJ02 2:02 – 2:17

LINE POSITIONS AND INTENSITIES FOR THE ν_3 BAND OF 5 ISOTOPOLOGUES OF GERMANE FOR PLANETARY APPLICATIONS

VINCENT BOUDON, TIGRAN GRIGORYAN, FLORIAN PHILIPOT, CYRIL RICHARD, *Laboratoire ICB, CNRS/Université de Bourgogne, DIJON, France*; F. KWABIA TCHANA, *LISA, CNRS, Universités Paris Est Créteil et Paris Diderot, Créteil, France*; LAURENT MANCERON, *Synchrotron SOLEIL, CNRS-MONARIS UMR 8233 and Beamline AILES, Saint Aubin, France*; ATHENA RIZOPOULOS, JEAN VANDER AUWERA, *Service de Chimie Quantique et Photophysique, Université Libre de Bruxelles, Brussels, Belgium*; THÉRÈSE ENCRENAZ, *LESIA, Observatoire de Paris / CNRS / UPMC, Meudon, France*.

Germane (GeH_4) is present in the atmospheres of the giant planets Jupiter and Saturn. The ongoing NASA mission Juno has renewed interest in its spectroscopy. The accurate modeling of which is essential for the retrieval of other tropospheric species. We present here the first complete analysis and modeling of line positions and intensities in the strongly absorbing ν_1/ν_3 stretching dyad region near 2100 cm^{-1}, for all five germane isotopologues in natural abundance[a]. New infrared spectra were recorded, absolute intensities were extracted through a careful procedure and modeled thanks to the formalism and programs developed in Dijon. A database of calculated germane lines, GeCaSDa, is available online through the *Virtual Atomic and Molecular Data Centre* (VAMDC) portal (http://portal.vamdc.org) and at http://vamdc.icb.cnrs.fr/PHP/gecasda.php. GeH_4 will integrate the HITRAN database as molecule number 50.

---

[a]V. Boudon, T. Grigoryan, F. Philipot, C. Richard, F. Kwabia Tchana, L. Manceron, A. Rizopoulos, J. Vander Auwera and T. Encrenaz, *J. Quant. Spectrosc. Radiat. Transfer* **205**, 174–183 (2018)

TJ03 2:19 – 2:34

INVESTIGATION OF THIOKETENE ISOMERS: MICROWAVE SPECTROSCOPY AND FORMATION CHEMISTRY OF HCCSH

<u>KELVIN LEE</u>, *Radio and Geoastronomy Division, Harvard-Smithsonian Center for Astrophysics, Cambridge, MA, USA*; MARIE-ALINE MARTIN-DRUMEL, *CNRS, Institut des Sciences Moleculaires d'Orsay, Orsay, France*; VALERIO LATTANZI, *The Center for Astrochemical Studies, Max-Planck-Institut für extraterrestrische Physik, Garching, Germany*; BRETT A. McGUIRE, *NAASC, National Radio Astronomy Observatory, Charlottesville, VA, USA*; MICHAEL C McCARTHY, *Atomic and Molecular Physics, Harvard-Smithsonian Center for Astrophysics, Cambridge, MA, USA*.

Because of their curious electronic properties, carbon-sulfide chains possess peculiar molecular structures and chemical reactivity. In this talk, we present the spectroscopic characterization of ethynethiol (HCCSH) - an isomer of thioketene (H_2CCS) and thiirene (c-H_2C_2S) - and its isotopologues using Fourier-transform microwave cavity and millimeter-wave absorption spectroscopies. HCCSH is produced in an electrical discharge of HCCH and H_2S. In our cavity spectrometer (10 - 40 GHz), the $J = 1 - 0, 2 - 1, 3 - 2$ a-type transitions have been observed. $HCC^{34}SH$ was detected in natural abundance while the deuterated isotologues required HCCD and D_2S for DCCSH and HCCSD respectively. Based on *ab initio* predictions, the fundamental b-type transitions were also measured using double resonance (150 - 300 GHz), allowing the A constant to be determined approximately and guiding subsequent surveys into the millimeter-wave (280 - 655 GHz). Based on the information derived from the isotopologues, we will also discuss the molecular structure and inferred formation chemistry, in addition to its relative abundance to thioketene.

TJ04 2:36 – 2:51

THE MILLIMETER/SUBMILLIMETER-WAVE SPECTRUM OF F_2SO ($\tilde{X}^1A'$)

JOHN P KEOGH, *Department of Chemistry and Biochemistry, University of Arizona, Tucson, AZ, USA*; <u>DeWAYNE T HALFEN</u>, *Department of Chemistry and Biochemistry, Department of Astronomy, The University of Arizona, Tucson, AZ, USA*; LUCY M. ZIURYS, *Department of Chemistry and Biochemistry; Department of Astronomy, Arizona Radio Observatory, University of Arizona, Tuscon, AZ, USA*.

The millimeter/submillimeter spectrum of F_2SO ($\tilde{X}^1A'$) has been measured using direct absorption techniques in the frequency range 271 – 508 GHz. Thionyl fluoride was created in the process of searching for a number of metal-containing fluoride molecules. This species was serendipitously produced from SF_6 as the fluorine source with residual water in the presence of a DC discharge. Multiple rotational transitions in the range $J = 16$ to $J = 30$ were recorded, each consisting of a c-type asymmetric top pattern, due to the large dipole moment along the $\hat{c}$ molecular axis $\mu_c = 1.62$ D. The data were analyzed using an asymmetric top Hamiltonian and rotational and centrifugal distortion constants were established. This work considerably expands the spectroscopic characterization of F_2SO. Previous microwave data consisted of measurements below 77 GHz.

TJ05 2:53 – 3:08

FLUORINATION EFFECT ON HYDROGEN BOND TOPOLOGIES IN WATER ADDUCTS OF FLUOROPYRIDINES

<u>JUAN WANG</u>, XIAOLONG LI, GANG FENG, QIAN GOU, *School of Chemistry and Chemical Engineering, Chongqing University, Chongqing, China*.

The rotational spectra of a series of 1:1 adducts of water-fluoropyridines have been investigated by using pulsed jet Fourier transform microwave spectroscopy. Depending on the fluorination sites, the hydrogen bond topologies are quite different from each other. The water links with the N n-obital of 2,3-difluoropyridine and 2,6-difluoropyridine, respectively. through an O-H...N bond. In addition, one weak C-H...O bond might be contributable to the stabilization of the complex 2,3-difluoropyridine-water, while an O-H...F bond plausibly formed in the complex 2,6-difluoropyridine-water in which water acts as a double proton donor. Additional fluorination of the third site of 2,6-difluoropyridine creates a new active site ready to interact with water: the isomer stabilized by one C-H...O and one O-H...F hydrogen bond is proved by the experimental evidences to be the global minimum. This hydrogen topology is quite different from the all former rotational studied complexes involving pyridine and fluoropyridines.

Intermission

TJ06 3:44 – 3:59

TIPPING THE BALANCE BETWEEN ELECTROSTATICS AND STERIC EFFECTS: THE MICROWAVE SPECTRA AND MOLECULAR STRUCTURES OF 2-CHLORO-1,1-DIFLUOROETHYLENE–ACETYLENE AND CIS-1,2-DIFLUOROETHYLENE–ACETYLENE

HELEN O. LEUNG, MARK D. MARSHALL, *Chemistry Department, Amherst College, Amherst, MA, USA*.

We have found that the observed average structures of haloethylene-protic acid heterodimers result from an interplay between favorable electrostatic interactions and steric effects. For vinyl fluoride and 1,1-difluoroethylene complexes, steric effects predominate and the acid binds across the double bond ("top"), while for trifluoroethylene, favorable electrostatics forces the complexes to adopt a sterically strained structure with the acid at one end of the olefin ("side"). A relaxation of steric requirements for binding with a chlorine atom leads to different geometries being observed for each of the vinyl chloride complexes with hydrogen fluoride, hydrogen chloride, and acetylene. The side binding motif to chlorine persists in (Z)-1-chloro-2-fluoroethylene–acetylene despite the presence of the more electronegative fluorine atom. For 2-chloro-1,1-difluoroethylene ethylene–acetylene, the acetylene is presented with the option of top binding to fluorine versus side binding to chlorine, whereas with *cis*-1,2-difluoroethylene, the only option is side binding to fluorine. The structures of these two complexes are compared to reveal the balance between electrostatics and sterics.

TJ07 4:01 – 4:16

MICROWAVE SPECTRUM AND MOLECULAR STRUCTURE OF 2,3,3,3-TETRAFLUOROPROPENE–HYDROGEN CHLORIDE

HELEN O. LEUNG, MARK D. MARSHALL, MILES A. WRONKOVICH, *Chemistry Department, Amherst College, Amherst, MA, USA*.

Our systematic study of the structures of heterodimers of haloethylenes with protic acids has provided a wealth of information, along with a few surprises, regarding the sometimes cooperative and sometimes competing effects of electrostatic, steric, and dispersion forces that contribute to the binding of these species. We seek to apply this knowledge to larger systems with a wider variety of possible interactions and binding sites via the addition of a trifluoromethyl moiety to the olefin to form halopropenes. The microwave spectrum of the complex formed between 2,3,3,3-tetrafluoropropene, which can be considered a trifluoromethyl analogue of vinyl fluoride, and hydrogen chloride is obtained and analyzed to determine the molecular structure of this species.

TJ08 4:18 – 4:33

THE CONFORMATIONS OF PROTEINOGENIC AMINO ACID GLUTAMINE: MORE ACCURACY IS URGENTLY NEEDED IN THEORETICAL CALCULATIONS

IKER LEÓN, ELENA R. ALONSO, CARLOS CABEZAS, SANTIAGO MATA, JOSÉ L. ALONSO, *Grupo de Espectroscopia Molecular, Lab. de Espectroscopia y Bioespectroscopia, Unidad Asociada CSIC, Universidad de Valladolid, Valladolid, Spain*.

Glutamine is α-amino acid that is used in the biosynthesis of proteins. The large flexibility of the molecule has several potential conformational candidates. Among them, the three most stable isomers have been characterized both using a laser ablation chirped pulse Fourier-transform microwave spectrometer (LA-CP-FTMW) and a laser ablation molecular-beam Fourier-transform microwave spectrometer (LA-MB-FTMW).[a,b] Two of the conformers can be determined using the spectroscopic constants provided by the theoretical methods. On the other hand, the third conformer is not reproduced by conventional theoretical methods and one should use some tricks to characterize the conformer's structure. In addition, even the slightest different prediction in the site position of the nitrogen atoms makes the conformer's spectroscopic characterization very challenging because the nuclear quadrupole coupling interactions depend critically on the electronic environment, position and orientation of the ^{14}N nuclei. In this work we show several important conclusions: first, we present the three conformers of glutamine detected in gas phase; second, we highlight that more accurate theoretical methods are needed and that glutamine can be used as a test to benchmark the calculations; finally, we propose one trick to help the scientific community when the calculations are not sufficient to predict the structure and how to deal with the complicated hyperfine structure with no starting grounds.

---

[a] S. Mata, I. Peña, C. Cabezas, J. C. López and J. L. Alonso, J. Mol. Spectrosc., 2012, 280, 91–96.
[b] J. L. Alonso and J. C. López, in Gas-Phase IR Spectroscopy and Structure of Biological Molecules, eds. A. M. Rijs and J. Oomens, Topics in., 2015, vol. 364, pp. 335–402.

TJ09
SULFUR HYDROGEN BONDING: A COMPARISON OF THE DIMERS AND MONOHYDRATES OF THENYL AND FURFURYL ALCOHOLS AND MERCAPTANS

MARCOS JUANES, RIZALINA TAMA SARAGI, ALBERTO LESARRI, *Departamento de Química Física y Química Inorgánica, Universidad de Valladolid, Valladolid, Spain*; RUTH PINACHO, JOSÉ EMILIANO RUBIO, LOURDES ENRIQUEZ, MARTIN JARAIZ, *Departamento de Electrónica, ETSIT, University of Valladolid, Valladolid, SPAIN*.

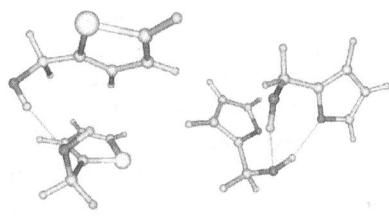

The dimers of thenyl alcohol (TA$_2$) and furfuryl alcohol (FA$_2$), and the monohydrates of both alcohols (TA$\cdots$H$_2$O, FA$\cdots$H$_2$O) and their corresponding mercaptans (TM$\cdots$H$_2$O, FM$\cdots$H$_2$O) were generated in a supersonic jet expansion and probed using both chirped-pulse and cavity Fourier transform microwave spectroscopy. The experimental results, supported by ab initio molecular orbital calculations, allow comparing the conformational preferences and the role of the sulfur and oxygen atoms in the O-H$\cdots$O/O-H$\cdots$S/S-H$\cdots$O hydrogen bonds (HBs) stabilizing the dimers. In the furfuryl monohydrates (FA$\cdots$H$_2$O, FM$\cdots$H$_2$O) water behaves as a proton donor to the ring oxygen, in competence with a second HB to the alcohol or thiol side chain[a]. Different behavior was observed when the ring oxygen is replaced by sulfur atom in the thenyl monohydrates (TA$\cdots$H$_2$O and TM$\cdots$H$_2$O) as the water molecule is binding to the side chain and the π electronic cloud of the ring. The water motion in both thenyl dimers is detected by tunneling splittings of the rotational transitions, denoting a weaker binding compared to the furfuryl compounds. The alcohol dimers (TA)$_2$ and (FA)$_2$ again reveal the different HB strengths in the furfuryl and thenyl groups. The alcohol dimer forms a network of two consecutive O-H$\cdots$O hydrogen bonds involving the oxygen ring, while in the thenyl dimer the alcohol group prefers binding to the π cloud than to the sulfur atom in the ring. Spectroscopic, structural and computational data will be reported.

[a]M. Juanes, A. Lesarri, R. Pinacho, E. Charro, J. E. Rubio, L. Enriquez, M. Jaraiz, *Chem. Eur. J.*, **2018**, 24, 1–9, in press

TJ10 — *Post-Deadline Abstract*
STRUCTURE DETERMINATION OF 5 MEMBERED SILANE RINGS USING MICROWAVE SPECTROSCOPY

FRANK E MARSHALL, AMANDA JO DUERDEN, NICOLE MOON, DAVID JOSEPH GILLCRIST, *Department of Chemistry, Missouri University of Science and Technology, Rolla, MO, USA*; IVAN SEDLACEK, *Chemistry, Missouri University of Science and Technology, Rolla, Missouri, United States*; GRIER JONES, THEODORE CARRIGAN-BRODA, GAMIL A GUIRGIS, *Department of Chemistry and Biochemistry, College of Charleston, Charleston, SC, USA*; G. S. GRUBBS II, *Department of Chemistry, Missouri University of Science and Technology, Rolla, MO, USA*.

Rotational spectra of 1,1-difluorosilacyclopent-3-ene, silacyclopent-3-ene, 1-(chloromethyl)-1-fluoro-silacyclopentane, 1,1-difluorosilacyclopentane, and 1,1-difluorosilacylopent-2-ene were observed in the 6 to 18 GHz range of the electromagnetic spectrum. The molecular structure for the parent and various isotopically substituted species were obtained from their respective spectra. The differences in structure between these similar molecules will be presented, showing how different functional groups and bond locations affect the overall structure and behavior of each system (ring puckering effects, etc.). Comparisons to similar known cyclopentane and silacyclopentane species will be presented.

TK. Large amplitude motions, internal rotation
Tuesday, June 19, 2018 – 1:45 PM
Room: B102 Chemical and Life Sciences

Chair: Jens-Uwe Grabow, Gottfried-Wilhelm-Leibniz-Universität, Hannover, NI, Germany

TK01　1:45 – 2:00
THE THERMAL SELF-POLYMERIZATION OF METHYL METHACRYLATE — ROTATIONAL CHARACTERIZATION OF THE METHYL METHACRYLATE DIMER (IT'S NOT A COMPLEX!)

SVEN HERBERS, DANIEL A. OBENCHAIN, KEVIN G. LENGSFELD, HENNING KUPER, JENS-UWE GRABOW, JÖRG AUGUST BECKER, *Institut für Physikalische Chemie und Elektrochemie, Gottfried-Wilhelm-Leibniz-Universität, Hannover, Germany.*

Structural data from microwave spectra of monomers and oligomers of the methyl methacrylate system compared to theoretical predictions allows for accurate predictions of the structure and physical properties of higher oligomers or even polymers. The idea behind this project is to start with small building blocks and to successively increase the size of the oligomers in order to obtain more and more accurate predictions. Following the previous analysis of the monomer, this contribution focuses on the dimer of methyl methacrylate.

In the dimer phase of methyl methacrylate, which was subjected to a thermal self-polymerization process, the linear methyl methacrylate dimer was identified by means of rotational spectroscopy. The analysis was performed using the coaxially oriented beam-resonator arrangement Fourier-transform microwave (COBRA-FTMW) spectrometer. The dimer comprises three methyl rotors. Coupling of the methyl internal rotation to the overall rotation causes a complicated, challenging splitting of the rotational spectrum. The fact that only the two methoxymethyl groups contributed resolvable (>5kHz) splittings simplified the spectrum a little and a fit of molecular parameters to the experimental data was achieved with experimental accuracy utilizing the program XIAM.

TK02　2:02 – 2:17
TOWARDS THE DETECTION OF EXPLOSIVE TAGGANTS: MICROWAVE AND MILLIMETER-WAVE GAS PHASE SPECTROSCOPIES OF 3-NITROTOLUENE

ANTHONY ROUCOU, *Laboratoire de Physico-Chimie de l'Atmosphère, Université du Littoral Côte d'Opale, Dunkerque, France*; ISABELLE KLEINER, *Laboratoire Interuniversitaire des Systèmes Atmosphériques (LISA), CNRS et Universités Paris Est et Paris Diderot, Créteil, France*; MANUEL GOUBET, SABATH BTEICH, *Laboratoire PhLAM, UMR 8523 CNRS - Université Lille 1, Villeneuve d'Ascq, France*; GAËL MOURET, ROBIN BOCQUET, FRANCIS HINDLE, *Laboratoire de Physico-Chimie de l'Atmosphère, Université du Littoral Côte d'Opale, Dunkerque, France*; W. LEO MEERTS, *Institute for Molecules and Materials (IMM), Radboud University Nijmegen, Nijmegen, Netherlands*; ARNAUD CUISSET, *Laboratoire de Physico-Chimie de l'Atmosphère, Université du Littoral Côte d'Opale, Dunkerque, France.*

The monitoring of gas phase mononitrotoluenes is crucial for defence, civil security and environmental interests since they are used as taggant for TNT detection and in the manufacturing of industrial compounds such as dyestuffs. We have succeeded to measure and analyse at high resolution the room temperature rotationally resolved millimeter-wave spectrum of 3-nitrotoluene (3-NT). Experimental and theoretical difficulties have been overcome, in particular, those related to the low vapour pressure of 3-NT and to the internal rotation of a CH_3 in almost free rotation regime (V_3=6.7659(24) cm^{-1}). Rotational spectra have been recorded in the microwave and millimeter-wave ranges using a supersonic jet Fourier Transform microwave spectrometer ($T_{rot} < 10$ K) [a] and a millimeter-wave frequency multiplication chain ($T = 293$ K) [b], respectively. Spectral analysis of pure rotation lines in the vibrational ground state and in the first torsional excited state supported by quantum chemistry calculations permits to characterise the rotational energy of the molecule, the hyperfine structure due to the ^{14}N nucleus and the internal rotation of the methyl group.[c]

---

[a] M. Tudorie, et al., J. Chem. Phys. **134**, (2011), 074314
[b] G. Mouret, et al., IEEE Sens. J. **13**, (2013), 133-138.
[c] A. Roucou et al., ChemPhysChem (2018)

TK03   2:19 – 2:34

INTERNAL ROTATION ANALYSIS OF THE FTMW AND MILLIMETER WAVE SPECTRA OF FLUORAL (CF_3CHO)

CELINA BERMÚDEZ[a], R. A. MOTIYENKO, L. MARGULÈS, *UMR 8523 CNRS - Université de Lille, Laboratoire PhLAM, Villeneuve d'Ascq, France*; CARLOS CABEZAS, YASUKI ENDO, *Department of Applied Chemistry, National Chiao Tung University, Hsinchu, Taiwan*; J.-C. GUILLEMIN, *ISCR - UMR6226, Univ. Rennes. Ecole Nationale Supérieure de Chimie de Rennes, Rennes, France.*

To protect atmosphere, hydrofluorocarbons (HFCs) are the current substituents of the dangerous CFCs and other ODS (ozone depleting substances).[b] Although HFCs are not ODS, they are potent greenhouse gases and, thus, they would be harmful to climate. Consequently, there is a keen interest on monitoring their reaction and the decomposition products in order to measure their effects. Fluoral (trifluoroacetaldehyde, CF_3CHO) is one of the stable decomposition products of several families of ODS substituents. Monitoring it in the atmosphere is hampered by the few spectroscopic data available in the literature. The rotational spectrum of fluoral from 8-40 GHz was measured previously,[c] however the performed analysis of the spectrum was rather limited due to difficulties in theoretical description. These difficulties reside mainly in the hindered internal rotation of the CF_3 group. Compared to acetaldehyde, in fluoral, the CF_3 group represents the major part of the molecular mass. Therefore, there is a strong coupling between the overall molecular rotation and the internal rotation of the top $\rho = 0.92$. As such, previously used principal axis method, where the axes remain unaffected by the large amplitude motion, is not fully suitable for the analysis. We present the analysis of new high resolution microwave and millimeter wave spectra of fluoral in the ranges 6-26 and 50-305 GHz, respectively, employing rho-axis method implemented in RAM36 program. The rotational distortional and internal rotational parameters that reproduce the spectral at experimental accuracy were determined for the ground state and several lowest excited torsional states.

---

[a] This work was supported by the CaPPA project funded by the French National Research Agency (ANR-10-LABX-005) through the PIA
[b] Burkholder J.B. *et al.*, Chem.Rev., 2015, **115**, 3704
[c] Woods R.C., 1967, J. Chem. Phys., **46**, 4789

TK04   2:36 – 2:51

GLOBAL FIT OF O-FLUOROTOLUENE TORSIONAL STATES FROM WAVEGUIDE CP-FTMW SPECTROSCOPY

J. H. WESTERFIELD, STEVEN SHIPMAN, *Department of Chemistry, New College of Florida, Sarasota, FL, USA.*

The microwave spectrum of o-fluorotuene has been investigated at -12 °C from 8.7-26.5 GHz with waveguide chirped-pulse Fourier transform microwave spectroscopy (CP-FTMW). This molecule has a measured V_3 barrier of 238.3 cm^{-1}. The low barrier height resulted in some challenges when fitting the excited states in XIAM. This work improves on our previous fit by extending into the 18-26.5 GHz frequency range and by switching to use the RAM36 fitting software instead. Based on newly collected data and our previous assignments of excited torsional states, a global fit of the ground and first two excited states has been conducted in RAM36. Additionally, scans were taken at 3 °C, 19 °C, and 35 °C to increase the population of the excited torsional modes. Details of the fit including improvements from previous work will be discussed in the talk.

TK05 2:53 – 3:08

SPECTROSCOPIC CHARACTERIZATION OF THE ELUSIVE *GAUCHE*-ISOPRENE BY HIGH RESOLUTION MICROWAVE SPECTROSCOPY

JESSIE P PORTERFIELD, *Atomic and Molecular Physics, Harvard Smithsonian Center for Astrophysics, Cambridge, MA, USA*; J. H. WESTERFIELD, *Department of Chemistry, New College of Florida, Sarasota, FL, USA*; BRYAN CHANGALA, *JILA, National Institute of Standards and Technology and Univ. of Colorado Department of Physics, University of Colorado, Boulder, CO, USA*; THANH LAM NGUYEN, *Department of Chemistry, University of Florida, Gainesville, FL, USA*; JOSHUA H BARABAN, *Chemistry, Ben-Gurion University of the Negev, Beer-Sheva, Israel*; STEVEN SHIPMAN, *Department of Chemistry, New College of Florida, Sarasota, FL, USA*; MICHAEL C McCARTHY, *Atomic and Molecular Physics, Harvard-Smithsonian Center for Astrophysics, Cambridge, MA, USA*.

The microwave spectrum of isoprene has been investigated in a cryogenic (4-7 K) He buffer gas cell (12-26.5 GHz) and in a room temperature waveguide (8.7-26.5 GHz). Rotation about the single C-C bond converts the lower-energy *trans* rotamer to the higher-energy *gauche*-conformer *via* an unstable, planar, *cis* transition state. As in butadiene, it is counterintuitive that *gauche* is more stable than *cis* because it has long been believed that planarity is required for π electron delocalization, a factor that often imparts greater stability. In standard jet experiments, observation of the higher-energy *gauche* conformer has proven challenging. However in the buffer gas cell, collisions with cold He result in rapid but gentle conformational cooling, allowing for straightforward observation of the *gauche*-isoprene rotamer. The rotational spectrum of *gauche* is complex owing to the combined effects of rotational line splitting (0^+ / 0^- from *gauche-gauche* inversion) and A/E splitting of the methyl rotor. On the basis of new theoretical calculations, steric hindrance of methyl rotation is predicted to be lower in the *gauche* than in the *trans* conformer (V_3 barrier approximately 653 cm^{-1} or 7.8 kJ/mol). In addition to the on-going spectroscopic analysis, efforts are now underway to better characterize the energetics of C-C bond rotation in isoprene, with the goal of ultimately understanding the factors that result in the greater stability of the *gauche* relative to the *cis* isoprene rotamer.

TK06 3:10 – 3:25

HYPERFINE SPLITTINGS OF METHANOL IN THE FIRST EXCITED TORSIONAL STATE

<u>LI-HONG XU</u>, *Department of Physics, University of New Brunswick, Saint John, NB, Canada*; JON T. HOUGEN, *Sensor Science Division, National Institute of Standards and Technology, Gaithersburg, MD, USA*; G YU GOLUBIATNIKOV, SERGEY BELOV, ALEXANDER LAPINOV, *Microwave Spectroscopy, Institute of Applied Physics, Nizhny Novgorod, Russia*; E. A. ALEKSEEV, *Quantum Radiophysics Department, Kharkiv National University and Institute of Radioastronomy of NASU, Kharkov, Ukraine*; IGOR KRAPIVIN, *Radiospectrometry Department, Institute of Radio Astronomy of NASU, Kharkov, Ukraine*; L. MARGULÈS, R. A. MOTIYENKO, *Laboratoire PhLAM, UMR 8523 CNRS - Université Lille 1, Villeneuve d'Ascq, France*; STEPHANE BAILLEUX, *Laboratoire PhLAM, Université de Lille - Sciences et Technologies, Villeneuve d'Ascq, France*.

Hyperfine splittings in the ground state of CH_3OH have recently been studied by several groups [JCP(2015)143_044304, (2016)145_024307, (2016)145_244301]. In our work [JCP(2016)145_024307], we treated splittings in the Lamb-dip sub-mm-wave transitions between some torsion-rotation states of E symmetry. These doublets increase nearly linearly with J, and we attributed them to the effect of torsionally mediated spin-rotation interaction of the methyl protons. Hyperfine doublets of this type have so far been observed only in methanol. The focus of this talk is on hyperfine doublet, "triplet" and quartet splittings observed in the first excited E torsional state of CH_3OH from three laboratories. Four series of lines dominate the available data. Measurements are: (i) from Kharkov/Lille, K = 6 ← 7, Q branch, quartets, with $7 \leq J \leq 15$; (ii) from NNOV, $K = 3 \leftarrow 2$, Q branch, with $3 \leq J \leq 18$, where the series starts as quartets, changes to doublets at $J = 7$, and then finally to singlets at $J = 17$; (iii) from NNOV, K = -2 ← -3, P branch, doublets, with $8 \leq J \leq 12$; (iv) from Kharkov/Lille, K = 8 ← 7, Q branch, with $8 \leq J \leq 24$, where the series starts as triplets and becomes doublets at $J = 15$. We have ignored the central features of the triplets, since we believe they might be due to unusual double-N crossover resonances; and (v) a few measurements that don't form branches.

We have empirically modeled these hyperfine splittings with spin-torsion, spin-rotational and spin-spin terms for the two $I = \frac{1}{2}$ spin systems arising from the OH and CH_3 protons, respectively. Work is in progress to understand the deeper physical meaning of these fitting parameters and compare them with ab initio calculations.

TK07 3:27 – 3:42

FURTHER PROGRESS IN FITTING 13000 TORSION-WAGGING-ROTATIONAL MW AND IR $v_t = 0,1$ TRANSITIONS IN CH_3NH_2 USING THE HYBRID (TUNNELLING + INTERNAL ROTATION) PROGRAM

ISABELLE KLEINER, *Laboratoire Interuniversitaire des Systèmes Atmosphériques (LISA), CNRS et Universités Paris Est et Paris Diderot, Créteil, France*; JON T. HOUGEN, *Sensor Science Division, National Institute of Standards and Technology, Gaithersburg, MD, USA.*

A few years ago, the authors wrote a hybrid program to fit rotational levels in molecules with one CH_3 internal-rotation large-amplitude motion, one NH_2 inversion large-amplitude motion, and symmetry described by the G_{12} PI group. This program was applied with success to the MW spectrum of 2-methylmalonaldehyde, but the rather small data set for this molecule did not provide a stringent test of the model. More challenging is the application of the hybrid program to CH_3NH_2, since this molecule has a much larger data set, containing both MW and IR transitions, as well as having a more extensive v_t, J, and K quantum number coverage. In our ISMS talk this year we will first give an overview of our best least-squares fit to date: The data set contains slightly more than 2500 MW and 11000 IR transitions with $J \leq 32$ and $K \leq 14$, which are fit to a weighted standard deviation of 1.64 using 71 parameters. Next, we present an assessment of this fit's strong points (e.g., significantly less parameters, ability to predict spectra in higher torsional states) and weak points (e.g., somewhat larger standard deviation, greater parameter correlation) when compared to the best all-tunneling-model fit in the literature. Based on this assessment, we believe that our fit, as well as the predictive abilities of the program, are sufficiently good that we can now begin considering collaborations with measurement and assignment campaigns of $v_t = 1$ MW data and $v_t = 2, 3$ IR data already underway in other laboratories. Finally, we will present a slightly modified ordering scheme for the operators in this hybrid program, and describe the need for devising a contact transformation treatment to specify determinable parameters in the hybrid Hamiltonian, in order to reduce parameter-correlation problems during the trial-and-error fitting process. A knowledge of determinable parameters would be particularly useful here, since there is little previous experience to guide the choice of "a good set" of higher-order constants to float when carrying out large fits having over 10000 lines, $J_{max} = 40$, $K_{max} = 15$, $v_t = 0$ and 1 torsional states, $A_1, A_2, B_1, B_2, E_1, E_2$ symmetry species, and nearly 100 parameters.

Intermission

TK08 4:18 – 4:33

A NEW MULTI-STATE VIBRATION-TORSION-ROTATION FITTING PROGRAM FOR MOLECULES WITH A C_{3v} TOP AND C_s FRAME: APPLICATION TO THE ν_{10} BAND OF ACETALDEHYDE

V. ILYUSHIN, E. A. ALEKSEEV, OLGA DOROVSKAYA, *Radiospectrometry Department, Institute of Radio Astronomy of NASU, Kharkov, Ukraine*; L. MARGULÈS, R. A. MOTIYENKO, MANUEL GOUBET, *Laboratoire PhLAM, UMR 8523 CNRS - Université Lille 1, Villeneuve d'Ascq, France*; OLIVIER PIRALI, *AILES beamline, Synchrotron SOLEIL, Saint Aubin, France*; SIGURD BAUERECKER, CHRISTOF MAUL, CHRISTIAN SYDOW, *Institut für Physikalische und Theoretische Chemie, Technische Universität Braunschweig, Braunschweig, Germany*; GEORG CH. MELLAU, *Physikalisch Chemisches Institut, Justus Liebig Universitat Giessen, Giessen, Germany*; ISABELLE KLEINER, *CNRS et Universités Paris Est et Paris Diderot, Laboratoire Interuniversitaire des Systèmes Atmosphériques (LISA), Créteil, France*; JON T. HOUGEN, *Sensor Science Division, National Institute of Standards and Technology, Gaithersburg, MD, USA.*

A new program is described for fitting several isolated small-amplitude fundamentals embedded in a pure torsional bath in molecules like acetaldehyde, in which the frame has C_s symmetry and the methyl top has C_{3v} symmetry. The program is based on the Longuet-Higgins group theoretical ideas and uses the Rho-axis method. In the talk the basic ideas, the structure of the program as well as the strategy for checking the program will be discussed. Also we present the first results [a] of application of the new program to the spectrum of the ν_{10} vibrational state of acetaldehyde, CH_3CHO, near 509 cm^{-1}. The analysis of the 509 cm^{-1} band is accompanied by the analysis of microwave spectrum of the ν_{10} vibrational state and $v_t = 3,4$ torsional states of acetaldehyde.

---

[a] This work was done under support of the Volkswagen foundation. The assistance of Science and Technology Center in Ukraine is acknowledged (STCU partner project P686).

TK09

FIRST RESULTS FOR ETHYLPHOSPHINE, $CH_3CH_2PH_2$, FROM AN EFFECTIVE ROTATIONAL HAMILTONIAN FOR TWO-ROTOR SYSTEMS WITH SYMMETRIC AND ASYMMETRIC INTERNAL ROTORS (LIKE ETHANOL)

PETER GRONER, *Department of Chemistry, University of Missouri - Kansas City, Kansas City, MO, USA.*

Spectra of molecules with a 3-fold internal rotor become much more interesting in the presence of another large-amplitude motion (LAM) that leads to tunneling between equivalent asymmetric forms which may also tunnel to a different conformer. An effective rotational Hamiltonian has been derived for such a system of which ethanol, CH_3CH_2OH, is a typical example [a]. For isolated vibrational states of molecules with two symmetric rotors with sufficiently "high" barriers, the ERHAM code[b] works well. Modifications were explored to find out whether ERHAM can be coaxed to treat ethanol-type systems, using "ancient" unpublished microwave data from vibrational ground and excited states of ethylphospine, $CH_3CH_2PH_2$, as test data. For gauche ethylphosphine, the splitting between the a-type Coriolis-coupled ground states is 5.215(6) MHz whereas it is 229.9(2) MHz in the ν_{24} state (PH_2 torsion). The tunneling energy coefficients ϵ_{01} for the methyl internal rotation are -0.63(2) MHz and 2.93(5) MHz (sign undeterminable), respectively. These results look promising; however, up to now, sets of assigned frequencies had to be omitted from fits to experimental uncertainty of 25 kHz: (a) for the ground state, all c-type transitions $J_{4,J-3} - J_{3,J-3}$ ($41 < J < 48$) for systematic large deviations (reason unknown); (b) for the ν_{24} state, half of the quartets of the $J_{3,J-2} - J_{2,J-2}$ series ($28 < J < 32$) because of interactions with a state of the trans conformer) and some of the $K_a = 1, 2$ low-J transitions (incorrect assignments or unknown reasons). Analyses of data for the ν_{23} (CH_3 torsion) and ν_{22} (CCP deformation) states are in progress.

---

[a]J.C. Pearson et al., J. Mol. Spectrosc. 251 (2008) 394
[b]P. Groner, J. Mol. Spectrosc. 278 (2012) 52–67

TK10

TORSIONAL SPLITTING AND FOUR-FOLD BARRIER TO INTERNAL ROTATION: THE ROTATIONAL SPECTRA OF VINYLSULFUR PENTAFLUORIDE

W. ORELLANA, SUSANNA L. STEPHENS, WALLACE C. PRINGLE, STEWART E. NOVICK, *Department of Chemistry, Wesleyan University, Middletown, CT, USA*; PETER GRONER, *Department of Chemistry, University of Missouri - Kansas City, Kansas City, MO, USA*; S. A. COOKE, *Natural and Social Science, Purchase College SUNY, Purchase, NY, USA.*

The rotational spectra of vinylsulfur pentafluoride, and three isotopologues (S-34 and both C-13's) have been recorded in the frequency region of 6 GHz to 20 GHz. Measurements were made using both cavity and chirped pulse Fourier transform microwave spectrometers. The four-fold barrier to the internal rotation of the $-SF_5$ group against the vinyl group has been approximated from the spectral data which is possible due to the observation of easily resolved pure rotational transitions in each of the A, B, and doubly degenerate E torsional substates. All transitions were successfully fit simultaneously using the ERHAM code. We note that this work, we believe, represents the first use of pure rotational spectroscopy to characterize a four-fold barrier internal rotation problem. Rotational constants, structure, and the internal rotation barrier height will be presented and compared to results from quantum chemical calculations.

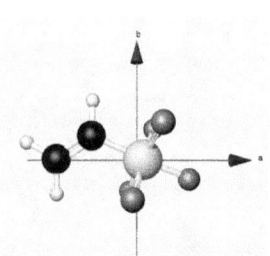

TK11   5:09–5:24

FURTHER STUDIES OF A FOUR-FOLD BARRIER TO INTERNAL ROTATION: THE ROTATIONAL SPECTRA OF PROPEN-1-YLSULFUR PENTAFLUORIDE AND BUTEN-1-YLSULFUR PENTAFLUORIDE

W. ORELLANA, SUSANNA L. STEPHENS, STEWART E. NOVICK, *Department of Chemistry, Wesleyan University, Middletown, CT, USA*; S. A. COOKE, *Natural and Social Science, Purchase College SUNY, Purchase, NY, USA.*

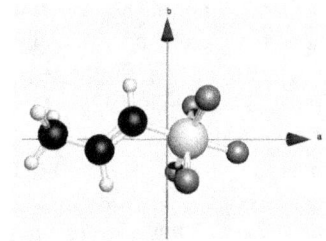

The rotational spectra of the two title molecules and several isotopologues have been recorded in the frequency region of 6 GHz to 20 GHz using Fourier transform microwave spectroscopy. For propen-1-ylsulfur pentafluoride, triplets of rotational transitions were observed appropriate for the A, B, and doubly degenerate E torsional substates arising from the four-fold barrier to internal rotation of the –SF_5 group against the propen-1-yl frame. However, the observed splittings, which are on the order of tens of kHz, were considerably smaller in magnitude than those analogous splittings observed in the spectra of the vinylsulfur pentafluoride, which were on the order of thousands of kHz. For the buten-1-ylsulfur pentafluoride, for which two conformers have been observed, at the resolution of the chirped pulse FTMW spectrometer used, splittings were not observable and the observed spectrum could be fit using the Hamiltonian of a semi-rigid rotor. Further experiments using a cavity FTMW spectrometer are underway. Constants from the spectral analyses, together with the results of quantum chemical calculations have allowed for the alkene-SF_5 barrier to internal rotation to be examined as a function of alkene chain length and the results will be presented.

TK12   5:26–5:41

THE ROTATIONAL STUDY OF THE VITAMINE B6 FORM PYRIDOXINE

ELENA R. ALONSO, IKER LEÓN, JOSÉ L. ALONSO, *Grupo de Espectroscopia Molecular, Lab. de Espectroscopia y Bioespectroscopia, Unidad Asociada CSIC, Universidad de Valladolid, Valladolid, Spain.*

Vitamin B6, like the rest of vitamins, is an important compound involved in numerous biological functions. Concretely, takes part in brain and nervous system health, in the metabolism of carbohydrates to produce energy and in the process of removing unwanted chemicals from our blood, among others. Vitamine B6 is found in a variety of forms, and here we present the rotational study of the pyridoxine form. Pyridoxine has been brought into gas-phase by means of laser ablation and probed by broadband LA-CP-FTMW microwave spectroscopy in the range 2-8 GHz. The presence of a methyl group in the structure offer us a nice internal rotation problem reflected in the spectrum, that together with the hyperfine structure due to the ^{14}N atom, makes this study very challenging. The high resolution of LA-MB-FTMW[a] spectroscopy has been crucial to overcome this problematic.

---

[a]C. Bermúdez, S. Mata, C. Cabezas and J. L. Alonso, Angew. Chemie - Int. Ed., 2014, 53, 11015–11018.

TL. Astronomy
Tuesday, June 19, 2018 – 1:45 PM
Room: 2079 Natural History

Chair: Jay A Kroll, University of Colorado at Boulder, Boulder, CO, USA

TL01   1:45 – 2:00

THE ROTATIONAL SPECTRUM OF PROTONATED ETHYL CYANIDE

<u>HARSHAL GUPTA</u>, *Division of Astronomical Sciences, National Science Foundation, Alexandria, VA, USA*; KELVIN LEE, *Radio and Geoastronomy Division, Harvard-Smithsonian Center for Astrophysics, Cambridge, MA, USA*; SVEN THORWIRTH, OSKAR ASVANY, STEPHAN SCHLEMMER, *I. Physikalisches Institut, Universität zu Köln, Köln, Germany*; MICHAEL C McCARTHY, *Atomic and Molecular Physics, Harvard-Smithsonian Center for Astrophysics, Cambridge, MA, USA*.

Ethyl cyanide (CH_3CH_2CN) is a well-known constituent of interstellar clouds and has recently been detected in the atmosphere of Titan.[a] It is so abundant in some interstellar clouds that its doubly substituted carbon-13 isotopologues, as well as highly excited vibrational satellites have been detected there.[b] Because of the high abundance and high proton affinity of CH_3CH_2CN, protonated ethyl cyanide ($CH_3CH_2CNH^+$) is a plausible intermediate in the chemistry of interstellar clouds and planetary atmospheres. Here we report the detection of $CH_3CH_2CNH^+$ by Fourier transform microwave spectroscopy of a supersonic molecular beam. Thirteen a-type rotational transitions have been observed between 8 and 44 GHz, some with partially resolved nitrogen hyperfine structure. This data set allows determination of all three rotational constants, as well as several of the leading centrifugal distortion constants to high accuracy. The derived rotational constants and those calculated at the CCSD(T) level of theory agree to better than 0.2%. Nitrogen hyperfine structure in the lower rotational transitions is so compact that only the quadrupole coupling tensor element along the a-inertial axis (χ_{aa}) could be determined. With accurate laboratory data in hand, a radio astronomical search for $CH_3CH_2CNH^+$ in publicly available spectral line surveys as well as through dedicated observations can now be undertaken with high confidence.

---

[a]Cordiner, M. A., Palmer, M. Y., Nixon, C. A. et al. 2015, ApJL, 800, L14.
[b]Margulès, L., Belloche, A., Müller, H. S. P. et al. 2016, A&A, 590, A93 and references therein.

TL02   2:02 – 2:17

TOWARDS LABORATORY LINE LISTS TO SEARCH FOR CH_3OD and $^{13}CH_3OD$ IN SPACE

<u>LI-HONG XU</u>, RONALD M. LEES, *Department of Physics, University of New Brunswick, Saint John, NB, Canada*; OLENA ZAKHARENKO, HOLGER S. P. MÜLLER, FRANK LEWEN, STEPHAN SCHLEMMER, *I. Physikalisches Institut, Universität zu Köln, Köln, Germany*; KARL M. MENTEN, *Millimeter- und Submillimeter-Astronomie, Max-Planck-Institut für Radioastronomie, Bonn, NRW, Germany*.

Understanding how, when, and where complex organic and potentially prebiotic molecules are formed is a fundamental goal of astrophysics and astrochemistry and an integral part of origins-of-life studies. The recent images from the Atacama Large Millimeter/submillimeter Array (ALMA) of a potentially planet-forming disk around a young star with an age of only 0.5-1 Myr (million years) have highlighted the importance of the physics and chemistry of the early protostellar stages. A particular focus of ALMA observations is the search for complex molecules in regions of high- and low-mass star formation.

Methanol is among the most abundant molecules in star-forming regions, to such an extent that it is considered as an interstellar "weed". Thus, its numerous isotopologues must be accounted for in search for new complex molecules as yet unknown. Here, **CalculatusEliminatus** approach – "... can help an awful lot. The way to find a missing something is to find out where it's not!"[a] In other words, if we establish all the knowns, then we can focus on the remaining unknowns.

This talk will report progress in our current effort in our on-going journey to pursue the weeding of the cosmos, focusing on CH_3OD and $^{13}CH_3OD$. While CH_3OD is of great interest to a considerable fraction of the radio astronomical community, the significance of $^{13}CH_3OD$ originates from Protostellar Interferometric Line Survey (PILS)[b] of the low-mass protostars IRAS 16291-2422 with ALMA. Here, many lines of CH_3OH, $^{13}CH_3OH$ and even CH_2DOH are optically thick, so that the ratio of CH_3OD to $^{13}CH_3OD$ may serve as a valuable probe of the $^{12}C/^{13}C$ ratio of methanol. New laboratory measurements obtained at Universität zu Köln will be described for: (i) sample mixture of $^{13}CH_3OH$ and CH_3OD yielding enriched $^{13}CH_3OD$ features from approximately 134 GHz to 510 GHz and (ii) CH_3OD from approximately 154 GHz to 510 GHz and from 1.12 THz to 1.34 THz.

---

[a]The Cat in the Hat (Video), Dr. Seuss.
[b]J. K. Jørgensen et al., Astron. Astrophys. 595 (216) 117.

TL03 2:19–2:34

PREBIOTIC MOLECULES IN INTERSTELLAR SPACE: ROTATIONAL SPECTROSCOPY OF CYANOMETHANIMINE AND ETHANIMINE

<u>CRISTINA PUZZARINI</u>, *Dep. Chemistry 'Giacomo Ciamician', University of Bologna, Bologna, Italy*; LORENZO SPADA, *Scuola Normale Superiore, Scuola Normale Superiore, Pisa, Italy*; MATTIA MELOSSO, LUCA DORE, *Dept. Chemistry "Giacomo Ciamician", University of Bologna, Bologna, ITALY*; VINCENZO BARONE, *Scuola Normale Superiore, Scuola Normale Superiore, Pisa, Italy*.

Ethanimine and cyanomethanimine are possible precursors of amino acids, and thus they are considered important prebiotic molecules that may play an important roles in the formation of biological building blocks in the interstellar medium. In addition, their identification in Titan's atmosphere would be important for understanding the abiotic synthesis of organic species. For both molecules, an accurate computational characterization of the molecular structure, energetics, and spectroscopic properties of the E and Z isomers has been carried out by means of a composite scheme based on coupled-cluster techniques. By combining the computational results with new millimeter-wave measurements, up to 300 GHz for ethanimine and to 420 GHz for cyanomethanimine, the rotational spectra of both isomers can be accurately predicted up to 500 GHz for ethanimine and 700 GHz for cyanomenthanimine. For the latter, spectral features have been searched in the mm-wave range using the high-sensitivity and unbiased spectral surveys obtained with the IRAM 30-m antenna in the ASAI context, thus sampling the earliest stages of star formation from starless to evolved Class I objects.

TL04 2:36–2:51

SEARCH FOR THE ROTATIONAL SPECTRUM OF THE β-CYANOVINYL RADICAL

<u>SOMMER L. JOHANSEN</u>, KYLE N. CRABTREE, *Department of Chemistry, The University of California, Davis, CA, USA*.

A fundamental question in the field of astrochemistry is whether the molecules essential to life originated in the interstellar medium and, if so, how they were formed. Nitrogen-containing heterocycles are of particular interest because of their role in biology. The discoveries of these molecules on meteorites provide evidence to support an interstellar origin. Yet, while many N-containing species have been identified in the interstellar medium, N-heterocycles have not, perhaps due to their susceptibility to UV photolysis. Recently, the β-cyanovinyl radical (HCCHCN) was implicated in the low temperature formation of N-heterocycles. While neutral vinyl cyanide (H_2CCHCN) has been rotationally characterized and detected in the interstellar medium, HCCHCN has not. In order to understand how this radical contributes to interstellar chemistry, further study is needed. We have launched a search for the rotational spectrum of HCCHCN using both cavity and broadband FTMW spectrometers, the current status of which will be discussed in this talk. Rotational characterization of this radical will enable a search in the interstellar medium and further experimental work on the low temperature formation of N-heterocycles.

TL05 2:53–3:08

INDIRECT ROTATIONAL SPECTROSCOPY OF THE D_2H^+ MOLECULAR ION

<u>CHARLES R. MARKUS</u>, PHILIP A. KOCHERIL, *Department of Chemistry, University of Illinois at Urbana-Champaign, Urbana, IL, USA*; BENJAMIN J. McCALL, *Departments of Chemistry and Astronomy, University of Illinois at Urbana-Champaign, Urbana, IL, USA*.

The partially deuterated isotopologues of H_3^+ have proven to be valuable probes of interstellar environments. In low temperature regions (< 20 K) such as dark molecular clouds, deuterium fractionation leads the ratios of D_2H^+/H_3^+ and H_2D^+/H_3^+ to be orders of magnitude greater than that of HD/H_2.[a] Unlike H_3^+, D_2H^+ and H_2D^+ have permanent dipole moments and their allowed rotational transitions can be used to probe interstellar conditions and give an indirect method of detection for H_3^+.

There are still unobserved transitions of D_2H^+ that are within the coverage of observatories such as the Stratospheric Observatory for Infrared Astronomy (SOFIA) which have relatively large uncertainties (8–16 MHz). Recently, we have measured over 20 rovibrational transitions of D_2H^+ with MHz-level uncertainty using the technique Noise-Immune Cavity-Enhanced Optical Heterodyne Velocity Modulation Spectroscopy (NICE-OHVMS).[b] These new measurements can be used to improve predicted frequencies of pure rotational transitions using combination differences and an improved fit to an effective Hamiltonian.

---

[a]H. Roberts, E. Herbst, and T. J. Millar, *Astrophys. J.*, **591**, L41 (2003).
[b]J. N. Hodges, A. J. Perry, P. A. Jenkins II, B. M. Siller, and B. J. McCall, *J. Chem. Phys*, **139**, 164201 (2013).

TL06 3:10 – 3:25

HIGH-RESOLUTION MICROWAVE SPECTROSCOPY OF RADIOACTIVE MOLECULES: MASS-INDEPENDENT STUDIES OF AlO, TiO, AND FeO

ALEXANDER A. BREIER, *Institute of Physics, University Kassel, Kassel, Germany*; BJÖRN WAßMUTH, *Institute of Physics, University of Kassel, Kassel, Germany*; THOMAS BÜCHLING, *Institute of Physics, University Kassel, Kassel, Germany*; GUIDO W FUCHS, *Physics Department, University of Kassel, Kassel, Germany*; THOMAS GIESEN, *Institute of Physics, University Kassel, Kassel, Germany*; JÜRGEN GAUSS, *Institut für Physikalische Chemie, Universität Mainz, Mainz, Germany*.

Astrophysical observations of radioactive isotopes, like ^{26}Al, ^{44}Ti, or ^{60}Fe, provide insight into the nucleosynthesis of stellar cores. The detection of characteristic γ-photons, which are released during radioactive decay, is used to map their spatial distribution on large scale. In general, the assignment to certain stellar objects fails due to limited sensitivity.

An alternative approach is the observation of molecules containing radioactive isotopes. Radio-telescope facilities, like *ALMA*, can identify these species via their rotational transitions. In the outer atmosphere of late-type stars, the molecular condensation starts with simple diatomic particles containing oxides of refractory elements. The astrophysical detection of diatomic radioactive molecules requires highly accurate rotational transition frequencies, which can be obtained from laboratory measurements of stable isotopologues using mass-independent Dunham parameters.

In this work, systematic studies are presented for ^{26}AlO, ^{44}TiO, and ^{60}FeO, as most promising tracers of nucleosynthesis in stellar environments, based on high-resolution measurements on the rotational transitions of their abundant stable isotopologues. Experiments were performed when a solid target (Al, Ti, Fe) is evaporated by a pulsed laser into an oxygen-rich buffer gas to form simple metal oxides. An adiabatically planar expansion of the gas into a vacuum chamber cools the gas to a few tens of Kelvin and subsequently, Doppler-free rotational absorption spectra are recorded in the frequency range up to 400 GHz. A global data analysis, which also includes results from the literature, reveals the molecular structure beyond the Born-Oppenheimer (BO) limit, resulting in experimentally derived BO correction coefficients of these species for the first time. Based on this analysis, the rotational transitions of the radioactive molecules are determined with high accuracy at the sub-MHz level, which enables their unambiguous identification in stellar environments.

Intermission

TL07 4:01 – 4:16

THE PURE ROTATIONAL SPECTRUM OF THE T-SHAPED AlC$_2$ RADICAL ($\tilde{X}^2A_1$)

DeWAYNE T HALFEN, *Department of Chemistry and Biochemistry, Department of Astronomy, The University of Arizona, Tucson, AZ, USA*; LUCY M. ZIURYS, *Department of Chemistry and Biochemistry; Department of Astronomy, Arizona Radio Observatory, University of Arizona, Tuscon, AZ, USA*.

The pure rotational spectrum of the AlC$_2$ radical ($\tilde{X}^2A_1$) has been recorded for the first time using Fourier transform microwave/millimeter-wave (FTMmmW) techniques in the frequency range 21 – 65 GHz. AlC$_2$ was produced in a supersonic jet from the reaction of aluminum, generated by laser ablation, with a mixture of CH$_4$ or HCCH, diluted in argon, with a DC discharge. Three transitions ($N_{K_a, K_c} = 1_{01} \to 0_{00}$, $2_{02} \to 1_{01}$, and $3_{03} \to 2_{02}$) were measured, consisting of multiple fine/hyperfine components, resulting from the unpaired electron and the aluminum-27 nuclear spin ($I = 5/2$). The data were analyzed using an asymmetric top Hamiltonian and rotational, fine structure, and hyperfine constants were established. These results are in excellent agreement with previous theoretical calculations and optical spectra. An r_0 structure of AlC$_2$ was determined to be r(Al–C) = 1.924 Å, r(C–C) = 1.260 Å, and θ(C–Al–C) = 38.2°. The data are consistent with a T-shaped geometry with Al$^+$C$_2^-$ bonding. A search for AlC$_2$ in the circumstellar envelope of IRC+10216 is currently underway.

TL08 4:18–4:33

TOWARDS UNRAVELLING THE FORMATION OF ICE GRAINS: THE PHENANTHRENE-WATER COMPLEX

<u>DONATELLA LORU</u>, SÉBASTIEN GRUET, *FS-SMP, Deutsches Elektronen-Synchrotron (DESY), Hamburg, Germany*; AMANDA STEBER, *CUI, The Hamburg Centre for Ultrafast Imaging, Hamburg, Germany*; CRISTOBAL PEREZ, MELANIE SCHNELL, *FS-SMP, Deutsches Elektronen-Synchrotron (DESY), Hamburg, Germany*.

Polycyclic aromatic hydrocarbons (PAHs) are believed to act as catalysts in ice grains formation. The formation of interstellar ice can thus be described as an aggregation process of gaseous water starting on PAH surfaces.

The structural investigation of PAH-H_2O clusters, therefore, represents a first and important step to undertake in order to shed light on potential ice grains formation pathways. Previous studies have focused on mimicking the initial stage of this aggregation process in laboratories by forming complexes between PAH and water molecules in the gas phase and investigating them by microwave spectroscopy. As the outcome of these studies, precise information on the structure and intermolecular interactions were obtained [1,2].

Herein we take this approach further and present preliminary data on the structural investigation of the complex of phenanthrene, a PAH molecule featuring three fused benzene rings, and water. Its pure rotational spectrum was recorded using a chirped pulse Fourier transform microwave spectrometer (CP-FTMW) operating in the region 2-8 GHz, and structural information on the respective water clusters can be obtained.

[1] Pérez, C.; Steber, A. L.; Rijs, A. M.; Temelso, B.; Shields, G. C.; Lopez, J. C.; Kisiel, Z.; Schnell, M., Phys. Chem. Chem. Phys. 2017, 19, 14214-14223. [2] Steber, A. L.; Pérez, C.; Temelso, B.; Shields, G. C.; Rijs, A. M.; Pate, B. H.; Kisiel, Z.; Schnell, J. Phys. Chem. Lett. 2017, 8, 5744-5750.

TL09 4:35–4:50

LABORATORY DETECTION OF VIBRATION-ROTATION TRANSITIONS OF $^{12}CH^+$ AND $^{13}CH^+$ AND IMPROVED MEASUREMENT OF THEIR ROTATIONAL TRANSITION FREQUENCIES

<u>JOSÉ LUIS DOMÉNECH</u>, *Instituto de Estructura de la Materia, (IEM-CSIC), Madrid, Spain*; PAVOL JUSKO[a], STEPHAN SCHLEMMER, OSKAR ASVANY, *I. Physikalisches Institut, Universität zu Köln, Köln, Germany*.

The elusive C-H vibration-rotation transitions of the fundamental ions $^{12}CH^+$ and $^{13}CH^+$ have been observed for the first time in the laboratory. The technique of state-dependent attachment of He atoms to these ions in a cryogenic trap[b] has been used to obtain high-resolution rovibrational data. The excitation source is an IR OPO whose frequency is measured using a frequency comb. In addition, the lowest frequency rotational transitions of $^{12}CH^+$, $^{13}CH^+$ and $^{12}CD^+$ have been revisited[c,d,e] using a synthesizer and a multiplier chain with the same ion trap, leading to a significant improvement of their rest frequency values.

[a]Current affiliation: Institut de Recherche en Astrophysique et Planétologie (IRAP), Université de Toulouse (UPS), CNRS, CNES, 9 Av. du Colonel Roche, 31028 Toulouse Cedex 4, France.
[b]O. Asvany, S. Brünken, L. Kluge, and S. Schlemmer 2014, Appl. Phys. B, 114, 203-211.
[c]S. Yu, B. J. Drouin, J. C. Pearson and T. Amano 2015, in Contribution RD06, 70th International Symposium on Molecular Spectroscopy.
[d]T. Amano 2010, Astrophys. J. Lett., 716, L1.
[e]S. Brünken, L. Kluge, A. Stoffels, J. Pérez-Ríos, and S. Schlemmer 2017, J. Mol. Spec. 332, 67.

TL10 4:52 – 5:07

LABORATORY INVESTIGATION OF ASTRONOMICAL REACTIVE SPECIES: THE VIBRATIONAL SATELLITES OF c-C_3H_2 RE-VISITED

<u>MARIE-ALINE MARTIN-DRUMEL</u>, *CNRS, Institut des Sciences Moleculaires d'Orsay, Orsay, France*; BRYAN CHANGALA, *JILA, National Institute of Standards and Technology and Univ. of Colorado Department of Physics, University of Colorado, Boulder, CO, USA*; HARSHAL GUPTA, *Division of Astronomical Sciences, National Science Foundation, Alexandria, VA, USA*; J. H. WESTERFIELD, *Department of Chemistry, New College of Florida, Sarasota, FL, USA*; OLIVIER PIRALI, *AILES beamline, Synchrotron SOLEIL, Saint Aubin, France*; SVEN THORWIRTH, *I. Physikalisches Institut, Universität zu Köln, Köln, Germany*; JOSHUA H BARABAN, *Chemistry, Ben-Gurion University of the Negev, Beer-Sheva, Israel*; JOHN F. STANTON, *Physical Chemistry, University of Florida, Gainesville, FL, USA*; MICHAEL C McCARTHY, *Atomic and Molecular Physics, Harvard-Smithsonian Center for Astrophysics, Cambridge, MA, USA*.

Cyclopropenylidene (c-C_3H_2) is one of the few polyatomic hydrocarbons ubiquitous in our galaxy, despite its reactive carbene nature (see e.g. [1]). Because it is so widely distributed in space, and because its ^{13}C, D, and D_2 isotopologues have also been detected (see e.g. [2]), c-C_3H_2 is an ideal probe of the physical conditions in various astrophysical objects. It is surprising though that its vibrational satellites have yet to be detected in the interstellar medium.

To enable the interstellar detection of vibrationally excited c-C_3H_2, and observe for the first time the elusive ν_4 and ν_9 vibrational modes, we have undertaken an extensive investigation of its spectrum from the centimeter to the submillimeter wavelengths, resulting in the observation of many new vibrational satellites in a promising spectral region for astronomical observations. Our measurements are supported by anharmonic rovibrational calculations using a high-quality ab initio potential energy surface, with particular attention paid to the ν_4/ν_9 Coriolis interaction.

[1] S. Spezzano *et al.*, *The Astrophysical Journal Supplement Series* **200**, 1 (2012)
[2] S. Spezzano *et al.*, *The Astrophysical Journal Letters*, **769**, L19 (2013)

TL11 5:09 – 5:24

SPECTROSCOPY OF NEW IMINE ASTROPHYSCICS TARGET: METHYLIMINO-ACETONITRILE ($CH_3N=CHCN$)

<u>L. MARGULÈS</u>, *Laboratoire PhLAM UMR 8523, Université de Lille, 59655 Villeneuve d'Ascq, FRANCE*; R. A. MOTIYENKO, *UMR 8523 CNRS - Université de Lille, Laboratoire PhLAM, Villeneuve d'Ascq, France*; J.-C. GUILLEMIN, *ISCR – UMR6226, Université de Rennes, 35000 Rennes, FRANCE*.

There are to date about 200 molecules that have been detected in the interstellar medium or circumstellar shells. Among these molecules, several tens are the methylated derivatives of compounds previously detected. For several years, we have been studying molecules belonging to the imine family. Following the detection of the dimer of HCN, the cynaoethanimine, its methylated derivative, methylimino-acetonitrile $CH_3N=CHCN$ appears as a privileged target. Methylimino-acetonitrile has two isomers E and Z. According to quantum chemical calculations, the E isomer is more stable than Z by about 1.5 kJ/mol. There was no spectrosocpic data allowing detection without ambiguity of this molecule in the interstellar medium. We recorded and analyzed the spectra of methylimino-acetonitrile up to 660 GHz. This compound is not stable in laboratory conditions, it was produced in-situ by pyrolisis and introduced in a 1 m long pyrex cell in a flow mode. The E isomer represents an interesting case from spectroscopic point of view. Even if the barrier to internal rotation of the methyl top is quite high 714 cm^{-1}, some A-E tunneling splittings were observed. This is due to quite high ρ value: 0.274, just slightly smaller than the acetaldehyde value of 0.329. The analysis is performed using the RAM36 code[a]. The spectroscopic results will be presented.

Acknowledgements: These results were supported by the Programme National PCMI of CNRS/INSU with INC/INP co- funded by CEA and CNES, the French National Research Agency (ANR-13-BS05-0008 "IMOLABS"

[a] Ilyushin, V.V. et al; *J. Mol. Spectrosc.* **259**, (2010) 26

WA. Mini-symposium: Frequency-Comb Spectroscopy

Wednesday, June 20, 2018 – 8:30 AM

Room: 116 Roger Adams Lab

Chair: Masatoshi Misono, FUKUOKA UNIVERSITY, Fukuoka, Japan

WA01   *INVITED TALK*   8:30 – 9:00

DUAL FREQUENCY COMB METHANE LEAK DETECTION AT OPERATIONAL OIL AND GAS FACILITIES

<u>GREGORY B RIEKER</u>, SEAN COBURN, CAROLINE ALDEN, ROBERT WRIGHT, *Department of Mechanical Engineering, University of Colorado Boulder, Boulder, CO, USA*; ALEX RYBCHUK, *Department of Mechanical Engineering, University of Colorado, Boulder, CO, USA*; KULDEEP PRASAD, *Fire Research Division, NIST, Gaithersburg, MD, USA*; KEVIN C COSSEL, ESTHER BAUMANN, IAN CODDINGTON, *Applied Physics Division, NIST, Boulder, CO, USA*.

We recently demonstrated a field-deployed dual frequency comb laser spectrometer capable of locating and sizing methane sources down to 1.6 grams/minute (which is equivalent to approximately one quarter of the human breathing rate) from a distance of 1 km. The system couples open-path methane concentration measurements over long distances together with wind information in a Bayesian inversion framework to locate sources within the monitoring region. We are now applying the technology for leak detection at operational oil and gas facilities. We will discuss the evolution of the project from laboratory proof-of-concept to controlled field testing to initial implementation in an industrial setting. We will also discuss the challenges of field deployment in real environments, which include remotely operating stabilized mode-locked frequency combs and maintaining a sensing network through rain, snow, and fog.

WA02   9:04 – 9:19

DUAL-COMB SPECTROSCOPY OF GREENHOUSE GAS BASED ON AN ERBIUM DUAL-COMB FIBER LASER

SIYAO YIN, JIE CHEN, <u>XIN ZHAO</u>, TING LI, ZHENG ZHENG, *School of Electronic Information Engineering , Beihang University, Beijing, China.*

Dual-comb spectroscopy holds the promise as real-time, high-resolution spectroscopy tools. It had been applied to measure the spectral features of a wide variety of samples. With the help of nonlinear optical spectral broadening schemes, gases with absorption in different spectral windows can be monitored using the dual-comb method[a]. Among them, methane is one that attracts much attention due to its important role in the greenhouse effect. Dual-comb spectroscopy based on a single mode-locked fiber laser has attracted more attention due to its simplification compared to the dual lasers systems. High-resolution optical spectroscopy had been demonstrated around the Erbium (Er) gain window by measuring the absorption of acetylene gas cell and other devices[b]. It had been suggested that certain degree of mutual coherence between two combs without stabilization, which is important for dual-comb applications, could exist due to some common-mode noise cancellation mechanisms.

In this work, dual-comb spectroscopy based on the nonlinear spectral broadening of a single laser dual comb source at the 1550 nm Er window is further investigated in ultralength (UL) wavelength band where the absorption of methane can be observed. The absorption line of methane around 1648 nm, which is 100 nm away from the lasing wavelengths of our seed laser, can be clearly obtained. The spectral signal to noise ratio improves significantly with the increase of the number of averaged interferograms. Despite the extra noise introduced by the nonlinear spectral broadening process, the absorption line of methane can still be resolved with good quality.

---

[a] K. Cossel, et al., Optica, 4, 724 (2017).
[b] X. Zhao, et al. Opt. Express 24, 21833 (2016).

WA03   9:21 – 9:36
DUAL FREQUENCY COMB SPECTROSCOPY FOR DEVELOPMENT AND TESTING OF HIGH PRESSURE, HIGH TEMPERATURE ABSORPTION MODELS

RYAN K. COLE, *Department of Mechanical Engineering, University of Colorado Boulder, Boulder, CO, USA*; PAUL JAMES SCHROEDER, *Chemical Sciences Division, National Oceanic and Atmospheric Administration, Boulder, Colorado, USA*; ANTHONY D. DRAPER, *Department of Mechanical Engineering, University of Colorado Boulder, Boulder, CO, USA*; MATTHEW J. CICH, BRIAN DROUIN, *Jet Propulsion Laboratory, California Institute of Technology, Pasadena, CA, USA*; GREGORY B RIEKER, *Department of Mechanical Engineering, University of Colorado Boulder, Boulder, CO, USA*.

The development of accurate absorption models for high pressure, high temperature environments is complicated by the increased relevance of higher order collisional phenomena on the absorption lineshape (e.g. line mixing, collision-induced absorption, finite duration of collisions). Accurate reference spectroscopy at these conditions is important for the study of combustion systems and remote sensing of dense planetary atmospheres. We present a new high pressure, high temperature absorption spectroscopy facility at the University of Colorado Boulder. This facility is coupled with a dual frequency comb absorption spectrometer to record broadband (~1500cm^{-1}), high resolution (~0.0066cm^{-1}) spectra in a controlled environment at high pressures and temperatures. Measurements of the NIR spectrum of carbon dioxide will be compared to modeled spectra extrapolated from the HITRAN 2016 database as well as other published models that include line mixing corrections. This comparison gives insight into the effectiveness of existing absorption models in the high pressure, high temperature limit as well as the improvements required to accurately model absorption spectra in harsh systems.

WA04   9:38 – 9:53
DYNAMIC REGIONAL AND CITY SCALE SENSING OF GHG'S USING A DUAL-COMB SPECTROMETER

ELEANOR WAXMAN, KEVIN C COSSEL, FABRIZIO GIORGETTA, GAR-WING TRUONG[a], MICHEAL CERMAK, WILLIAM C SWANN, *Applied Physics Division, NIST, Boulder, CO, USA*; DANIEL HESSELIUS, *College of Engineering, University of Colorado Boulder, Boulder, CO, USA*; GREGORY B RIEKER, *Department of Mechanical Engineering, University of Colorado Boulder, Boulder, CO, USA*; NATHAN R. NEWBURY, IAN CODDINGTON, *Applied Physics Division, NIST, Boulder, CO, USA*.

The output of a laser frequency comb is composed of 100,000+ perfectly spaced, discrete wavelength elements or comb teeth, which act as a massively parallel set of single frequency (CW) lasers with highly stable, well-known frequencies. In dual-comb spectroscopy (DCS), two such frequency combs are interfered on a single detector yielding absorption information for each individual comb tooth. This approach combines the strengths of both CW laser spectroscopy and broadband spectroscopy providing high spectral resolution and broad optical bandwidths, all with a single-mode, high-brightness laser beam and a simple, single photodetector, detection scheme. Inter comparisons of DCS instruments in the 1.55-1.7um region have shown that atmosphperic CO2 and CH4 concentrations can be retrieved with precisions of 0.14% and 0.35% respectively making this an attractive source for quantifying greenhouse gas emissions[b]. Here we show that DCS can be employed for dynamic regional monitoring using an unmanned aerial systems (UAS) to identify and quantify methane leaks[c]. Additionally, we will show that much larger scale (multi-kilometer) fixed path measurements can be used for continuous monitoring of city scale CO2 emissions. A preliminary demonstration of this technique in Boulder Colorado shows reasonable agreement with the city's own bottom up emission projections.

---
[a]currently at Crystalline Mirror Solutions
[b]E. M. Waxman, et al. "Intercomparison of open-path trace gas measurements with two dual-frequency-comb spectrometers." Atmos Meas Tech 1,3295–3311 (2017)
[c]K. C. Cossel, et al. "Open-path dual-comb spectroscopy to airborne retroreflector." Optica 4, 724–728 (2017)

Intermission

WA05
10:29 – 10:44

FREQUENCY COMB VERNIER SPECTROSCOPY OF METHANE IN THE MID-IR WITH TEMPORAL RETRIEVAL OF COMB LINES

JAMES R BOUNDS, FENG ZHU, ALEXANDER KOLOMENSKII, HANS A SCHUESSLER, *Department of Physics and Astronomy, Texas A&M University, College Station, TX, USA*; PAOTAI LIN, JUNCHAO ZHOU, *Electrical and Computer Engineering, Texas A&M University, College Station, USA.*

We develop a spectroscopic method combining the broadband spectral coverage and frequency resolution of a frequency comb with the optical path enhancement of a resonant optical cavity. The method requires only one frequency comb laser. We present measurements of the methane absorption spectrum performed in ambient air in the mid-IR from 3μm to 4μm using a comb derived from difference frequency generation. The resonant cavity provides the dual purpose of optical path enhancement along with Vernier filtering of the comb modes, allowing individual comb modes to be resolved with a grating. As the resonant cavity length is scanned, the transmitted comb lines in the whole spectral range of the cavity reflectivity are continuously recorded with a mid-IR camera, yielding an absorption spectrum in seconds. We show how this method can be realized with fewer moving parts and with broader spectral coverage than similar techniques reported in the literature, along with showing how the technique can be modified to suit different mid-IR detectors.

This work was supported by Robert A. Welch Foundation, grant No. A1546, the Qatar Foundation, grant NPRP 8-735-1-154.

WA06
10:46 – 11:01

DIRECT FREQUENCY COMB SPECTROSCOPY WITH AN 8.5 μm OPO

KANA IWAKUNI, *JILA, National Institute of Standards and Technology and Univ. of Colorado Department of Physics, University of Colorado, Boulder, CO, USA*; THINH QUOC BUI, JUSTIN NIEDERMEYER, *JILA, National Institute of Standards and Technology and Univ. of Colorado Department of Physics, University of Colorado, Boulder, Boulder, CO, USA*; BRYAN CHANGALA, MARISSA L. WEICHMAN, *JILA, National Institute of Standards and Technology and Univ. of Colorado Department of Physics, University of Colorado, Boulder, CO, USA*; TAKASHI SUKEGAWA, *Optical Products Operations, CANON, Utsunomiya, Japan*; JUN YE, *JILA, National Institute of Standards and Technology and Univ. of Colorado Department of Physics, University of Colorado, Boulder, CO, USA.*

Direct frequency comb spectroscopy provides high-resolution spectra over a broad bandwidth. Its high sensitivity has also enabled real time detection for gas sensing and chemical reaction kinetics[a]. Previous work has focused in the near-infrared or mid-infrared (1 - 5 μm), but there are stronger absorption lines in the >5 μm wavelength region. In addition, at longer wavelengths spectral congestion is significantly reduced owing to the decreasing strength of intramolecular vibrational energy redistribution. We have developed a new frequency comb spectrometer within 8.5 – 9.5 μm. The light source is a synchronously pumped optical parametric oscillator (OPO)-based frequency comb using a 2 μm Tm fiber comb as the pump wave. In direct frequency comb spectroscopy, several options exist to read out the spectrum, such as FTIR or highly dispersive optics like a virtually-imaged phased array (VIPA)[b]. In this work, an immersion grating and a reflective grating are used as cross dispersers and each comb mode is mapped to a 2D image in the same way as a VIPA spectrometer. Immersion gratings have been applied in astronomy and have resolving power ($\lambda/\Delta\lambda$) exceeding 10^5, which is suitable for high-resolution real-time comb spectroscopy. We report work done with this new spectrometer.

[a] T. Q. Bui, B. J. Bjork, P. B. Changala, T. L. Nguyen, J. F. Stanton, M. Okumura, J. Ye, Direct measurements of DOCO isomers in the kinetics of OD+CO, Science Advances, 4, eaao4777 (2018)

[b] L. Nugent-Glandorf, T. Neely, F. Adler, A. J. Fleisher, K. C. Cossel, B. Bjork, T. Dinneen, J. Ye, S. A. Diddams, Mid-infrared virtually imaged phased array spectrometer for rapid and broadband trace gas detection, Opt. Lett. 37, 3285 (2012)

WA07 11:03 – 11:18

MID-INFRARED FREQUENCY COMB SPECTROSCOPY USING A VIRTUALLY IMAGED PHASED ARRAY

<u>ADAM J. FLEISHER</u>, *Chemical Sciences Division, National Institute of Standards and Technology, Gaithersburg, MD, USA.*

Here we present a new mid-infrared frequency comb system for rapid spectral acquisition using a virtually imaged phased array (VIPA) spectrometer.[a] A difference-frequency generation comb, tuneable from 4.4 μm to 4.7 μm, was used to interrogate a single-pass absorption cell containing either N_2O or CO dilute in either N_2 or air. Precision molecular spectroscopy capabilities at timescales of less than 1 ms will be presented, and progress toward cavity-enhanced and time-resolved comb spectroscopies[b] will be discussed.

[a]L. Nugent-Glandorf et al., *Opt. Lett.* **37**, 3285 (2012)

[b]A.J. Fleisher et. al., *J. Phys. Chem. Lett.* **5**, 2241 (2014)

WA08 *Post-Deadline Abstract* 11:20 – 11:35

HIGH-RESOLUTION AND ULTRA-BROADBAND DIRECT-COMB ABSOLUTE-SPECTROSCOPY BY MEANS OF THE SCANNING MICRO-CAVITY RESONATOR (SMART) TECHNIQUE

<u>ALESSIO GAMBETTA</u>, EDOARDO VICENTINI, YUCHEN WANG, NICOLA COLUCCELLI, PAOLO LAPORTA, *Dipartimento di Fisica, Politecnico di Milano, Milano, Italy*; GIANLUCA GALZERANO, *Institute for photonics and nanotechnologies, National Research Council, Milano, Italy.*

By exploiting scanning Fabry-Pérot micro-cavity resonator (SMART) we developed a simple and compact spectrometer capable of resolving the mode structure of an optical frequency comb, with a frequency resolution limited only by the comb tooth linewidth.

The SMART approach can be adopted in any spectral window from UV to THz and represents an easy and completely "locking-free" approach to direct-comb spectroscopy, drastically reducing the system complexity when compared to the more conventional methods like the VIPA, dual- comb and Vernier. Furthermore, high-speed/high-sensitivity detection and straightforward absolute calibration of the optical-frequency axis are still granted by the SMART spectrometer, together with an ultimate resolution limited only by the optical linewidth of the frequency comb source adopted for the measurement.

We present an application to broadband and high-precision spectroscopy of acetylene at 1.54 μm. Also, by means of an auxiliary 2500-finesse cavity exploited as a sample with narrow- transmission features, we show the ability of the SMART approach to resolve the 400 kHz resonances of the auxiliary cavity, demonstrating a final resolution well below than the 20 MHz linewidth of the transmission-modes of the SMART micro-resonator employed for optical- detection.

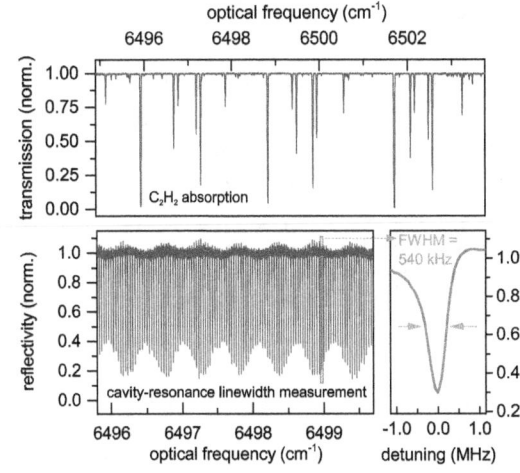

WB. Mini-symposium: New Ways of Understanding Molecular Spectra

Wednesday, June 20, 2018 – 8:30 AM

Room: 100 Noyes Laboratory

Chair: Taylor Smart, University of Virginia, Charlottesville, Virginia, USA

WB01   *INVITED TALK*   8:30 – 9:00

MOLECULAR SPECTROSCOPY FROM FIRST PRINCIPLES

SERGEI N. YURCHENKO, *Department of Physics and Astronomy, University College London, Gower Street, London WC1E 6BT, United Kingdom.*

Over the past years there have been a rapid improvement in nuclear motion approaches to solving spectroscopic problems, which has been described as the fourth age of quantum chemistry.[a] The methodology which is commonly attributed to the spectroscopy from first principles is in fact a combination of high level *ab initio* (electronic structure) calculations, high level nuclear motion (variational) calculations and empirical refinement to the highly accurate experimental data (e.g. line positions).[b] In this talk, I will discuss the current state-of-the-art of the theoretical molecular spectroscopy, which shows that this methodology is increasingly competitive with experiment and allows in many cases a more reliable determination of various molecular data.[c] Our variational program TROVE[d] is one of the modern computational tools successfully used to generate huge lists of transitions which provide the input for models of atmospheric absorption. I will review the methodology used by TROVE and other variational programs for accurate solution of the nuclear motion Schrödinger equation for general medium size polyatomic molecules, show examples of successful applications and discuss cases, which are still challenging for the modern *ab initio* methods.

---

[a] A. G. Császár, C. Fabri, T. Szidarovszky, E. Mátyus, T. Furtenbacher, and G. Czakó, *Phys. Chem. Chem. Phys.* **14**, 1085 (2012).

[b] J. Tennyson and S.N. Yurchenko , *Int. J. Quant. Chem.* **117**, 92, (2017).

[c] J. Tennyson, *J. Chem. Phys.* **145**, 120901 (2016).

[d] S.N. Yurchenko, W. Thiel, P. Jensen, *J. Mol. Spectrosc.* **245**, 126–140 (2007).

WB02   9:04 – 9:19

SPECTRA AND ASSIGNMENTS OF HOT METHANE UP TO 1000 K IN THE 1–2 μm REGION

ANDY WONG, PETER F. BERNATH, *Department of Chemistry and Biochemistry, Old Dominion University, Norfolk, VA, USA*; MICHAEL REY, *Groupe de Spectrométrie Moléculaire et Atmosphérique, UMR CNRS 7331, Université de Reims, Reims Cedex 2, France*; ANDREI V. NIKITIN, *Atmospheric Spectroscopy Div., Institute of Atmospheric Optics, Tomsk, Russia*; VLADIMIR TYUTEREV, *Groupe de Spectrométrie Moléculaire et Atmosphérique, UMR CNRS 7331, Université de Reims, Reims Cedex 2, France.*

Infrared absorption spectra of hot methane up to 1000 K were recorded with a high-resolution Fourier transform spectrometer in the 5200–9300 cm^{-1} spectral region. The experimental observations were compared to the predictions of variational calculations. Preliminary quantum number assignments were made for the observed features. Generally good agreement was found between observations and calculations particularly in the Tetradecad region, from 2.1 to 1.6 μm. Spectra in the Icosad (1.6–1.3 μm) and Triacontad (1.25–1.1 μm) regions suffered from some interference from a hot water impurity.

WB03　　　9:21 – 9:36

NEURAL NETWORK VS GUASSIAN PROCESS FITTING FOR REPRESENTING POTENTIAL ENERGY SURFACES

SERGEI MANZHOS, *Department of Mechanical Engineering, National University of Singapore, Singapore, China*; TUCKER CARRINGTON, *Department of Chemistry, Queen's University, Kingston, ON, Canada*; ROMAN KREMS, RODRIGO HERNANDEZ, *Departments of Chemistry, Physics and Astronomy, University of British Columbia, Vancouver, BC, Canada.*

Many methods have been proposed for fitting potential energy surfaces. Unfortunately, there are few comparative studies. In this paper, we compare neural networks (NN) with Gaussian process (GP) regression. We re-fit an accurate PES of formaldehyde and compare PES errors on the entire point set used to solve the vibrational Schrödinger equation, i.e. the only error that matters in quantum dynamics calculations. We also compare the vibrational spectra computed on the underlying reference PES and the NN and GP potential surfaces. The NN and GP surfaces are constructed with exactly the same points and the corresponding spectra are computed with the same points and the same basis. The GP fitting error is lower and the GP spectrum is more accurate. The best NN fits to 625/1250/2500 symmetry unique potential energy points have global PES root mean square errors (RMSE) of 6.53/2.54/0.86 cm-1, whereas the best GP surfaces have RMSE values of 3.87/1.13/0.62 cm-1, respectively. When fitting 625 symmetry unique points, the error the first 100 vibrational levels is only 0.06 cm-1 with the best GP fit, whereas the spectrum on the best NN PES has an error of 0.22 cm-1, with respect to the spectrum computed on the reference PES. This error is reduced to about 0.01 cm-1 when fitting 2,500 points with either NN or GP. We also find that the GP surface produces a relatively accurate spectrum when obtained based on as few as 313 points.

WB04　　　9:38 – 9:53

PolyMLR: AN ANALYTIC MODEL FOR POLYATOMIC POTENTIALS WITH FEWER UNPHYSICAL PARAMETERS. APPLICATION TO CO_2.

NIKESH S. DATTANI[a], *Department of Chemistry, Kyoto University, Kyoto, Japan.*

One has to calculate thousands or millions of *ab initio* points for potential energy surfaces even for molecules with only a few atoms. For diatomics, the MLR (Morse/long-range)[b,c] model has been very successful, making it possible to represent the entire curve accurately with just a few *ab initio* points, or a few spectral lines. With the MLR model it is also possible to extrapolate and interpolate in a way that allows successful predictions of energy level locations several thousand cm^{-1} away from the data region[d].

However no analogous model has existed yet for the intramolecular potentials of polyatomic molecules. A simple model is presented which accurately describes some small molecules with far fewer parameters than previous models, and can be extended to larger molecules too. The benefit of having a good model function is orders of magnitude greater for polyatomics than for diatomics since the amount of data needed for an accurate potential is reduced in each dimension. For example if the calculation of 100 *ab initio* points is reduced to 10 in a diatomic molecule, we may estimate that this factor of 10 reduction in cost becomes at least 10^{10} for a molecule whose potential depends on 10 radial coordinates.

As an example, an analytic potential for CO_2 is built, which requires fewer parameters than the previous state-of-the-art analytic potential, and obeys the theoretical long-range behavior more closely than all previous potentials, including inclusion of the Axelrod-Teller three-body interaction. The model is based on accurate diatomic potentials representing all atom-atom pairwise interactions, and for CO_2, a three-body correction representing the rest of the energy. This emphasizes the value of accurate molecular spectroscopy for simple diatomics, which is sometimes considered to be less interesting than research involving large molecules. Diatomic potentials are valuable as building blocks for large-molecule potentials.

An open-source computer program for building PolyMLR potentials for polyatomic molecules is introduced.

---

[a]nik.dattani@gmail.com
[b]Dattani N. S., Le Roy R. J., Ross A., Linton C. (2008) Proceedings of the 63rd Annual International Symposium on Molecular Spectroscopy. **p301**
[c]Le Roy R. J., Dattani N. S., Coxon J. A., Ross A. J., Crozet P., Linton C. (2009) *J. Chem Phys.* **131**, 204309
[d]Dattani N. S., Le Roy R. J. (2011) *J. Mol. Spec.*. **268**, 199-210.; Semczuk M., et al. (2013) *Phys. Rev. A* **87**, 052505

WB05   9:55 – 10:10

VMS-ROT: A NEW MODULE OF THE VIRTUAL MULTIFREQUENCY SPECTROMETER FOR SIMULATION, INTERPRETATION, AND FITTING OF ROTATIONAL SPECTRA

DANIELE LICARI, NICOLA TASINATO, LORENZO SPADA, *Scuola Normale Superiore, Scuola Normale Superiore, Pisa, Italy*; CRISTINA PUZZARINI, *Dep. Chemistry 'Giacomo Ciamician', University of Bologna, Bologna, Italy*; VINCENZO BARONE, *Scuola Normale Superiore, Scuola Normale Superiore, Pisa, Italy*.

The Virtual Multifrequency Spectrometer (VMS) is a tool that aims at integrating a wide range of computational and experimental spectroscopic techniques with the final goal of disclosing the static and dynamic physical-chemical properties "hidden" in molecular spectra. VMS is composed of two parts, namely, VMS-Comp, which provides access to the latest developments in the field of computational spectroscopy, and VMS-Draw, which provides a powerful graphical user interface (GUI) for an intuitive interpretation of theoretical outcomes and a direct comparison to experiment. In the present work, we introduce VMS-ROT, a new module of VMS that has been specifically designed to deal with rotational spectroscopy. This module offers an integrated environment for the analysis of rotational spectra: from the assignment of spectral transitions to the refinement of spectroscopic parameters and the simulation of the spectrum. While bridging theoretical and experimental rotational spectroscopy, VMS-ROT is strongly integrated with quantum-chemical calculations, and it is composed of four independent, yet interacting units: (1) the computational engine for the calculation of the spectroscopic parameters that are employed as a starting point for guiding experiments and for the spectral interpretation, (2) the fitting-prediction engine for the refinement of the molecular parameters on the basis of the assigned transitions and the prediction of the rotational spectrum of the target molecule, (3) the GUI module that offers a powerful set of tools for a vis-a-vis comparison between experimental and simulated spectra, and (4) the new assignment tool for the assignment of experimental transitions in terms of quantum numbers upon comparison with the simulated ones. The implementation and the main features of VMS-ROT are presented, and the software is validated by means of selected test cases ranging from isolated molecules of different sizes to molecular complexes. VMS-ROT offers an integrated environment for the analysis of the rotational spectra, with the innovative perspective of an intimate connection to quantum-chemical calculations that can be exploited at different levels of refinement, as an invaluable support and complement for experimental studies.

Intermission

WB06   10:46 – 11:01

PHOTOPHYSICS AND ELECTRONIC STRUCTURE STUDIES OF PROTONATION OF QUINOLINE

HIRDYESH MISHRA, *Department of Physics, Banaras Hindu University, Varanasi, Uttar Pradesh, India*.

Study of the photophysics and electronic structure properties of quinoline and its derivatives have been the subject of considerable interest because of their commercial and pharmaceutical applications. Since the quinoline ring is the basic fluorophor unit in all its derivatives, it is important to understand the change in dynamics and electronic structure of quinoline in presence some external perturbation. Being isoelectronic with naphthalene, these molecules provide useful comparisons for checking the electronic and vibrational state assignments, ionization potentials, and other properties of the parent hydrocarbon. In addition, these molecules possess nonbonding electrons which give rise to n– π^* states. The location and characterization of these states are of both theoretical and practical significance. Further, solvents have an important influence on the fluorescence property of N-heterocyclic compounds. Experimentally, Quinoline shows vibronic absorption spectrum and corresponding large Stoke shifted broad fluorescence emission spectrum having very low quantum yield and dual decay time, however protonated quinoline shows red shifted fluorescence spectrum with increase in quantum yield and fluorescence decay become mono-exponential. To understand the vibronic structure of electronic absorption spectra and photophysics of protonation of quinoline, both vibronic and electronic structure studies of quinoline (Q) and protonated quinoline (QH+) were carried out along with vibrational calculations for absorption and fluorescence bands at B3LYP 6-311++G(d, p) level in ground and excited state by density functional methods (DFT) and (time-dependent density functional) TD-DFT methods respectively with the help of Gaussian 09 software. Normal mode mixing is taken into account by the Duschinsky transformation. The vibronic structure of strongly dipole-allowed transitions is calculated within the Franck–Condon approximation. Weakly dipole-allowed and dipole-forbidden transitions are treated within the Franck–Condon–Herzberg–Teller and Herzberg–Teller approximation, respectively. A good correlation between computational spectroscopic calculations and experiment results are found to understand the photo-physics of protonation of quinoline.

WB07 11:03 – 11:18

EXPERIMENTAL AND THEORETICAL INVESTIGATIONS OF THE THRESHOLD PHOTOELECTRON SPECTRUM OF THE CH_2 RADICAL

B. GANS, F. HOLZMEIER, <u>L. H. COUDERT</u>, *Institut des Sciences Moléculaires d'Orsay, Université Paris-Sud, Orsay, France*; J.-C. LOISON, *Institut des Sciences Moléculaires, Université de Bordeaux, Talence, France*; G. A. GARCIA, *L'Orme des Merisiers; Saint Aubin BP 48, Synchrotron SOLEIL, Gif sur Yvette, France*; C. ALCARAZ, *Laboratoire de Chimie Physique, Université Paris-Sud, Orsay, France*.

The methylene cation CH_2^+ is spectroscopically poorly characterized as it is difficult to produce in large amounts. It is subject to the Renner-Teller effect giving rise to ground $\widetilde{X}^{+\,2}A_1$ and excited $\widetilde{A}^{+\,2}B_1$ electronic states. Photoelectron spectroscopy of the methylene radical CH_2 allows us to gain information about both CH_2 and its cation. The former is also theoretically challenging as it is a very non-rigid species characterized by a barrier to linearity of less than 2000 cm^{-1} in its ground $\widetilde{X}\,^3B_1$ electronic state. The first photoelectron spectra of CH_2 were investigated using pulsed-field-ionization zero-kinetic-energy spectroscopy.[a] A rotationally resolved spectrum containing $\widetilde{X}^{+\,2}A_1 \leftarrow \widetilde{X}\,^3B_1$ transitions was recorded from 83600 to 84070 cm^{-1} and analyzed in terms of CH_2^+ rotational constants.

The threshold photoelectron spectrum of CH_2 has been recorded from 9.8 to 12 eV (79040 to 96800 cm^{-1}) using a recently developed flow tube reactor[b] and VUV synchrotron radiation. This new spectrum spans a larger energy range than the previous ones,[a] but with less resolution. It displays narrow and broad features due respectively to the $\widetilde{X}^{+\,2}A_1 \leftarrow \widetilde{X}\,^3B_1$ and $\widetilde{A}^{+\,2}B_1 \leftarrow \widetilde{X}\,^3B_1$ ionizing transitions. Using new *ab initio* potential energy surfaces and available ones,[c] the photoelectron spectrum is currently being computed using two models. The first one accounts for the large amplitude bending mode and the rotation only; the second one, also accounts for the stretching modes. The experimental and theoretical spectra will be discussed in the paper.

---

[a]Willitsch *et al.*, *J. Chem. Phys.* **117** (2002) 1939; and Willitsch & Merkt, *ibid.* **118** (2003) 2235
[b]Garcia *et al.*, *J. Chem. Phys.* **142** (2015) 164201
[c]Jensen & Bunker, *J. Chem. Phys.* **89** (1988) 1327; and Jensen, Brumm, Kraemer & Bunker, *J. Molec. Spectrsoc.* **172** (1995) 194

WB08 11:20 – 11:35

0.06 cm^{-1} DISCREPANCY FOR $Li_2 \rightarrow 2Li$ AND 0.994 cm^{-1} FOR $C \rightarrow C^+$ BETWEEN LABORATORY AND COMPUTER SPECTROMETERS.

<u>NIKESH S. DATTANI</u>[a], *Department of Chemistry, Kyoto University, Kyoto, Japan.*

The energy at the empirical bond length of $Li_2(1^3\Sigma_u^+)$ of 4.1700Å [b] was obtained at all-electron FCI level with an aug-cc-pCV5Z-NR basis set, all-electron CCSDT(Q) with aug-cc-pCV7Z-NR, and all-electron CCSD(T) with aug-cc-pCV8Z-NR; along with corrections due to special relativity converged with respect to electron correlation and basis set size using the spin-free Dirac-Coulomb Hamiltonian, and further such corrections at the Hartree-Fock level using the Breit and Gaunt Hamiltonians. Corrections to the point-size nucleus approximation were calculated but found to be negligible. The result was compared to the lowest energy of the best empirical potentials[b] with the empirical Born-Oppenheimer breakdown corrections removed, making it essentially an infinite-mass to infinite-mass comparison. The discrepancy between the energy obtained from laboratory spectroscopy and the energy obtained completely by the computer was only 0.06 cm^{-1}, which is of the same order of magnitude as the uncertainty on the empirical value, which is ± 0.007 cm^{-1} before including the added uncertainty coming from the Born-Oppenheimer breakdown parameter u_0 which itself has an uncertainty of 0.01 cm^{-1}. It is discussed what is necessary for the computer spectrometer to outperform the laboratory spectrometer.

The ionization energy of the carbon atom was calculated at all-electron FCI level with aug-cc-pCV8Z-NR and aug-cc-pCV7Z-NR basis sets (the latter only for basis set extrapolation); along with corrections due to special relativity converged with respect to electron correlation and basis set size using the $1e^-$ X2C Hamiltonian, further corrections using state-averaged Dirac-Fock for the contribution from the Breit Hamiltonian and some QED contributions; along with DBOC corrections to the clamped nucleus approximation converged with respect to electron correlation and basis set size. Again, corrections to the point-size nucleus approximation were calcualted but found to be negligible. The final energy was compared to the very recent experimental value published by NIST[c] with the experimental spin-orbit lowering of 12.672508 cm^{-1} removed. The discrepancy was 0.994 cm^{-1} compared to the ± 0.009 cm^{-1} uncertainty in the laboratory value.

---

[a]nik.dattani@gmail.com
[b]Dattani N. S., Le Roy R. J. (2011) *J. Mol. Spec.*. **268**, 199-210.; Semczuk M., *et al.* (2013) *Phys. Rev. A* **87**, 052505
[c]Haris K., Krimada A. E., (2017) arXiv:1704.07474.

WC. Mini-symposium: Far-Infrared Spectroscopy

Wednesday, June 20, 2018 – 8:30 AM

Room: 1024 Chemistry Annex

Chair: Olivier Pirali, Synchrotron SOLEIL, Gif-sur-Yvette, France

---

WC01   ***INVITED TALK***   8:30 – 9:00

LINE INTENSITIES AND BROADENING COEFFICIENTS FROM HIGH RESOLUTION FAR INFRARED SPECTRA

<u>JEAN VANDER AUWERA</u>, *Service de Chimie Quantique et Photophysique, Université Libre de Bruxelles, Brussels, Belgium.*

Molecular lines observed in high resolution far infrared spectra are associated with pure rotation transitions or belong to low-energy vibrational bands. Together with other parameters characterizing them, the intensities and broadening coefficients of these lines are required for example to analyze spectra of planetary atmospheres or to gain insight into the physics of the studied molecules or intermolecular interactions. Pure rotation line intensities can also be used to determine the particle density of chemically unstable species, allowing to obtain line intensities in another spectral range as was for example done for hypochlorous acid[a] and ozone.[b]

This lecture will deal with the measurement of the intensities and broadening coefficients of molecular lines observed in high resolution far infrared absorption spectra. It will skim over measurements carried out using THz spectroscopy and focus on Fourier transform spectroscopy. This latter technique is now commonly associated with synchrotron radiation, the high brightness and highly collimated nature of which being big advantages at low energies over conventional sources such as mercury lamps.[c] The lecture will present and discuss some recent and ongoing measurements carried out relying on Fourier transform far infrared spectra recorded using synchrotron radiation, highlighting some aspects specific to these retrievals.

---

[a] J. Vander Auwera, J. Kleffmann, J.-M. Flaud, G. Pawelke, H. Bürger, D. Hurtmans, R. Pétrisse, J. Mol. Spectrosc. 204 (2000) 36–47.

[b] B.J. Drouin, T.J. Crawford, S. Yu, J. Quant. Spectrosc. Radiat. Transf. 203 (2017) 282–292.

[c] A.R.W. McKellar, J. Mol. Spectrosc. 262 (2010) 1–10.

WC02   9:04 – 9:19

HIGH RESOLUTION IR SPECTROSCOPY AND ANALYSIS OF THE BENDING DYAD OF RuO_4

SÉBASTIEN REYMOND-LARUINAZ, *Département de Physico-chimie, CEA/Saclay, CEA, DEN, Gif-sur-Yvette, France*; MBAYE FAYE, , *LISA CNRS et Universités Paris Est et Paris Diderot , Creteil, France*; VINCENT BOUDON, *Laboratoire ICB, CNRS/Université de Bourgogne, DIJON, France*; DENIS DOIZI, *Département de Physico-chimie, CEA/Saclay, CEA, DEN, Gif-sur-Yvette, France*; <u>LAURENT MANCERON</u>, *AILES beam line, Synchrotron Soleil, Gif-sur-Yvette, France.*

RuO_4 is a heavy tetrahedral molecule of interest in several fields. Due to its chemical toxicity and radiological impact of its 103 and 106 isotopologues, the possible remote sensing of this compound in the atmosphere in case of possible severe nuclear accident has renewed interest in its spectroscopic properties. We investigate here, for the first time at high resolution, the bending modes region in the far infrared. High resolution FTIR spectra have been recorded near room temperature, using a specially constructed cell and an isotopically pure sample of $^{102}RuO_4$. New assignments and effective Hamiltonian parameter fits for the main isotopologue ($^{102}RuO_4$) have been performed, treating the whole ν_2-ν_4 bending mode dyad. We provide precise effective Hamiltonian parameters, including band centers and Coriolis interaction parameters [1].

[1] S. Reymond-Laruinaz, M. Faye, V. Boudon, D. Doizi, L. Manceron, "High-resolution Infrared Spectroscopy and analysis of the ν_2-ν_4 bending dyad of Ruthenium Tetroxide", J. Mol. Spectrosc. 336 (2017) 29.

WC03　9:21 – 9:36

HIGH RESOLUTION STUDY OF THE ν_2 AND ν_5 ROVIBRATIONAL FUNDAMENTAL BANDS OF THIONYL CHLORIDE : INTERPLAY OF AN EVOLUTIONARY ALGORITHM AND A LINE-BY-LINE ANALYSIS

ANTHONY ROUCOU, GUILLAUME DHONT, ARNAUD CUISSET, *Laboratoire de Physico-Chimie de l'Atmosphère, Université du Littoral Côte d'Opale, Dunkerque, France*; MARIE-ALINE MARTIN-DRUMEL, *CNRS, Institut des Sciences Moleculaires d'Orsay, Orsay, France*; SVEN THORWIRTH, *I. Physikalisches Institut, Universität zu Köln, Köln, Germany*; DANIELE FONTANARI, *Laboratoire de Physico-Chimie de l'Atmosphère, Université du Littoral Côte d'Opale, Dunkerque, France*; W. LEO MEERTS, *Institute for Molecules and Materials (IMM), Radboud University Nijmegen, Nijmegen, Netherlands*.

The ν_2 and ν_5 fundamental bands of thionyl chloride (SOCl$_2$) were measured in the 420 cm^{-1} - 550 cm^{-1} region using the FT-Far-IR spectrometer exploiting synchrotron radiation on the AILES beamline at SOLEIL. A straightforward line-by-line analysis is complicated by the high congestion of the spectrum due to both the high density of SOCl$_2$ rovibrational bands and the presence of the strong ν_2 fundamental band of sulfur dioxide produced by hydrolysis of SOCl$_2$ with residual water. To overcome this difficulty, our assignment procedure for the two isotopologues ^{32}S^{16}O^{35}Cl$_2$ and ^{32}S^{16}O^{35}Cl^{37}Cl alternates between a direct fit of the spectrum, via a global optimization technique, and a traditional line-by-line analysis. The global optimization, based on an evolutionary algorithm [a], produces rotational constants and band centers that serve as useful starting values for the subsequent spectroscopic analysis. This work also helped to identify the pure rotational submillimeter spectrum of ^{32}S^{16}O^{35}Cl$_2$ in the v$_2$ = 1 and v$_5$ = 1 vibrational states. A global fit gathering all the data of SOCl$_2$ from the microwave, submillimeter, and far-infrared spectral regions [b,c] has been performed [d], showing that no major perturbation of rovibrational energy levels occurs for the main isotopologue of the molecule.

[a] W. Leo Meerts and Michael Schmitt, Int. Rev. Phys. Chem. **25** (2006) 353-406
[b] M. A. Martin-Drumel et al. J. Mol. Spectrosc. **315**, (2015), 30-36
[c] M. A. Martin-Drumel et al., J. Chem. Phys., **144**(8), (2016), 084305
[d] A. Roucou et al. J. Chem. Phys. **147**, (2017), 054303

WC04　9:38 – 9:53

FAR-INFRARED AND MICROWAVE SPECTROSCOPY OF HCOOCH$_3$

KAORI KOBAYASHI, RYO OHYAMA, *Department of Physics, University of Toyama, Toyama, Japan*; NOBUKIMI OHASHI, , *Kanazawa University, Kanazawa, Japan*; DENNIS W. TOKARYK, *Department of Physics, University of New Brunswick, Fredericton, NB, Canada*; BRANT E. BILLINGHURST, *EFD, Canadian Light Source Inc., Saskatoon, Saskatchewan, Canada*.

Methyl formate (HCOOCH$_3$), is an important interstellar molecule, first detected about 40 years ago in a giant molecular gas cloud SGR B2. [a] More than 1000 rotational transitions including those of its isotopologues and in the torsional excited states have been observed from several astrophysical sources. The laboratory spectra of methyl formate (HCOOCH$_3$) exhibit many unassigned transitions and many of them would be due to rotational transitions in the low-lying excited states. Previous astronomical identification of the torsional excited states indicates that rotational transitions in the more excited vibrational states can also be observed in astrophysical sources. In laboratory, two new series of transitions have been identified in the rotational data, and based on intensity, they lie about 300 cm^{-1} and 450 cm^{-1} above the ground state. [b]

There are many candidate vibrational excited states in this region and therefore, we decide to observe high-resolution far-infrared spectra to identify the responsible vibrational excited state and also to provide a feedback to the assignment of the microwave spectra.

The experiment was performed on the Far-Infrared Beamline of the Canadian Light Source synchrotron. Methyl formate at a pressure of about 2-8 mTorr was admitted into a 2-m-long White cell cooled to 198K. The cell was set to provide 36 transits of the far-infrared synchrotron radiation, for a total path length of 72 m. Spectra were obtained with both a Si:bolometer and Cu:Ge detector at full resolution (0.00096 cm^{-1}). Very dense spectra of the C-O-C deformation and C-O torsional modes were obtained with high signal-to-noise ratio between 300-360 cm^{-1}. More than 30000 transitions were observed and the detail of analysis will be reported.

[a] R. D. Brown, J. G. Crofts, P. D. Godfrey, F. F. Gardner, B. J. Robinson, & J. B. Whiteoak, *Astrophys. J. Lett.* **197**, L29 (1975).
[b] Y. Sakai, K. Kobayashi, M. Tsukamoto, M. Fujitake, & N. Ohashi, *International Symposium on Molecular Spectroscopy, 67th meeting.* **RF05** (2012).

Intermission

WC05   10:29 – 10:44
ON THE IMPORTANCE OF FAR-INFRARED SPECTROSCOPY FOR NON-POLAR SPHERICAL-TOP MOLECULES

VINCENT BOUDON, *Laboratoire ICB, CNRS/Université de Bourgogne, DIJON, France*; OLIVIER PIRALI, *AILES beamline, Synchrotron SOLEIL, Saint Aubin, France*; LAURENT MANCERON, *Synchrotron SOLEIL, CNRS-MONARIS UMR 8233 and Beamline AILES, Saint Aubin, France*; MBAYE FAYE, , *LISA CNRS et Universités Paris Est et Paris Diderot , Creteil, France*.

Highly-symmetric molecules like spherical-top possess no permanent dipole moment. Thus, at least in first approximation, their pure rotation spectrum is forbidden (or even strictly forbidden in the case of centrosymmetric species). It may thus seem useless to consider their far-infrared or THz spectrum. Nevertheless, this spectral region can provide invaluable information for these molecules. Firstly, in the case of tetrahedral species of type XY_4, pure rotation lines in the ground or in excited vibrational states can be induced through centrifugal distortion. Secondly, the strict selection rules for spherical-top molecules make some fundamental levels inaccessible though direct absorption. Here again, far-infrared studies can help to reach them through the study of low-lying difference bands. Thirdly, some larger and/or heavier species possess weak bands at low wavenumbers. In this talk, we will summarize some recent studies performed on the AILES beamline of the SOLEIL Synchrotron facility that illustrate these different cases with CH_4, CF_4, OsO_4, RuO_4, $C_{10}H_{16}$, $C_6N_4H_{10}$, C_8H_8 and SF_6.

WC06   10:46 – 11:01
IMPROVED FAR-INFRARED AMMONIA INTENSITY FROM EMPIRICAL HAMILTONIAN MODEL

JOHN PEARSON, SHANSHAN YU, KEEYOON SUNG, JENIVEVE PEARSON, BRIAN DROUIN, *Jet Propulsion Laboratory, California Institute of Technology, Pasadena, CA, USA*; OLIVIER PIRALI, *AILES beamline, Synchrotron SOLEIL, Saint Aubin, France*.

In the 2016 meeting we reported our experimental linelist of NH_3 in 50-660 cm^{-1} (See Paper FE08). The retrieved line positions and intensities were used as standards to validate HITRAN 2012 database and our empirical Hamiltonian models (Yu et al. 2010; Pearson et al. 2016). While the line position comparisons with HITRAN and our Hamiltonian models were excellent, the intensity comparisons were less satisfactory. During the past two years, we have updated our Hamiltonian model to improve the intensity prediction. In this presentation, we will report our significant improvement on intensity predictions, especially for the $\Delta K = 3$ forbidden transitions. We will also report comparisons of HITRAN 2016 with our existing experimental spectra.

WC07   11:03 – 11:18

THE JET-COOLED HIGH-RESOLUTION FAR-IR SPECTRUM OF FORMIC ACID CYCLIC DIMER

SABATH BTEICH, MANUEL GOUBET, <u>THERESE R. HUET</u>, *UMR 8523 - PhLAM - Physique des Lasers Atomes et Molécules, University of Lille, CNRS, F-59000 Lille, France*; OLIVIER PIRALI, *Institut des Sciences Moléculaires d'Orsay, Université Paris-Sud, Orsay, France*; PASCALE SOULARD, PIERRE ASSELIN, *CNRS, De la Molécule aux Nano-Objets: Réactivité, Interactions, Spectroscopies, MONARIS, Sorbonne Université, PARIS, France*; ROBERT GEORGES, *IPR UMR6251, CNRS - Université Rennes 1, Rennes, France.*

The cyclic conformation of the formic acid dimer $(HCOOH)_2$ (FACD) is an elementary system under study to understand the concerted hydrogen transfer through equivalent hydrogen bonds. To this end, high-resolution molecular spectroscopy coupled to quantum chemical calculations is a powerful technique. So far molecular parameters have been reported for the ground state, the ν_{22}, ν_{21} and $\nu_{12}+\nu_{14}$ vibrational levels, from spectra recorded using laser-based techniques in the 7.2 μm region [a].

Last year we reported the spectra associated with six rotationally resolved IR bands of FADC, recorded under jet-cooled conditions with the FTIR JET-AILES apparatus at synchrotron SOLEIL, and a QCL spectrometer at MONARIS.

A special attention was paid to the analysis of the far-IR ν_{24} fundamental band, associated with the intermolecular in-plane bending mode. Splittings due to vibration-rotation-tunneling motions were clearly observed and assigned. We will present the results of the analysis, using the model recently proposed by Zhang *et al* ,[a] which is based on the inclusion of a c-type Coriolis-like coupling term to reproduce the rotation-tunneling interaction. This model was successfully used in the 7.2 μm region.

The authors gratefully acknowledge the staff of the AILES beamline at synchrotron SOLEIL. The present work was funded by the French ANR Labex CaPPA through the PIA (contract ANR-11-LABX-0005-01), by the Regional Council Hauts de France, and by the European Funds for Regional Economic Development (FEDER).

---

[a] M. Ortlieb and M. Havenith, *J. Phys. Chem. A* **111**, 7355 (2007) ; K. G. Goroya, Y. Zhu, P. Sun, and C. Duan, *J. Chem. Phys.* **140**, 164311 (2014) ; Y. Zhang, W. Li, W. Luo, Y. Zhu, and C. Duan, *J. Chem. Phys.* **146**, 244306 (2017).

WC08   11:20 – 11:35

THE STRUCTURE OF *gauche*-BUTADIENE: INSIGHTS FROM THE CENTIMETER, MILLIMETER, AND FIR-INFRARED HIGH RESOLUTION SPECTRA

<u>MARIE-ALINE MARTIN-DRUMEL</u>, *CNRS, Institut des Sciences Moleculaires d'Orsay, Orsay, France*; JOSHUA H BARABAN, *Chemistry, Ben-Gurion University of the Negev, Beer-Sheva, Israel*; BRYAN CHANGALA, *JILA, National Institute of Standards and Technology and Univ. of Colorado Department of Physics, University of Colorado, Boulder, CO, USA*; MATTHEW NAVA, *Department of Chemistry, MIT, Cambridge, MA, USA*; JESSIE P PORTERFIELD, *AMP Division, Harvard-Smithsonian Center for Astrophysics, Cambridge, MA, USA*; BARNEY ELLISON, *Department of Chemistry and Biochemistry, University of Colorado, Boulder, CO, USA*; OLIVIER PIRALI, *AILES beamline, Synchrotron SOLEIL, Saint Aubin, France*; JOHN F. STANTON, *Physical Chemistry, University of Florida, Gainesville, FL, USA*; MICHAEL C McCARTHY, *Atomic and Molecular Physics, Harvard-Smithsonian Center for Astrophysics, Cambridge, MA, USA.*

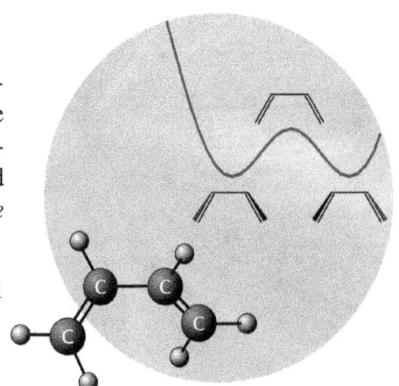

Recent investigation of the centimeter spectrum of *gauche*-butadiene has unambiguously established a non-planar conformation for this fundamental, archetypal diene for the Diels-Alder reaction [1]. We will present subsequent theoretical and experimental investigations aimed at determining a highly accurate molecular structure and the barrier height for interconversion between the two equivalent tunneling *gauche* forms.

[1] J. H. Baraban, M.-A. Martin-Drumel, *et al.*, *Angewandte Chemie*, **57**, 1821 (2018)

WC09

IMPROVE THE PREDICTION ACCURACY OF ISOTOPOLOGUE MICROWAVE SPECTRA BY COMBINING AMES-296K SO_2 IR LISTS WITH EXPERIMENTAL MODELS: A BENCHMARK STUDY

XINCHUAN HUANG, *Carl Sagan Center, SETI Institute, Moutain View, CA, USA*; DAVID SCHWENKE, *MS 258-2, NAS Facility, NASA Ames Research Center, Moffett Field, CA, USA*; TIMOTHY LEE, *Space Science and Astrobiology Division, NASA Ames Research Center, Moffett Field, CA, USA*.

Theoretical rovibrational IR line lists computed on the empirically refined potential energy surfaces (PES) have excellent isotopologue consistency and reliablity to push the ongoing pursuit of "Best Theory + reliable High-resolution Experiment" (BTRHE) strategy to a higher level of prediction accuracy. The SO_2 benchmark uses experimental (Expt) data based Effective Hamiltonian (EH) models of a few SO_2 isotopologues and Ames-296K IR line lists of 30 SO_2 isotopologues. For microwave (MW) intensity, the Einstein A_{21} coefficients demostrate isotopologue consistency better than 99.9%, which can help identify errors and inconsistencies in existing effective dipole moment (EDM) models or lab spectra analysis. For MW line position, the study goes from simple trial to systematic investigations on the convergence, uncertainties, higher order term effects, fixing EH parameters, mass coordinates, and other prediction scheme, etc. We confirm the feasibility of a two-orders-of-magnitude accuracy improvement over the original Ames IR line lists. By refining the rotational constants and quartic centrifugal distortion constatnts using the linear or quadratic extrapolations on their differences between the EH(Expt) and EH(Ames) IR list based parameter values, A_0 / B_0 / C_0 deviations can be as small as 0.01-0.02 MHz, and line position deviations can be reduced to 0-5 MHz for $J<30$, $K_a<10$-15 transitions.

We report a microwave line set consisting of 644,636 transitions with reliable 296K IR intensity and Einstein A_{21} coefficient for all 30 isotopologues of SO_2. The line position predictions are the best available, which will facilitate both the astronomical identification and lab MW analysis of those unobserved minor isotopologues. The procedure can be easily extended onto rovibrational bands and other molecular systems, while data precision higher than 0.003-0.03 MHz, or 1E-6 - 1E-7 cm^{-1}, is preferred.

WD. Radicals
Wednesday, June 20, 2018 – 8:30 AM
Room: 217 Noyes Laboratory

Chair: Neil J Reilly, University of Massachusetts Boston, Boston, MA, USA

WD01   8:30 – 8:45

SUB-DOPPLER INFRARED SPECTROSCOPY OF JET COOLED HCCL DIRADICAL: THE CH STRETCH AND VIBRATIONAL COUPLING IN THE GROUND ELECTRONIC STATE

ANDREW KORTYNA, PRESTON G. SCRAPE, *JILA, National Institute of Standards and Technology and Univ. of Colorado, Boulder, CO, USA*; DANIEL LESKO, *Department of Chemistry and Biochemistry, University of Colorado, Boulder, CO, USA*; DAVID NESBITT, *JILA, National Institute of Standards and Technology and Univ. of Colorado, Boulder, CO, USA*.

Diradical carbenes have long been recognized as important intermediates in a range of chemical processes, with the carbene's chemical reactivity being sensitive to the particular ground-state electronic structure. We have undertaken an investigation of chlorocarbene (HCCl) by seeding $CHCl_3$ into a Ne/He/H_2 mixture and passing this mixture through a pulsed slit discharge. In the discharge environment, the $CHCl_3$ undergoes a double Cl atom removal process through a combination of electron dissociative attachment and hydrogen abstraction. The subsequent jet expansion cools the HCCl diradical to a 32 K rotational temperature. With the goal of assisting the search for HCCl chemistry in interstellar molecular clouds, the rotational constants for the ground singlet state of both the ^{35}Cl and ^{37}Cl isotopologues are determined through least-squares fits of ground-state combination differences to an asymmetric top Watson Hamiltonian. A Watson Hamiltonian is also used to extract rotational constants for the nominally (100) vibrationally excited state, with a highly mixed combination band (nominally (012), one quantum of H-C-Cl bend plus two quanta of C-Cl stretch) of comparable intensity found within a few wavenumbers (cm^{-1}) of the CH stretch band origin. The proximity of these two bands and the comparable infrared intensities of both combination and fundamental bands points towards a highly mixed state with strong anharmonic coupling between these two zeroth order modes. We quantify the anharmonic coupling in a 2x2 matrix deperturbation treatment and find it to be similar in magnitude to the previously measured spin-orbit coupling constants between the singlet and nearby lying triplet manifold of states.

WD02   8:47 – 9:02

2C-R4WM SPECTROSCOPY OF JET COOLED NO_3 (II)

MASARU FUKUSHIMA, TAKASHI ISHIWATA, *Information Sciences, Hiroshima City University, Hiroshima, Japan*.

We have generated NO_3 in a supersonic free jet expansion, and observed laser induced fluorescence (LIF) and two-color resonant four-wave mixing (2C-R4WM) signals. We have measured dispersed fluorescence (DF) spectra from single vibronic levels. Among the vibrational levels observed in the DF spectrum from the vibration-less level, the ν_1 and ν_3 fundamental regions ($\sim$1050 and $\sim$1500 cm^{-1} regions, respectively) are now active for discussion, and thus we have tried to measure the rotationally resolved 2C-R4WM spectra[a]. The 2C-R4WM spectrum of the ν_3 fundamental region is consistent with a previous infra-red investigation[b], and that of ν_1 leads to the identification of the $K = 0$ and $N = 1$ level of the ν_1 fundamental for the first time. We have found an additional level near ν_1[c], and the 2C-R4WM spectrum of the level shows two rotational transitions separated by 0.27 cm^{-1}. Although the 0.27 cm^{-1} separation is about 10 times larger than the spin splitting, $\sim$0.025 cm^{-1}, of the $K = 0$ and $N = 1$ levels at the other a'_1 levels with $l = 0$, such as vibration-less and ν_1 (the latter value of which, 0.025 cm^{-1}, cannot be resolved under our instrumental resolution), the two transitions are thought to correspond to those terminating to spin sub-levels, $J = 0.5$ and $= 1.5$, at the present. We have assigned the additional level to $3\nu_4$ (a'_1) with $l = \pm 3$. For Σ vibronic levels with $K = 0$, such as $v_d = 1$ and $l = 1$, of a $^2\Pi$ electronic state, it is well known that $^2\Sigma^{(+)}$ and $^2\Sigma^{(-)}$ vibronic levels have relatively large Ω- or ρ-type doubling due to non-zero Λ, in spite of the Σ vibronic levels[d]. It is thought that the unexpectedly large spin splitting, 0.27 cm^{-1}, is induced by spin-vibration interaction, which has been discussed for degenerate vibronic levels of non-degenerate electronic states, $^2\Sigma$ and $^3\Sigma$, of linear polyatomic molecules[e].

---

[a]M. Fukushima and T. Ishiwata, 71st ISMS, paper RF01 (2016).
[b]K. Kawaguchi, et al., J. Mol. Spectrosco. 268, 85 (2011).
[c]M. Fukushima and T. Ishiwata, 68th ISMS, paper WJ03 (2013).
[d]J. Hougen, J. Chem. Phys. 36, 519 (1964)
[e]A. J. Merer and J. M. Allegretti, Can. J. Phys. 49, 2859 (1971).

WD03 — 9:04–9:19

VIBRONIC EMISSION SPECTROSCOPY OF JET-COOLED CHLORO-SUBSTITUTED BENZYL-TYPE RADICALS PRODUCED BY CORONA DISCHARGE

SANG LEE, *Department of Chemistry, Pusan National University, Pusan, Korea.*

Whereas benzyl radical, a prototype of aromatic free radicals, had attracted much attention from spectroscopists for the subject of large molecular radicals, chloro-substituted benzyl–type radicals have been little studied, presumably due to the difficulties associated with production in corona discharge from precursors. The weak C-Cl bond can be easily dissociated in high voltage corona discharge, [a] leading to the cleavage of benzene ring. During past years, we have concentrated on the spectroscopic observation of chloro-substituted methylbenzyl radicals in a technique of corona excited supersonic expansion using a pinhole-type glass nozzle which has been well developed in this lab. From the experiments, we could succeed the observation of vibronic emission spectra of chloro-substituted methylbenzyl radicals from 3- and 4-chloro-o-xylenes. From the analysis of the spectra observed, we can identify the radical species produced in corona discharge and determine the electronic energies of the $D_1 \rightarrow D_0$ transition. The variation of the electronic transition energies with the positions and types of substituents have been clearly explained by means of the additivity rule [b] and shape of the unoccupied lowest molecular orbitals (LUMO) which corresponds to the upper state of the electronic transition. In this presentation, the observation scheme of the chloro-substituted benzyl-type radicals and analysis of the spectra for the identification of the radical species generated will be discussed, together with the introduction of the method for the explanation of substituent effect [c] on the electronic transition energy of the benzyl-type radicals.

[a] S. Y. Chae, M. Lim, and S. K. Lee, *Chem. Phys. Lett.* **664**, 242-245 (2016).
[b] Y. W. Yoon, S. Y. Chae, and S. K. Lee, *Chem. Phys. Lett.* **644**, 167-170 (2016).
[c] C. Branciard-Larcher, E. Migirdicyan, and J. Baudet *Chem. Phys.* **2**, 95-106 (1973).

WD04 — 9:21–9:36

HIGH RESOLUTION SPECTRA OF THE SIMPLEST CRIEGEE INTERMEDIATE CH_2OO BETWEEN 880 AND 932 cm^{-1}

PEI-LING LUO, YASUKI ENDO, *Department of Applied Chemistry, National Chiao Tung University, Hsinchu, Taiwan*; YUAN-PERN LEE, *Applied Chemistry, National Chiao Tung University, Hsinchu, Taiwan, Institute of Atomic and Molecular Sciences, Academia Sinica, Taipei, Taiwan.*

The Criegee intermediates (CI) play critical roles in atmospheric chemistry. CH_2OO is the simplest CI and its characterization is important for investigations of reaction mechanisms and molecular structure. In this work, high-resolution spectra of the OO-stretching (ν_6) mode of CH_2OO in the range of 880–932 cm^{-1} have been recorded using a quantum cascade laser (QCL) system coupled with a multi-pass Herriott cell. The CH_2OO was produced from the reaction of $CH_2I + O_2$ in a flowing mixture of CH_2I_2/O_2 (1/213) at 3.2 Torr upon irradiation at 248 nm with an excimer laser. The spectrum was recorded by step-scanning the QCL with a step size of 0.0016 cm^{-1}; its wavelength was calibrated with a C_2H_4 reference cell and a germanium etalon. Over one thousand lines were assigned and used for fitting of molecular constants of CH_2OO. Furthermore, the rotational perturbations on the high-J levels of $K_a = 3$, $K_a = 6$, and $K_a \geq 11$ were observed.

WD05 — 9:38–9:53

MILLIMETER-WAVE SPECTROSCOPY OF KO: ESTABLISHING THE ELECTRONIC GROUND STATE

MARK BURTON, *Department of Chemistry and Biochemistry, University of Arizona, Tucson, AZ, USA*; BENJAMIN RUSS, PHILLIP M. SHERIDAN, *Department of Chemistry and Biochemistry, Canisius College, Buffalo, NY, USA*; MATTHEW BUCCHINO, LUCY M. ZIURYS, *Department of Chemistry and Biochemistry, University of Arizona, Tucson, AZ, USA.*

The ground electronic state of potassium monoxide (KO) has yet to be conclusively assigned, despite both experimental and theoretical investigations of this species. The ground state is either $^2\Pi_i$ (as for LiO and NaO) or $^2\Sigma^+$ (as for RbO and CsO), both of which are predicted to lie close in energy for KO. To solve this problem, we have conducted millimeter-wave direct absorption spectroscopy of KO. This species was synthesized via the reaction of potassium vapor, generated by a Broida-type oven, with nitrous oxide. We have found patterns that we have identified as the $\Omega = 3/2$ and $1/2$ ladders of a $^2\Pi_i$ state, as well as a $^2\Sigma^+$ state. Rotational and fine structure constants have been accurately determined assuming the $^2\Pi_i$ and $^2\Sigma^+$ assignments.

WD06   9:55 – 10:10

LASER SPECTROSCOPIC DETECTION OF THE JET-COOLED SnCH$_2$ MOLECULE

TONY SMITH, *New Product Development (NPD), Ideal Vacuum Products, ALBUQUERQUE, New Mexico, USA*; MOHAMMED GHARAIBEH, *Department of Chemistry, University of Jordan, Amman, Jordan*; DENNIS CLOUTHIER, *Department of Chemistry, University of Kentucky, Lexington, KY, USA.*

The stannylidene (SnCH$_2$) molecule has been detected for the first time in the gas phase by supersonic expansion/laser spectroscopy. This transient molecule was produced in an electric discharge through a dilute mixture of tetramethyltin [(CH3)$_4$Sn] in high-pressure argon and studied by laser induced and dispersed fluorescence through the $\tilde{B}\ ^1B_2$ - $\tilde{X}\ ^1A_1$ transition. The vibronic energy levels of the ground and excited states have been measured for both SnCH$_2$ and SnCD$_2$. The observed vibrational frequencies, partially resolved rotational band contours, deuterium isotope shifts, and electronic excitation energies are in accord with our predictions from ab initio calculations. This novel species has an unusual tin-carbon double bond in the ground state. It is the third in the series of X=CH$_2$ (X = Si, Ge and Sn) group IVA vinylidene species we have been able to produce and study in the gas phase.

Intermission

WD07   10:46 – 11:01

HIGH RESOLUTION LASER SPECTROSCOPY OF THE JET-COOLED SiCF FREE RADICAL

GRETCHEN K ROTHSCHOPF, TONY SMITH, *New Product Development, Ideal Vacuum Products, Albuquerque, NM, USA*; DENNIS CLOUTHIER, *Department of Chemistry, University of Kentucky, Lexington, KY, USA.*

The SiCF radical was produced in an electric discharge through a dilute mixture of trimethyl(trifluoromethyl)slilane (CH$_3$)$_3$SiCF$_3$ in high-pressure argon. Using our high-resolution pulse amplified ring dye laser system, the laser induced fluorescence of the 0-0 band of the $\tilde{A}^2\Sigma^+$ - $\tilde{X}^2\Pi_i$ transition has been rotationally resolved (linewidths 0.015 cm^{-1}) for the first time. The subsequent rotational analysis paired with previous ab initio calculations[a] allowed the determination of the ground and excited state SiC bond lengths. We find a Si-C double bond in the ground state and an unusual Si-C triple bond in the excited state. Finally, further low-resolution spectra were obtained to determine better values for the excited state vibrational frequencies.

[a] C. J. Evans and D. J. Clouthier, J. Chem. Phys., 117, 6439-6445 (2002)

WD08   11:03 – 11:18

PROBING SPIN-ORBIT COUPLING OF ORGANOCERIUM RADICALS FORMED IN Ce ATOM REACTIONS WITH ALKYLAMINES.

SILVER NYAMBO, YUCHEN ZHANG, DONG-SHENG YANG, *Department of Chemistry, University of Kentucky, Lexington, KY, USA.*

Ce atom reactions with alkyamines are carried out in a pulsed-laser ablation molecular beam source and characterized by mass-analyzed threshold ionization (MATI) spectroscopy. The MATI spectra of CeNR (R = CH$_3$, C$_2$H$_5$, and C$_3$H$_7$) formed by Ce reactions with H$_2$NR exhibit two band systems, separated by 78, 74, and 72 cm^{-1}, respectively. In contrast, the MATI spectrum of CeNC$_2$H$_5$ formed in the Ce + HN(CH$_3$)$_2$ reaction show two band systems with a much larger separation, 130 cm^{-1}. These separations are attributed to the spin-orbit (SO) splitting from the Ce 4f^1 electron. The different splittings between CeNR from the reactions of primary amines and CeNC$_2$H$_5$ from the reaction of secondary amine are due to their different structures. The CeNR complexes from the primary amines have acyclic structures with Ce double bonding to the N atom, whereas CeNC$_2$H$_5$ from the dimethylamine has a cyclic structure with Ce bonding to the N atom and one of the C atoms. The considerably smaller SO splittings in the CeNR species suggests that N coordination has a stronger quenching effect on the SO coupling of the Ce 4f electron than the C coordination.

WD09   11:20 – 11:35
ELECRONIC STRUCTURE OF ALKOXY RADICAL ISOMERS FROM ANION PEI SPECTROSCOPY

KELLYN M. PATROS, JENNIFER MANN[a], CAROLINE CHICK JARROLD, *Department of Chemistry, Indiana University, Bloomington, IN, USA.*

Anion photoelectron imaging spectra of two butenoxyl (3-buten-1-oxyl and 3-buten-2-oxyl) radical isomers are presented. The neutral electron affinities are comparable to those measured for saturated alkoxy radicals [Ramond et al., J. Chem. Phys. 112, 1158 (2000)], and the measured term energies for the Ã ^{2}A state of both isomers is approximately 0.1 eV. However, spectra of the two isomers exhibit distinct differences, particularly in the low electron binding energy signal that may be due to the presence of structural isomers. The experimental spectra are analyzed with supporting MP2 calculations and Franck-Condon simulations. Overall, the results underscore how the electronic properties vary with subtle changes in alkoxy radical structure, which may have implications for atmospheric photochemistry, as alkoxy radicals are key intermediates of the tropospheric oxidation of volatile organic compounds.

[a]Now at Physical Electronics, Chanhassen, MN, USA

WD10   11:37 – 11:52
ANION PHOTOELECTRON IMAGING OF 2-PROPENOL

MARISSA DOBULIS, KELLYN M. PATROS, JENNIFER MANN[a], CAROLINE CHICK JARROLD, *Department of Chemistry, Indiana University, Bloomington, IN, USA.*

Saturated and unsaturated alcohols are released into the atmosphere by vegetation and industrial activities and become radicals in the environment. These radical species are highly reactive toward other atmospheric species and are key intermediates of the oxidation of volatile organic compounds in the troposphere. In this talk we will investigate anion photoelectron imaging (PEI) of the 2-propenol radical at photon energies of 2.33 eV and 3.49 eV as an example of these radical species. DFT (B3LYP) and *ab initio* (MP2) calculations will be used to further elaborate on the transitions of these species and compare theory to experimental results.

[a]Now at Physical Electronics, Chanhassen, MN, USA

WD11   11:54 – 12:09
SUB-DOPPLER INFRARED SPECTROSCOPY OF JET COOLED CH_2I RADICAL: CH_2 STRETCH VIBRATIONS AND "CHARGE-SLOSHING" INTENSITY DYNAMICS

ANDREW KORTYNA, *JILA, National Institute of Standards and Technology and Univ. of Colorado, Boulder, CO, USA*; DANIEL LESKO, *Department of Chemistry and Biochemistry, University of Colorado, Boulder, CO, USA*; PRESTON G. SCRAPE, DAVID NESBITT, *JILA, National Institute of Standards and Technology and Univ. of Colorado, Boulder, CO, USA.*

Iodomethyl radical (CH_2I) is relevant to atmospheric chemistry, especially marine boundary layer dynamics, with recent attention arising from its use as novel precursor for Criegee intermediates (CH_2OO). As a first step towards the spectroscopic investigation of a Criegee intermediate, we have pursued high resolution characterization of the CH_2I radical in our slit jet discharge spectrometer. The methyl iodide radical is generated by seeding CH_2I_2 into a Ne/He/H_2 mixture in a pulsed slit discharge, produced through either electron dissociative attachment to form iodine anions or hydrogen abstraction of iodine, with subsequent cooling in a supersonic expansion to 16 K. Infrared absorption in the CH symmetric stretch vibrational band is observed at high single-to-noise ratio (S/N = 25:1), yielding a symmetric stretch band origin at $3046.9527 \pm 0.0006 \, cm^{-1}$. The sub-Doppler rotational structure is fitted to a rigid-rotor Hamiltonian with spin-rotation coupling, generating principal rotational constants and the spin-orbit coupling tensor for the vibrationally excited state. Interestingly, an extensive search for the asymmetric stretch mode yielded null results, despite simple bond-dipole model predictions of three-fold larger absorption intensities for the asymmetric vs. symmetric stretch band. We conclude that the asymmetric stretch absorption intensity must be at least a factor of 25 below that of the symmetric stretch. *Ab initio* calculations indicate that enhancement of the symmetric vs. asymmetric stretch intensity arises from "charge sloshing" motion of electrons in the highly polar carbon-iodine bond of the correct A_1 symmetry.

WE. Clusters/Complexes

Wednesday, June 20, 2018 – 8:30 AM

Room: B102 Chemical and Life Sciences

Chair: Nasser Moazzen-Ahmadi, University of Calgary, Calgary, AB, Canada

---

WE01   8:30 – 8:45

STRUCTURE OF MICROSOLVATED VERBENONE DETERMINED BY MICROWAVE FOURIER TRANSFORM SPECTROSCOPY AND QUANTUM CHEMICAL CALCULATIONS

MHAMAD CHRAYTEH, ANNUNZIATA SAVOIA, PASCAL DRÉAN, THERESE R. HUET, *UMR 8523 - PhLAM - Physique des Lasers Atomes et Molécules, University of Lille, CNRS, F-59000 Lille, France.*

Verbenone ($C_{10}H_{14}O$) is a bicyclic ketone terpene. It is one of the products of oxidation of α-pinene in the troposphere. It may have a significant role in the formation of secondary organic aerosols, in particular through its ability to interact with water molecules. Verbenone is almost insoluble in water so it is therefore important to understand how this type of molecules interacts with water.

The rotational spectrum of verbenone and the determination of its r_s and r_0 molecular structures were recently investigated [a]. This work deals with the study of its hydrates. Water is expected to form a primary hydrogen bond with the carbonyl group of verbenone, and the hydrogen atoms of the -CH_3 or -CH_2 groups may form weak interactions with the lone pairs of the water oxygen to stabilize different hydrates. The structures of two monohydrates, two dihydrates and four trihydrates of verbenone were optimized at the DFT B3LYP-D3BJ / def2-TZVP and *ab initio* MP2 / 6-311++G(d,p) levels, before searching for their rotational signatures, using a supersonic expansion coupled to a cavity-based Fourier transform microwave spectrometer working in the 2 - 20 GHz frequency range. We were able to analyse the spectra of the expected two mono- and two dihydrates, and of the lowest energy conformer of the trihydrate. We also analysed the spectra of the water-^{18}O substituted species using ^{18}O labeled water. For each hydrate, the sets of rotational constants were used to calculate the substitution coordinates of the water oxygen atoms and an effective r_0 structure of the water arrangements of water around the molecule of verbenone.

The present work was funded by the French ANR Labex CaPPA through the PIA (contract ANR-11-LABX-0005-01), by the Regional Council Hauts de France, by the European Funds for Regional Economic Development, and by the French Ministère de l'Enseignement Supérieur et de la Recherche. It is a contribution to the CPER research Project CLIMIBIO.

---

[a] F. E. Marshall, G. Sedo, C. West, B. H. Pate, S. M. Allpress, C. J. Evans, P. D. Godfrey, D. McNaughton and G. S. Grubbs, *J. Mol. Spectrosc.* **342**, 109 - 115 (2017).

WE02   8:47 – 9:02

MICROSOLVATION COMPLEXES OF ETHYL CARBAMATE STUDIED BY MICROWAVE SPECTROSCOPY.

PABLO PINACHO, JUAN CARLOS LOPEZ, *Departamento de Química Física y Química Inorgánica, Universidad de Valladolid, Valladolid, Spain*; ZBIGNIEW KISIEL, *ON2, Institute of Physics, Polish Academy of Sciences, Warszawa, Poland*; SUSANA BLANCO, *Departamento de Química Física y Química Inorgánica, Universidad de Valladolid, Valladolid, Spain.*

The rotational spectra of ethyl carbamate-$(H_2O)_n$ (n = 1, 2, 3) generated in a supersonic expansion have been studied using both a chirped-pulse and a molecular beam Fourier transform microwave spectrometer. Ethyl carbamate presents in the gas phase an equilibrium between two structures close in energy with a low interconversion barrier.[a,b] The observation of these structures and their complexes strongly depends on the carrier gas used due to collisional relaxation in the supersonic jet. Using argon, only the most stable form and its water complexes are observed. Using neon, both forms and their corresponding complexes are observed. The structures of the complexes have been characterized and show water closing sequential cycles with the H-N-C=O amide group. They show structural and dynamical features similar to those observed, for example, in formamide-$(H_2O)_n$ clusters.

---

[a] K.-M. Marstokk, H. Møllendal, *Acta Chem. Scand.*, 1999, **53**, 329-334.

[b] M. Goubet, R. A. Motiyenko, F. Real, L. Margulés, T. R. Huet, P. Asselin, P. Soulard, A. Krasnicki, Z. Kisiel, E. A. Alekseev, *Phys. Chem. Chem. Phys.*, 2009, **11**, 1719-1728.

WE03   9:04–9:19
STUDYING CO_2 SOLVENT PROPERTIES BY MICROWAVE SPECTROSCOPIC INVESTIGATION OF FLUOROETHYLENE...CO_2...CO_2 TRIMERS

PRASHANSA KANNANGARA, REBECCA A. PEEBLES, SEAN A. PEEBLES, *Department of Chemistry, Eastern Illinois University, Charleston, IL, USA*; BROOKS PATE, *Department of Chemistry, The University of Virginia, Charlottesville, VA, USA.*

Supercritical carbon dioxide (*sc*-CO_2) is an increasingly common green solvent, so it is important that its physical properties are well understood. In the present study, chirped-pulse Fourier-transform microwave (CP-FTMW) spectroscopy was used to study weak hydrogen bonding interactions in complexes of fluoroethylene (FE) with CO_2. Previous investigations of 1:1 dimers of CO_2 with FE observed two isomers for FE...CO_2. Our current focus is analysis of weakly bound trimers, and FE...CO_2...CO_2 was recently observed in the 2 – 8 GHz range using the CP-FTMW spectrometer at the University of Virginia. Four structures were optimized at the MP2/6-311++G(2d,2p) level. As with FE...CO_2 dimer, spectra of two trimer isomers were observed experimentally, corresponding to the two lowest energy ab initio structures, which lie within 25 cm^{-1} of each other. Although only planar forms of the isolated FE...CO_2 dimer were observed experimentally, both trimer structures trap a nonplanar dimer fragment, with one CO_2 molecule located above the plane of FE. Current work involves analysis of isotopic data to allow detailed structural comparisons, as well as searching for larger CO_2 clusters to explore structural changes as the number of solvating CO_2 molecules increases. Extended cross correlation and other techniques are being applied to assist assignment of the thousands of lines remaining in the scan and to suggest the carrier of a recently identified spectrum in the FE/CO_2 mixture.

WE04   9:21–9:36
π-π STACKING IN COMPETITION WITH HYDROGEN BONDING IN THE 1-NAPHTOL DIMER: A CP-FTMW SPECTROSCOPY STUDY

NATHAN A. SEIFERT, ARSH HAZRAH, WOLFGANG JÄGER, *Department of Chemistry, University of Alberta, Edmonton, AB, Canada.*

Present in a wide spectrum of chemical systems, π-π stacking and hydrogen bonding are intermolecular forces critical to the formation and stabilization of various chemical structures. However, these forces can be found to be competitive interactions in stabilizing model systems. This competition is clearly exhibited in the dimer of phenol, where hydrogen bonding is preferred over π-π stacking.[a] However, it is unclear how this competitive relationship will evolve as a function of molecular shape. To explore this, we use 1-naphthol, a naphthalene analogue of phenol, as a model to further understand the complex interplay between π-π stacking and hydrogen bonding.

Using chirped-pulse Fourier transform microwave (CP-FTMW) spectroscopy in the 2-6 GHz band[b], we observed a spectrum that is size-consistent with a dimer of 1-naphthol, in addition to the pure rotational spectra of two conformers of 1-naphthol as well as weakly-bound 1-naphthol complexes with neon and H_2O. We present an experimental analysis supplemented with a structure search enabled by dispersion corrected DFT and corroborated by interaction energies at the CCSD(T) level of theory, that provides the identity of a likely molecular carrier for the observed dimer spectrum. This analysis suggests that the dimer structure of 1-naphthol is not at all like that of the phenol dimer; conversely, the spectrum is consistent with a structure that stabilizes nearly exclusively through a π-π stacking interaction. This is in contradiction to previous observations using IR dip spectroscopy[c], which assigns the dimer vibrational spectrum to a hydrogen bonded structure.

[a] Seifert, N. A.; Steber, A. L.; Neill, J. L.; Pérez, C.; Zaleski, D. P.; Pate, B. H.; Lesarri, A. *Phys. Chem. Chem. Phys.* **2013**, *15*, 11468–11477.
[b] Pérez, C.; Lobsiger, S.; Seifert, N. A.; Zaleski, D. P.; Temelso, B.; Shields, G. C.; Kisiel, Z.; Pate, B. H. *Chem. Phys. Lett.* **2013**, *571*, 1–15.
[c] Saeki, M.; Ishiuchi, S.; Sakai, M.; Fujii, M. *J. Phys. Chem. A* **2007**, *111*, 1001–1005.

Intermission

WE05 10:12–10:27

CHARACTERIZATION OF MICROSOLVATED 15C5 CROWN ETHER FROM BROADBAND ROTATIONAL SPECTROSCOPY

JUAN CARLOS LOPEZ, SUSANA BLANCO, *Departamento de Química Física y Química Inorgánica, Universidad de Valladolid, Valladolid, Spain*; CRISTOBAL PEREZ, MELANIE SCHNELL, *FS-SMP, Deutsches Elektronen-Synchrotron (DESY), Hamburg, Germany.*

15-crown-5 ether (15C5) and its complexes with water generated in a supersonic jet have been studied using broadband Fourier transform microwave spectroscopy. The most stable form of the crown ether not previously reported, to complete a total of nine isolated forms, has been detected. In addition, two 1:1 and two 1:2 clusters have been observed. The clusters structures have been unambiguously identified through the observation of water ^{18}O isotopologue spectra and a detailed analysis of the rotational parameters. The structures of all the clusters show that at least one water molecule, located close to the axis of the ring, interacts through two simultaneous hydrogen bonds to the endocyclic oxygen atoms. This interaction reshapes the 15C5 ring to reduce its rich conformational panorama to only two open structures, related to those found in complexes with Li$^+$ or Na$^+$ ions. In the most intense 1:2 form, the two water molecules repeat the same interaction scheme in both sides of the ring while in the second one the water molecules lie on the same side of the ring.

WE06 10:29–10:44

CHARACTERIZATION OF SO_3-SO_2 BY MICROWAVE SPECTROSCOPY AND COMPUTATIONAL CHEMISTRY

BECCA MACKENZIE, ANNA HUFF, KEN LEOPOLD, *Chemistry Department, University of Minnesota, Minneapolis, MN, USA.*

The rotational spectrum for the complex formed between sulfur trioxide and sulfur dioxide has been observed by chirped-pulse and cavity Fourier transform microwave spectroscopy. Spectra were recorded for five isotopologues that include single substitution of ^{34}S and ^{33}S on each sulfur atom. Nuclear hyperfine structure was resolved for both ^{33}S isotopologues and their corresponding quadrupole coupling constants were obtained. *Ab initio* calculations predict a pair of structures which differ in energy by only 0.1 kcal/mol. In both structures, one oxygen of the SO_2 approaches the sulfur of the SO_3, but the two forms differ in the angular orientation of the SO_2. Despite the small calculated energy difference, only one set of spectra was identified. Isotopic substitution on SO_3 does not clearly distinguish between the two structures, but the experimental isotopic shifts and quadrupole coupling constants obtained for the SO_2 substituted isotopologues are more clearly consistent with those predicted for the lower energy form.

WE07 10:46–11:01

FORMAMIDE, WATER, AND THEIR COMPLEXES: A MICROWAVE SPECTROSCOPY STUDY

SUSANA BLANCO, JUAN CARLOS LOPEZ, *Departamento de Química Física y Química Inorgánica, Universidad de Valladolid, Valladolid, Spain*; CHANNING WEST, MARTIN S. HOLDREN, BROOKS PATE, *Department of Chemistry, The University of Virginia, Charlottesville, VA, USA.*

The rotational spectra of formamide and water mixtures have been recorded in the 2-8 GHz frequency region using a chirped-pulse Fourier transform microwave spectrometer. Samples of ^{14}N and ^{15}N of formamide have been used in this work. The ^{14}N quadrupole coupling hyperfine structure is a tool to identify the structure of the observed complexes; the ^{15}N isotopologue is of great help to explore the conformational panorama of complexes with several formamide units. In this work we present the detection and characterization of complexes of formamide and formamide-water, as F_3 and F-$(H_2O)_4$, which show interesting structural features.

WE08   11:03 – 11:18

DOES THE STRUCTURE OF THE POLYCYCLIC AROMATIC HYDROCARBON IMPACT THE AGGREGATION OF WATER ON ITS SURFACE? FLUORENE VS ACENAPHTHENE

AMANDA STEBER, *The Centre for Ultrafast Imaging (CUI), Universität Hamburg, Hamburg, Germany*; SÉBASTIEN GRUET, CRISTOBAL PEREZ, *FS-SMP, Deutsches Elektronen-Synchrotron (DESY), Hamburg, Germany*; BERHANE TEMELSO, *Department of Chemistry, Furman University, Greenville, SC, USA*; JANA MEISER, *Institute of Physikalische Chemie, Gottfried-Wilhelm-Leibniz-Universität, Hannover, Germany*; GEORGE C SHIELDS, *Department of Chemistry, Furman University, Greenville, SC, USA*; MELANIE SCHNELL, *FS-SMP, Deutsches Elektronen-Synchrotron (DESY), Hamburg, Germany*.

As polycyclic aromatic hydrocarbons (PAHs) are of interest to many communities, including astronomers, it is important to understand the interactions that they may be a part of. As water is ubiquitous in astronomical environments and PAHs are thought to form ice grains, the PAH-water interactions are of specific interest. In this investigation we used chirped pulse Fourier transform microwave (CP-FTMW) spectroscopy from 2-8 GHz to study the fluorene monomer and its complexes with water. While the monomer has previously been studied [1], our use of the COMPACT [2] instrument allowed us to observe not only new transitions for the monomer but also transitions for ^{13}C species. This allowed for a structural analysis of the monomer to be presented. This structural information is important when we move to complexes of PAHs with compounds such as water. We have previously studied the interactions of up to four water molecules clustered with the PAH acenaphthene (Ace)[3]. As in the Ace-water study, we have observed up to three water molecules complexed with fluorene and obtained isotopic data for the complexes. In this talk we will present these findings and the structural differences between the two PAH-water systems.

[1] Thorwirth, S., Theulé, P., Gottlieb, C.A., McCarthy, M.C., Thaddeus, P. *Astrophys. J.*, 662, 1309-1314, **2007**.
[2] Schmitz, D., Shubert, V.A., Betz, T., Schnell, M. *J. Mol. Spectro.*, 280, 77-84, **2012**.
[3] Steber, A.L., Pérez, C., Temelso, B., Shields, G.C., Rijs, A.M., Pate, B.H., Kisiel, Z., Schnell, M. *J. Phys. Chem. Lett.*, 8, 5744–5750, **2017**.

WE09   11:20 – 11:35

A ROTATIONAL STUDY OF 2-METHOXYBENZOIC ACID AND ITS WATER COMPLEXES

ALBERTO MACARIO, PABLO PINACHO, SUSANA BLANCO, JUAN CARLOS LOPEZ, *Departamento de Química Física y Química Inorgánica, Universidad de Valladolid, Valladolid, Spain*.

The 2-methoxybenzoic acid (*o*-anisic acid) and its complexes with water have been studied in the 2-12 GHz frequency region combining chirped-pulse Fourier transform microwave spectroscopy (CP-FTMW) and molecular Fourier transform microwave spectroscopy (MB-FTMW). *o*-Anisic acid has been vaporized using a heating nozzle were it partially reacts through a recombination reactions to give a series of products which have been all identified from its microwave spectra. Apart from these species, three different conformations for *o*-anisic acid, with distinct dispositions of carboxylic group, and two for the 1:1 water complex have been observed. For the lowest energy conformer of the monomer and the most abundant water complex, the spectra of various isotopologues have been measured and the molecular structures have been determined. Relative intensity measurements have been done to provide a better understanding of the complex formation in the supersonic expansion.

WE10 11:37 – 11:52

MICROWAVE SPECTROSCOPY OF 2-METHOXYETHYLAMINE-WATER: STRUCTURAL CHANGES DUE TO HYDROGEN BONDING NETWORKS

NATHAN HARPER[a], *Department of Chemistry, Emory University, Atlanta, GA, USA*; BRITTANY BASENBACK, RANIL GURUSINGHE, MICHAEL TUBERGEN, *Department of Chemistry and Biochemistry, Kent State University, Kent, OH, USA.*

2-Methoxyethylamine (2MEA) exists in trans and gauche conformations, each with at least one intramolecular hydrogen bond from the amine to the methoxy oxygen. Rotational spectra of 2MEA, recorded by Caminati et al., exhibit splittings arising from methyl internal rotation tunneling (V_3 = 822.6 cm^{-1} and 1102cm^{-1} respectively).[b,c] We report the rotational spectra of the three ^{13}C isotopologues for each 2MEA conformer. The rotational transitions, including resolved tunneling splittings and nuclear quadrupole hyperfine components, were fit with XIAM.[d] Ab initio calculations (MP2/6-311++G(d,p)) were used to model stable structures of 2MEA-water complexes. The most stable structure of the 2MEA-water complex was found to have an intermolecular hydrogen bond from the water to the amine and an intramolecular hydrogen bond from the amine to the methoxy oxygen. Twenty-five rotational transitions of 2MEA-water were recorded (A = 2922.74(3) MHz, B = 1696.72(2) MHz, and C = 1379.80(2) MHz) and assigned to the lowest-energy structure of the complex. Methyl internal rotation tunneling splittings were not resolved in the spectrum of 2MEA-water. The formation of hydrogen bonding networks in the 2MEA-water complex was found to alter the configuration of 2MEA within the complex.

---

[a] Summer 2017 REU student at Kent State University
[b] W. Caminati and E. B. Wislon, *J. Mol. Spectrosc.* **81**, 356-372 (1980).
[c] W. Caminati, *J. Mol. Spectrosc.* **121**, 61-68 (1987).
[d] H. Hartwig and H. Dreizler, *Z. Naturforsch, A: Phys. Sci.* **51**, 923-932 (1996).

WF. Metal containing
Wednesday, June 20, 2018 – 8:30 AM
Room: 2079 Natural History

Chair: Lindsay N. Zack, Austin College, Sherman, TX, USA

WF01   8:30–8:45

SPECTROSCOPIC INVESTIGATION OF A SERIES OF CERIUM-DOPED BORON CLUSTERS

JARRETT MASON, JOSEY E TOPOLSKI, CAROLINE CHICK JARROLD, *Department of Chemistry, Indiana University, Bloomington, IN, USA.*

Rivaled, perhaps, only by its neighbor carbon, boron encompasses a unique and diverse swath of chemistry that lends itself to the adoption of extraordinary characteristics when complexed to metals. In recent years, metal-boride clusters have garnered attention for their potential application in a variety of fields including hydrogen storage and high energy density fuels. As inorganic ligands, these boron clusters tend to prefer multiply aromatic (σ- and π-) electronic configurations which may contribute to their unusual stability and unique properties. In the present study, a series of cerium-boride clusters were interrogated using anion photoelectron spectroscopy as a means of reconciling the presence of a series of mass coincident CeB_x^- (x = 5-7) and $CeO_2B_y^-$ (y = 2-4) clusters as well as elucidating the electronic structure of the anionic and neutral species. Unlike previously studied cerium-oxide clusters, which typically have binding energies around 0.7-1.2 eV, the boride clusters have exhibited greater electron affinities between 1.0-2.0 eV. Moreover, the spectra collected show pronounced vibrational progressions that aid in the analysis of the clusters' molecular structure.

WF02   8:47–9:02

ROTATIONAL AND ISOTOPIC STUDY OF THE ZnBr RADICAL ($^2\Sigma^+$)

MARK BURTON, *Department of Chemistry and Biochemistry, University of Arizona, Tucson, AZ, USA*; LUCY M. ZIURYS, *Department of Astronomy, University of Arizona, Tucson, AZ, USA.*

The pure rotational spectrum of ZnBr ($^2\Sigma^+$) has been recorded using millimeter-wave direct absorption spectroscopy. This species was generated in the gas phase via the reaction of zinc vapor with CH_3Br in the presence of a DC discharge. Multiple rotational transitions were measured for 6 isotopologues ($^{64}Zn^{79}Br$, $^{64}Zn^{81}Br$, $^{66}Zn^{79}Br$, $^{66}Zn^{81}Br$, $^{68}Zn^{79}Br$, and $^{68}Zn^{81}Br$) in the frequency range of 270-300 GHz, each of which consisted of spin-rotation splittings. Furthermore, transitions originating in the v = 1 through 3 excited vibrational states for certain isotopologues were obtained. The equilibrium rotational constant for $^{64}Zn^{79}Br$ (B_e) was calculated to be near 2780 MHz, resulting in an equilibrium bond length (r_e) of 2.25 Å.

WF03   9:04–9:19

INFRARED SPECTRA OF THE Pd_nCO (n=2-5) MOLECULES ISOLATED IN SOLID ARGON AND NEON BETWEEN 100 AND 4000 cm^{-1}

BENOÎT TREMBLAY, *MONARIS, Sorbonne Université, Paris, France*; SIDI M.O. SOUVI, *PSN-RES/SAG/LETR, Institut de Radioprotection et de Sûreté Nucléaire (IRSN), St Paul-les-Durance, France*; ESMAÏL ALIKHANI, *MONARIS, Sorbonne Université, Paris, France.*

The Pd+CO reaction has been reinvestigated using deposition of ground state reagents in solid argon and neon and the formation of Pd_nCO (n=2-5) is evidenced by strong absorption in the range 2015-1650 cm^{-1}. Various isotopic data ($^{12}C/^{13}C$, ^{16}O/^{18}O, natural isotopes for the palladium) and number of two quantum transitions have been measured in the near- and far-infrared regions. In argon, selective irradiation in visible leads to conversion between two Pd_2CO isomers distinguished by the stretching frequency of the diatomic CO: bridged T-shaped (ν_{CO}= 1856 cm^{-1}) and side on (ν_{CO}= 2015 cm^{-1}). DFT calculations of the geometrical and electronic properties of Pd_nCO complexes are also presented and compared to the experimental values.

WF04   9:21 – 9:36

AN ELECTRONIC SPECTROSCOPIC STUDY OF A MOLECULAR BEAM SAMPLE OF YbOH[a]

<u>TIMOTHY STEIMLE</u>, *School of Molecular Sciences, Arizona State University, Tempe, AZ, USA*; NICKOLAS PILGRAM, NICHOLAS R HUTZLER, *Division of Physics, Mathematics and Astronomy, California Institute of Technology, Pasadena, CA, USA.*

    Ytterbium monofluoride, YbF, has long been used as a venue in attempts to measure the electron electric dipole moment (eEDM)[b,c]. In addition to the molecular EDM resulting from the eEDM contribution, the ^{173}Yb(16.1%, I=5/2) isotopic form of Yb-containing molecules are also expected to have an EDM caused by an interaction of a nuclear magnetic quadrupole moment (NMQM)[d] with the electrons. As pointed out by Kozyryev and Hutzler[e], certain energy levels of Yb-polyatomic molecules (e.g. YbOH, YbCCH, YbCH$_3$, and YbOCH$_3$) are expected to exhibit enhanced sensitivity for EDM measurements, relative to YbF, largely due to their ease of polarization. The properties of such molecules are poorly characterized. Here we report on our initial molecular beam studies of the known[f] $A^2\Pi_{1/2}$(000) - $X^2\Sigma^+$(000) transition of YbOH. The high-resolution (30 MHz) laser induced fluorescence (LIF) spectrum in the 17320 cm^{-1} to 17326 cm^{-1} range was recorded both field-free and in the presence of a static electric field. Stark spectra were analyzed to determine the molecular frame permanent electric dipole moments, μ_{el}, for the $A^2\Pi_{1/2}$ and $X^2\Sigma^+$ states. The dispersed fluorescence resulting from the excitation of rotationally resolved branch features has been analyzed to produce fluorescence branching ratios. Implications for planned EDM measurements will be presented.

---

[a] Funded by a grant from the Heising-Simons Foundation.

[b] Hudson, J. J.; Sauer, B. E.; Tarbutt, M. R.; Hinds, E. A., Measurement of the Electron Electric Dipole Moment Using YbF Molecules. Phys. Rev. Lett. 2002, 89 (2), 023003/1-023003/4.

[c] Tarbutt, M. R.; Sauer, B. E.; Hudson, J. J.; Hinds, E. A., Design for a fountain of YbF molecules to measure the electron's electric dipole moment. New J. Phys. 2013, 15 (May), 053034/1-053034/17.

[d] Lackenby, B. G. C.; Flambaum, V. V., Weak quadrupole moment, quadrupole distribution of neutrons and Lorentz invariance violation in deformed nuclei. arXiv.org, e-Print Arch., Nucl. Theory 2017, 1-7.

[e] Kozyryev, I.; Hutzler, N. R., Precision measurement of time-reversal symmetry violation with laser-cooled polyatomic molecules. arXiv.org, e-Print Arch., Phys. 2017, 1-11.

[f] Melville, T. C.; Coxon, J. A., The visible laser excitation spectrum of YbOH: The $A^2\Pi_{1/2}$(000) - $X^2\Sigma^+$(000) transition. J. Chem. Phys. 2001, 115 (15), 6974-6978.

Intermission

WF05 10:12 – 10:27

ALKALINE EARTH MONOALKOXIDE FREE RADICALS AS CANDIDATES FOR LASER COOLING OF POLYATOMIC MOLECULES

ANAM C. PAUL, MD ASMAUL REZA, *Department of Chemistry, University of Louisville, Louisville, KY, USA*; KETAN SHARMA, TERRY A. MILLER, *Department of Chemistry and Biochemistry, The Ohio State University, Columbus, OH, USA*; JINJUN LIU, *Department of Chemistry, University of Louisville, Louisville, KY, USA*.

Alkaline earth monoalkoxide free radicals, e.g., $CaOCH_3$, $CaOCH_2CH_3$, and $CaOCH(CH_3)_2$, have been proposed recently as candidates for future laser cooling of polyatomic molecules.[a] Their $\tilde{A} \leftarrow \tilde{X}$ and $\tilde{B} \leftarrow \tilde{X}$ electronic transitions correspond to promotion of the unpaired electron in the $4s$ orbital of Ca^+ to each of the three components of its $4p$ orbital perturbed by the presence of the alkoxy group. These electronic transitions are limited to unbonding orbitals, from which the existence of quasiclosed transition loops for laser cooling has been surmised. Moreover, molecules suitable for Doppler cooling with lasers must feature (quasi-)diagonal Franck-Condon (FC) matrices for transitions between vibronic levels involved in the closed transition cycles, which is expected for alkaline earth monoalkoxides thanks to their bonding scheme. Laser-induced fluorescence (LIF) and dispersed-fluorescence (DF) spectra can provide valuable information to guild future laser cooling experiments. Experimentally obtained vibronic transition frequencies and intensities have been used to benchmark ab initio calculations carried out using both complete active space self-consistent field (CASSCF) and coupled cluster (CC) methods. Although FC factor calculations using the harmonic oscillator approximation reproduce major peaks in the spectra, it has been found that Jahn-Teller (JT) and pseudo-Jahn-Teller (pJT) effects introduce certain new transitions. Furthermore, The $\tilde{A}$ state of alkaline earth monoalkoxides is split by the spin-orbit interaction and, for those free radicals with symmetry lower than C_{3v}, the difference potential between two nearly degenerate electronic states. The rotational and fine structure of the involved electronic states as well as line intensities for rotational transitions between these states are predicted using a newly proposed spectroscopic model. The implications of the present experimental and computational investigations to future laser cooling experiments will be discussed.

---

[a] Kozyryev, L. Baum, K. Matsuda and J. M. Doyle, ChemPhysChem 17, 3641 (2016).

WF06 10:29 – 10:44

LASER-INDUCED FLUORESCENCE AND DISPERSED-FLUORESCENCE SPECTROSCOPY OF JET-COOLED CALCIUM MONOALKOXIDE RADICALS

ANAM C. PAUL, MD ASMAUL REZA, *Department of Chemistry, University of Louisville, Louisville, KY, USA*; PRANOY DEB SHUVRA, *Electrical and Computer Engineering, University of Louisville, Louisville, KY, USA*; JINJUN LIU, *Department of Chemistry, University of Louisville, Louisville, KY, USA*.

Laser-induced fluorescence (LIF) and dispersed-fluorescence spectroscopy (DF) spectroscopic investigations of $\tilde{A} \leftarrow \tilde{X}$ transitions of a series of calcium monoalkoxides, including $CaOCH_3$, $CaOCH_2CH_3$, and $CaOCH(CH_3)_2$, have been carried out. The free radicals were produced by laser ablation of a calcium rod in the presence of alcohols under jet-cooled conditions. Dominant transitions in the vibrationally resolved LIF and DF spectra obtained by pumping the orgin bands are reproduced using Franck-Condon (FC) factors calculated by complete active space self-consistent field (CASSCF) as well as coupled cluster (CC) methods. DF spectra obtained by pumping other vibronic bands in the LIF spectra provide valuable information about the FC matrices and aid the assignment of vibronic transitions. The (pseudo-)Jahn-Teller effects introduce transitions that are not predicted using the harmonic oscillator approximation.

WF07 10:46 – 11:01

COUPLED-CLUSTER CALCULATIONS FOR LOW-LYING ELECTRONIC STATES OF HEAVY-METAL CONTAINING MOLECULES

LAN CHENG, *Department of Chemistry, Johns Hopkins University, Baltimore, MD, USA.*

Coupled-cluster calculations of low-lying electronic states for heavy-metal containing diatomic molecules (e.g., PtH, ThO^+, ThN, BaO^+, CsF^+) are reported. Recently-developed relativistic quantum-chemical techniques have been used, including an atomic mean-field approach for efficent construction of spin-orbit integrals [1], a perturbative approach for treating spin-orbit coupling within exact-two-component equation-of-motion coupled-cluster methods [2], and a new implementation of two-component coupled-cluster methods for non-perturbative treatments of spin-orbit coupling [3]. Bond lengths, vibrational frequencies, and dipole moments of these molecules containing heavy metals are compared with experimental data to assess the accuracy and usefulness of the computational methods.

References

[1] J. Liu and L. Cheng, J. Chem. Phys. submitted.

[2] L. Cheng, F. Wang, J. F. Stanton, and J. Gauss, J. Chem. Phys. **148**, 044108 (2018).

[3] J. Liu, Y. Shen, A. Asthana, and L. Cheng, J. Chem. Phys. **148**, 034106 (2018).

WF08 11:03 – 11:18

HIGH RESOLUTION SPECTROSCOPY OF THE $[18.0]^2\Pi_{3/2}$ - $X^2\Sigma^+$ TRANSITION OF THORIUM NITRIDE, ThN[a]

ANH T. LE, DUC-TRUNG NGUYEN, TIMOTHY STEIMLE, *School of Molecular Sciences, Arizona State University, Tempe, AZ, USA*; LAN CHENG, *Department of Chemistry, Johns Hopkins University, Baltimore, MD, USA.*

Serious draw backs to nuclear power include long-term nuclear waste storage and amelioration of the existing waste. The 4% enriched uranium fuel used in a typical light water reactor is converted to spent nuclear fuel (SNF) made up of 3% fission products, of which ∼30% are lanthanides (Ln), and 1% transuranium actinide (Ac) elements Np, Pu, Am and Cm. Partitioning of the Ln from the Ac present in the SNF by developing element-specific ligands for solvent extraction is a one of the most challenging facet of nuclear waste processing.[b] Systematic experimental and theoretical studies of simple Ac and Ln containing molecules is one avenue for garnering insight into element-specific ligation.[c] As part of an effort to establish trends in Th-X bonding, a combined experimental and theoretical study of ThN has been undertaken. High-resolution (∼30MHz) LIF spectroscopy, both field-free and in the presence of static magnetic and electric fields, were recorded. A strong band near 555 nm, which was not previously detected via REMPI spectroscopy [b] has been assigned to a $[18.0]^2\Pi_{3/2}$ - $^2\Sigma^+$ transition. The determined fine structure parameters, electric dipole moments, and magnetic g-factors will be discussed in terms of the present, and previous[b], ab initio predictions.

---

[a]Supported by the United States Department of Energy (DOE) under the Grant. No. DE-SC0018241
[b]Leoncini, A.; Huskens, J.; Verboom, W., Ligands for f-element extraction used in the nuclear fuel cycle. Chem. Soc. Rev. 2017, 46 (23), 7229-7273.
[c]Heaven, M. C.; Barker, B. J.; Antonov, I. O., Spectroscopy and Structure of the Simplest Actinide Bonds. J. Phys. Chem. A 2014, 118 (46), 10867-10881.

WF09   11:20–11:35

HIGH RESOLUTION SPECTROSCOPY OF THE [18.2]1.5 - $X^2\Delta_{3/2}$ TRANSITION OF THORIUM MONOCHLORIDE, ThCl[a].

COLAN LINTON, *Department of Physics, University of New Brunswick, Fredericton, NB, Canada*; DUC-TRUNG NGUYEN, TIMOTHY STEIMLE, *School of Molecular Sciences, Arizona State University, Tempe, AZ, USA*.

A systematic experimental and theoretical studies of simple Ac and Ln containing molecules is one avenue for garnering insight into element-specific ligation[b]. Here we report on the high resolution (∼30 MHz) laser induced fluorescence (LIF) spectra of supersonic cooled molecular beam of ThCl produces in the reaction of laser ablated Th with an Ar/CCl$_4$ mixture. The present work builds on the recent LIF, dispersed fluorescence, and REMPI study of the Heaven and Peterson groups[c]. Analysis of a band near 550 nm has been assigned as the [18.2]1.5-$X^2\Pi_{3/2}$ transition. Observed doubling the lines has shown to be caused Ω-doubling in the upper state. No ^{35}Cl(I=3/2) hyperfine splitting was observed. Progress on recording the electric dipole moments and magnetic g-factors will be reported. Interpretation of the spectrum is based, in part, upon previously published electronic structure prediction and a simple molecular orbital correlation diagram.

[a] Supported by the United States Department of Energy (DOE) under the Grant. No. DE-SC0018241.
[b] Heaven, M. C.; Barker, B. J.; Antonov, I. O., Spectroscopy and Structure of the Simplest Actinide Bonds. J. Phys. Chem. A 2014, 118 (46), 10867-10881.
[c] Van Gundy, R. A.; Bartlett, J. H.; Heaven, M. C.; Battey, S. R.; Peterson, K. A., Spectroscopic and theoretical studies of ThCl and ThCl$^+$. J. Chem. Phys. 2017, 146 (5), 054307/1-054307/8.

WG. Mini-symposium: Frequency-Comb Spectroscopy

Wednesday, June 20, 2018 – 1:45 PM

Room: 116 Roger Adams Lab

Chair: Adam J. Fleisher, National Institute of Standards & Technology, Gaithersburg, MD, USA

WG01   1:45 – 2:00

MID-INFRARED OPTICAL FREQUENCY COMB VIA COHERENT SUPERCONTINUUM PROCESSES IN NANO-PHOTONIC WAVEGUIDES [a]

HAIRUN GUO, WENLE WENG, TOBIAS J. KIPPENBERG, *SB-IPHYS-LPQM, École polytechnique fédérale de Lausanne, Lausanne, Switzerland.*

Mid-infrared (Mid-IR) optical frequency combs [b] are of significant interest for molecular spectroscopy. Recent work has also highlighted the potential to generate mid-IR optical frequency combs from coherent supercontinuum process [c], particularly in chip-based nano-photonic waveguides.

Here we demonstrate the ability to synthesize mid-IR frequency combs in the range $2.5 - 4\ \mu$m [d], directly from an erbium-fiber based femtosecond laser in the telecom-band (at $1.55\ \mu$m), based on mid-IR dispersive wave generation, and using Si_3N_4 nano-photonic waveguides, cf. Figure 1. We further demonstrate that the dispersive wave inherits a high level of coherence from the seed laser and therefore serves as a mid-IR frequency comb. This approach has certain advantages such as being fully compatible with planar fabrication techniques and with compact telecom-band femtosecond fiber laser, and can be readily extended for mid-infrared dual-comb spectroscopy.

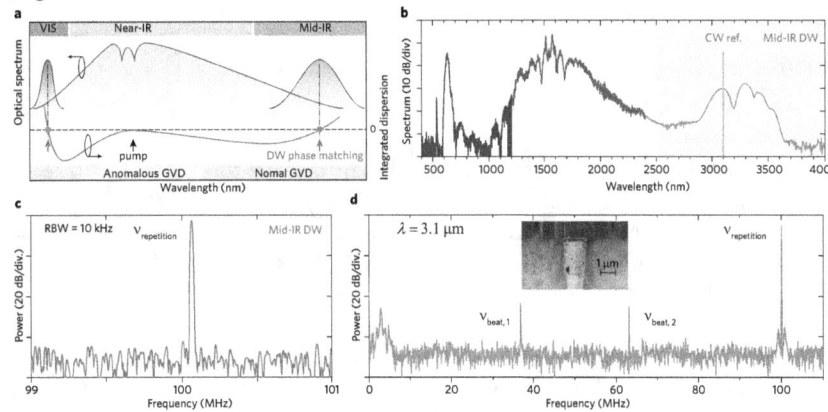

Figure 1: (a) Schematic representation of soliton induced dispersive wave generation; (b) Spectrum of the generated supercontinuum in a large-cross-section Si_3N_4 waveguide, in which the mid-IR wave spans $2.9 - 3.6\ \mu$m; (c) The repetition beatnote of the mid-IR wave which is filtered by a long-pass edge-filter (cut-on wavelength at $2.5\ \mu$m), resolution bandwidth 10 kHz; (d) The heterodyne beatnote of the mid-IR wave with a CW reference laser (wavelength at $\sim 3.1\ \mu$m; inset shows the corss-section SEM picture of Si_3N_4 waveguides, in which the height of the waveguide is as large as $\sim 2.3\ \mu$m.

---

[a] This work is supported by DARPA (SCOUT: W31P4Q-16-1-0002); Air Force (No. FA9550-15-1-0099); Marie Sklodowska-Curie IF grant (No. 709249).
[b] A. Schliesser, N. Picque, and T. W. Hansch, "Mid-infrared frequency combs," Nat. Photon. 6, 440-449 (2012).
[c] J. M. Dudley, G. Genty, and S. Coen, "Supercontinuum generation in photonic crystal fiber," Rev. Mod. Phys. 78, 1135-1184 (2006).
[d] C. Herkommer, et al. "Mid-infrared dispersive wave generation in silicon nitride nano-photonic waveguides," arXiv:1704.02478, 2017.

WG02　　　　　　　　　　　　　　　　　　　　　　　　　　　　　　　　　　　　2:02 – 2:17

HARMONIC FREQUENCY COMB COVERING THE MID-INFRARED MOLECULAR FINGERPRINT REGION

CHRISTIAN GAIDA, MARTIN GEBHARDT, TOBIAS HEUERMANN, *Institute of Applied Physics, Abbe Center of Photonics, Friedrich-Schiller-Universität Jena, Jena, Germany*; THOMAS BUTLER, DANIEL GERZ, CHRISTINA HOFER, LENARD VAMOS, FERENC KRAUSZ, *Laboratory for Attosecond Physics, Max Planck Institute for Quantum Optics, Garching, Germany*; JENS LIMPERT, *Institute of Applied Physics, Abbe Center of Photonics, Friedrich-Schiller-Universität Jena, Jena, Germany*; IOACHIM PUPEZA, *Laboratory for Attosecond Physics, Max Planck Institute for Quantum Optics, Garching, Germany*.

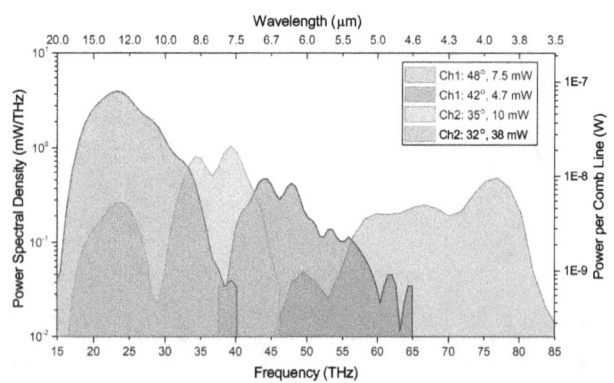

We present a multi-channel harmonic frequency comb covering the mid-infrared spectral range between 15 and 85 THz (or 3.5 - 20 μm, or 500 to 2860 cm^{-1}) with a record 1-mW/THz-level power spectral density. An Er-fiber-based oscillator is wavelength-shifted to a central wavelength of 1960 nm and a chirped-pulse Tm-fiber amplifier provides a 50-MHz-repetition-rate train of 250-fs pulses with 120 W of average power. Nonlinear self-compression in two fused-silica fibers results in two channels, yielding 11-fs pulses with 4.5 W (Channel 1) and 25-fs pulses with 25 W (Channel 2). Subsequent intrapulse difference-frequency generation (DFG) in 1-mm-thin GaSe crystals results in a coverage of the entire molecular fingerprint region with only two phase matching angles for each channel (see Figure). DFG inherently provides phase-stable pulses, leading to a harmonic frequency comb. The 120-W average power of the near-infrared frontend suffices for the parallel implementation of multiple channels, facilitating broadband spectroscopy.

WG03　　　　　　　　　　　　　　　　　　　　　　　　　　　　　　　　　　　　2:19 – 2:34

HIGH-POWER MID-IR COMB GENERATION FOR CAVITY-ENHANCED 2DIR SPECTROSCOPY

MYLES C SILFIES, *Department of Physics, Stony Brook University, Stony Brook, NY, USA*; YUNING CHEN, *Department of Chemistry, Stony Brook University, Stony Brook, NY, USA*; HENRY TIMMERS, ABIJITH S KOWLIGY, ALEX LIND, SCOTT DIDDAMS, *Time and Frequency Division, National Institute of Standards and Technology, Boulder, CO, USA*; THOMAS K ALLISON, *Department of Chemistry, Stony Brook University, Stony Brook, NY, USA*.

Using frequency combs and optical cavities, we have previously demonstrated ultrafast transient absorption measurements with a detection limit of ΔOD $= 1 \times 10^{-9}/\sqrt{\text{Hz}}$, enabling work in dilute molecular beams.[a] Similar methods can be applied to multidimensional spectroscopy as well.[b] Since molecules undergoing ultrafast dynamics have broad spectral features, cavity-enhanced ultrafast spectroscopy then demands broadband and widely tunable frequency combs. Here we present a frequency conversion setup for the generation of high power mid infrared frequency combs in the 3-10 μm region. The initial comb is generated using an Er:fiber oscillator with 100 MHz repetition rate. After nonlinear amplification, the comb is shifted in a highly nonlinear fiber (HNLF) to 1 μm and amplified to 10 W in a home built, multi-stage Yb:fiber amplifier. We have measured the output comb tooth linewidth to be less than 10 kHz and the pulse duration is 120 fs. This laser is then used as a pump for several nonlinear difference frequency generation stages seeded by additional HNLF-shifted combs. Cavity-enhanced mid-infrared combs in the 3-5 μm region will be applied to studying ultrafast dynamics of hydrogen-bonded clusters.

[a] M. A. R. Reber, Y. Chen, and T. K. Allison, Optica **3**, 311 (2016).
[b] T. K. Allison, J. Phys. B: At. Mol. Opt. Phys. **50**, 044004 (2017).

WG04   2:36 – 2:51
COMB-REFERENCED MOLECULAR BEAM SPECTROSCOPY OF POLYCYCLIC HYDROCARBONS

MASATOSHI MISONO, *Applied Physics, Fukuoka University, Fukuoka, Japan*; AKIKO NISHIYAMA, *Japan Science and Technology Agency (JST), ERATO MINOSHIMA Intelligent Optical Synthesizer (IOS) Pro, Tokyo, Japan*; MASAAKI BABA, *Division of Chemistry, Graduate School of Science, Kyoto University, Kyoto, Japan.*

We have studied the electronic excited states of aromatic hydrocarbons such as, benzene or naphthalene by high-resolution spectroscopy.[a] In the excited electronic states of these molecules, there are various interesting interactions such as intramolecular vibrational energy redistribution (IVR), intersystem crossing (ISC), and internal conversion (IC).

In the present study, we observe high-resolution spectra of larger polycyclic hydrocarbons such as perylene. In our experiment, we use a frequency-doubled single mode Ti:Sapphire laser as a light source. Sub-Doppler spectra are obtained with a supersonic molecular beam, which crosses the laser light at right angles. A GPS-disciplined Er-doped fiber optical frequency comb is used as a frequency ruler to decide transition frequencies at the uncertainty of 10 kHz.

[a] A. Nishiyama, K. Nakashima, A. Matsuba, and M. Misono, J. Mol. Spectrosc. **318**, 40 (2015).

WG05   2:53 – 3:08
PRIMARY THERMOMETRY FROM A CO_2 OVERTONE LINE VIA COMB-ASSISTED CAVITY-RING-DOWN SPECTROSCOPY

RICCARDO GOTTI, *Dipartimento di Fisica, Politecnico di Milano, Milano, Italy*; LUIGI MORETTI, *Mathematics and Physics, Second University of Naples, Caserta, Italy*; DAVIDE GATTI, *Dipartimento di Fisica, Politecnico di Milano, Milano, Italy*; ANTONIO CASTRILLO, *Mathematics and Physics, Second University of Naples, Caserta, Italy*; GIANLUCA GALZERANO, *Institute for photonics and nanotechnologies, National Research Council, Milano, Italy*; PAOLO LAPORTA, *Dipartimento di Fisica, Politecnico di Milano, Milano, Italy*; LIVIO GIANFRANI, *Mathematics and Physics, Second University of Naples, Caserta, Italy*; MARCO MARANGONI, *Dipartimento di Fisica, Politecnico di Milano, Milano, Italy.*

We provide the most accurate absolute temperature measurement ever performed on an atomic or molecular sample with a Doppler-Broadening-Thermometry approach. Specifically, the absorption profile of the $P_e(12)$ line of the (30012) - (00001) band of a CO_2 sample at thermodynamic equilibrium is accurately measured at 1.578 μm by a comb-assisted cavity-ring-down spectrometer that combines an extremely dense frequency axis (3000 points over 4.2 GHz) with an acquisition time as low as a few seconds. The Doppler width is extracted from a refined multi-spectrum fitting procedure accounting for the speed dependence of the relaxation rates, which were found to play a role even at the very low pressures explored, from 1 to 7 Pa. The thermodynamic gas temperature is retrieved with relative uncertainties of $8 \cdot 10^{-6}$ (type A) and $11 \cdot 10^{-6}$ (type B), which rank the system at the first place among optical methods. Thanks to a measurement time of only 5 h, the technique represents a promising pathway towards the optical determination of the thermodynamic temperature with a global uncertainty at the 10^{-6} level [a]. An additional element of interest derives from the forthcoming redefinition of the unit Kelvin [b], in 2018, which calls for primary thermometers that are capable to operate over a large part of the temperature scale with very high accuracy.

[a] Gotti R., Moretti L., Gatti D., Galzerano G., Castrillo A., Laporta P., Gianfrani L., and Marangoni M., Phys. Rev. A 97, 12512 (2018)
[b] J. Fischer, Phil. Trans. R. Soc. A 374, 20150038 (2016)

Intermission

WG06 3:44 – 3:59

TOWARD QUANTUM STATE RESOLVED INFRARED FREQUENCY COMB SPECTROSCOPY OF THE C_{60} FULLERENE

BRYAN CHANGALA, MARISSA L. WEICHMAN, *JILA, National Institute of Standards and Technology and Univ. of Colorado Department of Physics, University of Colorado, Boulder, CO, USA*; KEVIN LEE, MARTIN FERMANN, *Laser Research, IMRA AMERICA, Inc, Ann Arbor, MI, USA*; JUN YE, *JILA, National Institute of Standards and Technology and Univ. of Colorado Department of Physics, University of Colorado, Boulder, CO, USA.*

In this talk, we report on progress toward high resolution infrared frequency comb spectroscopy of buckminsterfullerene, C_{60}. A rotationally resolved spectrum of C_{60} has to date remained elusive, despite the very intense research into this molecule's chemical and physical properties since its discovery in 1985. Our approach utilizes cyrogenic buffer gas cooling of the output of a 1000 K effusive oven to prepare cold gas phase C_{60} molecules. We subsequently probe these with a difference frequency generation-based frequency comb tuned to the 8.5 μm IR active fundamental. The combination of a high finesse absorption enhancement cavity and Fourier transform interferometry read-out provide sensitive, broadband detection, while retaining the high spectral resolution of the frequency comb light. We will discuss our preliminary results, which tentatively suggest successful ground state vibrational cooling and observation of resolved rotational fine structure, as well as experimental modifications that we expect to improve the C_{60} number density and internal state cooling efficiency.

WG07 4:01 – 4:16

CO_2 LINE PARAMETER RETRIEVAL BEYOND THE VOIGT PROFILE USING COMB-BASED FOURIER TRANSFORM SPECTROSCOPY

ALEXANDRA C JOHANSSSON, ANNA FILIPSSON, LUCILE RUTKOWSKI, *Department of Physics, Umea University, Umea, Sweden*; PIOTR MASLOWSKI, *Institute of Physics, Faculty of Physics, Astronomy and Informatics, Nicolaus Copernicus University, Torun, Poland*; ALEKSANDRA FOLTYNOWICZ, *Department of Physics, Umea University, Umea, Sweden.*

Mechanical Fourier transform spectrometers (FTS) based on optical frequency combs (OFC) allow precise measurement of comb line intensities when the nominal resolution of the FTS is precisely matched to the repetition rate of the comb[a,b]. Under this condition, the resolution and frequency scale accuracy are given by the narrow comb lines rather than the spectrometer, and the measured spectra are not influenced by the instrumental lineshape. Here we use an FTS based on an Er:fiber frequency comb to perform high-precision direct absorption measurements of the $3\nu_1+\nu_3$ absorption band of pure CO_2 at 1.57 μm. The sample is held inside a multipass cell, and a spectrum of the entire band with signal to noise ratio (SNR) up to 1000 is acquired in $\sim$27 min at one pressure. We retrieve the parameters of individual lines using multiline fitting with the Voigt profile (VP) and the speed-dependent Voigt profile (SDVP), with significantly improved residuals for the SDVP (quality factors close to the SNR for the SDVP). Moreover, the SDVP fits agree with the measured profiles much better than in previous measurements of the same molecular band[c]. The transition frequencies have precision much better than those stated in the HITRAN2016 database[d] and at a similar level as those obtained with continuous wave cavity ring-down spectroscopy[e]. Thus comb-based FTS is a perfect tool for spectroscopy of entire absorption bands with precision beyond the Voigt profile and for retrieval of molecular line parameters for improved spectroscopic databases.

---

[a] Masłowski, P., et al., Phys. Rev. A 93, 021802 (2016).
[b] Rutkowski, L., et al., J. Quant. Spectrosc. Radiat. Transf. 204, 63-73 (2018).
[c] Larcher, G., et al., J. Quant. Spectrosc. Radiat. Transf. 164, 82-88 (2015).
[d] Gordon, I. E., et al., J. Quant. Spectrosc. Radiat. Transf. 203, 3-69 (2017).
[e] Long, D.A., et al., J. Quant. Spectrosc. Radiat. Transf. 161, 35-40 (2015).

WG08   4:18 – 4:33

PREDICTING *PARA-ORTHO* CONVERSION IN AMMONIA

> GUANG YANG, *Center for Free-Electron Laser Science (CFEL), Deutsches Elektronen-Synchrotron (DESY), Hamburg, Germany*; CHRISTOPH HEYL, INGMAR HARTL, *Deutsches Elektronen-Synchrotron DESY, Deutsches Elektronen-Synchrotron DESY, Hamburg, Germany*; ANDREY YACHMENEV, JOCHEN KÜPPER, *Center for Free-Electron Laser Science (CFEL), Deutsches Elektronen-Synchrotron (DESY), Hamburg, Germany.*

We present a combined theoretical and experimental study of the hyperfine-resolved spectrum of ammonia and its deuterated isotopologues. The calculations have been performed using the variational approach TROVE, a new spectroscopically determined potential energy surface, and *ab initio* quadrupole, spin-spin, and spin-rotation coupling surfaces. The computed spectroscopic line lists cover transitions between levels with rotational excitations $J = 0...20$ and vibrational band centers with up to 8000 cm^{-1} above the zero-point-energy level.

For the spectroscopic observation of the *para-ortho* interconversion we use mid-infrared frequency comb spectroscopy in both ammonia vapour and a cold molecular beam. Furthermore, its modulation by external electric field is discussed. Our theoretical model, i.e., the underlying potential energy surface will be refined using the experimentally observed transitions.

WG09   4:35 – 4:50

SEARCH FOR INVERSION SPLITTING OF PHOSPHINE

> SHOKO OKUDA, HIROYUKI SASADA, *Department of Physics, Faculty of Science and Technology, Keio University, Yokohama, Japan.*

Inversion splitting of phosphine molecules has been one of open questions in molecular spectroscopy. A recent calculation predicted that the splitting is 300 kHz and 3 MHz in the $v_2 = 3$ and 4 states [1], where v_2 is the vibrational quantum number of the ν_2 mode. We have observed three Q-branch transitions in the $3\nu_2$ band of phosphine using a comb-referenced sub-Doppler resolution spectrometer [2]. The spectrometer consists of a difference-frequency-generation source and a cavity-enhanced absorption cell with large beam spot size at beam waist to reduce transit-time broadening. The observed spectral linewidths are 150 kHz, but no inversion splitting has been observed. We now try to observe the $4\nu_2 - \nu_2$ hot band. [1] C. Sousa-Silva, J. Tennyson, S. N. Yurchenko, J. Chem. Phys. **145**, 091102 (2016). [2] S. Okuda, H. Sasada, J. Mol. Spectrosc., in press (2018).

WG10   4:52 – 5:07

ONLINE GAS MONITORING USING A MID-INFRARED OPO BASED DUAL COMB SPECTROMETER

> FRANS HARREN, *Molecular and Laser Physics, Radboud University, Nijmegen, Netherlands.*

A dual-frequency comb-based spectrometer for the mid-infrared (3-5 micrometer) wavelength region will open many opportunities for spectroscopic applications. Non-linear conversion provides a wide spectral coverage in the mid-infrared using Optical Parametric Oscillators, keeping the optical properties of the well-established near infrared frequency combs with good frequency accuracy, high spectral resolution at seconds time scale. A number of challenges remain when OPOs are used for optical conversion, because the OPO cavity generate frequency and intensity fluctuation in the combs. Here, we present how these variations are measured and used, in real-time, to correct the recorded broadband spectrum. By monitoring the frequency and amplitude variations of a single absorption line in a reference gas cell, each individual spectrum is normalized in amplitude and corrected by an offset-frequency. As such, real-time averaging is achieved over minutes with minor losses in spectral resolution or degradation, leading to an improvement in spectral resolution. A high-to-noise ratio of about 2400 is achieved with such spectral resolution, demonstrating the efficiency of the proposed method.

WH. Small molecules
Wednesday, June 20, 2018 – 1:45 PM
Room: 100 Noyes Laboratory

Chair: Stephen T Gibson, Australian National University, Canberra, ACT, Australia

WH01 1:45 – 2:00

THE DICARBON BONDING PUZZLE

BENJAMIN A LAWS, STEPHEN T GIBSON, *Research School of Physics and Engineering, Australian National University, Canberra, ACT, Australia.*

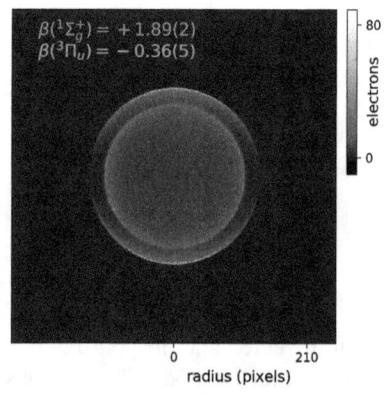

At first glance dicarbon, C_2, would appear to be a very simple homonuclear diatomic molecule. However the bonding structure of C_2 has long been a topic of debate, with different qualitative theories predicting a C–C bond order of 2, 3, or even 4[a]. Recent evidence for a quadruply bonded dicarbon has been provided by multiconfigurational *ab-initio* calculations[b]. However, the interpretation of these calculations has caused debate, with some research preferring the notion of a double, triple, or quasi double-triple bond, while other studies note that there is not enough evidence to clearly define the bonding nature of C_2[c].

In this work, photoelectron spectra of the C_2^- anion are measured using a high resolution photoelectron imaging (HR-PEI) spectrometer. The electron anisotropy of the detachment reveals the character of the parent anion orbital. Detachment to both the ground $\tilde{X}^1\Sigma_g^+$ and first excited $\tilde{a}^3\Pi_u$ electronic states is observed, identifying the character of two orbitals: the diffuse detachment orbital of the anion, and the HOMO of the neutral. The measurements show that electron detachment occurs from a pure s-like orbital ($3\sigma_g$) and a dominant p-like orbital ($1\pi_u$), that is inconsistent with the predictions of strongly mixed (50:50) sp orbitals required for the high bond order models, a result compatible only with the predictions of a C=C double bonding scheme.

[a]R. M. Macrae, *Sci. Prog.* **99**, 1 (2016)
[b]S. Shaik, D. Danovich, W. Wu, P. Su, H. Rzepa, P. Hiberty, *Nat. Chem.* **4**, 195 (2012)
[c]R. Zhong, M.Zhang, H. Xu, Z. Su, *Chem. Sci.* **7**, 1028 (2016)
Research supported by the Australian Research Council Discovery Project Grant DP160102585.

WH02 2:02 – 2:17

ANOMALOUS Q BRANCH INTENSITY IN THE 2+1 REMPI SPECTRUM OF THE $^1\Pi$-$^1\Sigma^+$ TRANSITION IN HIGHLY ROTATIONALLY EXCITED CO PHOTOFRAGMENTS FROM OCS PHOTODISSOCIATION AT 215 NM

CAROLYN E. GUNTHARDT, COLIN J. WALLACE, *Department of Chemistry, Texas A & M University, College Station, TX, USA*; GREGORY HALL, *Division of Chemistry, Department of Energy and Photon Sciences, Brookhaven National Laboratory, Upton, NY, USA*; SIMON NORTH, *Department of Chemistry, Texas A & M University, College Station, TX, USA.*

Nascent CO (X $^1\Sigma^+$) photoproducts formed in the dissociation of OCS at 215 nm were probed using 2+1 resonance enhanced multiphoton ionization (REMPI) through the E $^1\Pi$ state. This photodissociation produces a highly rotationally excited CO distribution, with fragment rotational levels ranging from J=48 to J=77. The resulting REMPI spectrum contains a prominent Q branch, despite negligible line strength factors for high J, two-photon, Π-Σ, Q branch transitions. The presence of a Q branch in the spectrum is explained by intensity borrowing from the nearby C $^1\Sigma^+$ state, as coupling between the C and E states is well documented, and two photon, Σ-Σ, Q branch transitions are intense for high J states. The observed relative intensities of the Q and S branch lines are well described by extrapolation to high J of the J-dependent mixing of C and E states inferred from the E state lambda doublet splittings at lower J. Improved D_e and D_f constants have been derived through the incorporation of this high J data.

WH03　2:19 – 2:34

BRIDGING THE GAP - NEWLY OBSERVED VIBRATIONAL LEVELS OF A AND B STATES OF CaH

KYOHEI WATANABE, IORI TANI, TAKUMI NAMEKATA, KAORI KOBAYASHI, FUSAKAZU MATSUSHIMA, YOSHIKI MORIWAKI, *Department of Physics, University of Toyama, Toyama, Japan*; STEPHEN CARY ROSS, *Department of Physics, University of New Brunswick, Fredericton, NB, Canada.*

The electronic spectrum of CaH has been studied for over 90 years. The first laboratory spectroscopy of CaH was carried out in 1925 on $C^2\Sigma^+ - X^2\Sigma^+$ transitions in the near-UV region[a] Our primary interest is the B state and how it is affected by other nearby states. The B state has a double-minimum potential energy function. For this state we can identify three energy regimes. The lowest is the energy range between the minimum of the inner well and the minimum of the higher lying outer well. In this lower energy region the B-X and A-X spectra were recently investigated by Shayestech et al.[b] The high energy range is that lying above the potential energy barrier between the two wells. Our previous laser induced fluorescence report was on the vibrational states in this higher energy range.[c] In that previous report we were able to confirm a strong irregularity in the vibrational energy spacings that had been previously predicted by the *ab initio* study of Carlsund-Levin et al.. This irregularity is due to interaction between the B and D states.[d]

In the current study we have investigated the intermediate energy regime. These are energies starting from somewhat below the minimum of the higher lying outer well and continuing up to somewhat above the potential energy barrier between the two wells. In this intermediate energy range we have identified the A-X(4,0) and B-X(3 or 5, 0) bands. We present evidence for possible interactions between these vibrational levels. We will report on the current status of our work as we continue our program of delineating the vibrational levels of the B state over a full energy range: starting at the energy of the minimum of the inner well, progressing through the energy of the minimum of the outer well, the energy of the barrier, and on towards the dissociation limit.

---

[a]R. S. Mulliken, *Phys. Rev.* **25**, 509 (1925).
[b]A. Shayesteh, R.S. Ram, P.F. Bernath, *J. Mol. Spectrosc.* **288**, 46 (2013).
[c]K. Watanabe, N. Yoneyama, K. Uchida, K. Kobayashi, F. Matsushima, Y. Moriwaki, S. C. Ross, *Chem. Phys. Lett.* **657**, 1 (2016).
[d]C. Carlsund-Levin, N. Elander, A. Nunez, A. Scrinzi, *Phys. Scripta* **65**, 306 (2002).

WH04　2:36 – 2:51

FLUOROCARBONS IN SATELLITE PLUMES: THE PHOTOSYNTHESIS AND FLUORESCENCE FROM TRIFLUOROMETHYL RADICAL.

JUSTIN W. YOUNG, *Institute for Scientific Research, Boston College, Boston, MA, USA*; CHRISTOPHER ANNESLEY, *Space Vehicles Directorate, Air Force Research Lab, Kirtland AFB, NM, USA.*

Detailed information on the absorption and emission spectra of satellite plume constituents is lacking. Fluorocarbons are used as a space vehicle propellant in cold gas and pulsed plasma thrusters. Consequently, these uses likely produce small fluorocarbon radicals through pyrolysis and solar driven photolysis. Specifically, it's been shown that certain VUV wavelengths can produce electronically excited CF_3 radicals from parent fluorocarbons,[a] but it has not been shown if ground state CF_3 radicals may be produced as well. Furthermore, the CF_3 radical's laser induced fluorescence spectrum has never been reported. Here we investigate the photosynthesis of CF_3 radical from VUV excitation of fluorocarbons, the direct fluorescence from the CF_3 radical, and the radical's relevance in the space environment.

---

[a]Washida, N., Suto, M., Nagase, S., Nagashima, U., Morokuma, K., *J. Chem Phys.* 78, 1025, (1983).

WH05 2:53 – 3:08
COMPUTING SPECTRA OF OPEN-SHELL DIATOMIC MOLECULES WITH DUO

SERGEI N. YURCHENKO, *Department of Physics and Astronomy, University College London, Gower Street, London WC1E 6BT, United Kingdom*; JONATHAN TENNYSON, JAMES R. ASHFORD, HENG YING LI, ELIZAVETA PYATENKO, *Department of Physics and Astronomy, University College London, Gower Street, London WC1E 6BT, United Kingdom*; MAIRE N. GORMAN, *Department of Physics, Aberystwyth University, Aberystwyth, UK.*

DUO is a program designed to solve a coupled Schrödinger equation for the motion of nuclei of a given diatomic molecule characterized by an arbitrary set of electronic states.[a] DUO is capable of both refining potential energy curves (by fitting data to experimental energies or transition frequencies) and producing line lists. Our most recent results of applying DUO to produce hot line lists for open-shell diatomic molecules include NO,[b] SiH,[c] PS and PO,[d] C_2,[e] SN and SH,[f] and AlH.[g] The published version of DUO only considers truly bound states. We are now working on extending DUO to treat quasi-bound or resonance states, or indeed the continuum itself, using the stabilization method.[h] As an illustration, we present simulations of spectra of the quasi-bound system $A\ ^1\Pi - X\ ^1\Sigma^+$ of AlH and of the continuum system $A\ ^1\Pi - X\ ^1\Sigma^+$ and $B\ ^1\Sigma^+ - X\ ^1\Sigma^+$ of NaCl.

[a] S.N. Yurchenko, L. Lodi, J. Tennyson, and A.V. Stolyarov, *Comput. Phys. Commun.* **202**, 262 (2016).
[b] A. Wong, S. N. Yurchenko, P. Bernath, H. S. P. Mueller, S. McConkey, and J. Tennyson, *Mon. Not. R. Astron. Soc.* **470**, 882 (2017).
[c] S. N. Yurchenko, F. Sinden, L. Lodi, C. Hill, M. N. Gorman, and J. Tennyson, *Mon. Not. R. Astron. Soc.* **473**, 5324 (2018)
[d] L. Prajapat, P. Jagoda, L. Lodi, M. N. Gorman, S. N. Yurchenko, and J. Tennyson, *Mon. Not. R. Astron. Soc.* **472**, 3648 (2017).
[e] S. N. Yurchenko, J. Tennyson, and et al, *Mon. Not. R. Astron. Soc.* in preparation (2018).
[f] S. N. Yurchenko, W. Bond, M. N. Gorman, L. Lodi, L. K. McKemmish, W. Nunn, R. Shah, and J. Tennyson, *Mon. Not. R. Astron. Soc.* submitted (2018)
[g] H. Williams, P. C. Leyland, L. Lodi, S. N. Yurchenko, and J. Tennyson, *Mon. Not. R. Astron. Soc.* in preparation (2018).
[h] A.U. Hazi, H.S. Taylor, *Phys. Rev. A* **1**, 1109 (1970)

WH06 3:10 – 3:25
MICROWAVE SPECTRUM AND THEORETICAL INVESTIGATION OF TRIFLUOROACETIC SULFURIC ANHYDRIDE

ANNA HUFF, BECCA MACKENZIE, CJ SMITH, KEN LEOPOLD, *Chemistry Department, University of Minnesota, Minneapolis, MN, USA.*

Trifluoroacetic sulfuric anhydride, CF_3COOSO_2OH, has been produced under supersonic jet conditions from the reaction of sulfur trioxide and trifluoroacetic acid. The rotational spectra for both the parent and deuterated isotopologues have been recorded, but were notably weaker than similar spectra obtained for other carboxylic sulfuric anhydrides. The spectra were readily fit to a Watson A-reduced Hamiltonian with no evidence of internal rotation. M06-2X/6-311++G(3df,3pd) calculations indicate that the formation of CF_3COOSO_2OH proceeds through a $\pi_2 + \pi_2 + \sigma_2$ cycloaddition mechanism analogous to that previously established for other carboxylic sulfuric anhydrides. The barrier to formation for CF_3COOSO_2OH calculated at the CCSD(T)/CBS(D-T) level is slightly positive (0.7 kcal/mol), in contrast to the slightly negative value obtained for the formation of its acetic acid analog (acetic sulfuric anhydride). The possible role of internal rotation in the formation of both systems will be discussed.

Intermission

WH07   4:01–4:16
VERY DIFFERENT CH$_3$ INTERNAL ROTATION BARRIERS IN THE SYN- AND ANTI- FORMS OF THIOACETIC ACID: MICROWAVE MEASUREMENTS AND ENERGY DECOMPOSITION ANALYSIS

<u>CJ SMITH</u>, ANNA HUFF, KEN LEOPOLD, *Chemistry Department, University of Minnesota, Minneapolis, MN, USA*; HUAIYU ZHANG, *College of Chemistry and Material Science, Hebei Normal University, Shijiazhuang, China*; YIRONG MO, *Department of Chemistry, Kalamazoo College, Kalamazoo, MI, USA.*

Rotational spectra of two conformers of thioacetic acid (CH$_3$COSH) have been observed by pulsed-nozzle Fourier transform microwave spectroscopy. Spectroscopic constants are reported for both the syn- and anti- forms of the parent species, as well as for five isotopologues which include ^{34}S and ^{13}C substitution on the methyl and carboxyl atoms. Spectra were fit using two different internal rotation fitting programs, XIAM and BELGI-Cs, and comparisons between their performances will be discussed. The experimental internal rotation barriers for the parent syn- and anti-thioacetic acid obtained from BELGI-Cs are 69.3(10) and 435.2(22) cm^{-1}, respectively, and compare favorably with the computed values of 83 and 344 cm^{-1} at the M06-2X/6-311+G(d,p) level of theory. An energy decomposition analysis using the block localized wavefunction method indicates that the steric (including both Pauli and electrostatic) repulsion between the -SH and CH$_3$ groups, which is further enhanced by the pi-conjugation from the -SH to the carbonyl group, is largely responsible for the large difference in the internal rotation barrier between the syn- and anti- forms.

WH08   4:18–4:33
LINE INTENSITY MEASUREMENTS AND ANALYSIS IN THE ν_3 BAND OF RUTHENIUM TETROXIDE

<u>JEAN VANDER AUWERA</u>, *Service de Chimie Quantique et Photophysique, Université Libre de Bruxelles, Brussels, Belgium*; SÉBASTIEN REYMOND-LARUINAZ, *Département de Physico-chimie, CEA/Saclay, CEA, DEN, Gif-sur-Yvette, France*; VINCENT BOUDON, *Laboratoire ICB, CNRS/Université de Bourgogne, DIJON, France*; DENIS DOIZI, *Département de Physico-chimie, CEA/Saclay, CEA, DEN, Gif-sur-Yvette, France*; LAURENT MANCERON, *Synchrotron SOLEIL, CNRS-MONARIS UMR 8233 and Beamline AILES, Saint Aubin, France.*

Ruthenium tetroxide (RuO$_4$) is a heavy tetrahedral molecule characterized by an unusual volatility near ambient temperature. Because of its chemical toxicity and the radiological impact of its ^{103}Ru and ^{106}Ru isotopologues, the possible remote sensing of this compound in the atmosphere has renewed interest in its spectroscopic properties. In a recent study, the strong fundamental band associated with the excitation of the infrared active stretching mode ν_3 of ^{102}Ru^{16}O$_4$, observed near 10 μm, was re-investigated at high-resolution (0.001 cm^{-1}) with the help of a ^{102}Ru isotopically pure sample.[a] Building upon that work, the present contribution is the first investigation dealing with high-resolution line-by-line intensity measurements for the ν_3 fundamental band of ^{102}Ru^{16}O$_4$. It relies on high resolution Fourier transform infrared spectra specifically recorded at room temperature at the AILES beam line of SOLEIL using synchrotron radiation, a specially constructed cell and an isotopically pure sample of ^{102}Ru^{16}O$_4$. Relying on an effective Hamiltonian and associated effective dipole moment,[a] the measured line intensities were assigned and dipole moment parameters determined. A HITRAN-formatted frequency and intensity line list was generated.

---

[a]S. Reymond-Laruinaz, V. Boudon, L. Manceron, L. Lago, D. Doizi, J Mol Spectrosc 315 (2015) 46–54.

WH09 4:35–4:50
MOLECULAR LINE INTENSITIES OF CARBON DIOXIDE IN THE 1.6 μm REGION DETERMINED BY CAVITY RINGDOWN SPECTROSCOPY

ZACHARY REED, DAVID A. LONG, JOSEPH T. HODGES, *Material Measurement Laboratory, National Institute of Standards and Technology, Gaithersburg, MD, USA.*

Here we present some recent advances in frequency stabilized cavity ring-down spectroscopy (FS-CRDS) measurements of molecular line intensities of carbon dioxide in the (30012)←(00001), the (30013)←(00001), and the (30014)←(00001) bands near 1.6 μm.

These measurements were performed near 296K using a frequency stabilized cavity ringdown spectrometer [1]. Additional independent measurements were performed on a frequency agile rapid scanning (FARS) CRDS [2].

We have compared the line intensities obtained from Hartmann Tran Profile (HTP) fits of the measured spectra to several spectroscopic databases, including UCL (ie, HITRAN2016) [3]. The overall agreement between these results and the ab initio calculations of Zak et al is excellent [3], although some individual transitions show deviations of up to 1%. The intensities for the (30012)←(00001) show average agreement at the 0.1% level. Preliminary measurements on the (30013)←(00001), and the (30014)←(00001) bands in this region also show good agreement with the ab initio of Zak et al for the (30013)←(00001), but considerably poorer agreement for the (30014)←(00001) band. No significant J-dependence is observed for any of the three bands.

This work demonstrates significant improvement in experimental determination of important CO_2 line intensities in the 1.6 μm region. It also demonstrates that it may be feasible for ab initio theory to provide sufficiently accurate results for global determinations of line intensities in the near future.

[1] H. Lin, Z. D. Reed, V. T. Sironneau, and J. T. Hodges, J. Quant. Spectrosc. Radiat. Transfer 161, 11-20 (2015).

[2] G. W. Truong, K. O. Douglass, S. E. Maxwell, R. D. van Zee, D. F. Plusquellic, J. T. Hodges, and D. A. Long, Nat. Photonics 7, 532-534 (2013).

[3] E. J. Zak, J. Tennyson, O. L. Polyansky, L. Lodi, N. F. Zobov, S. A. Tashkun, and V. I. Perevalov, J. Quant. Spectrosc. Radiat. Transfer 189, 267-280 (2017).

WH10 4:52–5:07
FIRST HIGH RESOLUTION IR SPECTRA OF 2-D_1-PROPANE. THE ν_9 (A_1) B-TYPE BAND NEAR 367.2389 cm^{-1}.

STEPHEN J. DAUNT, ROBERT GRZYWACZ, *Department of Physics & Astronomy, The University of Tennessee-Knoxville, Knoxville, TN, USA;* WALTER LAFFERTY, *Optical Technology Division, National Institute of Standards and Technology, Gaithersburg, MD, USA;* JEAN-MARIE FLAUD, *LISA, CNRS, Universités Paris Est Créteil et Paris Diderot, Créteil, France;* BRANT E. BILLINGHURST, *EFD, Canadian Light Source Inc., Saskatoon, Saskatchewan, Canada.*

This is a further report in a project to record high resolution IR data of the ^{13}C and D substituted isotopologues of propane (see talks FA04, FA05 and TK08 at 2017 ISMS). Initially in CLS Cycle 23 (Jan-Jun, 2015) we recorded spectra of the ν_{26} (B_2) C-Type band whose corresponding band in C_3H_8 is observed in Titan's Atmosphere. That band and others seen in the 550-950 cm^{-1} region were too perturbed by complex torsional splittings for analysis at this time. In this talk will give details on the first high resolution ($\Delta\nu = 0.00096$ cm^{-1}) IR investigation of the spectrum in the Far-IR region. We recorded spectra during Cycle 25 (Jan-Jun, 2017) of the ν_9 (A_1) CCC skeletal bending mode near 367.2389 cm^{-1}. This has a B-type band structure and appears unperturbed. Spectra were recorded at pressures of 0.014, 0.056, 3.995 & 8.087 Torr in a 72m optical path at room temperature. We used the Bruker IFS-125HR on the Far-IR beamline of the CLS. The spectra were assigned both traditionally and with the aid of the PGOPHER program of Colin Western.[a] We were able to assign over 8100 lines with up to K = 35 and J = 60 using both the 4 and 8 Torr data sets. The only available MW data on this molecule are the seven K = 0, J = 0-6 lines from Lide.[b] We therefore had to use the present data to determine a new set of ground state constants that included centrifugal distortion terms for this molecule. We compare these experimentally determined values with both Lide's A, B, C values and the recent calculated *ab initio* values of Villa, Senent & Carvajal.[c] Upper state constants have also been been derived that provide a good simulation of the observed spectra. The hope is that this data will be useful in identifying isotopic propane lines in Titan and other astrophysical objects.

[a] C. Western, J. Quant. Spectrosc. & Rad. Transf. **186**, 221 ff. (2017).
[b] Lide, J.Chem. Phys. **33**, p.1514ff. (1960).
[c] Villa, Senent & Carvajal, PCCP **15**, 10258 (2013).

WH11

FIRST FAR-IR SPECTRA OF 2,2-D$_2$-PROPANE: THE ν_9 (A$_1$) B-TYPE BAND NEAR 365.3508 cm^{-1}. THE DETERMINATION OF GROUND AND UPPER STATE CONSTANTS.

<u>DANIEL GJURAJ</u>, *Department of Physics, Iona College, New Rochelle, NY, USA*; STEPHEN J. DAUNT, ROBERT GRZYWACZ, *Department of Physics & Astronomy, The University of Tennessee-Knoxville, Knoxville, TN, USA*; WALTER LAFFERTY, *Optical Technology Division, National Institute of Standards and Technology, Gaithersburg, MD, USA*; JEAN-MARIE FLAUD, *CNRS, Universités Paris Est Créteil et Paris Diderot, LISA, Créteil, France*; BRANT E. BILLINGHURST, *EFD, Canadian Light Source Inc., Saskatoon, Saskatchewan, Canada*.

Only old IR and no MW data exist for this molecule.[a] Last year we reported (2017 ISMS, TK08) on the ν_{20} (B$_1$) A-type band recorded in CLS Cycle 23 (Jan-May, 2016). This was the only band with easily assignable lines. Other bands were perturbed and not assignable at present. We assigned most of ν_{20} and rotational constants were determined for this molecule. Our ν_{20} data was limited in K and J values due to the pressure used and time limitations in that cycle. Also some small torsional perturbations may have affected the derived constants. From our recent studies ν_9 bands appear unperturbed for the other ^{13}C and D isotopologues. Therefore we recorded that band for 2,2-D$_2$-Propane in the Far-IR at higher pressures in new experiments. The spectrum of the ν_9 (A$_1$) band (CCC bend) was recorded on the Far-IR beamline during CLS Cycle 27 (Jun-Dec, 2017). Spectra were recorded at 4.055 Torr ($\Delta\nu$ = 0.00096 cm-1) and 7.691 Torr ($\Delta\nu$ = 0.0020 cm-1) to see higher K and J transitions. An optical path of 72 m and a cell temperature of 265.75K were used. We assigned over 5900 lines with both traditional methods and the aid of the PGOPHER program of Colin Western.[b] Lines up to K = 37 and J = 55 were assigned by using both pressure data sets. Improved rotational constants including the inertial and centrifugal distortional constants will be reported. This varied isotopic data should improve the r$_0$ structure.

[a] Friedman & Turkevich, J. Chem. Phys. **17**, 1012 ff. (1949); McMurry, Thornton & Condon, J. Chem. Phys. **17**, 918 ff. (1949); McMurry & Thornton, J. Chem. Phys. **19**, 1014 ff. (1951).; Gayles & King, Spectrochim. Acta **21**, 543 ff. (1965); Kondo & Saeki, Spectrochim. Acta **29A**, 735 ff. (1973).
[b] C. Western, JQSRT **186**, 221 ff. (2017)

WI. Mini-symposium: Far-Infrared Spectroscopy

Wednesday, June 20, 2018 – 1:45 PM

Room: 1024 Chemistry Annex

Chair: Brian Drouin, California Institute of Technology, Pasadena, CA, USA

WI01   *INVITED TALK*   1:45 – 2:15

PHOTONICS-BASED TERAHERTZ SOURCES FOR MOLECULAR SPECTROSCOPY

<u>JEAN-FRANÇOIS LAMPIN</u>, *UMR CNRS 8520, Institut d'Electronique de Microélectronique et de Nanotechnologie, Villeneuve d'Ascq, France.*

Since twenty years photonics-based terahertz (THz) sources have made great progress. They are now usable for low- and high-resolution molecular spectroscopy in the 0.1-4 THz range (3-130 cm^{-1}).

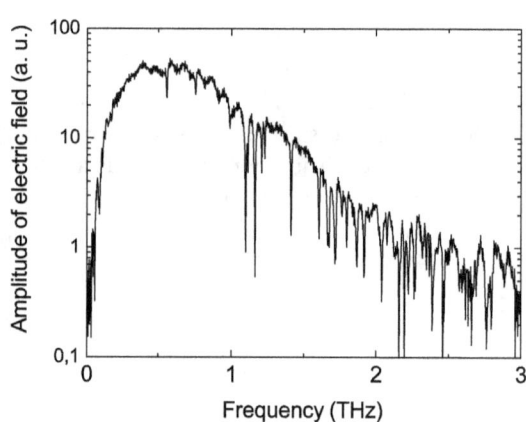

We will present the principles of near-infrared laser-based THz set-ups: time-domain spectroscopy (see figure showing ambient pressure H$_2$O lines) and frequency-domain photomixing. Then we will review the different types of high bandwidth semiconductors photodetectors used to convert laser beams into THz beams. Two families are mainly used: photoconductors and photodiodes. Advantages and drawbacks of each will be presented. The subject of THz antennas will be also discussed and a state of the art will be given including the devices developed at IEMN. Some example of molecular spectroscopy measurements using these devices at various frequencies and resolutions will be given. Finally new ways for photonics-based THz sources will be exposed.

WI02   2:19 – 2:34

THZ HETERODYNE SPECTROSCOPY ON THE AILES BEAMLINE OF SOLEIL FACILITY USING THE SYNCHROTRON RADIATION EMITTED BY THE MULTIBUNCH OPERATION MODE

<u>OLIVIER PIRALI</u>[a], *AILES beamline, Synchrotron SOLEIL, Saint Aubin, France*; GAËL MOURET, FRANCIS HINDLE, ARNAUD CUISSET, *Laboratoire de Physico-Chimie de l'Atmosphère, Université du Littoral Côte d'Opale, Dunkerque, France*; JEAN-FRANÇOIS LAMPIN, SOPHIE ELIET, JOAN TURUT, *Institut d'Electronique de Microélectronique et de Nanotechnologie, Université de Lille 1, Villeneuve d'Ascq, France*; P. ROY, *AILES beam line, Synchrotron Soleil, Gif-sur-Yvette, France.*

The goal of our project is to develop a new high-resolution spectrometer on the AILES beamline of synchrotron SOLEIL. The spectrometer will record absorption spectra in the 1–5 THz range with sub-MHz resolution using a heterodyne detection scheme. We recently performed a test experiment using the far-IR continuum produced by synchrotron radiation in the multibunch mode, a multiplication chain as local oscillator, and a hot electron bolometer (HEB) as heterodyne mixer. The set-up allowed the detection of one D$_2$O absorption line centered at about 782 GHz (^{R}Q(4)$_{0,4}$) with a resolution better than 1 MHz. We will present the details of the experiment and some instrumental developments in progress.

---

[a]*Institut des Sciences Moléculaires d'Orsay, UMR8214, CNRS, Université Paris-Saclay, Univ. Paris-Sud, France*

WI03 2:36 – 2:51

GAS-PHASE INFRARED SPECTROSCOPY OF METAL-LIGAND REDOX PAIRS

<u>MUSLEH UDDIN MUNSHI</u>, *FELIX Laboratory, Radboud University, Nijmegen, The Netherlands*; GIEL BERDEN, JONATHAN K MARTENS, *Institute for Molecules and Materials (IMM), Radboud University Nijmegen, Nijmegen, Netherlands*; JOS OOMENS, *FELIX Laboratory, Radboud University, Nijmegen, The Netherlands*.

Transition metal organometallic complexes are known for their catalytic activation of small molecules. For example, CO_2 activation by Ni^{1+}(cyclam) complexes has recently been reported in the gas-phase[a], where the oxidation state of the metal appeared to be the key for the electrochemical reduction of CO_2. In this contribution, we show that metal(II) complexes, including Ni-cyclam, generated via electrospray ionization can be charge reduced inside a quadrupole ion trap mass spectrometer[b] (QIT MS) via an electron transfer reduction (ETR) reaction using fluoranthene radical anions as reagents[c]. Moreover, their IR spectra can be recorded by action spectroscopy using the FELIX free electron laser, which enables their structural characterization. The complex in either of two charge states is isolated by its m/z, stored and irradiated with tunable IR light in the range of 500-1800 cm^{-1}. IR induced dissociation occurs upon resonant vibrational excitation of the trapped ions, so that an IR spectrum of the (charge reduced) complexes can be obtained by plotting the dissociation yield as a function of IR laser frequency. Experimental spectra are compared with theoretical spectra computed at the density functional theory (DFT) level to obtain detailed structural information of the complexes undergoing charge reduction, eventually providing new insights in their catalytic capabilities.

---

[a] F. S. Menges, S. M. Craig, N. Tötsch, A. Bloomfield, S. Ghosh, H. J. Krüger and M. A. Johnson, *Angewandte Chemie International Edition*, 2016, **55**, 1282-1285.

[b] J. Martens, G. Berden, C. R. Gebhardt and J. Oomens, *Review of Scientific Instruments*, 2016, **87**, 103108.

[c] M. U. Munshi, S. M. Craig, G. Berden, J. Martens, A. F. DeBlase, D. J. Foreman, S. A. McLuckey, J. Oomens and M. A. Johnson, *The Journal of Physical Chemistry Letters*, 2017, **8**, 5047-5052

WI04 2:53 – 3:08

HIGH-RESOLUTION TERAHERTZ GAIN SPECTRA OF MID-INFRARED PUMPED NH_3

MARTIN MICICA, *Institut d'électronique de microélectronique et de nanotechnologie, Université de Lille 1, Villeneuve d'Ascq, France*; SOPHIE ELIET, *Institut d'Electronique de Microélectronique et de Nanotechnologie, Université de Lille 1, Villeneuve d'Ascq, France*; A. PIENKINA, *Laboratoire PhLAM, UMR 8523 CNRS - Université Lille 1, Villeneuve d'Ascq, France*; R. A. MOTIYENKO, *UMR 8523 - PhLAM - Physique des Lasers Atomes et Molécules, University of Lille, CNRS, F-59000 Lille, France*; L. MARGULÈS, *Laboratoire PhLAM UMR 8523, Université de Lille, 59655 Villeneuve d'Ascq, FRANCE*; MATHIAS VANWOLLEGHEM, *UMR CNRS 8520, Institut d'Electronique de Microélectronique et de Nanotechnologie, Villeneuve d'Ascq, France*; KAMIL POSTAVA, *IT4Innovations, VSB - Technical University of Ostrava, Ostrava - Poruba, Czech Republic*; JEAN-FRANÇOIS LAMPIN, *UMR CNRS 8520, Institut d'Electronique de Microélectronique et de Nanotechnologie, Villeneuve d'Ascq, France*.

Inversion of population in the terahertz (THz) range can be obtained thanks to the optical pumping of polar molecules in the mid-infrared range. Generally it is done with CO_2 lasers but recently we have demonstrated the first molecular laser pumped by a quantum cascade laser (QCL). It is based on the optical pumping of the NH_3 molecule in the ν_2=1 state. The gain is obtained by the stimulated emission on pure inversion transitions of NH_3 (large amplitude motions) around 1 THz that are not accessible to continuous-wave (CW) CO_2 lasers. We present here CW high-resolution gain measurements of two strong lines: the (3,3) around 1.073 THz and the (4,4) around 1.083 THz. The measurements are done with a THz multiplication chain and an InSb bolometer. The gain profiles are recorded at different pressure and different QCL frequencies as for an IR/THz double resonance experiment. The highest gain at the best conditions are obtained with the (3,3) line: 10 dB/m for a pump power of about 40 mW. To our knowledge this gain is highest measured in the THz range for a CW-pumped molecule. These measurements will help the understanding and the design of our NH_3 lasers. This kind of laser will find applications in THz molecular spectroscopy/astronomy as a source or as a local oscillator for heterodyne detection.

Intermission

WI05 INVITED TALK 3:44–4:14

EXPLORING THE SOLID STATE PHASE TRANSITION IN DL-NORVALINE WITH TERAHERTZ SPECTROSCOPY

JENS NEU, COLEEN T. NEMES, KEVIN P. REGAN, *Department of Chemistry, Yale University, New Haven, CT, USA*; MICHAEL R. C. WILLIAMS, *Center for Integrated Nanotechnologies, Los Alamos National Laboratory, Los Alamos, NM, USA*; CHARLES A. SCHMUTTENMAER, *Department of Chemistry, Yale University, New Haven, CT, USA*.

DL-Norvaline is a molecular crystal at room temperature and it undergoes a phase transition when cooled below 190 K. This phase transition is believed to be Martensitic. We investigate this phase transition by measuring its terahertz (THz) spectrum over a range of temperatures. Temperature-dependent THz time-domain spectroscopy (THz-TDS) measurements reveal that the transition temperature ($T_{\beta \to \alpha}$) is 190 K. The influence of nucleation seeds was analyzed by determining the $T_{\beta \to \alpha}$ of molecular crystals with varying grain size. Grains of 5 μm or less result in a lower transition temperature ($T_{\beta \to \alpha}$ = 180 K) compared to larger grains of 125–250 μm ($T_{\beta \to \alpha}$ = 190 K). Additionally, we gain insight into the physical process of the phase transition *via* temperature-dependent THz-TDS spectra of doped and mixed molecular crystals. The addition of molecular dopants, which differ from DL-norvaline only at the end of the side chain which resides in the hydrophobic layers of the crystal, decreases $T_{\beta \to \alpha}$. This is consistent with a solid-solid phase transition in which the unit cell shifts along this hydrophobic layer, and it leads us to believe that the phase transition in DL-norvaline is Martensitic in nature.

WI06 4:18–4:33

PULSE-ECHO MILLIMETER WAVE *IN SITU* SENSOR WITH 65 nm CMOS TRANSMITTER AND HETERODYNE RECEIVER ELECTRONICS

DEACON J NEMCHICK, BRIAN DROUIN, ADRIAN TANG, YANGHYO KIM, *Jet Propulsion Laboratory, California Institute of Technology, Pasadena, CA, USA*; GABRIEL VIRBILA, M.-C. FRANK CHANG, *Electrical Engineering, University of California - Los Angeles, Los Angeles, CA, USA*.

Cavity enhanced pure rotational spectroscopy has long been a potent laboratory tool for the elucidation of structure and dynamics in isolated molecular systems where sensitive pulsed-echo techniques are routinely performed up to frequencies as high ∼50 GHz. Although the associated narrow linewidths (∼800kHz), wide-bandwidth (often >10 GHz), and long optical path lengths have long been identified as a desirable combination for sensitive and specific gas sensing, the unaccommodating size and power requirements of traditional microwave optics/electronics are unsuitable for the stringent demands required for *in situ* deployment. Additionally, efforts to drive pulsed-echo techniques into millimeter and submillimeter wavelength regimes, where the size of optics can be reduced without suffering large diffraction losses, have failed largely due to inefficiencies of injecting radiation into the resonant optical cavity.

Recent pursuits at the Jet Propulsion Laboratory to realize compact, low-power devices capable of *in situ* chemical detections on extra-terrestrial objects have found success in calling upon novel transmitter and receiver elements built from CMOS architectures commonly employed in the high-speed communications industry. These low-power integrated circuit chipsets can be embedded directly into quasi-optical devices allowing for the realization of cavity based instruments where all source and detection electronics are hosted by a single 16 in^2 printed circuit board. The current talk will present a full system description of this miniaturized CMOS-based pulse-echo rotational spectrometer,[a] which has an operational bandwidth of 90-105 GHz, along with experimental trials taken in bulk gas flows and seeded molecular beam environments.

---

[a]D. J. Nemchick *et al.*, "A 90-102 GHz CMOS Based Pulsed-Echo Fourier Transform Spectrometer: New Approaches for *In Situ* Chemical Detection and Millimeter-Wave Cavity-Based Molecular Spectroscopy," *Rev. Sci. Inst.*, In Submission.

WI07   4:35 – 4:50

FOURIER TRANSFORM MILLIMETER-WAVE SPECTROMETER WITH ORIGINAL DESIGN

R. A. MOTIYENKO, L. MARGULÈS, *UMR 8523 CNRS - Université de Lille, Laboratoire PhLAM, Villeneuve d'Ascq, France*; E. A. ALEKSEEV, *Radiospectrometry Department, Institute of Radio Astronomy of NASU, Kharkov, Ukraine.*

Direct digital synthesizers (DDS) have a number of advantages, especially such as the high precision and rate of frequency adjustment. In addition, these synthesizers possess a unique property to allow changing the frequency from one value to another with continuous phase. Few years ago, we built fast-scan absorption spectrometer with Schottky diode frequency multiplication chains as a radiation source.[a] The rapid frequency scan in the spectrometer is provided by RF synthesizer based on up-conversion of the DDS signal. Owing to the capability of fast frequency switching, the same up-converted RF synthesizer may generate short pulses to polarize molecules and subsequently may be used as a local oscillator in the heterodyne detection of free induction decay. This feature simplifies the spectrometer design, as it allows using only one radiation source to polarize molecular sample, and to detect molecular signal. Using this principle, we built a Fourier transform spectrometer in the millimeter wave range. The spectrometer covers the frequency range between 150 and 220 GHz. In the current design, the RF synthesizer allows generation of frequency pulses with a bandwidth inverse proportional to pulse duration, as well as chirped pulses with a bandwidth of about 350 MHz. The performances of the spectrometer will be presented and discussed.

[a] A. Pienkina, R.A. Motiyenko, L. Margulès et al. ISMS, 71st symposium (2016), FB05

WI08   4:52 – 5:07

THE ROTATIONAL SPECTRUM OF THE METHANETHIOL ISOTOPOLOG $CH_3^{34}SH$

OLENA ZAKHARENKO, FRANK LEWEN, STEPHAN SCHLEMMER, HOLGER S. P. MÜLLER, *I. Physikalisches Institut, Universität zu Köln, Köln, Germany*; V. ILYUSHIN, E. A. ALEKSEEV, IGOR KRAPIVIN, *Radiospectrometry Department, Institute of Radio Astronomy of NASU, Kharkov, Ukraine*; LI-HONG XU, RONALD M. LEES, *Department of Physics, University of New Brunswick, Saint John, NB, Canada*; ROBIN T. GARROD, *Departments of Chemistry and Astronomy, The University of Virginia, Charlottesville, VA, USA*; ARNAUD BELLOCHE, KARL M. MENTEN, *Millimeter- und Submillimeter-Astronomie, Max-Planck-Institut für Radioastronomie, Bonn, NRW, Germany.*

Methanethiol, CH_3SH, has been found in the warm and dense parts of high as well as low mass star-forming regions.[a] The molecule is also of fundamental interest because of the large amplitude internal rotation of the CH_3 group whose effects are somewhat less pronounced than in its lighter homolog CH_3OH. In the course of our ongoing study of $CH_3^{32}SH$,[b] we have recorded new spectra which currently cover 49−510 GHz.[c] These spectra, as well as existing ones covering part of the 1.1−1.5 THz region,[d] were inspected for lines of $CH_3^{34}SH$. We made extensive assignments of $\Delta K = 0$ transitions in $v_t = 0$ to 2. Numerous assignments of $\Delta K = \pm 1$ transitions were made in $v_t = 0$ and to a lesser extent in the two higher torsional states. We will present results of modeling these data with the RAM36 program and of searches for this isotopolog in our 3 mm ALMA data of Sagittarius B2(N).

[a] See, e.g., H. S. P. Müller at al., Astron. Astrophys. 587 (2016) A92; M. N. Drozdovskaya et al., Mon. Not. R. Astron. Soc., accepted, arXiv: 1802.02977; and references therein.
[b] V. Ilyushin et al., TI04 at the 72nd ISMS, 2017
[c] The work in Kharkov was done under support of the Volkswagen foundation. The assistance of the Science and Technology Center in the Ukraine is acknowledged (STCU partner project #P686). The work in Cologne was supported by the Deutsche Forschungsgemeinschaft (DFG) via grant SFB 956, project B3 and via the Gerätezentrum "Cologne Center for Terahertz Spectroscopy".
[d] L.-H. Xu at al., J. Chem. Phys. 137 (2012) 104313.

WJ. Conformers and isomers

Wednesday, June 20, 2018 – 1:45 PM

Room: 217 Noyes Laboratory

Chair: Timothy S. Zwier, Purdue University, West Lafayette, IN, USA

WJ01   1:45 – 2:00

THE ROTATIONAL STUDY OF DOPAC, A NEURAL METABOLITE

<u>ELENA R. ALONSO</u>, IKER LEÓN, LUCIE KOLESNIKOVÁ, CARLOS CABEZAS, JOSÉ L. ALONSO, *Grupo de Espectroscopia Molecular, Lab. de Espectroscopia y Bioespectroscopia, Unidad Asociada CSIC, Universidad de Valladolid, Valladolid, Spain.*

DOPAC (3,4-Dihydroxyphenylacetic acid) is an intermediate neural metabolite in the cycle of catecholamines,[a] concretely in the metabolic reaction of dopamine (DA) to produce homovanillic acid (HVA), a substance that being present in urine in a large quantity, is indicative of health problems. In this work, we present a rotational study to explore the conformational preferences of DOPAC. It has been brought into gas phase avoiding decomposition by laser ablation LA of the solid sample, and the conformers present in the supersonic expansion have been probed by CP-FTMW spectroscopy.[b] As we will show, several conformers have been detected and characterized. All the data obtained is presented in this communication.

[a] C. Cabezas, I. Peña, J. C. López and J. L. Alonso, Seven conformers of neutral dopamine revealed in the gas phase, J. Phys. Chem. Lett., 2013, 4, 486–490 and references therein.

[b] I. Peña, S. Mata, A. Martín, C. Cabezas, A. M. Daly and J. L. Alonso, Conformations of D-xylose: the pivotal role of the intramolecular hydrogen-bonding., Phys. Chem. Chem. Phys., 2013, 15, 18243–8.

WJ02   2:02 – 2:17

MM-WAVE AND AB INITIO STUDIES OF THE CONFORMATIONAL LANDSCAPE OF METHOXYPHENOLS IDENTIFIED AS SOA PRECURSORS

<u>ATEF JABRI</u>, ANTHONY ROUCOU, DANIELE FONTANARI, CÉDRIC BRAY, FRANCIS HINDLE, GAËL MOURET, ROBIN BOCQUET, ARNAUD CUISSET, *Laboratoire de Physico-Chimie de l'Atmosphère, Université du Littoral Côte d'Opale, Dunkerque, France.*

Methoxyphenols are emitted in the atmosphere from biomass burning and recent works have shown the potential role of these oxygenated aromatic species in the formation of secondary organic aerosols [a]. In fact, these semi-volatile polar aromatic species are produced from the pyrolysis of wood lignin and mainly consist in Guaiacol (2-methoxyphneol), syringol (2,6-dimetoxyphenols) and their derivatives.

In this work, we carried out a complete conformational landscape of the ortho-, meta- and para- isomers of methoxyphenol (known also as 2-methoxyphenol or guaiacol, 3-methoxyphenol and 4-methoxyphenol or Mequinol, respectively) using quantum chemical calculations with *GAUSSIAN* program as well as room temperature millimeter wave spectrometer of the LPCA Laboratory in Dunkirk [b].

The 70 - 330 GHz rotational spectrum is measured in this study for each isomer and analysis of their rotational signatures for all the stable conformers in ground and low-energy vibrationally excited states will be detailed and discussed.

The main goal of the study is to understand the isomeric influence on the conformational landscape of the methoxy-aromatic compounds.

[a] A. Lauraguais, C. Coeur-Tourneur, A. Cassez, K. Deboudt, M. Fourmentin, and M. Choel., Atmospheric reactivity of hydroxyl radicals with guaiacol (2-methoxyphenol), a biomass burning emitted compound: Secondary organic aerosol formation and gas-phase oxidation products. Atmos. Environ., 86:155-163, 2014

[b] G. Mouret, M. Guinet, A. Cuisset, L. Croize, S. Eliet, R. Bocquet, and F. Hindle. Versatile sub-thz spectrometer for trace gas analysis. IEEE Sens. J., 13(1):133-138, 2013

WJ03 2:19 – 2:34
CONFORMATIONAL STUDY OF SECONDARY ORGANIC AEROSOL PRECURSORS CONTAINING INTERNAL ROTATIONS: CASES OF METHYL ANISOLE ISOMERS

ATEF JABRI, DANIELE FONTANARI, ANTHONY ROUCOU, GUILLAUME DHONT, GAËL MOURET, ARNAUD CUISSET, *Laboratoire de Physico-Chimie de l'Atmosphère, Université du Littoral Côte d'Opale, Dunkerque, France*; WOLFGANG STAHL, *Institute for Physical Chemistry, RWTH Aachen University, Aachen, Germany*; HA VINH LAM NGUYEN, ISABELLE KLEINER, *CNRS et Universités Paris Est et Paris Diderot, Laboratoire Interuniversitaire des Systèmes Atmosphériques (LISA), Créteil, France.*

Methyl anisole (MA), also known as methoxytoluene, exists as three isomers with the ring methyl group in ortho, meta, and para position relative to the methoxy group. Despite the similarity of their chemical properties, the effect of methyl internal rotation which often depends on the steric and electronic surroundings can be completely different. In the other hand, This species is very important as a secondary organic aerosol precursor.

In our present study, we focus on the millimeter wave studies on ortho, meta and para-methyl anisole in the 70-330 GHz frequency domain in order to complete the previous microwave studies between 2-40 GHz on the ground state [a, b, c] and to determine internal rotation higher order parameters by analysis of the rotational structures in the excited states. Our millimeter wave spectra are measured at room temperature, which allow observing rotational structures in the ground and low-energy vibrationally excited states. Thus, we could determine precisely the V_6 potential barriers and evaluate their contributions.

[a] L. Ferres, H Mouhib, W. Stahl, and H. V. L. Nguyen, Methyl Internal Rotation in the Microwave Spectrum of o-Methyl Anisole. ChemPhysChem. 18, 1855–1859 (2017)

[b] L. Ferres, W. Stahl, and H. V. L. Nguyen, Conformational Effects on the Torsional Barriers in m-Methylanisole Studied by Microwave Spectroscopy, accepted in J. Chem. Phys.

[c] L. Ferres, W. Stahl, I. Kleiner, H. V. L. Nguyen, The effect of internal rotation in p-methyl anisole studied by microwave spectroscopy. J. Mol. Spectrosc. 343, 44–49 (2018)

WJ04 2:36 – 2:51
HIGH RESOLUTION SPECTROSCOPY OF 3-METHYLBUTYRONITRILE BETWEEN 2 AND 400 GHZ

NADINE WEHRES, MARIUS HERMANNS, OLIVIA H. WILKINS[a], KIRILL BORISOV, FRANK LEWEN, *I. Physikalisches Institut, Universität zu Köln, Köln, Germany*; JENS-UWE GRABOW, *Institut für Physikalische Chemie und Elektrochemie, Gottfried-Wilhelm-Leibniz-Universität, Hannover, Germany*; STEPHAN SCHLEMMER, HOLGER S. P. MÜLLER, *I. Physikalisches Institut, Universität zu Köln, Köln, Germany.*

We present high-resolution rotational spectroscopy of the two conformers of 3-methylbutyronitrile. Spectra were taken between 2-24 GHz by means of Fourier transform microwave spectroscopy. Spectra between 36 and 402 GHz were recorded by means of frequency modulated (FM) absorption spectroscopy. The analysis yields precise rotational constants and higher order distortion constants, as well as a set of ^{14}N nuclear electric quadrupole coupling parameters. In addition, quantum chemical calculations were performed assisting the assignments. Frequency calculations yield insight into the vibrational energy structure from which partition functions and vibrational correction factors are determined. These are used to determine experimentally and computationally the energy difference between the conformers, which is revealed to be negligeable. Overall, this study provides precise spectroscopic constants for the search of 3-methylbutyronitrile in the interstellar medium. In particular, this molecule is an interesting testcase to check our knowledge of (branched) molecule formation in space.

[a] Current Address: Division of Chemistry and Chemical Engineering, California Institute of Technology, Pasadena, California 91125, USA

WJ05 2:53 – 3:08
HIGH-RESOLUTION SPECTROSCOPY OF TWO CONFORMERS OF 2-CYANOBUTANE BETWEEN 10 AND 400 GHZ

MARIUS HERMANNS, NADINE WEHRES, FRANK LEWEN, HOLGER S. P. MÜLLER, STEPHAN SCHLEMMER, *I. Physikalisches Institut, Universität zu Köln, Köln, Germany.*

We present high-resolution rotational spectroscopy of two out of three conformers of 2-cyanobutane. Spectra were taken between 10-26 GHz by means of chirped-pulse spectroscopy. Spectra between 36 and 402 GHz were recorded by means of frequency modulated (FM) absorption spectroscopy. The analysis yields precise rotational constants and higher order distortion constants, as well as a set of ^{14}N nuclear electric quadrupole coupling parameters. In addition, quantum chemical calculations were performed assisting the assignments. Calculations of vibrational frequencies yield insight into the vibrational energy structure from which partition functions and vibrational correction factors are determined. These are used to determine experimentally and computationally the energy difference between the conformers. Overall, this study provides precise spectroscopic constants for the search of 2-cyanobutane in the interstellar medium. In particular, this molecule appears as an interesting case to test our knowledge of (branched) molecule formation in space.

WJ06 3:10 – 3:25
CONFORMATIONAL ISOMERISM OF 1-IODOPENTANE

SUSANNA L. STEPHENS, *Department of Chemistry, Wesleyan University, Middletown, CT, USA*; JOSHUA A. SIGNORE, *Chemistry , Wesleyan University , Middletown, USA*; DANIEL A. OBENCHAIN, *Institut für Physikalische Chemie und Elektrochemie, Gottfried-Wilhelm-Leibniz-Universität, Hannover, Germany*; ROBERT KARL BOHN, *Department of Chemistry, University of Connecticut, Storrs, CT, USA*; STEWART E. NOVICK, *Department of Chemistry, Wesleyan University, Middletown, CT, USA*; S. A. COOKE, *Natural and Social Science, Purchase College SUNY, Purchase, NY, USA.*

The rotational spectrum of 1-iodopentane was measured over the 7-13 GHz frequency range with a chirped pulse Fourier transform microwave spectrometer revealing rotational transitions from a number of conformers. This continues the group's work on how a large substituent, in this case an iodine atom, at the terminal position will affect the dihedral angles of the alkyl carbon backbone and what influence it will exert with continuing chain length. In keeping with last year's study of 1-iodobutane,[a] we find that the corresponding GAA conformer is the most abundant, and that while the nuclear quadrupole coupling tensor is poorly predicted by direct *ab initio* methods, scaling methods[b] allow very reasonable predictions to be obtained.

---

[a] Arsenault E.A.; Obenchain, D.A.; Blake, T.A.; Cooke, S.A.; Novick, S.E; *J. Mol. Spectrosc.*, **2017** *335* 17-22.
[b] Anticipated future communication with W. C. Bailey.

Intermission

WJ07　　　　　　　　　　　　　　　　　　　　　　　　　　　　　　　　　　　　4:01 – 4:16
ISOMER-SPECIFIC SPECTROSCOPY OF ETHYL NAPHTHALENE DERIVATIVES: SPECTROSCOPIC FOUNDATION FOR UNDERSTANDING ETHYL-BRIDGED DINAPHTHYLS

VICTORIA M. BOULOS, DANIEL M. HEWETT, TIMOTHY S. ZWIER, *Department of Chemistry, Purdue University, West Lafayette, IN, USA.*

The incomplete combustion of gasoline under fuel-rich conditions leads to soot formation, an unwanted pollutant and energetically wasteful product. While it is known that soot is composed of π-stacked graphitic-like structures, the incipient stages of soot formation are still something of a mystery, with conflicting evidence as to whether π-stacked polycyclic aromatic hydrocarbon (PAH) aggregates can form and survive at combustion temperatures still unresolved. The mechanism by which the initial aggregation of PAH precursors occurs is still under active investigation. This talk will describe studies of two isomeric naphthalene derivatives containing short alkyl chains: 1-ethyl naphthalene and 2-ethyl naphthalene. The structural isomers differ in the position of substitution of the ethyl side-chain on the naphthalene framework. These molecules are not only components of diesel fuel and of early stages of PAH formation in flames, but are also close analogs of a series of ethyl-bridge dinaphthyl compounds studied by our group. We will present the jet-cooled LIF excitation, dispersed fluorescence, and fluorescence-dip infrared spectra of these molecules, cooled in a supersonic free jet. The vibronic spectroscopy of these molecules is similar to that in naphthalene, modified only subtly by the ethyl side chain. Vibronic coupling between S_1 and S_2 states is pervasive, leading to textbook evidence of Herzberg-Teller coupling. We have also observed the onset of the S_0-S_2 transition near 288 nm. These results will be compared with calculations of the excited states, setting a foundation for understanding the fascinating vibronic spectroscopy of the ethyl-bridged dinaphthyl compounds.

WJ08　　　　　　　　　　　　　　　　　　　　　　　　　　　　　　　　　　　　4:18 – 4:33
THE ORIGINATION OF SOOT FORMATION: A STUDY ON ETHYL-LINKED NAPHTHALENE DIMERS

DANIEL M. HEWETT, VICTORIA M. BOULOS, TIMOTHY S. ZWIER, *Department of Chemistry, Purdue University, West Lafayette, IN, USA.*

Combustion is the main source of power worldwide. This makes understanding combustion pathways and byproducts vital to optimizing power output and minimizing the unwanted byproducts. One of the main byproducts of the combustion process is soot, which can be accurately described as a large aggregate of polyaromatic hydrocarbons (PAHs). While the general process by which soot particles grow in size have been laid out, the incipient steps by which individual PAH molecules begin to aggregate is still a mystery. Current models of soot formation require individual PAH molecules to dimerize and π stack in PAHs significantly smaller than are thermodynamically capable of doing so. One largely untested possibility is that PAHs linked together by short chemical linkages could be responsible for the initial steps of aggregation. In particular, we propose that resonance-stabilized benzylic-like radicals could recombine to form ethyl-linked PAH dimers that are responsible for the initial stages of aggregation and π stacking. These covalently-linked dimers could withstand the high energies present in the combustion flame and provide a seed for soot aggregation. This brings us to the following question: At what sized PAH will these ethyl-linked aromatics be capable of π stacking, and if they can, will conformations leading to π stacking compete with other, more extended geometries? We begin to answer this question by studying a series of ethyl-linked naphthalene dimers in which the ethyl linkage bridges the two unique sites of substitution (shown below). These molecules are brought into the gas phase by heating, and cooled in a supersonic expansion. LIF excitation, IR-UV holeburning, fluorescence-dip infrared spectroscopy, and dispersed fluorescence are used to record conformer-specific UV and IR spectra. These dimers also have fascinating electronic spectroscopy associated with the presence of two UV chromophores that are in identical or nearly identical environments, leading to extensive vibronic coupling. The experimental results will be compared with a multi-mode theoretical model of near-resonant vibronic coupling.

WJ09 4:35–4:50
STUDYING THE FOLDING PROPENSITY OF ASPARAGINE-CONTAINING PEPTIDES IN A MOLECULAR BEAM

KARL N. BLODGETT, JOSHUA L. FISCHER, *Department of Chemistry, Purdue University, West Lafayette, IN, USA*; DEWEI SUN, *Department of Chemistry, Purdue University, Valparaiso, IN, USA*; TIMOTHY S. ZWIER, *Department of Chemistry, Purdue University, West Lafayette, IN, USA*.

Asparagine (Asn) and Glutamine (Gln) are the only two naturally-occuring amino acids that incorporate an amide group in the side chain, differing only by a single methylene group that lengthens the chain in Gln relative to Asn. These two amino acids appear with unusual abundance in the prion forming domain of proteins that are implicated in neurodegenerative disease pathogenesis. In a bottom-up approach towards understanding nature's preference for Asn and Gln in prion domains, we build off of our previous study of Gln-containing peptides by elucidating the inherent folding propensities of three Asn-containing peptides: Ac-Asn-NHBn, Ac-Ala-Asn-NHBn, and Ac-Asn-Asn-NHBn. These molecules are brought into the gas phase by laser desorption and cooled in a supersonic expansion before interrogation via resonant two-photon ionization (R2PI), IR-UV holeburning, and resonant ion-dip infrared (RIDIR) spectroscopies. This talk will describe the conformation-specific IR and UV spectroscopy of these three peptides, concentrating on the fundamental question of whether side-chain/side-chain, backbone/backbone, or side-chain/backbone interactions are preferred in each case. Assignment of the observed conformations are deduced from a comparison of the experimental IR spectra with the predictions of calculations. Two unique conformational isomers are assigned to the capped Asn amino acid while only one conformational isomer is present in both dipeptides. These structures will be compared with those found in the analogous glutamine-containing peptides. Best fit calculated structures reveal intriguing incipient secondary structure formation at the first possible instance.

WJ10 4:52–5:07
INSIGHTS INTO PROTON TRANSFER MECHANISMS IN HALOGEN-SUBSTITUTED MALDI MATRICES

CHELSEA N BRIDGMOHAN, KRISTOPHER M KIRMESS, LICHANG WANG, *Department of Chemistry and Biochemistry, Southern Illinois University Carbondale, Carbondale, IL, USA*.

In MALDI mass spectrometry, the role of a matrix is to transfer charge to the analyte so that it can be detected and quantified. The fundamental process of how this charge transfer actually occurs, however, is still largely unknown. Experimental evidence suggests that InterSystem Crossing (ISC) to the triplet excited state (T1) of the matrix is a crucial step in effective charge transfer to the analyte. To encourage ISC via spin-orbit coupling, we have utilized the heavy atom substitution effect, which suggests that the addition of a heavy atom to an otherwise "dead" matrix should increase the rate of ISC to the T1 state. With the addition of a halogen atom, experiments showed a visible decay in singlet lifetime and an increase in triplet lifetime, as well as improved matrix performance. Our system of interest in this study is 2,4–dihydroxybenzoic acid, which will be compared with two of its heavily substituted, halogenated (F,Cl, Br, and I) partners. Using *Gaussian09* software and Density Functional Theory (DFT), all structural isomers were identified and ranked by stability using the functional B3LYP and basis set 6-31g+(d,p). Zero-Point Energies (ZPE) and IR spectra were compared for S0, S1, and T1 states of both the substituted and unsubstituted isomers. Additionally, Time Dependent (TD)-DFT calculations generated UV-Vis and fluorescence spectra. Results showed that Proton Affinity (PA) and Gas Phase Acidity (GPA) values improved with heavy atom substitution in both the 5th and 6th ring positions of 2,4-dihydroxybenzoic acid. Comparison of the S0, S1, and T1 vibrational spectra of the heavily substituted isomers revealed the weakening of bonds between the molecule and its protons, which may be partially responsible for more effective charge transfer to the analyte.

WJ11

CONFORMERS OF L-GLUTAMIC ACID: MATRIX ISOLATION FTIR AND *AB-INITIO* STUDIES.[a]

PANKAJ DUBEY[b], K S VISWANATHAN, *Chemical Science, Indian Institute of Science Education and Research, MOHALI, PUNJAB, India.*

L-glutamic acid is most abundant free amino acid in brain and it is the major excitatory neurotransmitter of the vertebrate central nervous system and known to play an important role in neural differentiation process of a developing brain. Study of the rich conformational landscape of L-glutamic acid can serve as basis to understand interactions of this amino acid with other biomolecules and receptors present in central nervous system.[c] L-glutamic acid was trapped in an inert gas matrix by employing a heated nozzle to provide an effusive molecular beam and the various conformers of the amino acid trapped in the matrix were then characterized by FTIR spectroscopy. *Ab-initio* calculations were also performed, using MP2/6-311++G(d,p) and M06-2X/6-311++G(d,p) level of theories, to corroborate with experimental observations. A comprehensive scan of the potential energy surface was performed to arrive at the various conformers of L-glutamic acid, which were further classified based on their backbone structure. The tendency of lower energy conformers to adopt certain backbone structures has been pictorially represented using a 'conformational dartboard'. Factors such as intramolecular H-bonding, delocalized orbital interactions and entropy were found to determine conformational preferences in L-glutamic acid, which will be discussed.

---
[a] Authors acknowledge IISER Mohali, India, for research facilities.
[b] PD thanks MHRD, India, for research fellowship.
[c] Tapiero, H.; Mathe, G.; Couvreur, P.; Tew, K. D. Biomed. Pharmacother. 2002, 56, 446-457.

WK. Structure determination
Wednesday, June 20, 2018 – 1:45 PM
Room: B102 Chemical and Life Sciences

Chair: Helen O. Leung, Amherst College, Amherst, MA, USA

WK01   1:45 – 2:00
DESIGN AND APPLICATIONS OF A MASS-CORRELATED BROADBAND MICROWAVE SPECTROMETER

<u>SEAN FRITZ</u>, ALICIA O. HERNANDEZ-CASTILLO, BRIAN M HAYS, TIMOTHY S. ZWIER, *Department of Chemistry, Purdue University, West Lafayette, IN, USA.*

Chirped pulse Fourier transform microwave (CP-FTMW) spectroscopy serves as a faster-than-ever technique for structure determination that is being utilized in an ever-widening range of applications. Vacuum ultraviolet (VUV) time-of-flight mass spectrometry (TOFMS) provides a gentle and general ionization scheme for measurement of the masses, and hence molecular formulae, of the components of a gas mixture, as long as these components have ionization potentials below the threshold for single-photon photoionization (118 nm = 10.5 eV in this case). By combining these two methods, a powerful tool is created for determining the structures of the components in a mixture. By following the correlation between the microwave transition intensities and the mass-resolved signals in the TOF mass spectrum, the molecular formulae of the carriers of the microwave transitions can be determined. In this sense, the combination of CP-FTMW with VUV TOFMS produces mass-correlated broadband microwave spectra. Herein, we discuss the design, construction and initial applications of a new chamber that was designed to acquire simultaneous mass and rotational spectra. Proof of concept will be demonstrated by examining a gaseous mixture of 2-methoxyfuran, furan and furfural. The new spectrometer is also being used to characterize resonance stabilized radicals formed by a pyrolysis source. Application in this context will also be discussed.

WK02   2:02 – 2:17
STRUCTURAL CHARACTERIZATION OF PHENOXY RADICAL USING A MASS-CORRELATED BROADBAND MICROWAVE SPECTROMETER

<u>ALICIA O. HERNANDEZ-CASTILLO</u>, CHAMARA ABEYSEKERA, *Department of Chemistry, Purdue University, West Lafayette, IN, USA*; JOHN F. STANTON, *Physical Chemistry, University of Florida, Gainesville, FL, USA*; TIMOTHY S. ZWIER, *Department of Chemistry, Purdue University, West Lafayette, IN, USA.*

We have combined the high-resolution provided by chirped pulse Fourier transform microwave (CP-FTMW) spectroscopy with a vacuum ultraviolet (VUV) time-of-flight mass spectrometry (TOF-MS) that allows us to find optimal conditions to detect reactive intermediates. This unique setup was used to structurally characterize the phenoxy radical. This is an important intermediate in the oxidation of many aromatic compounds. The decomposition of this radical has as primary product cyclopentadienyl radical, which is an important species in the formation of polynuclear aromatic hydrocarbons. Phenoxy radical was generated through the pyrolysis of anisole and allyl phenyl ether. The work was carried out over a 300-1500 K temperature range, using a high-temperature flash pyrolysis micro-reactor coupled with a supersonic expansion. The 2-18 GHz pure rotational spectrum was obtained concurrently with the mass spectrum and determined the resonant-stabilized radical molecular parameters. For further structural analysis spectra of ^{13}C isotopomers were obtained using segmented chirps. The new structural insights and the pyrolysis setup coupled with the CP-FTMW/TOF spectrometer will be discussed.

WK03   2:19 – 2:34
AN ARGON-OXYGEN COVALENT BOND IN THE ArOH$^+$ MOLECULAR ION

<u>J. PHILIPP WAGNER</u>, *Department of Chemistry, University of Georgia, Athens, GA, USA*; DAVID C McDONALD, *Chemistry, University of Georgia, Athens, GA, USA*; MICHAEL A DUNCAN, *Department of Chemistry, University of Georgia, Athens, GA, USA.*

Although the OH$^+$ cation is decidedly a triplet ($^3\Sigma^-$) being over 50 kcal mol^{-1} more stable than the corresponding singlet ($^1\Delta$), binding to an argon atom can reverse this situation. The noble gas forms a strong donor-acceptor bond to the excited state singlet cation with a bond strength of 66.4 kcal mol^{-1} at the CCSDT(Q)/CBS level of theory. This makes the singlet 3.9 kcal mol^{-1} more stable than the most favorable triplet Ar–HO$^+$ complex. In a cold molecular beam experiment we have prepared both, singlet and triplet, isomers of this molecular ion depending on the employed ion source. Photodissociation spectroscopy in combination with messenger atom tagging reveals that the two observed spin isomers exhibit completely different spectral signatures in the infrared and the O–H stretching fundamentals differ by about 900 cm^{-1}. These findings might encourage the search for a new potential interstellar noble gas molecule.

WK04   2:36 – 2:51

PROTON IN A DOUBLE-WELL POTENTIAL AS SEEN FROM MICROWAVE AND CORE LEVEL PHOTOEMISSION SPECTROSCOPY

LUCA EVANGELISTI, WEIXING LI, ASSIMO MARIS, SONIA MELANDRI, *Dipartimento di Chimica G. Ciamician, Università di Bologna, Bologna, Italy.*

The extended nature of the proton wave function, the shape of the ground and final state potentials in which the proton is located has been investigated in gaseous acetylacetone and three of its derivatives, benzoylacetone, dibenzoylmethane[a] and 3,5-heptanedione by quantum chemical calculations, microwave spectroscopy and core level photoemission study. These molecules show intramolecular hydrogen bonds, in which a proton is located in a double well potential, whose barrier height is different for the four compounds, allowing us to examine the effect of the shape of double well on photoemission and rotational spectra. For all of them, two distinct O 1s core hole peaks are observed, previously assigned to two chemical states of oxygen in the ground state. We provide an alternative assignment by taking full account of the finite temperature of the samples based on quantum chemical calculations and symmetry consideration.

---

[a]Vitaliy Feyer, Kevin C. Prince, Marcello Coreno, Sonia Melandri, Assimo Maris, Luca Evangelisti, Walther Caminati, Barbara M. Giuliano, Henrik G. Kjaergaard, and Vincenzo Carravetta, J. Phys. Chem. Lett. 9 (2018) 521–526. DOI: 10.1021/acs.jpclett.7b03175

WK05   2:53 – 3:08

HETERO-OLIGOMERS OF DIFLUOROMETHANE AND 1,1-DIFLUOROETHANE: CONFORMATIONAL EQUILIBRIA, MOLECULAR STRUCTURE AND WEAK HYDROGEN BONDS

TAO LU, JUNHA CHEN, QIAN GOU, GANG FENG, *School of Chemistry and Chemical Engineering, Chongqing University, Chongqing, China.*

The hetero-oligomers formed by difluoromethane (CH_2F_2) and 1,1-difluoroethane (CH_3CHF_2) were investigated by pulsed jet Fourier transform microwave spectroscopy and theoretical calculations. For the hetero-dimer of CH_2F_2-CH_3CHF_2, three most stable conformers predicted at MP2/6-311++G(d,p) level were observed. Experimental results, ab initio calculations and quantum theory of atoms in molecules (QTAIM) analyses indicate that all the observed conformers are stabilized through a net of three weak C-H⋯F-C interactions. Rotational measurements have also been extended to three ^{13}C isotopologues in natural abundance for each observed conformer, which allowed precisely structural determinations. The relative populations of these three conformers in the jet were estimated by relative intensity measurements. For the hetero-trimer of $(CH_2F_2)_2$-CH_3CHF_2, one conformer has been observed. The conformer is stabilized through eight C-H⋯F-C interactions.

WK06   3:10 – 3:25

STRUCTURE DETERMINATION, CONFORMATIONAL EQUILIBRIA AND WEAK HYDROGEN BONDS IN HETERODIMERS OF FREONS: THE ROTATIONAL STUDY OF CH_2F_2-CF_3CH_2F

TAO LU, JUNHA CHEN, QIAN GOU, GANG FENG, *School of Chemistry and Chemical Engineering, Chongqing University, Chongqing, China.*

The rotational spectra of the 1:1 heterodimer of difluoromethane and 1,1,1,2-tetrafluoroethane has been investigated by pulsed jet Fourier transform microwave spectroscopy and quantum chemical calculations. Three most stable conformers predicted at MP2/6-311++G(d,p) level were observed. Experimental results, ab initio calculations and quantum theory of atoms in molecules (QTAIM) analyses indicate that all the observed conformers are stabilized through a net of three weak C-H...F-C interactions.The measurements have also been extended to three or two ^{13}C isotopologues in natural abundance for the conformers II and III, respectively, which allowed precisely structural determinations of these two conformers. The relative populations of these three conformers in the jet estimated to be 50/7/1.

WK07 3:27–3:42

THE CONFORMATIONAL PREFERENCE OF A STRONG O-H•••O HYDROGEN BONDED COMPLEX IS DETERMINED BY WEAK C-H•••π INTERACTION: A LIF STUDY OF BINARY COMPLEXES OF P-FLUOROPHENOL WITH 2,5-DIHYDROFURAN AND TETRAHYDROFURAN

<u>DEB PRATIM MUKHOPADHYAY</u>, SOUVICK BISWAS, TAPAS CHAKRABORTY, *Physical Chemistry, Indian Association for the Cultivation of Science, Kolkata, India.*

Hydrogen bonded binary complexes of p-fluorophenol (pFP) with two cyclic ethers, tetrahydrofuran (THF) and 2,5-dihydrofuran(2,5-DHF), have been studied using laser induced fluorescence spectroscopy in jet expansion condition. For both the complexes, a nearly linear type geometry should be preferred as the complexes are bound via conventional O-H•••O type hydrogen bonding. However, for pFP-2,5-DHF complex, two nearly isoenergetic conformers are detected in experimental condition by identifying two discreet origin bands in vibrationally resolved fluorescence excitation spectrum. By measuring dispersed fluorescence spectra and using electronic structure calculations these two complexes are unequivocally assigned. Though THF complex has the conventional linear geometry, the lowest energy conformer of 2,5-DHF complex prefers a folded form where substantial dispersion interaction present between ortho-H and the ethylenic bond. The other higher energy conformer has a similar structure of pFP-THF complex. Apart from the structural aspects, these conformers also show different behavior of vibrational energy relaxation (VER) in S_1 state. When excited at first intramolecular band ($6a^1$) the folded form displays the selective and restricted VER whereas the conventional forms of both THF and 2,5-DHF complex show structureless, broad emission corresponds to unrestricted VER in the excited state.

Intermission

WK08 4:18–4:33

TUNING OF NON-COVALENT INTERACTIONS IN MOLECULAR COMPLEXES OF FLUORINATED AROMATIC COMPOUNDS

<u>SONIA MELANDRI</u>, LUCA EVANGELISTI, ASSIMO MARIS, IMANOL USABIAGA GUTIERREZ, WEIXING LI, *Dipartimento di Chimica G. Ciamician, Università di Bologna, Bologna, Italy*; CAMILLA CALABRESE, *Departamento de Química Física, Universidad del País Vasco (UPV-EHU), Bilbao, Spain*; LAURA B. FAVERO, *Istituto per lo Studio dei Materiali Nanostrutturati, Consiglio Nazionale delle Ricerche (ISMN-CNR), Bologna, Italy.*

The rotational spectra of pentafluoropyridine-water[a] and hexafluorobenzene-water[b] have shown unambiguously that substitution by fluorine atoms on the ring strongly influences the binding abilities of the aromatic ligand. Differently from their non-substituted counterparts, which form hydrogen bonds with water, both molecules interact with water forming a lone-pair-π-hole interaction between the water oxygen and the ring. We report on a series of rotational spectroscopy studies performed with a Molecular Beam Fourier Transform Microwave spectrometer in which we have tested the binding abilities of a series of fluorine substituted pyridines, namely 2,4-fluoropyridine, 3,5-fluoropyridine and 2,4,6-fluoropyridine with water. In the complexes formed by the di-substituted pyridines, the water moiety lies in the aromatic plane and forms a hydrogen bond with the heterocyclic nitrogen atom and this is by far the most stable conformation. In the tri-substituted pyridine-water complexes two isomers are possible, but the lp-π-hole form is the one observed in the rotational spectrum. More studies involving the pentafluoropyridine and hexafluorobenzene molecules and several ligands such as NH_3 and CO have also been performed and the interaction with the π-hole has been characterized.

---

[a]Camilla Calabrese, Qian Gou, Assimo Maris, Walther Caminati, Sonia Melandri, Probing the Lone Pair-π-Hole Interaction in Perfluorinated Heteroaromatic Rings: The Rotational Spectrum of Pentafluoropyridine-Water, J. Phys. Chem. Lett. 2016, 7, 1513-1517 http://dx.doi.org/10.1021/acs.jpclett.6b00473

[b]Luca Evangelisti, Kai Brendel, Heinrich Maeder, Walther Caminati, and Sonia Melandri, Rotational Spectroscopy Probes Water Flipping by Full Fluorination of Benzene, Angew. Chem. Int. Ed. 2017, 56, 13699–13703 https://doi.org/10.1002/anie.201707155

WK09   4:35–4:50

THE CYCLOHEXANOL DIMER AND THE MONOHYDRATE: INTERNAL ROTATION AND CONVERSION FROM TRANSIENT TO PERMANENT CHIRALITY

<u>MARCOS JUANES</u>, *Departamento de Química Física y Química Inorgánica, Universidad de Valladolid, Valladolid, Spain*; IKER LEÓN, *Grupo de Espectroscopia Molecular, Lab. de Espectroscopia y Bioespectroscopia, Unidad Asociada CSIC, Universidad de Valladolid, Valladolid, Spain*; RUTH PINACHO, JOSÉ EMILIANO RUBIO, *Departamento de Electrónica, ETSIT, University of Valladolid, Valladolid, SPAIN*; WEIXING LI, LUCA EVANGELISTI, WALTHER CAMINATI, *Dep. Chemistry 'Giacomo Ciamician', University of Bologna, Bologna, Italy*; ALBERTO LESARRI, *Departamento de Química Física y Química Inorgánica, Universidad de Valladolid, Valladolid, Spain*.

The conformational preferences, molecular structure and intramolecular dynamics of cyclohexanol (CHOL), the cyclohexanol dimer (CHOL$_2$) and cyclohexanol-water (CHOL$\cdots$H$_2$O) have been analyzed using rotational spectroscopy in a supersonic jet expansion. The monomer conformation is controlled by ring inversion (equatorial/axial) and alcohol internal rotation (gauche+/gauche- or trans). Both equatorial trans (rigid) and equatorial gauche (tunneling) isomers were observed, but axial cyclohexanol went undetected. The monohydrate CHOL$\cdots$H$_2$O exhibits a single hydrogen bonded equatorial gauche conformation, with tunneling splittings revealing a concerted motion of the hydroxyl and water molecules. The formation of the dimer CHOL$_2$ stabilizes the transient chirality of the gauche+/gauche- monomer conformations, as different chiral combinations donor-acceptor result in permanent chirality. This molecular dimer represent a larger level of complexity compared with previous studies [a,b], as 36 different donor-acceptor combinations are now possible and the relative orientation of the two rings generates multiple conformations. Finally, six different isomers were observed for the dimer. Accurate rotational parameters for the observed conformations and supporting ab initio and DFT calculations will be reported.

[a] M. S. Snow, B. J. Howard, L. Evangelisti, W. Caminati, *J. Phys. Chem. A* **2011**, 115, 47.
[b] A. K. King, B. J. Howard, *Chem. Phys. Lett.* **2001**, 348, 343.

WK10   4:52–5:07

ROTATIONAL SPECTRA AND GEOMETRIES OF FOUR CONFORMERS OF ISOLATED UROCANIC ACID; AND OF A COMPLEX OF UROCANIC ACID WITH WATER

<u>NICK WALKER</u>, *School of Natural and Environmental Sciences, Newcastle University, Newcastle-upon-Tyne, UK*; GRAHAM A. COOPER, CHRIS MEDCRAFT, EVA GOUGOULA, *School of Chemistry, Newcastle University, Newcastle-upon-Tyne, United Kingdom*.

The rotational spectra of four conformers of urocanic acid ((2E)-3-(1H-imidazol-4-yl)prop-2-enoic acid) have been measured between 6 and 18 GHz. The molecule was prepared for study through laser vaporisation of a solid target in the presence of argon gas undergoing supersonic expansion into a vacuum chamber. The solid target was composed of urocanic acid, copper powder and a small amount of polyvinyl acetate glue used as binder. Rotational constants, B_0, centrifugal distortion constants, D_J, D_{JK} and the nuclear quadrupole coupling constants of the nitrogen atoms, $\chi_{aa}(N)$ and $\chi_{bb-cc}(N)$, have been determined through analysis of the assigned broadband rotational spectra. The geometry of each conformer was identified by comparison of the experimental results with geometries calculated *ab initio*. Isotopic substitutions at carbon and hydrogen atoms allow further insight into the molecular geometries of the two lowest energy conformers. The geometry of a complex formed between urocanic acid and water will also be described.

WK11 5:09 – 5:24
ROTATIONAL SPECTROSCOPIC STUDIES ON THE CH_3CN-CO_2 COMPLEX

SHARON PRIYA GNANASEKAR, ELANGANNAN ARUNAN, *Department of Inorganic and Physical Chemistry, Indian Institute of Science, Bangalore, India.*

The CH_3CN-CO_2 complex was investigated using a Pulsed-Nozzle Fourier Transform Microwave Spectrometer. This complex offers the possibility of observing a 'carbon-bonded' structure. The ab initio calculations give three minima structures; a π-stacked structure, a T-shaped structure with the N end of CH_3CN interacting with the C atom of CO_2, and a linear structure with the O of CO_2 interacting with the tetrahedral C of CH_3CN. Thus, the T-shaped and the linear structures are bound with a 'carbon-bond'. The π-stacked and T-shaped structures have similar binding energies (-7.7 kJmol^{-1} and -7.6 kJmol^{-1}, respectively). Many lines were observed which depend on both CH_3CN and CO_2 concentrations. Six of these lines follow a nearly prolate, 'a'-type spectra. The K=0 (J = 3-2 to 8-7) rotational transitions have been observed. The B+C value obtained from fitting these transitions is consistent with the value predicted for the T-shaped geometry with an N-C interaction. All lines show hyperfine splitting due to quadrupole coupling of the nitrogen atom. Measurements with isotopic substitutions have been carried out to ascertain the assignment of the rotational transitions. The unassigned lines may belong to the other possible geometries of the CH_3CN-CO_2 complex.

WK12 5:26 – 5:41
CONFORMER SPECIFIC METHYL INTERNAL ROTATION AND OBSERVATION OF PHOTODISSOCIATION DYNAMICS: IS METHYL INTERNAL ROTATION COUPLED WITH TORSIONAL VIBRATION?

HEESUNG LEE, SO-YEON KIM, JEAN SUN LIM, JUNGGIL KIM, SANG KYU KIM, *Chemistry, Korea Advanced Institute of Science and Technology, Daejeon, Republic of Korea.*

The spectroscopic peak assignment gives barrier height of internal rotation as well as minimum angle displacement between S_1 and S_0 electronic states. The most well-known type is methyl internal rotation because most organic molecules have it. Conventional way to assign the spectra uses 1D hindered rigid rotor model. The model system is thioanisole because it engendered mode specific nonadiabatic photofragment in previous experiment[a]. Resonance enhanced multiphoton ionization (REMPI) followed by velocity-map imaging (VMI) made us be possible to assign the methyl internal rotation peaks and to see the dynamics of the photodissociation. Conformers were selectively sampled using different carrier gases in supersonic jet-expansion and their electronic transitions were confirmed by UV-UV hole-burning spectroscopy equipped with Stark-Deflector. In this presentation, *trans meta*-methylthioanisole (*m*MTA) manifests active progressions of methyl internal rotation ($0a_1$, 1e, 2e, $3a_1$, 4e, 5e, $6a_1$), whereas a few of them ($0a_1$, 1e, $3a_1$) for *cis m*MTA are tangible. For *cis m*MTA, the rotor progressions are combined with the torsional vibration. In case of *trans ortho*-methylthioanisole (*o*MTA), not only active combinations of methyl internal rotation and torsional vibration but also couplings of them seem to be exist in S_1-S_0 transitions[b]. VMI studies imply two distinguished photodissociation dynamics of *o*MTA and *m*MTA, respectively. The detailed comparison of them is to be presented.

---
[a] Jeong Sik Lim, Sang Kyu Kim, Nat. Chem. 2, 627-632(2010)
[b] *trans m*MTA also shows a coupling of **e** transition of methyl rotor and the vibronic transition of methylthio group at around 200 cm^{-1} above S_1 origin.

WL. Astronomy
Wednesday, June 20, 2018 – 1:45 PM
Room: 2079 Natural History

Chair: Cristina Puzzarini, University of Bologna, Bologna, Italy

WL01 1:45 – 2:00

EXPLOITING TUNABLE VACUUM ULTRAVIOLET PHOTOIONIZATION COMBINED WITH REFLECTRON TIME-OF-FLIGHT MASS SPECTROMETRY TO UNRAVEL THE NITROGEN CHEMISTRY OF COMPLEX ORGANICS IN THE INTERSTELLAR MEDIUM

ROBERT FRIGGE, ANDREW MARTIN TURNER, MATTHEW JAMES ABPLANALP, RALF INGO KAISER, *Department of Chemistry, University of Hawaii at Manoa, Honolulu, HI, USA.*

For more than half a century, gas-phase reaction networks of rapid ion-molecule and neutral-neutral reactions have played a fundamental role in aiding our understanding of the evolution of the interstellar medium (ISM). However, with about 200 molecules detected in interstellar and circumstellar environments, these models fail to explain the synthesis of ubiquitous complex organic molecules (COMs) – organics containing several atoms of carbon, hydrogen, nitrogen, and oxygen – predicting abundances which are lower by several orders of magnitude compared to observations toward hot molecular cores like Sagittarius $B_2(N)$. Here, we report that key COMs - methanimine (CH_2NH) and ethylenediamine ($NH_2CH_2CH_2NH_2$) along with n-methylformamide ($CH_3NHC(O)H$) can be synthesized within interstellar ices containing methylamine (CH_3NH_2) at temperatures as low as 5 K via an facile non-equilibrium chemistry initiated by energetic electrons initiated by galactic cosmic rays once penetrating interstellar ices. After the radiation exposure, the subliming molecules were analyzed isomer selectively after single photon vacuum ultraviolet ionization coupled with a reflectron time-of-flight mass spectrometer (PI-ReTOF-MS). This methodology has several advantages compared to traditional infrared spectroscopy of ices such as the possibility to identify structural isomers of complex organics. The underlying reaction mechanisms leading to methanimine, ethylenediamine, and n-methylformamide in methylamine bearing ices are compared to the chemistry of isoelectronic methanol ices studies previously in our laboratory, revealing exciting similarities, but also differences in the synthesis of complex organic molecules in the interstellar medium.

We thank the US National Science Foundation (AST-1505502) for support to conduct the experiments and data analysis. Furthermore, we thank the W. M. Keck Foundation for financing the experimental setup.

WL02 2:02 – 2:17

RADIO ASTRONOMY RECEIVERS AND A GAS REACTION CHAMBER FOR LABORATORY ASTROCHEMICAL SIMULATIONS.

JOSE CERNICHARO, JUAN R. PARDO, *Instituto de Fisica Fundamental, CSIC, Madrid, Spain*; JUAN DANIEL GALLEGO, PABLO DE VICENTE, *Centro Astronómico de Yebes, Observatorio Astronómico Nacional, Yebes, Guadalajara, Spain*; ISABEL TANARRO, VICTOR JOSE HERRERO, JOSÉ LUIS DOMÉNECH, RAMÓN J. PELÁEZ, *Instituto de Estructura de la Materia, (IEM-CSIC), Madrid, Spain.*

We present the current status of an experimental setup in which astronomical receivers and spectrometers are coupled to a reaction chamber to study the spectroscopy and chemical evolution of gas mixtures via their rotational emission lines. In a first proof of concept a small prototype reactor was placed in the beam path of the Aries 40 m radio telescope (Yebes, Guadalajara, Spain) facing the Q-band receiver operating in the 41-49 GHz frequency range providing 2 GHz bandwidth and 38 kHz resolution. Experiments with static samples or in flow mode, exposed to UV irradiation or an inductively coupled cold plasma were performed and the feasibility of the experiment demonstrated[a]. In a second phase, new receivers have been designed and built by the team of astronomers and engineers at the Yebes observatory, and are now coupled to a new larger reaction chamber in a dedicated laboratory. The new receivers cover the 31.5-50 GHz and 72-116 GHz bands quasi-simultaneously, with resolution ∼38 kHz. The performance and first results of the system will be discussed.

---
[a] I. Tanarro et al. Astron. & Astrophys. 609, A15 (2018)

WL03   2:19 – 2:34

O(^{1}D) INSERTION REACTIONS FOR THE PRODUCTION AND SPECTRAL ANALYSIS OF INTERSTELLAR ORGANIC MOLECULES

HAYLEY BUNN, SAMUEL ZINGA, CARSON REED POWERS, BRIAN SAVINO, MORGAN N McCABE, BRIAN M HAYS[a], SUSANNA L. WIDICUS WEAVER, *Department of Chemistry, Emory University, Atlanta, GA, USA.*

O(^{1}D) insertion reactions with stable precursors have proved an efficient way of producing important prebiotic molecules that are highly reactive and otherwise unstable under laboratory conditions. In 2015, Hays et al.[b] reported successful production of gaseous methanol and vinyl alcohol by exothermic O(^{1}D) insertion into methane and ethylene, respectively, and collected their rotational spectra in the millimeter/submillimeter region. Prior to this, in 2013 Hays et al.[c] reported a computational study predicting the formation of methanediol, methoxymethanol and aminomethanol, through O(^{1}D) insertion into methanol, dimethyl ether and methylamine, respectively. These species are all important prebiotic molecules and have been shown to be stable under interstellar conditions. We therefore seek to collect their spectra for comparison to interstellar observations. Here we will report experimental progress toward producing and characterizing the spectra of aminomethanol and methanediol using O(^{1}D) insertion reactions and millimeter/submillimeter spectroscopy.

[a] Current address Institut de Physique de Rennes, Département Physique Moléculaire
[b] B. M.Hays, N. Wehres, B. Alligood DePrince, A. A.M. Roy, J. C. Laas, S. L. Widicus Weaver, Chem.Phys. Lett., 630, 18 (2015)
[c] B. M. Hays, S. L. Widicus Weaver, J. Phys. Chem., 117, 7142 (2013)

WL04   2:36 – 2:51

COSMIC RAY-DRIVEN RADIATION CHEMISTRY IN COLD INTERSTELLAR ENVIRONMENTS

CHRISTOPHER N SHINGLEDECKER, *Department of Chemistry, The University of Virginia, Charlottesville, VA, USA*; JESSICA D. TENNIS, *Chemistry, University of Virginia, Charlottesville, VA, USA*; ROMANE LE GAL, ERIC HERBST, *Department of Chemistry, The University of Virginia, Charlottesville, VA, USA.*

The physiochemical impact of cosmic rays on interstellar regions is widely known to be significant [a]. Indeed, the cosmic ray-driven formation of H_3^+ via the ionization of H_2 was shown to be of key importance in even the first astrochemical models [b]. Later, cosmic rays were implicated in the collisional excitation of H_2, which leads to the production of internally produced UV photons that also have profound effects on the chemistry of molecular clouds [c]. Despite these key findings, though, attempts at a more complete consideration of interstellar radiation chemistry have been stymied by the lack of a general method suitable for use in astrochemical models and capable of preserving the salient macroscopic phenomena that emerge from a large number of discrete microscopic events.

Recently, we have developed a theoretical framework which meets these criteria and allows for the estimation of the decomposition pathways, yields, and rate coefficients of radiation-chemical reactions [d]. In this talk, we present preliminary results illustrating the effect of solid-phase radiation chemistry on models of TMC-1 in which we consider the radiolysis of the primary ice-mantle constituents of dust grains. We further discuss how the inclusion of this non-thermal chemistry can lead to the formation of complex organic molecules from simpler ice-mantle constituents, even under cold core conditions.

[a] Indriolo, N. & McCall, B. J.,*Chem. Soc. Rev.*, 42, 7763-7773, 2013
[b] Herbst, E. & Klemperer, W., *Ap.J.*, 185, 505-534, 1973
[c] Prasad, S. S. & Tarafdar, S. P.,*Ap.J.*, 267, 603-609, 1983
[d] Shingledecker, C. N. & Herbst, E., *Phys. Chem. Chem. Phys.*, 20, 5359-5367, 2018

WL05 2:53 – 3:08
PROBING THE PHOTOPRODUCTS OF INTERSTELLAR ICE ANALOGUES VIA LABORATORY SUBMILLIMETER SPECTROSCOPY

KATARINA YOCUM, HOUSTON H SMITH, *Department of Chemistry, Emory University, Atlanta, GA, USA*; STEFANIE N MILAM, *Astrochemistry, NASA Goddard Space Flight Center, Greenbelt, MD, USA*; SUSANNA L. WIDICUS WEAVER, *Department of Chemistry, Emory University, Atlanta, GA, USA*.

Studying the chemical evolution of the interstellar medium (ISM) is critical for understanding chemical processes which take place during the formation of stars and planetary systems. The gas-phase composition of interstellar space is revealed through remote observations employing high-resolution spectroscopy. It is believed that many complex organics found in the ISM, some of which are of prebiotic interest, first formed in the ices coating interstellar dust grains. The results of laboratory simulations of interstellar ices provide great insight into how complex organics form and/or evolve in the ISM. Previous experimental techniques have monitored the thermal and photoprocessing of relevant ices via infrared spectroscopy while studying the sublimated gases with mass spectrometry. Here we will discuss a new approach that uses noninvasive submillimeter spectroscopy to analyze the gas-phase reactions occurring above the ice during processing. New results and experimental improvements will be discussed.

WL06 3:10 – 3:25
SUB-DOPPLER INFRARED SPECTROSCOPY OF JET COOLED NDH_3^+: N–H STRETCH VIBRATIONS IN A KEY ASTROCHEMICAL ION

PRESTON G. SCRAPE, ANDREW KORTYNA, *JILA, National Institute of Standards and Technology and Univ. of Colorado, Boulder, CO, USA*; DANIEL LESKO, *Department of Chemistry and Biochemistry, University of Colorado, Boulder, CO, USA*; DAVID NESBITT, *JILA, National Institute of Standards and Technology and Univ. of Colorado, Boulder, CO, USA*.

The ammonium-*d* cation (NDH_3^+) is the singly-deuterated isotologue of ammonium (NH_4^+), an abundant species in terrestrial chemistry that has been proposed as a key species in the chemistry of interstellar objects. Unlike NH_4^+, NDH_3^+ has a nonzero rotational dipole transition moment, so it can be detected by its microwave emissions. To assist in the detection of this ion, we have collected rovibrationally resolved infrared spectra of its symmetric (ν_1, A_1) and asymmetric (ν_4, E) N–H stretching modes, with special attention to accurately determining its ground-state rotational constants. In this study, NDH_3^+ is generated by seeding NDH_2 in a H_2/Ne/He mixture through a pulsed slit discharge. Electron ionization of H_2 in the mixture produces H^+, which readily protonates other H_2 molecules to form H_3^+; this H_3^+ goes on to protonate NDH_2. The resulting NDH_3^+ ions are cooled in a slit jet supersonic expansion to a rotational temperature of 40 K. Rotational constants for the ground state and both singly-excited vibrational states, as well as the ν_1 and ν_4 band origins, are obtained by least-squares fitting of the sub-Doppler rotational structure to a symmetric top Hamiltonian. The fitted band origins agree with the best theoretical predictions to within 1 cm^{-1}.

WL07 3:27 – 3:42
ULTRAVIOLET AND INFRARED OSCILLATOR STRENGTHS FOR OH^+

JAMES NEIL HODGES, DROR M. BITTNER, PETER F. BERNATH, *Department of Chemistry and Biochemistry, Old Dominion University, Norfolk, VA, USA*.

OH^+ is an important astrophysical species. OH+ has been detected in the interstellar medium by UV and terahertz spectroscopy. Following the recent analysis of OH^+ emission spectra,[a] empirical potential energy surfaces have been calculated for the $A^3\Pi$ and $X^3\Sigma^-$ states using the RKR method. Ab initio transition and dipole moment functions were calculated and together with the potential energy surfaces have been used to compute oscillator strengths using Le Roy's LEVEL program.[b] The new oscillator strengths account for the Herman–Wallis effect, a rotational dependence in the vibrational wavefunction, and are now in good agreement with the measured lifetime.[c] The Herman–Wallis effect creates a 5% difference in UV oscillator strengths by J" = 15 and an 80% difference in oscillator strengths by J" = 10 in the IR. We recommend these new oscillator strengths be used to determine OH^+ column densities.

---
[a] Hodges, J. N., & Bernath, P. F. Astrophys. J., 840.2 (2017) 81
[b] Le Roy, R. J., J. Quant. Spectrosc. Radiat. Transf. 186 (2017) 167
[c] Möhlmann, G. R., et al., Chem. Phys. 31.2 (1978) 273

Intermission

WL08   4:18 – 4:33

INFRARED ABSORPTION CROSS SECTIONS OF HYDROCARBONS

<u>PETER F. BERNATH</u>, ANDY WONG, *Department of Chemistry and Biochemistry, Old Dominion University, Norfolk, VA, USA*; DOMINIQUE APPADOO, *800 Blackburn Road, Australian Synchrotron, Melbourne, Victoria, Australia*; BRANT E. BILLINGHURST, *EFD, Canadian Light Source Inc., Saskatoon, Saskatchewan, Canada.*

Absorption cross sections for a range of small hydrocarbons, from C2-C4, in the far and mid IR spectral regions are presented. Cross sections were obtained from high resolution spectra recorded at cold temperatures from experiments performed at two synchrotron facilities: the Australian Synchrotron (AS) and the Canadian Light Source (CLS), as well as at Old Dominion University (ODU). The experimental conditions that were sampled (pressure, composition and temperature) were chosen to mimic those found in the planetary atmospheres of Titan, Saturn and Jupiter. These cross sections can be used to determine molecular abundances from remote sensing observations.

WL09   4:35 – 4:50

INFRARED SPECTROSCOPY ON SMALL METAL-BEARING OXIDES

<u>DANIEL WITSCH</u>, *Institute of Physics, University of Kassel, Kassel, Germany*; ALEXANDER A. BREIER, *Institute of Physics, University Kassel, Kassel, Germany*; GUIDO W FUCHS, *Physics Department, University of Kassel, Kassel, Germany*; THOMAS GIESEN, *Institute of Physics, University Kassel, Kassel, Germany.*

Interstellar dust is an integral part of the interstellar medium and is important for star forming processes and the associated chemical evolution in these regions. However the formation of dust is still not well understood. Around oxygen rich late type stars, titanium oxides are thought of being an important seed molecule of the dust formation and has been observed at optical and radio wavelengths.

A new generation of high resolution infrared telescope instruments, like TEXES at Gemini North or EXES onboard SOFIA, allows the identification of astrophysical molecules by means of their rovibrational spectra, probing warm atmospheres of evolved late type stars in the mid-infrared.

In this talk we present high resolution laboratory spectra of titanium monoxide (TiO) and its isotopologues in the gas-phase around 1000cm^{-1} ($10 \mu \text{m}$). In a global fit we determine molecular parameters of the vibrational excited states with high accuracy. In our experiments, molecules are produced using high intense laser pulses to ablate a titanium sample in an atmosphere of nitrous oxide diluted in helium buffer gas. Guided through a reaction channel different molecules - including TiO - are formed. Subsequent adiabatic expansion in a supersonic jet cools down the molecules to rotational temperatures of around 30 K. To record a rotationally resolved infrared spectrum a quantum cascade laser is used.

WL10   4:52 – 5:07

HIGH ACCURACY THERMOCHEMISTRY AND KINETICS OF THE HCN/HNC SYSTEM

>KELVIN LEE, *Radio and Geoastronomy Division, Harvard-Smithsonian Center for Astrophysics, Cambridge, MA, USA*; MICHAEL C McCARTHY, *Atomic and Molecular Physics, Harvard-Smithsonian Center for Astrophysics, Cambridge, MA, USA.*

The relative abundance of HCN/HNC is a ubiquitous issue in astrochemistry. This ratio is largely governed by competition between thermodynamic stability and kinetics: in cold dense clouds, the thermodynamically unfavorable HNC is enhanced, while in hot cores and young stellar objects there is a much stronger preference for HCN. Efforts to develop a consistent and universally accepted set of reaction rates and thermochemical parameters involving both gas- and condensed-phase dynamics has proven challenging. Considerable interest has focused on accurate determinations of molecular properties, and many theoretical and experimental efforts, spanning several decades, have sought to provide the necessary rates and enthalpies. Despite much work, estimates of the uncertainty of enthalpies and rates vary substantially, particularly with respect to quantum chemical treatments that involve a plethora of basis sets and methods. To address this issue, we have undertaken a systematic study to calculate a consistent set of thermochemical quantities and rates involving gas-phase reactions presumed to be important in determining the branching between HCN and HNC in astrophysical environments. Using the HEAT345(Q) method, we have calculated the energetics of neutral and ion reactants and products. This method routinely achieves chemical accuracy (1 kJ/mol/120 K) without empirical corrections. We validate our thermochemical network by comparison with reliable databases such as the ATcT, and by doing so lends confidence into the species that are not yet included in databases. Finally, we report reaction rates for significant reactions from first principles (V)TST theory.

WL11   5:09 – 5:24

DETECTION OF INTERSTELLAR BENZONITRILE (c-C_6H_5CN)

>BRETT A. McGUIRE, *NAASC, National Radio Astronomy Observatory, Charlottesville, VA, USA*; ANDREW M BURKHARDT, *Department of Astronomy, The University of Virginia, Charlottesville, VA, USA*; SERGEI KALENSKII, *Astro Space Center, Lebedev Physical Institute, Russian Academy of Sciences, Moscow, Russia*; CHRISTOPHER N SHINGLEDECKER, *Department of Chemistry, The University of Virginia, Charlottesville, VA, USA*; ANTHONY REMIJAN, *ALMA, National Radio Astronomy Observatory, Charlottesville, VA, USA*; ERIC HERBST, *Department of Chemistry, The University of Virginia, Charlottesville, VA, USA*; MICHAEL C McCARTHY, *Atomic and Molecular Physics, Harvard-Smithsonian Center for Astrophysics, Cambridge, MA, USA.*

The Unidentified Infrared Bands are now widely believed to originate from the emission of large, aromatic molecules in high-energy environments. Despite this, no individual species has been identified as a carrier, and indeed the only five- or six-membered aromatic ring molecule reported in the ISM is benzene, which is seen in only a small handful of sources at infrared wavelengths. Here, I will discuss a dedicated laboratory, observational (GBT), and modeling effort which has resulted in the first definitive radio identification and quantification of a benzene-ring containing aromatic molecule: benzonitrile (c-C_6H_5CN). The results will shed light on the probable formation pathways for larger aromatic species, and have identified a successful methodology for future, comprehensive investigations.

WL12

SYNTHESIS OF INTERSTELLAR BENZONITRILE (c-C_6H_5CN): A MICROWAVE SPECTROSCOPIC STUDY

BRETT A. McGUIRE, *NAASC, National Radio Astronomy Observatory, Charlottesville, VA, USA*; KELVIN LEE, *Radio and Geoastronomy Division, Harvard-Smithsonian Center for Astrophysics, Cambridge, MA, USA*; MICHAEL C McCARTHY, *Atomic and Molecular Physics, Harvard-Smithsonian Center for Astrophysics, Cambridge, MA, USA.*

Benzonitrile (c-C_6H_5CN) has recently been detected in the interstellar medium - the first molecule containing a benzene ring to be observed by radio astronomy. Its detection in a cold, starless dark cloud affords one the opportunity to probe aromatic chemistry at the earliest stages of the star formation process. Here, we explore the formation chemistry of benzonitrile using a combination of laboratory microwave spectroscopic and quantum chemical approaches. We demonstrate the synthesis of benzonitrile from a variety simple, acyclic precursors (acetylene [HCCH], diacetylene [HC_4H], cyanoacetylene [HC_3N], and 1,3-butadiene [$CH_2(CH)_2CH_2$]), providing definitive evidence for facile bottom-up generation of aromatic carbon chemistry from small interstellar precursors. The results show that benzonitrile can already be used as a reliable proxy for the presence of benzene in the ISM, and that there may exist a much larger array of aromatic species that are 'hidden' just below the current sensitivity of spectral surveys.

RA. Plenary
Thursday, June 21, 2018 – 8:30 AM
Room: Foellinger Auditorium

Chair: Leslie Looney, University of Illinois at Urbana-Champaign, Urbana, IL, USA

RA01   8:30–9:10

VIBRATIONAL AND ROTATIONAL SPECTROSCOPY IN CRYOGENIC ION TRAPS

SANDRA BRÜNKEN, BRITTA REDLICH, *FELIX Laboratory, Radboud University, Nijmegen, The Netherlands*; PAVOL JUSKO, OSKAR ASVANY, STEPHAN SCHLEMMER, *I. Physikalisches Institut, Universität zu Köln, Köln, Germany*.

Reactive molecular ions play a central role in the chemistry of the interstellar medium and in planetary atmospheres. Spectroscopic studies of these often elusive ions yield fundamental insights on their geometrical and electronic structure, and provide vibrational and rotational signatures needed for their identification in space. Cryogenic ion traps have proven to be ideal tools for the development of sensitive spectroscopic schemes of mass-selected, cold, and isolated molecular ions. Recent progress on these so-called action spectroscopic methods allows not only to probe electronic and vibrational excitation processes, but also to record high-resolution purely rotational molecular spectra[a], which are a direct prerequisite for radio-astronomical detections of new species in space as will be demonstrated with selected examples. In addition, details of broadband infrared experiments on several astrophysically important hydrocarbon cations ranging in size from comparatively small systems (e.g., C_2H^+, $C_3H_2^+$, and C_3H^+) to PAH cations will be given, using the unique combination of a cryogenic ion trap instrument[b] interfaced to the free electron lasers at the FELIX Laboratory.

[a] S. Brünken, L. Kluge, A. Stoffels, O. Asvany, and S. Schlemmer, Astrophys. J. Lett., 783, L4 (2014); A. Stoffels, L. Kluge, S. Schlemmer, and S. Brünken, A&A 593, A56 (2016); S. Brünken, L. Kluge, A. Stoffels, J. Pèrez-Rios, and S. Schlemmer, J. Mol. Spectrosc., 332, 67 (2017)

[b] O. Asvany, S. Bünken, L. Kluge, and S. Schlemmer, Appl. Phys. B, 114, 203 (2014)

RA02   9:15–9:55

ULTRAFAST TRANSIENT ABSORPTION SPECTROSCOPY OF PHOTOCHEMICAL DYNAMICS IN SOLUTION

DAISUKE KOYAMA, RAVI KUMAR VENKATRAMAN, *School of Chemistry, University of Bristol, Bristol, United Kingdom*; HARVEY J A DALE, *School of Chemistry, University of Edinburgh, Edinburgh, United Kingdom*; ANDREW ORR-EWING, *School of Chemistry, University of Bristol, Bristol, United Kingdom*.

The methods of ultrafast UV-visible and mid-IR transient absorption spectroscopy are powerful probes of dynamical processes occurring in solution [1]. They can be used to determine the rates and mechanisms of chemical and photochemical reactions, identify short-lived reactive intermediates, and examine how the dynamics are modified by solute-solvent interactions. The mechanistic insights which derive from the application of transient absorption spectroscopy will be illustrated by recent studies of solvent effects on electronically non-adiabatic pathways in photoexcited molecules, and direct observation of the intermediates involved in radical reactions controlled by the use of organic photoredox catalysts [2].

[1] Taking the plunge: chemical reaction dynamics in liquids, A.J. Orr-Ewing, Chem. Soc. Rev. 2017, 46, 7597-7614. DOI: 10.1039/C7CS00331E. [2] Ultrafast observation of a photoredox reaction mechanism: photo-initiation in organocatalyzed atom-transfer radial polymerization, D. Koyama, H.J.A. Dale and A.J. Orr-Ewing, J. Am. Chem. Soc. 2018, 140, 1285-1293. DOI: 10.1021/jacs.7b07829.

Intermission

RAO AWARDS   10:35
Presentation of Awards by Gary Douberly, University of Georgia

2017 Rao Award Winners
Bryce Bjork, University of Colorado at Boulder
Anna Huff, University of Minnesota
Christopher Shingledecker, University of Virginia

MILLER PRIZE 10:45

Introduction by Michael Heaven, Emory University

RA03 *Miller Prize Lecture* 10:50 – 11:05

BOND INSERTION IN METAL–CARBON DIOXIDE ANIONIC CLUSTERS STUDIED BY INFRARED PHOTODISSOCIATION SPECTROSCOPY

<u>LEAH G DODSON</u>, *JILA and NIST, University of Colorado, Boulder, CO, USA*; MICHAEL C THOMPSON, J. MATHIAS WEBER, *JILA and the Department of Chemistry and Biochemistry, University of Colorado-Boulder, Boulder, CO, USA.*

C–O bond breaking is an important process in the activation of CO_2 that can be catalyzed by the presence of a metal. In this talk, we investigate the factors that lead to bond insertion in $[M(CO_2)_y]^-$ gas phase clusters, specifically addressing differences amongst the metals M = Ni, Fe, and Ti. Gas phase anionic clusters were generated using laser ablation of a metal target in the presence of a CO_2 expansion, and the infrared photodissociation spectra were measured from 950–2400 cm^{-1}. Metal carbonyl vibrational signatures were used to infer bond insertion, and computational chemistry simulations were used to assess the feasibility of bond breaking in these systems.

COBLENTZ AWARD 11:10

Presentation of Award by Linda Kidder, Coblentz Society

RA04 *Coblentz Society Award Lecture* 11:15 – 11:55

ATTOSECOND TIME-RESOLVED MOLECULAR SPECTROSCOPY

<u>HANS JAKOB WÖRNER</u>, *Laboratory of Physical Chemistry, ETH Zurich, Zürich, Switzerland.*

Attosecond time-resolved spectroscopy is beginning to provide experimental access to the most fundamental time scales of molecules, on which the electronic dynamics take place. A few recent experiments that access purely electronic dynamics, as well as coupled electronic and nuclear dynamics in molecules will be discussed. The theoretical developments that accompanied the experimental work will also be presented. The ionization of most molecules on the sub-femtosecond time scale prepares the molecular cation in a superposition of several electronic states that supports charge migration. Detailed measurements of the phase and amplitude of high-harmonic emission from spatially oriented iodoacetylene molecules have enabled the reconstruction of sub-femtosecond charge migration in the iodoacetylene cation (see figure)[a]. The ionization of molecules by attosecond pulses and a synchronized infrared field was used to measure photoionization time delays between the two highest-lying occupied valence orbitals of H_2O and N_2O. These measurements revealed delays of up to ∼160 as in the case of N_2O, which are characteristic of the transient trapping of the photoelectron by shape resonances[b]. Finally, the extension of attosecond spectroscopy to the soft-X-ray domain (water window) will be discussed. The broad spectral bandwidth available in this domain has been exploited to synthesize one of the shortest attosecond pulses to date (43 as)[c]. Transient absorption spectroscopy at the carbon K-edge has been used to study the photodissociation dynamics of CF_4^+, revealing the rearrangement of the electronic structure during this ultrafast (∼40 fs) process[d]. An outlook on attosecond spectroscopy of both isolated and solvated molecules will be given.

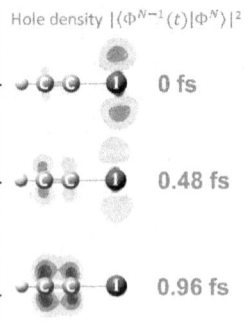

[a] P. M. Kraus *et al.*, *Science* **350**, 790 (2015)
[b] M. Huppert *et al.*, *Phys. Rev. Lett.* **117, 093001 (2016)**
[c] T. Gaumnitz *et al.*, *Opt. Exp.* **25**, 27506 (2017)
[d] Y. Pertot *et al.*, *Science* **355, 264 (2017)**

RG. Cold and ultracold molecules
Thursday, June 21, 2018 – 1:45 PM
Room: 116 Roger Adams Lab

Chair: Hideto Kanamori, Tokyo Institute of Technology, Tokyo, Japan

RG01   1:45 – 2:00
ELECTRONIC PHOTODISSOCIATION SPECTROSCOPY OF COLD NITROPHENOLATE IONS. PART I. ORTHO- AND PARA-NITROPHENOLATE

<u>WYATT ZAGOREC-MARKS</u>, *JILA and the Department of Chemistry and Biochemistry, University of Colorado-Boulder, Boulder, CO, USA*; LEAH G DODSON, *JILA and NIST, University of Colorado, Boulder, CO, USA*; J. MATHIAS WEBER, *JILA and the Department of Chemistry and Biochemistry, University of Colorado-Boulder, Boulder, CO, USA.*

Isomers of nitrophenolate can serve as models for flourophores commonly found in fluorescent proteins. Here we report electronic spectra for mass-selected 2- and 4-nitrophenolate ions prepared in a cryogenic ion trap, measured by photodissociation spectroscopy. The features in the spectra remain broad with no resolvable vibrational structure down to 25 K. We discuss the width of the experimental spectral features in the framework of excited state lifetime and spectral congestion, based on time-dependent density functional theory calculations.

RG02   2:02 – 2:17
ELECTRONIC PHOTODISSOCIATION SPECTROSCOPY OF COLD NITROPHENOLATE IONS. PART II. META-NITROPHENOLATE

WYATT ZAGOREC-MARKS, *JILA and the Department of Chemistry and Biochemistry, University of Colorado-Boulder, Boulder, CO, USA*; LEAH G DODSON, *JILA and NIST, University of Colorado, Boulder, CO, USA*; <u>J. MATHIAS WEBER</u>, *JILA and the Department of Chemistry and Biochemistry, University of Colorado-Boulder, Boulder, CO, USA.*

Isomers of nitrophenolate can serve as models for flourophores commonly found in fluorescent proteins. Here we report electronic spectra for mass-selected 3-nitrophenolate ions prepared in a cryogenic ion trap, measured by photodissociation spectroscopy. Different from the two other isomers, the spectrum shows sharp vibrational bands at low temperatures. We present a Franck-Condon analysis of the spectrum, and discuss the differences between the spectra of the different isomers.

RG03   2:19 – 2:34
ROTAMERS OF ISOPRENE: INFRARED SPECTROSCOPY IN HELIUM DROPLETS AND AB INITIO THERMOCHEMISTRY

<u>PETER R. FRANKE</u>, GARY E. DOUBERLY, *Department of Chemistry, University of Georgia, Athens, GA, USA.*

Isoprene (C_5H_8) is an abundant, reactive tropospheric hydrocarbon, derived from biogenic emissions. A detailed understanding of the spectroscopy of isoprene is therefore desirable. Isoprene monomer is isolated in helium droplets and its infrared spectrum is measured in the CH stretching region. Anharmonic frequencies are predicted by VPT2+K simulations employing CCSD(T) force fields with quadratic (cubic and quartic) force constants computed using the ANO1 (ANO0) basis set. The vast majority of the spectral features can be assigned to trans-isoprene on the basis of these computations. Some features of the higher energy gauche conformer are also assignable, by comparison to experiments using heated isoprene. Convergent ab initio thermochemistry is presented for the isomerization pathway, for which the partition function explicitly accounts for the eigenstates associated with separate, uncoupled one-dimensional potential surfaces for methyl torsion and internal rotation between rotamers. The respective 0 and 298.15 K trans/gauche energy differences are 2.82 and 2.52 kcal/mol, which implies a room temperature gauche population of 2.8%. Additionally, preliminary spectroscopic results for the OH–π complexes between hydroxyl radical and isoprene are presented.

RG04　　2:36–2:51
INFRARED SPECTRA OF PROPENE IN HELIUM NANODROPLETS AND SOLID PARA-HYDROGEN

GREGORY T. PULLEN, PETER R. FRANKE, GARY E. DOUBERLY, *Department of Chemistry, University of Georgia, Athens, GA, USA*; YUAN-PERN LEE, *Applied Chemistry, National Chiao Tung University, Hsinchu, Taiwan, Institute of Atomic and Molecular Sciences, Academia Sinica, Taipei, Taiwan.*

We report the infrared spectra of propene in the C–H stretching region measured in helium nanodroplets (HENDI) at 0.4 K and solid *para*-hydrogen (p-H_2) matrices at 3.2 K, in order to probe the effects of the matrix host environments on the experimental spectra. Propene is an ideal test molecule to study these matrix effects, due to the many anharmonic resonance polyads present in the C–H stretching region of the spectrum. We observe a $4 - 5$ cm^{-1} on average red-shift of the bands in p-H_2 relative to HENDI. Moreover, the choice of matrix environment influences the positions and intensity ratios of transitions within each resonance polyad, leading to qualitatively different spectra. To better understand the nuances involved, simulations were performed that capture the important resonance interactions in a VPT2+K effective Hamiltonian. Certain elements of the Hamiltonian were adjusted to model the impact that different matrix environments have on the anharmonic couplings. In addition, propene reacted with hydrogen atoms *via* electron bombardment of a p-H_2 matrix during sample deposition, producing propyl radicals. *i*-Propyl radicals were produced in greater proportion than *n*-propyl radicals, indicating that for hydrogen addition to the double bond, the rate of addition to the terminal carbon (*i*-propyl) is faster than the rate of addition to the center carbon (*n*-propyl). Because the barriers for addition are approximately 700 cm^{-1} – 1500 cm^{-1} (1000 K – 2000 K), the only available mechanism for reaction in the p-H_2 matrix (3.2 K) is tunneling. *Ab initio* calculations were used to compute the tunneling probabilities for the formation of the *n*-propyl and *i*-propyl radicals. The rate of addition to the terminal carbon (*i*-propyl) was calculated to be faster, in agreement with experiment.

RG05　　2:53–3:08
DETECTION AND SPECTROSCOPY OF POLYATOMIC MOLECULES INSIDE A CRYOGENIC BUFFER GAS CELL

THOMAS WALL, JULIA BIENIEWSKA, B. E. SAUER, MICHAEL TARBUTT, *Centre for Cold Matter, Imperial College London, London, United Kingdom*; BENOIT DARQUIE, *Laboratoire de Physique des Lasers, CNRS, Université Paris 13, Sorbonne Paris Cité, 93430 Villetaneuse, France*; TREVOR SEARS, *Department of Chemistry, Stony Brook University, Stony Brook, NY, USA.*

We are building a cryogenic source of polyatomic molecules that will be used for tests of fundamental physics[a,b]. The molecules are cooled by a buffer gas of 4 K He inside a copper cell mounted on the cold stage of a cryo-cooler. For the development of this source we are using 1,3,5-trioxane. Although a solid at room temperature, it has a high vapour pressure. We inject this vapour into the buffer gas cell through a room temperature tube. We probe the cooled molecules inside the cell using wavelength modulation (WM) spectroscopy, driving vibration-rotation transitions using 10.2 μm wavelength radiation from a quantum cascade laser.

I will present data from recent experiments in which we performed WM spectroscopy close to the Q-branch origin of the ν_5 vibrational fundamental band. We use these spectra to probe the temperature and density of the molecules inside the cell. Most recently we have performed sub-Doppler spectroscopy, recording Lamb dips that can be used to study collision rates inside the cell.

I will discuss our plans to perform sensitive detection of a slow, cold beam of trioxane produced by this source, including a multi-pass optical assembly and cavity enhanced absorption spectroscopy.

---

[a] B. Darquié *et al.*, Chirality **22**, 870-884 (2010)
[b] S.K. Tokunaga *et al.* New Journal of Physics **19**, 053006 (2017)

RG06 3:10–3:25

CHARACTERIZING MOLECULAR IONS FOR LASER CONTROL

<u>SRUTHI VENKATARAMANABABU</u>, *Physics, Northwestern University, Evanston, Illinois, USA*; PATRICK R STOLLENWERK, IVAN ANTONOV, BRIAN C. ODOM, *Department of Physics and Astronomy, Northwestern University, Evanston, IL, USA.*

Variation of fundamental constants would indicate physics beyond the Standard Model. Astronomical evidence and proposed theoretical models suggest a possible variation in the proton to electron mass ratio (μ). Detection of μ variation will require high precision measurements. Historically, the highest precision measurements have been performed on ultracold atoms. Atomic transitions, however, have limited sensitivity to μ compared to what is found in molecules. Unfortunately, the additional motional degrees of freedom in molecules that give them this sensitivity also lead to more complex internal structure, making it difficult to control them using powerful techniques such as optical pumping. To achieve laser control of the molecular degrees of freedom we need an accurate knowledge of transition energies, state lifetimes and radiative branching ratios. In addition, we need a method to reliably produce molecular ions. In this talk, I will discuss spectroscopic techniques developed in our lab to probe molecular ions.

Intermission

RG07 4:01–4:16

DETERMINATION OF THE SPIN-ROTATION FINE STRUCTURE OF He_2^+

<u>PAUL JANSEN</u>, LUCA SEMERIA, FREDERIC MERKT, *Laboratorium für Physikalische Chemie, ETH Zurich, Zurich, Switzerland.*

Measuring spin-rotation intervals in molecular cations is challenging, particularly so when the ions do not have electric-dipole-allowed rovibrational transitions. We present a method to determine the spin-rotational fine structure of molecular ions from the fine structure of high Rydberg states[a]. The method is illustrated by the determination of the so far unknown spin-rotation fine structure of the fundamentally important He_2^+ ion in the X^+ $^2\Sigma_u^+$ ground electronic state. The interaction that is responsible for the level structure in the high Rydberg states of He_2 that were probed in our experiment is the n-independent spin-rotation interaction of the ion core. As a consequence, the fine-structure splittings in He_2^+ can be related to the fine-structure of the Rydberg states by applying an angular-momentum basis transformation from Hund's case (e[b]) to Hund's case (d).

The experiment relies on the use of single-mode cw radiation to record spectra of high Rydberg states of He_2 from the a $^3\Sigma_u^+$ metastable state. Metastable helium molecules are produced by striking a discharge in a pulsed expansion of neat helium gas[b]. Cooling the valve body to a temperature of 10 K and using continuous-wave excitation results in an observed Doppler-limited linewidth of 25 MHz. The fine structure of Rydberg states of He_2 is determined from strict selection rules by comparing the observed splitting of the Rydberg spectrum with the spin-rotational intervals of the initial metastable state. The fine-structure splittings of the $v^+ = 0$, $N^+ = 1, 3$, and 5 levels of He_2^+ are 7.96(14)MHz, 17.91(32) MHz and 28.0(6) MHz, respectively.

---

[a]P. Jansen, L. Semeria, and F. Merkt, *Phys. Rev. Lett.* **120**, 043001 (2018).
[b]M. Motsch, P. Jansen, J.A. Agner, H. Schmutz, and F. Merkt, *Phys. Rev. A* **89**, 043420 (2014).

RG08   4:18–4:33

FINE STRUCTURE OF METASTABLE ^{4}He$_2$ USING ZEEMAN-DECELERATED MOLECULAR-BEAM RESONANCE SPECTROSCOPY

<u>LUCA SEMERIA</u>, PAUL JANSEN, JOSEF A. AGNER, HANSJÜRG SCHMUTZ, FREDERIC MERKT, *Laboratorium für Physikalische Chemie, ETH Zurich, Zurich, Switzerland.*

The $a\,^3\Sigma_u^+$ state of ^{4}He$_2$ is a metastable state with a lifetime of about 18 s. The spin-spin and spin-rotation interactions result in a splitting of each rotational level N into three components $J = N, N \pm 1$. The fine structure intervals of the $N = 1$, 3, 5, 7 – 11 and 25 - 29 have been measured by radio frequency (rf) spectroscopy[a][b][c][d] and were included in a global analysis of the $a\,^3\Sigma_u^+$ state [e].

A new measurement of the fine structure of all rotational levels between $N = 1$ and 21 of the $a\,^3\Sigma_u^+$ ($v = 0$) state will be presented. The $J = N$ fine-structure components, which are high-field seeking in magnetic fields, have been eliminated using a multistage Zeeman decelerator, and repopulated from the low-field-seeking $J = N \pm 1$ components using rf radiation prior to detection by excitation to Rydberg states followed by pulsed-field ionization. The low velocity of the Zeeman decelerated beam[f][g] enabled long interaction times of the molecules with the rf radiation and therefore a reduction of the transit-time broadening down to 10 kHz (FWHM), allowing the transition frequencies to be determined very accurately. The fine structure has been analyzed using an effective Hamiltonian to obtain improved values of the spin-spin and spin-rotation coupling constants for the $a\,^3\Sigma_u^+$ ($v = 0$) metastable state of ^{4}He$_2$, including centrifugal distortion corrections.

---

[a] W. Lichten, M.V. McCusker and T. L. Vierima, *J. Chem. Phys.*, **61**, 2200 (1974).
[b] W. Lichten and T. Wik, *J. Chem. Phys.*, **69**, 98 (1978).
[c] M. Kristensen and N. Bjerre, *J. Chem. Phys.*, **93**, 983 (1990).
[d] I. Hazell, A. Nørregaard and N. Bjerre, *J. Mol. Spectrosc.*, **172**, 135 (1995).
[e] C. Focsa, P. F. Bernath and R. Colin, *J. Mol. Spectrosc.*, **191**, 209, (1998).
[f] M. Motsch, P. Jansen, J. A. Agner, H. Schmutz and F. Merkt, *Phys. Rev. A*, **89**, 043420 (2014).
[g] P. Jansen, L. Semeria, L. E. Hofer, S. Scheidegger, J. A. Agner, H. Schmutz and F. Merkt, *Phys. Rev. Lett.*, **115**, 133202 (2015).

RG09   4:35–4:50

SPECTOSCOPY OF SiO AND SiO$^+$ IN SUPPORT OF ULTACOLD MOLECULE STUDIES

<u>IVAN ANTONOV</u>, PATRICK R STOLLENWERK, SRUTHI VENKATARAMANABABU, BRIAN C. ODOM, *Department of Physics and Astronomy, Northwestern University, Evanston, IL, USA.*

SiO$^+$ was proposed as a candidate for ultracold molecule experiments. Cooling schemes required to prepare SiO$^+$ in its ground state require knowledge of state energies, lifetimes and branching of selected SiO$^+$ transitions. Knowledge of dissociative transitions is needed to probe state populations of SiO$^+$ in the proposed experiments. Finally, efficient loading of SiO$^+$ into a trap by photoionization requires studying spectroscopy of neutral SiO. In this talk, I will discuss recent progress in study of SiO and SiO$^+$ spectroscopy in our lab and approaches used to address these studies.

RG10   4:52–5:07

TOWARDS STATE-RESOLVED ULTRACOLD CHEMICAL REACTIONS OF KRb MOLECULES

<u>DAVID GRIMES</u>, MING-GUANG HU, YU LIU, ANDREI GHEORGHE, KANG-KUEN NI, *Department of Chemistry and Chemical Biology, Harvard University, Cambridge, MA, USA.*

Chemical reactions at ultralow temperatures can proceed surprisingly efficiently due to their quantum mechanical nature. However, the detailed chemical physics to describe and predict the distribution of final states is still unclear. We will discuss a barrier-less, likely 4-center reaction, 2 KRb → K$_2$ + Rb$_2$, in the temperature regime below 1 μK. Due to the low exothermic energy of this reaction, ~ 10 cm^{-1}, we aim to resolve individual product quantum states through ionization detection. Our approach combines the physical chemistry techniques of REMPI spectroscopy and velocity-map imaging for ion and quantum state detection with AMO physics techniques for preparation of the ultracold molecular reagents.

RH. Spectroscopy as an analytical tool
Thursday, June 21, 2018 – 1:45 PM
Room: 100 Noyes Laboratory

Chair: Kyle N. Crabtree, University of California, Davis, CA, USA

RH01   1:45 – 2:00

MEASUREMENTS OF $N_2(A^3\Sigma_u^+,v)$ POPULATIONS IN A NANOSECOND PULSE DISCHARGE BY CAVITY RING-DOWN SPECTROSCOPY

ELIJAH R JANS, KRAIG FREDERICKSON, IGOR V. ADAMOVICH, *Department of Aerospace and Mechanical Engineering, The Ohio State University, Columbus, OH, USA.*

Time-resolved number densities of excited electronic state of nitrogen, $N_2(A^3\Sigma_u^+,v=0,1,2)$, have been measured in a nanosecond pulse discharge using Cavity Ring-Down Spectroscopy (CRDS). The CRDS spectrometer is operated using a 10 Hz Nd:YAG laser which pumps a narrowband tunable dye laser using a LDS765 dye, to produce output between 745 to 770 nm with a linewidth of 0.12 cm^{-1}. The ring-down cavity is a 10 mm x 22 mm rectangular cross section quartz channel 55 cm long, fused to two 1.5 inch diameter quartz tubes at both ends, with the total cavity length of 90 cm. The mirrors (reflectivity of 0.99995) are attached to the ends of the ring-down cavity using stainless steel adjustable mounts, for precision alignment. Two rectangular plate copper electrodes, 12 mm x 60 mm, are attached to the top and bottom walls of the quartz channel in the middle of the cavity, using silicone rubber adhesive. The electrodes are powered by a custom-built high-voltage pulse generator producing alternating polarity pulses with peak voltage up to 15 kV and pulse duration of approximately 100 ns FWHM. The pulser is operated in burst mode, with burst repetition rate of 10 Hz, pulse repetition rate of 10 kHz, and 10 pulses per burst, with coupled discharge energy of approximately 0.3 mJ/pulse. Spectra of $N_2(B^3\Pi_g \leftarrow A^3\Sigma_u^+,v)$ absorption bands are taken 25 μs after the last discharge pulse in the burst, with all absorption transitions identified. Time resolved CRDS data are taken from isolated rotational lines for each vibrational state to infer temporal evolution of absolute populations of vibrational levels of $N_2(A^3\Sigma_u^+)$ at t=25-1500 μs after the last discharge pulse. This diagnostics is being developed for measurements of excited metastable state populations of N_2 and O_2 in nonequilibrium plasmas and nonequilibrium high-speed flows.

RH02   2:02 – 2:17

ABSOLUTE NUMBER DENSITY MEASUREMENTS OF HYDROPEROXYL RADICAL IN A NANOSECOND PULSE DISCHARGE USING CAVITY RING-DOWN SPECTROSCOPY

KRAIG FREDERICKSON, *Department of Aerospace and Mechanical Engineering, The Ohio State University, Columbus, OH, USA*; TERRY A. MILLER, *Department of Chemistry and Biochemistry, The Ohio State University, Columbus, OH, USA*; IGOR V. ADAMOVICH, *Department of Aerospace and Mechanical Engineering, The Ohio State University, Columbus, OH, USA.*

A recently implemented cavity ring-down spectrometer has been used to perform absolute number density measurements of hydroperoxyl radical (HO_2) generated in a repetitive nanosecond-duration, pulse discharge sustained in a mixture of $H_2/O_2/Ar$. The probe source for the spectrometer is a custom-built, injection-seeded, optical parametric oscillator emitting an idler beam in the 1500 nm region accessing the first overtone ($2\nu_1$) of the O-H stretch. Water vapor was used as a standard species to characterize the spectrometer and provide estimates of the spectral linewidth, sensitivity, and noise level. A specially constructed ring-down cell, with the central portion consisting of rectangular quartz channel tubing and a pair of copper plate electrodes, was used to produce a repetitively pulsed discharge in a $H_2/O_2/Ar$ mixture. Narrow bandwidth cavity ring down spectra are acquired of a hydroperoxyl absorption feature composed of numerous closely spaced ro-vibrational lines centered at 6638.20 cm^{-1} and number density is determined from the resulting spectral line. This is believed to be the first detection and quantitative measurement of hydroperoxyl radical produced in a nanosecond pulse discharge. The measured number density is compared to the value predicted by the kinetic model of a nanosecond pulse discharge in a reacting $H_2/O_2/Ar$ mixture.

RH03 — 2:19–2:34
ACETONE AND METHANE DETECTION WITH WAVELENGTH MODULATION SPECTROSCOPY IN THE NEAR- AND MID-IR

JINBAO XIA, FENG ZHU, JAMES R BOUNDS, *Department of Physics and Astronomy, Texas A&M University, College Station, TX, USA*; SASA ZHANG, *School of Information Science and Engineering, Shandong University, Jinan, China*; ALEXANDER KOLOMENSKII, HANS A SCHUESSLER, *Department of Physics and Astronomy, Texas A&M University, College Station, TX, USA.*

A high sensitivity sensor, combining a multipass cell and wavelength modulation spectroscopy in the near-IR spectral region (1.651μm) was designed and implemented for trace gas detection. The sensor uses a DFB laser and software lock-in detection, realized with a LabVIEW code. The high sensitivity was achieved by combining the multipass cell having a long effective absorption length of 290 meters, the wavelength modulation spectroscopy, and noise suppression by using a dual beam scheme. The developed spectroscopic technique demonstrates an improved sensitivity for methane in ambient air and a relatively short detection time compared to previously reported sensors. The average methane concentration measured in ambient air was 2.01ppm with a relative error of $\pm 2.5\%$. With Allan deviation analysis, it was found that the methane detection limit of 1.2ppb was achieved in 650s. A modification of this scheme for acetone detection with a mid-IR distributed feedback interband cascade laser with the center wavelength around 3.367μm was also developed, achieving the detection limit was 0.58 ppm with 1s and down to 0.12 ppm with 60s signal averaging.

This work was supported by Robert A. Welch Foundation, grant No. A1546, the Qatar Foundation, grant NPRP 8-735-1-154.

RH04 — 2:36–2:51
SPECTROSCOPIC CHARACTERIZATION OF SMALL POLAR IMPURITIES IN GASOLINE

SYLVESTRE TWAGIRAYEZU, *Chemistry and Biochemistry, Lamar University, Beaumont, TX, USA*; ALEX MIKHONIN, MATT MUCKLE, JUSTIN L. NEILL, *BrightSpec Labs, BrightSpec, Inc., Charlottesville, VA, USA.*

Small polar compounds in gasoline have been identified using a BrightSpec Fourier Transform Microwave Rotational Resonance (FT-MRR) spectrometer in the 260-290 GHz band with Headspace Sampling Module. The design of this spectrometer is based on segmented Chirped Pulse Fourier Transform millimeter wave (CP-FTmmW) spectroscopy, which exploits recent advances in digital electronics to allow the measurement of broadband rotational spectra in a few minutes. As part of efforts to determine applications for rotational spectroscopy to petrochemical problems, FT-MRR has been employed to record rotationally resolved spectra of small polar compounds in gasoline. Preliminary analysis of the observed features using the BrightSpec spectral database reveals a rich, but interpretable, pattern, due to the sensitivity of FT-MRR to only polar compounds. The complex hydrocarbon matrix, which in many analytical instruments obscures the signals from low concentration impurities, is nearly invisible in FT-MRR. Spectroscopic and quantitative analyses of detected polar compounds are underway and will be given in this talk.

RH05
2:53 – 3:08

OPTICAL SENSING OF ENVIRONMENTALLY HAZARDOUS HEAVY METALS (Cr^{3+}, Pb^{2+}, Zn^{2+}) AND CANCER CELLS BY FUNCTIONALIZED CORE/SHELL QUANTUM DOTS

PAPIA CHOWDHURY, *DEPARTMENT OF PHYSICS AND MATERIAL SCIENCE, JAYPEE INSTITUTE OF INFORMATION TECHNOLOGY, NOIDA, UTTAR PRADESH, INDIA.*

Over the last few years, confined nanometric systems such as quantum wells, quantum wires and quantum dots (QDs) etc. have become most fascinating and promising research fields in view of their tremendous applications environmental safety [1]. When organic/inorganic material reduces to nano size, their electronic and optical properties drastically change from their bulk form. QDs are such nanocrystals. There are two main categories of QD: core type QDs and core-shell type QDs. Core-shell type QDs show less surface defects, enhanced luminescence efficiencies, photo stability and less toxicity compared to that of core type QDs. CdSeS/ZnS is one of such Core/shell type QD which shows all of the above mentioned advantages over its core structure CdSeS. Optical properties mainly high fluorescence with large quantum yield (QY) (up to 85Development of industries generates numerous heavy metal wastes that can cause direct or indirect harm to the environment and humans. Many hazardous heavy metals, such as copper (Cu), chromium (Cr), lead (Pb), zinc (Zn), nickel (Ni), iron (Fe), cadmium (Cd), mercury (Hg), tungsten (W) and silver (Ag) are toxic to living organism [2]. High percentage of Cr and Pb ions within a living organism may lead to various diseases, such as hypersensitivity, lung cancer, nasal cancer, and many other types of cancer. Therefore, the detections of hazardous metal ions like Cr, Zn and Pb are our prime focus. In the present work, we have synthesized functionalized CdSeS/ZnS core shell QDs using L-glutathione (L-GSH) in view of their application to detect hazardous metal ions and some cancer affected diseased cells. The surface modification of QDs with L-GSH make them available for interaction with the targeted materials, which can be used for the detection of hazardous ions and diseased cells present in water. Prepared functionalized CdSeS/ZnS QDs were characterized and tested with the help of several molecular spectroscopic techniques (UV-Vis spectroscopy and fluorescence spectroscopy) by their fluorescence signals. The present work opens a door to the study of new water soluble and biocompatible QDs by the use of their fluorescence sensing for the detection of hazardous metal ions and living cancer cells. References: [1]M. Ishikawa, V. Biju, Prog. Mol. Transl. Sci., 104 (2011) 53. [2] N. Singla, A. Tripathi, M. Rana, S. K Goswami, A. Pathak, P. Chowdhury, J of Luminescence, 165 (2015) 46-55.

RH06
3:10 – 3:25

ANALYSIS OF PEAR ESTER FLAVORING SAMPLES USING BROADBAND ROTATIONAL SPECTROSCOPY

CHANNING WEST, RACHEL BOCWINSKI, AISLING FOLEY, SASHA HOYT, SARAH JOHNSON, ALEXANDER KHLOPENKOV, JULIA MARKS, RACHEL SCHELLING, XUAYNE ZHU, JINBUM DUPONT, LIAM FINEMAN, *Department of Chemistry, The University of Virginia, Charlottesville, VA, USA*; JUSTIN L. NEILL, *BrightSpec Labs, BrightSpec, Inc., Charlottesville, VA, USA*; BROOKS PATE, *Department of Chemistry, The University of Virginia, Charlottesville, VA, USA.*

Pear ester (ethyl decadienoate, $C_{12}H_{20}O_2$) is a molecule used in perfumes and as a food flavoring. It can be obtained from both natural and synthetic sources. The motivation for this study was the interest of a local distillery (Vitae Spirits) in understanding differences in the composition of pear ester samples that might cause off flavors in their spirits. Different samples were analyzed by broadband rotational spectroscopy in the attempt to identify possible impurities. The measurement uses a head space sampling approach where the liquid sample is held in a reservoir and heated. The vapor pressure above the sample is entrained in inert neon gas, and this gas mixture is injected into the spectrometer using a pulsed valve. One challenge in the analysis of chemical mixtures using broadband rotational spectroscopy is that the measured spectrum contains overlapping rotational spectra of each mixture component, making it difficult to isolate the spectral pattern of a single chemical species. To aid the analysis, a version of temperature programmed spectroscopy was performed, where the head space spectrum was acquired for a series of sample reservoir temperatures. This measurement method produces characteristic intensity vs. temperature profiles for transitions from a single species. This makes it possible to separate the overall measurement into spectra arising from different species and is an analysis process that can be fully automated. The analysis of different pear ester samples will be summarized including the ability to identify impurity species, like n-hexanal. The identification of pear ester posed challenges in its own right due to the conformational flexibility of the molecule. The approach to obtaining accurate quantum chemistry estimates of the rotational spectrum parameters for pear ester, so the molecule can be identified in the broadband rotational spectrum, will also be described.

Intermission

RH07   4:01–4:16
MOLECULAR COMPOSITION OF GALLBLADDER STONE USING PHOTOACOUSTIC SPECTROSCOPY[a]

ZAINAB GAZALI, *Department of Physics, Allahabad University, Allahabad, India*; SURYA NARAYAN THAKUR, *Department of Physics, Banaras Hindu University, Varanasi, Uttar Pradesh, India*; AWADHESH KUMAR RAI[b], *Department of Physics, Allahabad University, Allahabad, India.*

Gallstone formation in gallbladder is common in India. Molecular composition of the gallstone in not well known till date. But the knowledge of molecular composition of different kinds of gallstones can provide a significant clue for its formation and treatment. Thus in the present paper molecular composition of gallbladder stone has been investigated by means of photoacoustic spectroscopy (PAS) as it does not require any sample preparation. The PA spectra of gallstone samples are recorded using PA spectrometer developed in our laboratory. The presence of cholesterol, calcium carbonate and bilirubine, in the photoacoustic spectrum, have been recognized and compared with its UV-Visible absorption spectrum. The results of this investigation show that PAS is more suitable to detect the presence of different molecular composition in gallstones as compared to conventional absorption spectroscopy.

---
[a]The author thankful to UGC for providing financial assistance .
[b]awadheshkrai@rediffmail.com

RH08   4:18–4:33
IN SITU CHEMICAL CHARACTERIZATION OF THE MOTILE TO SESSILE TRANSITION OF *PSEDOMONAS AERUGINOSA* COMMUNITIES

TIANYUAN CAO, *Department of Chemistry and Biochemistry, University of Notre Dame, Notre Dame, IN, USA*; NYDIA MORALES-SOTO, KRISTEN M. KRAMER, *Department of Civil and Environmental Engineering and Earth Sciences, University of Notre Dame, Notre Dame, USA*; NAMEERA F. BAIG, *Department of Chemistry and Biochemistry, University of Notre Dame, Notre Dame, IN, USA*; JOSHUA D. SHROUT, *Department of Civil and Environmental Engineering and Earth Sciences, University of Notre Dame, Notre Dame, USA*; PAUL W. BOHN, *Department of Chemistry and Biochemistry, University of Notre Dame, Notre Dame, IN, USA.*

Pseudomonas aeruginosa is a Gram-negative opportunistic pathogen which infects more than 50,000 people each year in the United States alone. Its abilities to move, colonize surfaces, and develop biofilms give rise to the high resistance to antimicrobial treatment. One type of motility employed by *P. aeruginosa* is swarming motility, where bacterial cells undergo physical and metabolic alterations. Swarming has been studied by many researchers but the knowledge on its chemical composition that relates to the transition between the motile and sessile biofilm stages are still lack. Here we apply confocal Raman microscopy (CRM) to examine *P. aeruginosa* wild-type PA14 (a virulent strain isolated from a burn wound) under swarming and biofilm conditions.

The comparison between the swarming and biofilm samples indicates different molecules linked to the motile to sessile transition, revealing their community-specific chemical features. While the *Pseudomonas* quinolone signal (PQS) is found in swarm colonies and biofilms, the *N*-oxide quinolines (4-hydroxy-2-heptylquinoline-*N*-oxide,2-nonyl-4-hydroxyquinoline, etc.) are present in higher abundance and are synthesized and secreted much earlier in swarm colonies. Moreover, a closer investigation spanning from the center to the edge of a swarm colony shows high abundance of PQS at the center while *N*-oxides dominate the edge of the colony. The results provide insights into the chemical profile change occurring during the motile to sessile transition in *P. aeruginosa*, and demonstrate the broad application of CRM in biomolecular imaging.

RH09   4:35 – 4:50

UTILISING DIFFUSE REFLECTANCE INFRA-RED SPECTROSCOPY TO MONITOR THE OXIDATION OF BITUMEN AND ASPHALT AS A RESULT OF ARTIFICIAL AND NATURAL AGEING

<u>HANNAH BOWDEN</u>, MATTHEW ALMOND, WAYNE HAYES, *Department of Chemistry, University of Reading, Reading, United Kingdom*; STUART McROBBIE, *Infrastructure, Transport Research Laboratory, Crowthorne, United Kingdom.*

At present the road surface condition in the UK is monitored visually for any defects. This system works well to identify any major issues; however there is a very short window of time between detecting the defects and the complete failure of the surface. The road then requires resurfacing. It is therefore of interest to be able to predict the failure of the road surface in order to ensure the success of rejuvenation techniques. Asphalt road surfaces are constructed with three main components. Bitumen, a semi-solid, hydrocarbon based tar-like substance; fine filler, commonly calcium carbonate which adds bulk to the bitumen and stone based aggregates. There are many different mechanisms for the degradation of the road surfaces that involve chemical and physical factors. The chemical oxidation of bitumen is a contributing factor to the ageing of the asphalt road surfaces. The increase in oxygen levels within the composition and the loss of the lower molecular weight volatile components increases the polarity of the bitumen and leads to an increase in stiffness. As the bitumen becomes more brittle it loses its cohesion and adhesion with the aggregates and the surface begins to deteriorate. Bitumen oxidation can be monitored with the use of FTIR spectroscopy. The evolution of oxidation product functional group absorbance bands, including carbonyl, carboxylic and sulphoxide bonds can be monitored. This phenomenon has been well documented for raw bitumen but is less well understood for real road surfaces. This work investigates the use of diffuse reflectance IR spectroscopy, a non-contact measurement, to monitor the oxidation of bitumen and asphalt and relate this to pavement degradation. A number of different bitumen and asphalt samples have been aged naturally and artificially. Reflectance spectra have been collected alongside standardised mechanical testing of the physical properties of the bitumen in order to determine a link between the chemical and physical degradation. Preliminary results from this work identify the presence of oxidation product absorbance bands in the reflectance spectra as a result of ageing alongside a decrease in mechanical cohesion and an increase in stiffness and viscosity at lower temperatures.

RI. Mini-symposium: Far-Infrared Spectroscopy

Thursday, June 21, 2018 – 1:45 PM

Room: 1024 Chemistry Annex

Chair: Sandra Brünken, Radboud University, Nijmegen, The Netherlands

RI01   *INVITED TALK*   1:45 – 2:15

POLAR RADIANT ENERGY IN THE FAR-INFRARED EXPERIMENT (PREFIRE)

<u>BRIAN DROUIN</u>, *Jet Propulsion Laboratory, California Institute of Technology, Pasadena, CA, USA*; TRISTAN S L'ECUYER, *Department of Atmospheric and Oceanic Sciences, University of Wisconsin - Madison, Madison, Wisconsin, USA.*

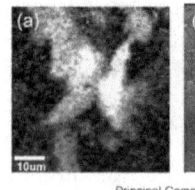
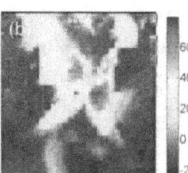
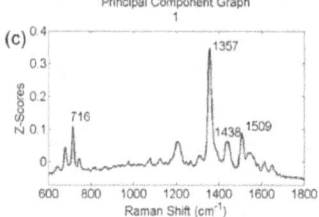

Figure 1. Analysis of 24hr swarming assays of PA14. Raman image intergrated over 1330-1380 cm^{-1} for representative regions of signal molecules (a); principal component heat map (b); Raman spectral loading plot (c).

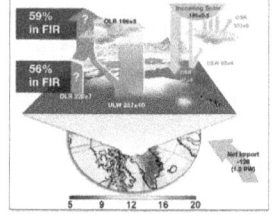

Much of the Far-infrared radiation (FIR) emitted by the earth surface is trapped primarily by the insulating greenhouse effect. At the poles, the greenhouse effect is minimized by the nominal cold and dry atmospheric state. This is how a significant amount of absorbed solar energy is vented back to space, acting like a thermostat. Under these conditions, the effects of surface emissivity become disproportionately large and have a significant impact on the radiative balance. Earth system models have consistently under-estimated the rapid warming occurring in the Arctic, perhaps due to poor assumptions about the nature of far-infrared spectral emissions.

The Polar Radiant Energy in the Far-InfraRed Experiment (PREFIRE) is a NASA Earth Ventures mission, currently in formulation, that would test the hypothesis that time-varying errors in FIR surface emissivity and atmospheric greenhouse effects bias the modeled energy balance that under-estimates Arctic warming. This presentation covers the processes involved in the energy balance, and how spectrally resolved measurements provide the means to extract critical information. We also discuss the instrument difficulties associated with remote measurements across the far-infrared, emphasizing the differing challenges associated with earth science vs. astrophysics. Finally, we provide an overview of the planned PREFIRE mission and how it would address these challenges.

RI02 2:19 – 2:34

PROGRESS ON THE FT-IR MEASUREMENTS OF WATER CONTINUUM IN THE FAR-INFRARED REGION AT 252 – 296 K

KEEYOON SUNG, BRIAN DROUIN, TIMOTHY J. CRAWFORD, *Jet Propulsion Laboratory, California Institute of Technology, Pasadena, CA, USA*; EDWARD H WISHNOW, *Space Sciences Laboratory and Department of Physics, University of California, Berkeley, CA, USA.*

Water is the strongest greenhouse gas in the Earth atmosphere, which plays a critical role in the energy balance of the earth atmosphere. It has long been observed particularly in the far-infrared that there is significant longwave continuum absorption due to water vapor (dimers or multimers), not attributable to the Lorentz line contribution within 25 cm^{-1} from the line center for individual water vapor lines. The MT_CKD model offers the water vapor continuum predictions, which are to be validated by a laboratory study in the far infrared. In order to directly measure this water vapor continuum absorption, we have obtained a series of spectra of water vapor broadened by Self, N_2, and O_2 in the 50 – 500 cm^{-1} (200 – 20 μm) at temperatures between 251 and 296 K. For this, we used a coolable White cell system (whose optics are optimized for the far-infrared spectrometry) with passive temperature control, configured to the Fourier transform spectrometer, Bruker IFS-125HR at the Jet Propulsion Laboratory (JPL). We have been analyzing the spectra to make direct measurement of the far-infrared water continuum in two steps; (1) we obtained their transmission spectra by ratioing the sample spectrum to their corresponding background spectrum, (2) we obtained the continuum part of the transmission by dividing the measured spectrum by a synthetic spectrum of the resonant lines calculated using the HITRAN database. As shown in Figure 1, it has revealed the underlying water-water, water-O_2, and water-N_2 continua in the temperature range, depending on the spectrum type. The preliminary results from this on-going analysis are presented along with their comparison with the MT_CKD (ver.3.5) model predictions. Temperature dependence of the water vapor continuum will be discussed as part of future work.[a]

---

[a]Government sponsorship acknowledged

RI03 2:36 – 2:51

MILLIMETER-WAVE CHIRALITY SPECTROMETER (CHIRALSPEC)

SHANSHAN YU, THEODORE J RECK, JOHN PEARSON, MICHAEL MALASKA, ROBERT HODYSS, *Jet Propulsion Laboratory, California Institute of Technology, Pasadena, CA, USA*; BROOKS PATE, *Department of Chemistry, The University of Virginia, Charlottesville, VA, USA.*

Life detection through chemical analysis requires nearly unambiguous detection of specific chemical biosignatures. The US Decadal Survey paper, Vision and Voyages for Planetary Science in the Decade 2013-2022, recommends "a detailed characterization of organics to search for signatures of biological origin, such as molecules with a preferred chirality or unusual patterns of molecular weights" as a key future investigation of life detection (page 240). While mass spectrometry has often been proposed for measuring the abundance patterns of molecular weights, it lacks the chirality detection capability required for chiral analyses of chiral molecules such as amino acids, and cannot uniquely identify specific structure-based isomers such as fatty acids.

In this presentation we will report the status of the development of a millimeter-wave chirality spectrometer (ChiralSpec). ChiralSpec advances key technologies to enable chirality detection and discrimination of structural isomers with a simple instrument. It is applicable to mission focus areas such as Enceladus, Europa, Titan, and Mars. It could be used on planetary in-situ probes to measure amino acids and other organic molecules in the gas phase or brought into the gas phase.

ChiralSpec employs an innovative microwave three-wave mixing technology for chirality detection and the cavity resonance technology for sensitivity enhancement. It can be operated under two modes: (1) survey mode, with the instrument acting as a traditional microwave spectrometer to characterize chemical composition and quantify abundance of planetary samples; and (2) chirality detection mode, with the instrument determining which enantiomer is in excess and how much it is in excess for each existing chiral molecule.

RI04 2:53 – 3:08

MICROWAVE SPECTRUM OF 1-ADAMANTANOL $C_{10}H_{15}$–OH

OLIVIER PIRALI, MARIE-ALINE MARTIN-DRUMEL, L. H. COUDERT, *Institut des Sciences Moléculaires d'Orsay, Université Paris-Sud, Orsay, France*; MANUEL GOUBET, *Laboratoire PhLAM, Université de Lille 1, Villeneuve de Ascq, France*; SÉBASTIEN GRUET, MELANIE SCHNELL, *CoCoMol, Max-Planck-Institut für Struktur und Dynamik der Materie, Hamburg, Germany.*

1-Adamantanol is a heavy non-rigid molecule consisting of 1-adamantyl and hydroxyl groups. Internal rotation about the 1-adamantyl 3-fold axis of symmetry was evidenced some time ago[a] leading to an estimated value of the A-E splitting of 10 cm^{-1}. The microwave spectrum of 1-adamantanol was recorded later[b] in the 8 to 40 GHz region. Even though individual rotational lines could not be assigned, a value of 410 cm^{-1} was obtained for V_3 the height of the barrier hindering the internal rotation.

A cold molecular beam and a room temperature submillimeter wave spectra of 1-adamantanol were recorded in the 2–12 and 140–220 GHz ranges, respectively. 1404 parallel a-type transitions have been assigned in both spectra. A line frequency analysis of this new data set and of the perpendicular b-type clusters previously observed[b] was carried out using an IAM approach.[c]

In the paper, the new data and the results of the analysis will be presented. As 1-adamantanol is a nearly symmetric top molecule with an asymmetry parameter[b] κ close to -0.99, asymmetry splittings could not be resolved in the new spectra and $B - C$ was set to zero. Owing to the fact that the moment of inertia of 1-adamantyl about the axis of internal rotation is 400 times larger than that of the OH group about the same axis, ρ the parameter describing the rotational dependence of the torsional splitting is 0.9975. The implication for the energy level diagram of a value so close to 1 for this parameter will be discussed. Work is still in progress and it is hoped that it will be possible to identify torsional subbands in the crowded submillimeter wave spectrum recorded at room temperature.

---
[a]Craven, *Spectrochim. Acta*, 29A (1973) 679
[b]Corbelli, Degli Esposti, Favero, and Lister, *J. Chem. Soc. Trans. 2*, **83** (1987) 2225
[c]Hougen, *J. Mol. Spectrosc.*, **114** (1985) 395; and Coudert and Hougen, *ibid*, **130** (1988) 86

Intermission

RI05 *INVITED TALK* 3:44 – 4:14

FAR-INFRARED SPECTROSCOPY OF SHORT-LIVED SPECIES

HIROYUKI OZEKI, *Department of Environmental Science, Toho University, Funabashi, Japan.*

Detection and characterization of short-lived species, or radicals, have been one of the main targets of high-resolution molecular spectroscopy. These kinetically unstable substances can be produced only in a very small amount under ordinary laboratory measurement conditions, it is essential to increase the sensitivity of the spectrometer and/or to improve the production efficiency of the molecules to be studied. Conventional microwave spectroscopy has taken an approach to raise the operating frequency to far-infrared region, expecting that effective absorption coefficient will increase. Sensitivity of the spectrometer in far-infrared region, or THz frequency region, has greatly improved thanks to succesful development of frequency multiplication techinques. Along with searching for an efficient production method of the short-lived species, many kinds of short-lived species can be possible to observe. Based on this situation, I would like to show several examples of far-infrared spectrosocpy of reactive species such as CH_2, NH_2, and CHF_2.

RI06   4:18 – 4:33
THZ SPECTROSCOPY OF SULFUR DERIVATIVES OF ASTROPHYSICAL INTEREST

<u>L. MARGULÈS</u>, S. BAILLEUX, R. A. MOTIYENKO, *Laboratoire PhLAM UMR 8523, Université de Lille, 59655 Villeneuve d'Ascq, FRANCE*; J.-C. GUILLEMIN, *ISCR – UMR6226, Université de Rennes, 35000 Rennes, FRANCE*; JOSE CERNICHARO, *Molecular Astrophysics, ICMM, Madrid, Spain*; ARNAUD BELLOCHE, *Millimeter- und Submillimeter-Astronomie, Max-Planck-Institut für Radioastronomie, Bonn, NRW, Germany*; BRETT A. McGUIRE, ANTHONY REMIJAN, *NAASC, National Radio Astronomy Observatory, Charlottesville, VA, USA*; OLGA DOROVSKAYA, V. ILYUSHIN, *Radiospectrometry Department, Institute of Radio Astronomy of NASU, Kharkov, Ukraine.*

About 200 molecules have thus far been detected in the interstellar medium. Twenty-two are sulfur-bearing chemical compounds (and analogues of oxygenated species), making sulfur the tenth most abundant element in the galaxy. We report here the sub-THz spectroscopic observations of two reactive species: thioacetaldehyde (CH_3CHS) and NS^+.[a] The latter new cation has been firmly detected for the first time towards many interstellar sources (cold molecular clouds, pre-stellar cores and shocks) using the IRAM-30m radiotelescope. Although a recent study of the chemistry of sulfur in cold dense clouds has been carried out [b] the formation pathways of the sulfur species are still misunderstood.

The rotational spectrum of CH_3CHS was previously recorded up to 40 GHz.[c] New measurements performed up to 660 GHz represent a significant extension in terms of frequency range and analysis. The final spectroscopic analysis, including the internal rotation treatment, and searches for it towards SgrB2 and other sources will be presented.

Acknowledgements: These results were supported by the Programme National PCMI of CNRS/INSU, the French National Research Agency (ANR-13-BS05-0008 "IMOLABS"), the CaPPA project (ANR-11-LABX-0005-01), the spanish MINECO (grants AYA2012-32032, AYA2016-75066-C2-1-P, CSD2009-00038 and RyC-2014-16277) and the European Research Council (grant ERC- 2013-SyG 610256, NANOCOSMOS). Kharkov group acknowledge support of the Volkswagen foundation and assistance of the Science and Technology Center in Ukraine (STCU partner project P686).

---

[a] Cernicharo J.; *et al.*, 2018, *ApJL* **L22**, 852
[b] Fuente A.; *et al. A&A* **593**, (2016) A94
[c] H. Kroto; *et al.*, 1976, *J. Mol. Spectrosc.* **62**, 346

RI07   4:35 – 4:50
WATER VAPOUR AND AMMONIA IN CIRCUMSTELLAR ENVELOPES OF C-RICH EVOLVED STARS

<u>MIROSLAW R. SCHMIDT</u>, *Department of Astrophysics I, NICOLAUS COPERNICUS ASTRONOMICAL CENTER, TORUN, Poland.*

HIFI survey for water vapor and ammonia in a sample of the carbon-rich Asymptotic Giant Branch (AGB) stars has shown that their presence in circumstellar envelopes is almost universal. Models for thermochemical equilibrium in the photospheres of the carbon-rich stars predict abundances of water and ammonia many orders below the observed values. Modeling of emission lines suggests, that both molecules should be formed very close to the photospheres of the central stars. The mechanism of formation of these molecules is uncertain, and the proposed hypotheses include shock chemistry, photochemistry driven by the UV radiation leaking into the inner part of clumpy envelope and formation on dust grains. We present the results of a detailed modeling of the lowest rotational transitions observed with the Herschel/HIFI instrument in the sample of C-rich AGB stars. The aim of this analysis is to constrain their abundances and, where possible, their formation radius. Excitation of both molecules is governed by the radiative pumping, mainly in their ν_2 vibrational modes. Models of molecules include a large number of levels in the ground and the first excited vibrational states. Ortho and para species of both molecules are treated separately. The transitions intensities are adopted from the BYTe (Yurchenko et al. 2011) and HITRAN (Rothman et al. 2009) databases for ammonia and water, respectively.

EXTENDED MEASUREMENTS AND AN EXPERIMENTAL ACCURACY EFFECTIVE HAMILTONIAN MODEL FOR THE $3\nu_2$ AND $\nu_2 + \nu_4$ STATES OF AMMONIA

JENIVEVE PEARSON, SHANSHAN YU, JOHN PEARSON, KEEYOON SUNG, BRIAN DROUIN, *Jet Propulsion Laboratory, California Institute of Technology, Pasadena, CA, USA*; OLIVIER PIRALI, *AILES beamline, Synchrotron SOLEIL, Saint Aubin, France.*

The infrared spectrum of ammonia has proven to be highly problematic for effective Hamiltonian analysis. As a result, previous studies failed to model the $3\nu_2$ and $\nu_2 + \nu_4$ bands of the spectrum close to experimental accuracy. To remedy this a global fit of the $3\nu_2$ and $\nu_2 + \nu_4$ bands has been undertaken using SPFIT. The analysis includes about 1000 newly assigned vibrational transitions in $3\nu_2$ to $2\nu_2$ as well as inversion transitions in $3\nu_2$ to $3\nu_2$. The spectra were a long path infrared absorption spectrum recorded with the Synchrotron light source at SOLEIL, with a path length of 180 m and a resolution of 0.0011 cm^{-1} at room temperature and 1 Torr of pressure, and a mid-infrared discharge spectrum recorded similarly at SOLEIL, with a path length of 0.7 m and resolution .004 cm^{-1} at 10 Torr and 900 K. Our fit has achieved experimental accuracy through the use of a number of terms that had not previously been in the Hamiltonian proving that ammonia is tractable to effective Hamiltonians despite previous beliefs.

RJ. Electronic structure, potential energy surfaces
Thursday, June 21, 2018 – 1:45 PM
Room: 217 Noyes Laboratory

Chair: Steve Alexandre Ndengue, Missouri University of Science & Technology, Rolla, MO, USA

RJ01   1:45 – 2:00

ROTATION-TUNNELING ANALYSIS OF PROTON-TRANSFER DYNAMICS IN ELECTRONICALLY EXCITED 6-HYDROXY-2-FORMYLFULVENE USING DEGENERATE FOUR-WAVE MIXING

<u>ZACHARY VEALEY</u>, LIDOR FOGUEL, PATRICK VACCARO, *Department of Chemistry, Yale University, New Haven, CT, USA.*

The multidimensional nature of classically hindered proton transfer is demonstrated clearly by the pronounced effects engendered by attendant electronic and vibrational degrees of freedom, the selective excitation of which can enhance or diminsh the rate of hydron migration greatly. To explore the provenance of such phenomena, the origin band (0_0^0) of the $\tilde{A}^1B_2$–$\tilde{X}^1A_1$ absorption system in 6-hydroxy-2-formylfulvene (HFF) has been probed under ambient bulk-gas conditions by using polarization-resolved degenerate four-wave mixing (DFWM) spectroscopy. The alleviation of rovibronic congestion and suppression of rotational-branch structure afforded by judicious selection of incident/detected polarizations for the DFWM interaction has enabled refined rotation-tunneling information to be extracted for the lowest-lying singlet excited manifold, $\tilde{A}^1B_2$ ($\pi^*\pi$). In contrast to the ultrafast dynamics ($\tau_{pt}\leq 120$fs) that characterize the $\tilde{X}^1A_1$ ground electronic state of HFF, the $\pi^*\leftarrow\pi$ electron promotion is found to impede intramolecular proton transfer markedly, leading to the near complete quenching of tunneling-induced spectral signatures. The intrinsic dependence of reaction coordinate and proton-transfer efficacy on the nuances of potential-surface topology and transition-state geometry will be discussed in light of these experimental results, with complementary quantum-chemical calculations serving to elucidate the dramatic impact that subtle changes in energy landscape can exert upon unimolecular dynamics.

RJ02   2:02 – 2:17

NEW ELECTRONIC STATES OF YO IN THE UV REGION

<u>ALLAN S.C. CHEUNG</u>, NA WANG, YUK WAI NG, *Department of Chemistry, The University of Hong Kong, Hong Kong, Hong Kong*; ANDREW CLARK, *LCPM-ISIC, EPFL, Lausanne, Switzerland*; WENLI ZOU, *Institute of Modern Physics, Northwest University, Xi'an, China.*

Laser excitation spectra of the yttrium monoxide (YO) molecule in the ultra violet region between the 280 and 320 nm have been recorded and studied using optical-optical double resonance (OODR) spectroscopy. The YO molecule was prepared by the reaction of laser ablated yttrium atom with oxygen under supersonic jet cooled conditions. Thirteen vibration bands have been observed via the intermediate $B^2\Sigma^+$ state from the $X^2\Sigma^+$ state. The excited states analyzed so far are generally in good case (c) coupling scheme. Besides the observation of excited $\Omega = 0.5$ and 1.5 sub-states, and $^2\Sigma^+$ state, we have also identified and studied a forbidden transition, the $[33.7]^4\Sigma^-$ - $B^2\Sigma^+$ transition. Molecular constants for the newly observed electronic states were determined by least squares fitting the measured rotational lines.

A number of low-lying Λ-S states and Ω sub-states of the YO molecule have been calculated using SA-CASSCF (state-averaged complete active space self-consistent field) followed by MS-CASPT2 (multi-state complete active space second-order perturbation theory). Since the active Y 5p shell is very important to get some low-lying electronic states with the correct principal configurations, the active space consists of 7 electrons in 12 orbitals corresponding to the Y 4d5s5p and O 2p shells. The molecular orbitals from Y 4s4p and O 2s are inactive but are also correlated, whereas the lower core-shells are relaxed only by SA-CASSCF and then kept frozen at the CASPT2 level. Spin-orbit coupling (SOC) is treated via the state-interaction (SI) approach with the one-center atomic mean field integral (AMFI) approximation for one- and two-electron spin-orbit integrals. In the SOC calculations of potential energy curves (PECs), the SA-CASSCF wavefunctions are adopted where the diagonal elements in the SOC matrix are replaced by the corresponding MS-CASPT2 energies calculated above. A comparison of the spectroscopic properties of electronic states determined experimentally and from calculations will be presented.

RJ03 2:19 – 2:34

STARK AND ZEEMAN EFFECT IN THE [18.5]$^2\Delta_{3/2}$ - X$^2\Delta_{3/2}$ TRANSITION OF THORIUM MONOFLUORIDE[a]

<u>DUC-TRUNG NGUYEN</u>, TIMOTHY STEIMLE, *School of Molecular Sciences, Arizona State University, Tempe, AZ, USA.*

Studies of the bonding and electronic structure of simple actinide compounds[b,c,d,e] are attractive because they provide insight into the chemistry of more complex molecules associated with radioactive waste. These molecules are the most effective venues for developing a synergism between theory and experiment. The primary goal of the present study is to understand and identify different levels of covalency in a series of gas phase actinides and lanthanides containing molecules via the determination of the permanent electric dipole moment, μ, and magnetic g-factors. The electronic spectrum of ThF has been investigated using: a) medium resolution two dimensional (2D)[f]; ultrahigh field free, Stark, and Zeeman spectroscopy of a supersonically cooled molecular beam sample. A strong band system near 540 nm was detected and the Stark shifts and splitting were analyzed to produce μ values of 1.426(18)D and 0.586(30)D, for the X$^2\Delta_{3/2}$ and [18.5] Ω=3/2 states, respectively. Zeeman splittings were analyzed to show that both the ground and excited [18.5] Ω=3/2 states are predominately $^2\Delta_{3/2}$ spin-orbit components. A molecular orbital correlation diagram will be used to rationalize the observed very small μ values, the electronic state distribution, and garner insight into the bonding mechanism.

[a] Supported by the United States Department of Energy (DOE) under the Grant. No. DE-SC0018241
[b] M. C. Heaven, B. J. Barker and I. O. Antonov, J. Phys. Chem. A, 2014, 118, 10867-10881.
[c] T. Steimle, D. L. Kokkin, S. Muscarella and T. Ma, J. Phys. Chem. A, 2015, 119, 9281-9285.
[d] C. Linton, A. G. Adam and T. C. Steimle, J. Chem. Phys., 2014, 140, 214305/214301-214305/214307.
[e] D. L. Kokkin, T. C. Steimle and D. DeMille, Phys. Rev. A: At., Mol., Opt. Phys., 2014, 90, 062503/062501-062503/062510.
[f] N. J. Reilly, T. W. Schmidt and S. H. Kable, J. Phys. Chem. A, 2006, 110, 12355-12359.

RJ04 2:36 – 2:51

LASER INDUCED FLUORESCENCE (LIF) SPECTROSCOPY OF JET COOLED ThO

<u>JOEL R SCHMITZ</u>, MICHAEL HEAVEN, *Department of Chemistry, Emory University, Atlanta, GA, USA.*

Knowledge of actinide bonding is essential for nuclear energy and reactor applications, including nuclear waste treatment. Due to its relatively simple electronic structure, thorium oxide (ThO) is an ideal molecule to study actinide bonding. Previous studies have reported visible and near UV band systems that were recorded under high temperature conditions. Spectral congestion significantly complicated the analyses of these data. In the present work we have examined ThO bands under conditions where jet cooling was used to reduce the rotational temperature to approximately 90 K. LIF spectra were recorded over the range 18,000-19,800 cm^{-1}. Several new vibronic bands have been characterized. Some extend the data range for known electronic transitions, while others belong to electronic states that have not been reported previously. Analysis of these data and models for the electronically excited states of ThO will be presented.

RJ05   2:53 – 3:08
INELASTIC COLLISIONS OF Ar AND O_3

<u>SANGEETA SUR</u>, ERNESTO QUINTAS SÁNCHEZ, STEVE ALEXANDRE NDENGUE, RICHARD DAWES, *Department of Chemistry, Missouri University of Science and Technology, Rolla, MO, USA.*

According to the Chapman cycle, during formation, metastable ozone can be stabilized by a third body (M) collision (buffer gas) in the atmosphere. The stabilization occurs through an energy transfer (ET) mechanism from O_3^* to M. The details of this ET are not well known and one of the reasons is the lack of an accurate potential energy surface (PES) including the collision partner. The PES of the O_3-Ar complex is a 6D problem in full-dimensionality, or 3D for rigid O_3. Here we present a global 3D PES for O_3 fixed at equilibrium, interacting with Ar. Highly accurate Davidson-corrected multi-reference configuration interaction (MRCI-f12) energies were computed at 2112 data points. The AUTOSURF code was used to construct the PES automatically, represented by a local interpolating moving least-squares (L-IMLS) method. A global RMS fitting error of 0.6 cm^{-1} was obtained. Symmetry equivalent minima with a well depth of -229 cm^{-1} are located above and below the plane of O_3. We present here bound vdW states of the O_3-Ar complex obtained by variational rovibrational calculations, as well as quantum scattering cross-sections for rotationally inelastic collisions.

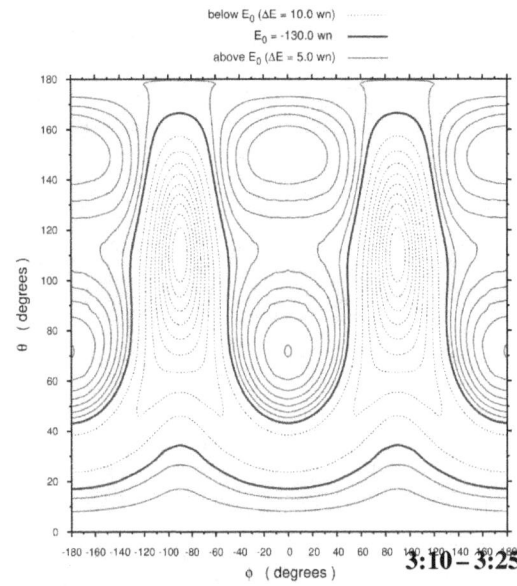

RJ06   3:10 – 3:25
METAL-AMMONIA COMPLEXES DISCLOSE A SECRET PERIODIC TABLE OF SOLVATED ELECTRON PRECURSORS

<u>EVANGELOS MILIORDOS</u>, *Chemistry and Biochemistry, Auburn University, Auburn, AL, 36849.*

Metal-ammonia complexes, $[M(NH_3)_4]^{0,\pm}$, are shown theoretically to have a $M(NH_3)_4^{x+}$ positively charged core with one, two, or three outer electrons orbiting in its periphery. Our results reveal a new class of molecular entities (solvated electron precursors) which host outer electrons resembling atoms. The observed electronic shell model (1s, 1p, 1d, 2s, 1f, 2p, 2d) differs from that of the hydrogen-like model and resembles the jellium or nuclear shell model. This fact is attributed to the different effective electrostatic potential experienced by the outer electrons. Multi-reference and propagator approaches combined with diffuse basis sets are employed to calculate accurate geometries, ionization energies, electron affinities and vertical excitation energies. Our results are expected to trigger the interest of the experimental spectroscopy community.

RJ07   3:27 – 3:42
EXTRA HIGH ACCURACY FITTING OF THE PES FOR SUB-PERCENT CALCULATION OF INTENSITIES

<u>OLEG L. POLYANSKY</u>, JONATHAN TENNYSON, *Department of Physics and Astronomy, University College London, Gower Street, London WC1E 6BT, United Kingdom*; VLADIMIR YU. MAKHNEV, ALEKSANDRA A. KYUBERIS, NIKOLAY F. ZOBOV, *Microwave Spectroscopy, Institute of Applied Physics, Nizhny Novgorod, Russia.*

Calculation of rotation-vibration line intensities with sub-percent accuracy has recently become a standard requirement for the applications in retrieval and monitoring of gases in the Earth's atmosphere and potentially in the atmospheres of exoplanets. A major factor in the accurate calculation of intensities is the requirement for a high accuracy *ab initio* Dipole moment surface (DMS) (e.g. references [a] and [b]). We demonstrate here that the change from the "good" potential energy surface (PES) to "excellent" PES, used for the intensity calculations is also important. By "good" we mean here, for example, the PES a standard deviation of 0.025 cm^{-1} and by "excellent" - the PES with the standard deviation 0.011 cm^{-1}. Details of studies on H_2O[c], O_3 [d], HCN and CO_2 molecules will be presented in the talk.

---

[a] L.Lodi, J. Tennyson and O.L. Polyansky, *Journal of Chemical Physics*, **135**, 034113, (2011)

[b] O.L. Polyansky, K. Bielska, M. Ghysels, L. Lodi, N.F.Zobov, J.T.Hodges, J. Tennyson *Physical Review Letters*, **114**, 243001, (2015)

[c] I.I Mizus, A.A. Kyuberis, N.F. Zobov, V.Y. Makhnev, O.L. Polyansky and J. Tennyson *Phil. Trans. R. Soc. A*, **376**, 20170149, (2018)

[d] O.L.Polyansky, N.F. Zobov, I.I Mizus, A.A. Kyuberis. L. Lodi and J. Tennyson *JQSRT*, **210**, 127-135 (2018)

Intermission

RJ08 4:18–4:33

AUTOSURF: A CODE FOR AUTOMATED CONSTRUCTION OF POTENTIAL ENERGY SURFACES

<u>ERNESTO QUINTAS SÁNCHEZ</u>, RICHARD DAWES, *Department of Chemistry, Missouri University of Science and Technology, Rolla, MO, USA.*

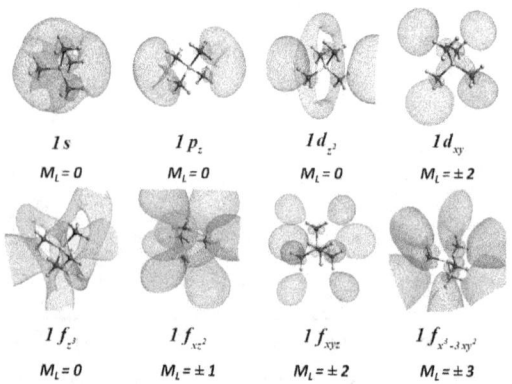

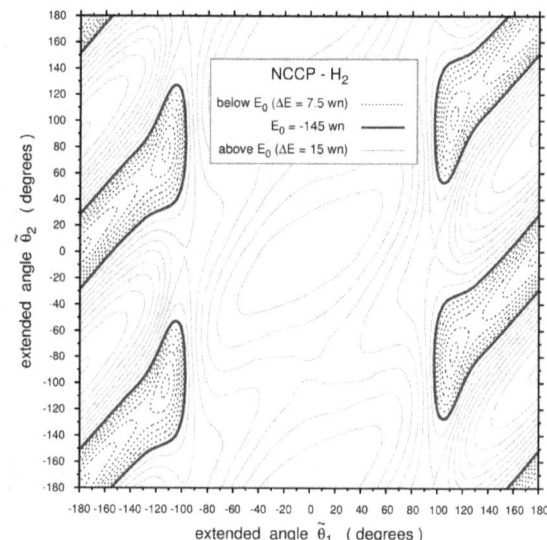

The potential energy surface (PES) of a molecular system constitutes a cornerstone for every theoretical study of spectroscopy and dynamics. We describe here our general code for the automated construction of PESs for van der Waals (vdW) systems composed of two (rigid) fragments. The AUTOSURF suite is designed to completely automate all of the steps and procedures that go into fitting various classes of PESs. The algorithms are based on the local interpolating moving least squares (L-IMLS) methodology, and have many advanced features such as options for data-point placement, and iterative refinement. We have interfaced this fitting approach to popular electronic structure codes such as Molpro and CFOUR to automatically generate ab initio PESs for 3D (atom - general molecule) and 4D (linear molecule - linear molecule) vdWs systems. The niche of these algorithms is to obtain an interpolative representation of high-level ab initio energies with negligible (arbitrarily small) fitting error, enabling a broad community of non-experts in PES fitting methods to bridge electronic structure calculations and spectroscopic and dynamics research.

The code is demonstrated here by presenting PESs and analysis of the corresponding rovibrational bound states for 7 highly anisotropic "heavy-light" systems: C_6H^--H_2, HC_2NC-H_2, HNC_3-H_2, HC_5N-H_2, C_4H^--H_2, $MgCCH$-H_2, $NCCP$-H_2.

RJ09 4:35 – 4:50
APPROXIMATIONS FOR HIGH-ACCURACY THEORETICAL THERMOCHEMISTRY

BRADLEY WELCH, RICHARD DAWES, *Department of Chemistry, Missouri University of Science and Technology, Rolla, MO, USA.*

The Active Thermochemical Tables (ATcT) approach by Ruscic[a] incorporates data for a large number of chemical species from a variety of sources (both experimental and theoretical) and derives a self-consistent network capable of making extremely accurate estimates of quantities such as temperature dependent enthalpies of formation. The network provides rigorous uncertainties, and since the values don't rely on a single measurement or calculation, the provenance of each quantity is also obtained. To expand and improve the network it is desirable to have a reliable protocol such as the HEAT approach[b] for calculating accurate theoretical data.

Anharmonic zero-point-energies are essential for accurate enthalpies of formation even at 0 K. Coupled Cluster based vibrational perturbation theory (VPT2) quickly becomes prohibitively expensive for larger, more chemically relevant molecules. Here we present benchmark work based upon testing the use of B3LYP and MP2 based VPT2 in-place of CCSD(T). The benchmark set includes some species from the original HEAT set[b], as well as some larger organic ($C_a H_b O_c$) species. We also consider scaled harmonic zero-point-energies based upon recent work[c] with the B2PLYP double hybrid functional and comment on its outlook for the same benchmark set as the DFT and MP2 VPT2 approach. The derived methods are implemented in a fully automated computational workflow.

[a] B. Ruscic, Active Thermochemical Tables (ATcT) values based on ver. 1.118 of the Thermochemical Network (2015); available at ATcT.anl.gov
[b] A.Tajti, P. G. Szalay, A. G. Császár, M. Kállay, J. Gauss, E. F. Valeev, B. A. Flowers, J. Vázquez, and J. F. Stanton. JCP 121, (2004):11599.
[c] M.Kesharwani, B.Brauer and J.M.L. Martin. J.Phys.Chem. A 119,(2015):1701

RJ10 4:52 – 5:07
VACUUM UV LABORATORY STUDY OF THE PHOTODISSOCIATION OF CS

ZHONGXING XU, YIH-CHUNG CHANG, KYLE N. CRABTREE, WILLIAM M. JACKSON, CHEUK-YIU NG, *Department of Chemistry, The University of California, Davis, CA, USA.*

Photodissociation of carbon monosulfide (CS) in UV-dominated regions, including diffuse interstellar medium and protoplanetary regions, may produce metastable carbon and sulfur in ^1S and ^1D states, which could contribute internal energy to gas-phase chemical reactions. However, unlike its isoelectronic CO molecule, little is known about Vacuum UV (VUV) photodissociation of CS. In the present study, we investigated the $C^1\Sigma^+ - X^1\Sigma^+$ band of CS. CS is generated by photolysis of CS_2 and then adiabatically expanded into a vacuum chamber. The two-independently-tunable-VUV photodissociation-photoionization spectroscopy coupled with velocity map imaging (VMI) detection was used to measure state-specific photodissociation cross sections and atomic state branching fractions. Our experiment is the first quantitative study of CS in the VUV spectral region.

RJ11

HIGH-SPIN ELECTRONIC STATES OF MOLECULAR OXYGEN

GABRIEL J. VÁZQUEZ, *Instituto de Ciencias Fisicas, Universidad Nacional Autonoma de Mexico (UNAM), Cuernavaca, Morelos, Mexico*; H. P. LIEBERMANN, *Fachbereich C-Mathematik und Naturwissenschaften, Universität Wuppertal, Wuppertal, Germany*; H. LEFEBVRE-BRION, *Institut des Sciences Moléculaires d'Orsay, Université Paris-Sud, Orsay, France.*

As a by-product of an ongoing rather comprehensive study of the electronic structure of the lowest valence and Rydberg states of O_2, chiefly singlets and triples, in this contribution we focus on high-spin electronic states, namely, quintets and septets. Although these latter states may be thought of as of pure academic interest, the current calculations show interesting features of their potential energy curves (PECs) which have not been studied and are actually unknown for most diatomics made up of first-row atoms (plus hydrogen and helium). Experimentally, there is essentially no information whatsoever, aside from some indirect evidence of the possible involvement of high-spin species in the spectroscopy or photodissociation processes. Theoretically, there are a few studies, but they are usually issued from early and modest calculations, so the accuracy is not good enough.

We report in this contribution an insight into the quintet and septet electronic states of molecular oxygen. Their PECs display a rich and complex structure, and interactions among states which could not be anticipated. We report PECs of valence, Rydberg and ion-pair quintet states as well as of various Rydberg septet states. Most PECs are repulsive, as expected, yet, a few of the high-spin states are bound. Excitation energies are tabulated for all states. Spectroscopic constants are given for the bound states. A case is also presented of a bound sextet state of O_2^+, along with potential curves of several sextet repulsive states of the cation.

RK. Clusters/Complexes

Thursday, June 21, 2018 – 1:45 PM
Room: B102 Chemical and Life Sciences

Chair: Haruki Ishikawa, Kitasato University, Sagamihara, Japan

RK01   1:45 – 2:00

A CONFORMATIONAL STUDY OF *META*-ANISIC ACID AND ITS COMPLEXES WITH FORMIC ACID BY MICROWAVE SPECTROSCOPY

<u>ALBERTO MACARIO</u>, SUSANA BLANCO, *Departamento de Química Física y Química Inorgánica, Universidad de Valladolid, Valladolid, Spain*; JAVIX THOMAS, YUNJIE XU, *Department of Chemistry, University of Alberta, Edmonton, AB, Canada*; JUAN CARLOS LOPEZ, *Departamento de Química Física y Química Inorgánica, Universidad de Valladolid, Valladolid, Spain.*

The *meta*-anisic acid and its complexes with formic acid have been studied in the 2-15 GHz frequency region using chirped-pulse Fourier transform microwave spectroscopy (CP-FTMW). For *meta*-anisic acid, four conformations have been identified. The four conformations resulting from the possible arrangements of the caboxylic and methoxy groups have been identified. In addition, the spectra of mixtures of *meta*-anisic acid and formic acid were investigated. Four species were found where formic acid is attached to the acid group through two O-H···O=C complementary hydrogen bonds. These complexes correspond to the four *meta*-anisic acid stable structures that seem to be not essentially modified after the complexation process.

RK02   2:02 – 2:17

CP-FTMW SPECTROSCOPY OF 2-CYANOACETIC ACID

<u>ERIKA JOHNSON</u>, STEVEN SHIPMAN, *Department of Chemistry, New College of Florida, Sarasota, FL, USA*; IKER LEÓN, LUCIE KOLESNIKOVÁ, SANTIAGO MATA, JOSÉ L. ALONSO, *Grupo de Espectroscopia Molecular, Lab. de Espectroscopia y Bioespectroscopia, Unidad Asociada CSIC, Universidad de Valladolid, Valladolid, Spain.*

2-Cyanoacetic acid (m.p. 67-73 °C), a molecule with astrochemical significance and structural similarities to glycine[a] has been transferred into the gas phase by heating and supersonic expansion. Possible conformational relaxation paths were explored by using different buffer gases such as neon, argon, and helium. Rotational constants were found using chirped-pulse Fourier transform microwave spectroscopy (CP-FTMW). 2-Cyanoacetic acid possesses one 14N nucleus with a nuclear quadrupole moment, giving rise to a complex hyperfine structure. To assist with spectral fitting, Autofit[b] was used. One stable conformer of the monomer and its complex with a water molecule have been identified on the basis of the experimental rotational constants in conjunction with DFT and *ab initio* predictions.

---
[a] I. D. Reva, S. G. Stepanian, L. Adamowicz and R. Fausto, J. Phys. Chem A 2003 107, 6351-6359
[b] N. A. Seifert, I. A. Finneran, C. Perez, D. P. Zaleski, J. L. Neill, A. L. Steber, R. D. Suenram, A. Lesarri, S. T. Shipman, and B. H. Pate J. Mol. Spec. 2015 312, 12-21

RK03   2:19 – 2:34

HIGH RESOLUTION MICROWAVE SPECTROSCOPY AND STRUCTURE OF THE WEAKLY BOUND Xe···OCS COMPLEX

<u>DANIEL A. OBENCHAIN</u>, SVEN HERBERS, PETER KRAUS, DENNIS WACHSMUTH, JENS-UWE GRABOW, *Institut für Physikalische Chemie und Elektrochemie, Gottfried-Wilhelm-Leibniz-Universität, Hannover, Germany.*

The rotational spectrum of the weakly bound complex between xenon and carbonyl sulfide has been measured using a coaxially oriented beam resonator arrangement (COBRA) Fourier transform microwave spectrometer in Hannover. There are nine naturally occurring isotopes of xenon in addition to the many possible isotopologues of carbonyl sulfide, which allows for a detailed analysis of the structure of the van der Waals complex and subsequent comparisons to other rare gas van der Waals complexes with carbonyl sulfide that have already been reported.

Of the nine isotopes of xenon, two have non-zero nuclear spins, ^{131}Xe (I=3/2) and ^{129}Xe (I=1/2). In the case of the ^{131}Xe, a hyperfine structure was observed in all transitions, revealing a non-zero field gradient at the xenon nucleus in the complex. High-level ab initio calculations were carried out to identify an accurate method for the prediction of the structure of the complex and the nuclear quadrupole coupling constants of ^{131}Xe. The van der Waals interaction energies of xenon and carbonyl sulfide will be discussed.

RK04 2:36 – 2:51
MICROWAVE SPECTRUM OF THE A INTERNAL ROTOR STATE OF Ar-CH_3I

ANNA HUFF, CJ SMITH, KEN LEOPOLD, *Chemistry Department, University of Minnesota, Minneapolis, MN, USA.*

The Ar-CH_3I complex has been observed by pulsed-nozzle Fourier transform microwave spectroscopy. The rotational spectrum is complicated by both the low internal rotation barrier of the methyl group and the large nuclear quadrupole coupling from the iodine. The identification of closed loops was essential to the successful analysis of 140 observed transitions which have been assigned to the A internal rotor state of the complex. An additional 102 observed transitions are likely due to the E internal rotor state. Associated computational work, as well as ongoing attempts to assign the E state spectrum, will be discussed.

RK05 2:53 – 3:08
MICROWAVE STUDY OF 2-PHENYLPYRIDINE AND THEIR WATER COMPLEXES

SUSANA BLANCO, ALBERTO MACARIO, JUAN CARLOS LOPEZ, *Departamento de Química Física y Química Inorgánica, Universidad de Valladolid, Valladolid, Spain.*

2-Phenylpyridine (2pp) is well known as an organic ligand in metal complexes, systems which present a relevant importance in fields such as inorganic photochemistry, biological electron-transfer dynamics or, for example, as chromophores in semiconductor assemblies in solar energy panels. The analysis of the rotational spectra of the parent and all the ^{13}C and ^{15}N isotopologues for the 2pp, split by the torsional vibration, have allowed us to determine the structure of 2pp. Furthermore, it has been possible to detect the spectra of the 1:1 and 1:2 water complexes. For the 1:1 adduct, the spectra of the parent and the same isotopologues as observed for the monomer have been measured. This has enabled direct structural comparison between the bare molecule and its water complex, determining how the water affects the structure of 2pp.

Intermission

RK06 3:44 – 3:59
INTRA AND INTERMOLECULAR DYNAMICS AND STRUCTURE IN THE FORMANILIDE-$(H_2O)_n$ (n=1,2) CLUSTERS

PABLO PINACHO, SUSANA BLANCO, JUAN CARLOS LOPEZ, *Departamento de Química Física y Química Inorgánica, Universidad de Valladolid, Valladolid, Spain.*

Formanilide (C_7H_7NO) is a molecule that mimics the -NH-CO- peptide bond. It adopts two different conformations, planar or non-planar, depending respectively on the *trans* or *cis* arrangement of the peptidic group. Formanilide offers a variety of hydrogen bond binding sites so its microsolvated clusters can be taken as good models to investigate the interaction of the peptide functional group with water. In this work, the rotational spectra of formanilide-$(H_2O)_n$ (n=1, 2) complexes have been studied in the 2-12.5 GHz frequency range using both a pulsed-chirp and a molecular beam Fourier transform microwave spectrometers. Three heterodimers, *cis*-1:1a *trans*-1:1b and *trans*-1:1c forms and one heterotrimer, *cis*-1:2a, have been observed. The rotational spectra of the parent, several D/H and ^{18}O/^{16}O isotopically substituted species have been measured for the adducts in order to investigate their structures. All species are characterized by the quadrupole coupling hyperfine structure due to the presence of a ^{14}N atom in formanilide. The rotational spectra of some species show small doublets attributable to either intramolecular motions within the formanilide subunit, as occur in *cis*-1:1a, or to intermolecular motions as the internal rotation of water in form *trans*-1:1c.

RK07   4:01–4:16
OH-π HYDROGEN BOND IN THE COMPLEX OF STYRENE-WATER: A ROTATIONAL STUDY

YANG ZHENG, JUNCHENG LEI, GANG FENG, QIAN GOU, *School of Chemistry and Chemical Engineering, Chongqing University, Chongqing, China.*

The rotational spectra of the styrene-water complex has been investigated by using the pulsed jet Fourier transform microwave spectroscopic technique. Styrene has two π systems which can act as the proton acceptor and link with water through the OH-π hydrogen bond. Ab initio calculations suggested that the vinyl π system is favored to form such a hydrogen bond. In contrast, the experimental evidences of four isotopologues pointed out that the water O-H group prefers to link to the benzene π system. The internal rotation of water around its symmetry axis splits all the rotational transitions into two component line with a relative intensity ration of 1:3.

RK08   4:18–4:33
ISOTOPIC SUBSTITUTIONS UNVEILED THE IDENTIFICATION OF THE MORE STABLE CONFORMER OF FENCHOL AND OF ITS WATER COMPLEX

ELIAS M. NEEMAN, THERESE R. HUET, *UMR 8523 - PhLAM - Physique des Lasers Atomes et Molécules, University of Lille, CNRS, F-59000 Lille, France.*

Fenchol $C_{10}H_{18}O$ was identified as one of the many products emitted by several plants, by pin radiata wood, and by fibers panels [a]. It is a monoterpene of spectroscopic interest in order to determine the more stable conformer. Indeed the fenchol molecule presents two stereoisomers depending on the position of the OH group and of the hydrogen atom, named endo- and exo-fenchol.

In the present work, we present the conformational landscape study of fenchol in order to identify in the gas phase the more stable conformer for each stereoisomer. A combination of theoretical calculations and Fourier transform microwave spectroscopy in a supersonic molecular jet was used. Because of a disagreement between the different calculation methods for the endo-fenchol stereoisomer, fenchol-D was used to identify the most stable conformer. The hyperfine structure signature was found very helpful.

Moreover the hydration of endo-fenchol was studied. The multi-isotopic substitution of deuterium has led to identify the observed conformer. Surprisingly the substituted structure of the hydrogen atoms shows that the observed complex is formed by hydrogen bonding between the high energy conformer of endo-fenchol - which is not observed in the jet - with the water molecule. The set of molecular parameters was adjusted using a Watson Hamiltonian in the A reduction to the experimental accuracy.

The present work was funded by the French ANR Labex CaPPA through the PIA (contract ANR-11-LABX-0005-01), by the Regional Council Hauts de France, by the European Funds for Regional Economic Development (FEDER), and by the French Ministère de l'Enseignement Supérieur et de la Recherche. It is a contribution to the CPER research Project CLIMIBIO.

---

[a] N. Yassa *et al*, *Atmos. Env.* **34**, 2809 (2000) ; A. G. McDonald *et al*, *Holz als Roh und Werkstoff* **64**, 291 (2004) ; M. G. D. Baumann, *Forest products journal* **50**, 75 (2000).

RK09
4:35 – 4:50

A ROTATIONAL STUDY OF THE METHYL CARBAMATE-$(H_2O)_n$ n=1,2 COMPLEXES: MICROWAVE SPECTRUM, INTERNAL ROTATION AND HYPERFINE STRUCTURE.

PABLO PINACHO, JUAN CARLOS LOPEZ, *Departamento de Química Física y Química Inorgánica, Universidad de Valladolid, Valladolid, Spain*; ZBIGNIEW KISIEL, *ON2, Institute of Physics, Polish Academy of Sciences, Warszawa, Poland*; SUSANA BLANCO, *Departamento de Química Física y Química Inorgánica, Universidad de Valladolid, Valladolid, Spain.*

The rotational spectrum of methyl carbamate has been recorded in the 2-8 GHz frequency region using a chirped-pulse Fourier transform microwave spectrometer. The carrier gas was seeded with water and methyl carbamate vapors for the formation of microsolvated complexes. Complexes with one and two molecules of water have been detected. Both spectra show the fine structure arising from the internal rotation of the methyl top together with the hyperfine structure due to the presence of one ^{14}N nucleus. The spectra were further registered in the 5-18 GHz frequency region by the higher resolution supersonic expansion cavity Fourier transform microwave spectroscopy to analyze more accurately the hyperfine structure. The determined rotational parameters provide the key for the identification of the complexes in the light of ab initio computations. Both the methyl group internal rotation barrier and the quadrupole coupling constants show interesting trends in going from isolation to the microsolvated complexes.

RK10
4:52 – 5:07

ROTATIONAL SPECTRUM OF THE ISOPRENE-WATER COMPLEX

BRANDON CARROLL, MICHAEL C McCARTHY, *Atomic and Molecular Physics, Harvard-Smithsonian Center for Astrophysics, Cambridge, MA, USA.*

Measuring the structure of molecular dimers is a stepping stone in the study of chemical interactions. These structures are normally highly sensitive to the two-body potential and therefore form a basis for interpreting structure, dynamics, and reactions in more complex systems. Studies of strong and weak hydrogen bonded clusters have provided important insight into the nature of these interactions. The structures of numerous dimers have been determined, however many complexes, especially those involving weakly polar species such as isoprene, have not been measured. Isoprene(C_5H_8), the simplest terpene, is the building block for terpenes and terpenoids, many of which are volatile compounds released by trees and shrubs. Isoprene accounts for 50% of biogenic non-methane hydrocarbon emission, and is therefore a major contributor to aerosol production. Furthermore, the composition of isoprene-derived aerosols varies significantly with relative humidity. The study of the structure of heterogeneous isoprene clusters is therefore of considerable interest for furthering our understanding of the interactions that drive isoprene chemistry.

Towards this end, we will present the pure rotational spectrum of the isoprene-water complex recorded with a chirped-pulse Fourier transform microwave spectrometer, and discuss the structure of the complex.

RL. Astronomy
Thursday, June 21, 2018 – 1:45 PM
Room: 2079 Natural History

Chair: Brett A. McGuire, National Radio Astronomy Observatory, Charlottesville, VA, USA

RL01 1:45 – 2:00

MOLECULAR COMPLEXITY IN PRESTELLAR CORES

VALERIO LATTANZI, PAOLA CASELLI, *The Center for Astrochemical Studies, Max-Planck-Institut für extraterrestrische Physik, Garching, Germany.*

Prestellar cores, such as L183 (= L134N), are the first phase of the star formation process and the nursery of chemical complexity. Despite the lack of embedded sources, whose energy input to the cloud can complicate the analysis by permitting a wider range of gas-phase reactions to occur, and can cause the release of molecules formed on grain surface back into the gas phase, cold dark clouds exhibit a complex molecular inventory and distribution.
Here we present single-pointing observations of L183 performed with the IRAM 30m telescope at 3mm wavelength. The focus of the present study is to compare the chemical abundances of the detected molecular species with those found previously in L1544, a very similar astronomical environment, to study similarities and differences between these two prestellar cores. Also, our results will be put in comparison with an astrochemical model recently developed, providing a valuable test for its validity in the context of pre-stellar cores.

RL02 *Post-Deadline Abstract* 2:02 – 2:17

MILLIMETER/SUBMILLIMETER-WAVE SPECTROSCOPY OF THE CrP RADICAL $^4\Sigma^-$

MARK BURTON, *Department of Chemistry and Biochemistry, University of Arizona, Tucson, AZ, USA*; DeWAYNE T HALFEN, *Department of Chemistry and Biochemistry, Department of Astronomy, The University of Arizona, Tucson, AZ, USA*; LUCY M. ZIURYS, *Department of Chemistry and Biochemistry; Department of Astronomy, Arizona Radio Observatory, University of Arizona, Tuscon, AZ, USA.*

The millimeter/sub-mm spectrum of the CrP radical in its $X^4\Sigma^-$ ground electronic state has been recorded using direct absorption techniques in the frequency range of 340-540 GHz. This study is the first pure rotational measurement of a metal-phosphide species. CrP was synthesized in an AC discharge by the reaction of gas-phase chromium, generated from $Cr(CO)_6$, with phosphorus vapor, in argon carrier gas. Twelve rotational transitions were measured, each consisting of four fine structure components. The data were analyzed using a Hund's case b Hamiltonian; rotational, spin-spin, and spin-rotation constants were determined, improving on previous optical work. Comparison with our previous study of CrN indicates differences between metal-nitrogen and metal-phosphorus bonds.

RL03 2:36 – 2:51

NEW CARBON-CHAIN MOLECULAR DETECTIONS IN TMC-1 WITH THE GREEN BANK TELESCOPE

ANDREW M BURKHARDT, *Department of Astronomy, The University of Virginia, Charlottesville, VA, USA*; CHRISTOPHER N SHINGLEDECKER, ERIC HERBST, *Department of Chemistry, The University of Virginia, Charlottesville, VA, USA*; SERGEI KALENSKII, *Astro Space Center, Lebedev Physical Institute, Russian Academy of Sciences, Moscow, Russia*; MICHAEL C McCARTHY, *Atomic and Molecular Physics, Harvard-Smithsonian Center for Astrophysics, Cambridge, MA, USA*; ANTHONY REMIJAN, BRETT A. McGUIRE, *NAASC, National Radio Astronomy Observatory, Charlottesville, VA, USA.*

The source of molecular complexity in the interstellar medium is strongly dependent on the build up of carbon-chain molecules. As such, it is crucial to develop a robust chemical inventory of the largest of these carbon-chain species and, in turn, constrain their formation mechanisms. The cold core TMC-1 has long been a source of new molecular detections, particularly for unsaturated carbon-rich molecules. Through deep observations with the Green Bank Telescope of TMC-1, we report 8 new isotopologues of HC_5N and HC_7N and an entirely new molecular family (HC_5O, HC_7O). These new detections provide crucial insights to the formation of PAHs and the underlying carbon-chain chemistry of dark clouds. In addition, we will also discuss preliminary results from the next stage of GBT chemical surveys toward TMC-1.

RL04

INVESTIGATING THE DISTRIBUTION OF COMPLEX MOLECULES AT LOW FREQUENCY USING THE KARL G. JANSKY VERY LARGE ARRAY IN SEARCH OF THE EXCITATION OF HNCNH

ANTHONY REMIJAN, BRETT A. McGUIRE, *NAASC, National Radio Astronomy Observatory, Charlottesville, VA, USA*; ANDREW M BURKHARDT, *Department of Astronomy, The University of Virginia, Charlottesville, VA, USA*; JOANNA F. CORBY, *Physics, University of South Florida, Tampa, FL, USA*.

In 2012, McGuire et al. identified carbodiimide (HNCNH), an isomer of the well-known interstellar species cyanamide (NH2CN) which was first detected in 1975 at millimeter wavelengths by Turner et al. The detection of HNCNH was done using the Robert C. Byrd Green Bank Telescope as part of the PRIMOS survey toward Sgr B2(N). Given the excitation of the detected transitions, it was concluded that NHCNH was a weak astronomical maser and the only way HNCNH could be detected was by those transitions which are amplified by masing. Many other species detected at centimeter wavelengths also have transitions amplified by masing which include, but is not limited to, cyanoacetylene (HC_3N), methyl formate (CH_3OCHO), methanol (CH_3OH) and more recently, formamide (NH_2CHO) and possibly methylamine (CH_3NH_2). The outstanding question remains as to whether the transitions are being enhanced by radiative or collisional processes. To try to ascertain the answer to this question, several low frequency transitions ($\sim$4 GHz) were observed with the Karl G. Jansky Very Large Array to determine the overall spatial distribution with respect to the background continuum sources and other sources of molecular emission. This talk will present the results of these observations and discuss the possible time variability of some of these low frequency, large molecule astronomical masers.

RL05

DETECTION OF CH_3CN IN DIFFUSE CLOUD TOWARD GALACTIC CENTER SGRB2(M)

MITSUNORI ARAKI, *Faculty of Science Division I, Tokyo University of Science, Shinjuku-ku, Tokyo, Japan*; SHURO TAKANO, *College of Engineering, Nihon University, Fukushima, Japan*; YOSHIAKI MINAMI, TAKAHIRO OYAMA, *Faculty of Science Division I, Tokyo University of Science, Shinjuku-ku, Tokyo, Japan*; NOBUHIKO KUZE, *Faculty of Science and Technology, Sophia University, Tokyo, Japan*; KAZUHISA KAMEGAI, , *National Astronomical Observatory of Japan, Tokyo, Japan*; KOICHI TSUKIYAMA, *Faculty of Science Division I, Tokyo University of Science, Shinjuku-ku, Tokyo, Japan*.

Organic molecules have been detected mainly in dense clouds. As diffuse clouds are a previous phase of dense clouds in evolutionary history of interstellar clouds, detection of organic molecules in diffuse clouds can reveal a longer history of organic molecules. Despite of its importance, emission lines from molecules in diffuse clouds are difficult to detect because of inactive excitation by collisions and active cooling by radiations. However, for CH_3CN, a rotation around a molecular axis cannot be cooled by radiations. Thus, this molecule can be detected by absorption having a relatively strong line from $J = K$ rotational levels, because these levels are well populated. In our previous work, this rotational behavior was formulated as "Hot Axis Effect" [1]. In this work, to detect this molecule in diffuse clouds that are more diffuse than the known diffuse clouds carrying organic molecules [2], we have been searched for absorption lines of the $J_K = 4_3-3_3$ transition of CH_3CN at 73 GHz toward the galactic center SgrB2(M) by using Nobeyama 45 m telescope. As a result, this transition was detected in the diffuse cloud of SgrB2 envelop. The column density was derived to be 2×10^{14} cm^{-2}. Therefore, we detected an organic molecule in the diffuse cloud that are more diffuse than the known diffuse cloud carrying CH_3CN [2] because this molecule detected shows the low excitation temperature of $\sim$ 3 K and the high kinetic temperature of $\sim$ 70 K.

[1] Araki et al. Astronomical Journal, 148, 87 (2014)
[2] Muller et al., A&A, 535, 103 (2011)

RL06 3:27 – 3:42

CONSTRAINING SULFUR ISOTOPE ABUNDANCES IN MOLECULAR CLOUDS: A METEORITIC PERSPECTIVE

JACOB BERNAL, *Department of Chemistry and Biochemistry, University of Arizona, Tucson, AZ, USA*; MAITRAYEE BOSE, *School of Earth and Space Exploration, Arizona State University, Tempe, AZ, USA*; LUCY M. ZIURYS, *Department of Chemistry and Biochemistry, University of Arizona, Tucson, AZ, USA*.

Carbonaceous chondrite class meteorites exhibit isotopic enrichments in deuterium, ^{13}C, ^{15}N, and possibly ^{33}S, implying interstellar origins. To study possible sulfur enrichments, we have been conducting observations of SO in molecular clouds to determine abundance ratios between ^{32}S, ^{34}S, and ^{33}S. The N=2-1 3mm transitions of SO have been measured towards W51M, DR21, G34.3, W3(OH), and NGC7538 using the ARO 12m telescope. The sulfur isotope ratios are currently being calculated and will be compared to meteoritic samples (Bose et al. 2017). The results presented will help to determine the connection between molecular cloud and meteoritic materials.

RL07 3:44 – 3:59

FIFI-LS FIR VIEW OF ORION: FINE STRUCTURE AND CO LINES

FRANKIE ENCALADA, *Astronomy, University of Illinois at Urbana-Champaign, Urbana-Champaign, IL, USA*; LESLIE LOONEY, *Department of Astronomy, University of Illinois at Urbana-Champaign, Urbana, IL, USA*; RANDOLF KLEIN, *SOFIA Science Center, NASA Ames Research Center, Moffett Field, CA, USA*; CHRISTIAN FISCHER, *Deutsches SOFIA Institut, Universität Stuttgart, Stuttgart, Germany*; SEBASTIAN COLDITZ, *5. Physikalisches Institut, Universität Stuttgart, Stuttgart, Germany*; DARIO FADDA, *SOFIA Science Center, NASA Ames Research Center, Moffett Field, CA, USA*; NORBERT GEIS, *Optical and Interpretative Astronomy, Max Planck Institute for Extraterrestrial Physics, Garching, Germany*; RAINER HÖNLE, CHRISTOF ISERLOHE, ALFRED KRABBE, *Deutsches SOFIA Institut, Universität Stuttgart, Stuttgart, Germany*; ALBRECHT POGLITSCH, *Infrared/Submillimeter Group, Max Planck Institute for Extraterrestrial Physics, Garching, Germany*; WALFRIED RAAB, *RSSD, ESA/ESTEC, Noordwijk, Netherlands*; WILLIAM VACCA, *SOFIA Science Center, NASA Ames Research Center, Moffett Field, CA, USA*.

The Orion Nebula is the closest massive star forming region, which allows us to study its physical conditions at high spatial resolution. We used the far infrared integral-field spectrometer, FIFI-LS, on-board the airborne observatory SOFIA to study the Orion Nebula's atomic and molecular gas.

We obtained large maps of fine structure and CO lines that span the nebula from the BN/KL-object to the bar. These maps allow us to study the conditions of the photon-dominated region and the interface to the molecular cloud.

A five-hundred-year-old violent explosion in the Orion Nebula has been stirring up the BN/KL region via wide-angled molecular outflows. We present maps of several high-J CO observations, allowing analysis of the heated molecular gas.

Intermission

RL08 4:35 – 4:50

THE TRANSITION FROM DIFFUSE ATOMIC CLOUDS TO DENSE MOLECULAR CLOUDS

JOHNATHAN S RICE, STEVEN FEDERMAN, *Physics and Astronomy, University of Toledo, Toledo, OH, USA*.

We explore the transition from diffuse to dense molecular gas by combining a variety of tracers for density and composition. Observations and chemical modeling of CH, CH$^+$, and CN absorption at visible wavelengths from McDonald Observatory and the European Southern Observatory are combined with ultraviolet observations of CO and H$_2$ absorption from the Far Ultraviolet Spectroscopic Explorer and the Hubble Space Telescope, as well as emission data from the GOT C+ survey with the Herschel Space Telescope. The selected tracers from visible, ultraviolet, and radio wavelengths allow the characterization of neutral diffuse gas, including CO-dark gas. Sight lines, such as those toward h and χ Persei or those toward Chamaeleon provide a opportunity to examine the transition from atomic to molecular gas and help to describe the nature of the gas.

RL09
4:52 – 5:07

DUST POLARIZATION IN THREE PROTOSTELLAR DISKS

RACHEL E. HARRISON, *Astronomy, University of Illinois at Urbana-Champaign, Urbana, IL, United States*; LESLIE LOONEY, ROBERT J HARRIS, *Department of Astronomy, University of Illinois at Urbana-Champaign, Urbana, IL, USA*; ZHI-YUN LI, *Department of Astronomy, The University of Virginia, Charlottesville, VA, USA*; IAN STEPHENS, *Radio and Geoastronomy Division, Harvard-Smithsonian Center for Astrophysics, Cambridge, MA, USA*; WOOJIN KWON, *Radio Astronomy, Korea Astronomy and Space Science Institute, Daejeon, Republic of Korea.*

We present 1.3 mm ALMA dust polarization observations of three T Tauri stars: DG Tauri, DL Tauri, and LkCA 15. All three sources show some degree of polarization at a resolution of 0.5". DL Tauri shows a polarization morphology that is consistent with polarization produced by dust scattering. DG Tauri and LkCa 15 have polarization morphologies that may be produced by dust grain alignment with the disk's radiation field and/or magnetic field. Dust grains can serve as a site for chemical reactions in protostellar disks. Observations of dust polarization can constrain the distribution and properties of dust within the disk, which would provide insight into the chemical evolution of protostellar disks.

RL10
5:09 – 5:24

IDENTITY OF THE CARRIER OF $\lambda 5797$ DIFFUSE INTERSTELLAR BAND and $\lambda 5800$ RED-RECTANGLE EMISSION BAND

KEIR ADAMS, *Department of Chemistry, The University of Chicago, Chicago, IL, USA*; TAKESHI OKA, *Astronomy and Astrophysics, Chemistry, The Enrico Fermi Institute, University of Chicago, Chicago, USA.*

The remarkable Red Rectangle nebula is well known for its emission bands (RRBs), excited by the central binary star HD 44179.[a] The proximity in wavelength between the strong $\lambda 5800$ RRB and the long known and intense Diffuse Interstellar Band $\lambda 5797$ DIB has led to speculation that these two bands originate from identical molecules. This speculation, however, has been challenged on the grounds that the peak wavelength of $\lambda 5800$ RRB fails to converge to 5797 Å when observed at large angular offsets from HD 44179. Consequently, $\lambda 5800$ RRB has been interpreted as being caused by various PAHs.[b]

We investigate the possibility that $\lambda 5800$ RRB and $\lambda 5797$ DIB originate from the same molecule. We speculate that absorption in the foreground gas causes the peak wavelength discrepancy, and that the red-shifting of the $\lambda 5800$ RRB peaks is a combined effect of the extended tail toward the red (ETR)[c] resulting from the high radiative temperature near HD 44179 and the foreground gas absorption. We use the temperatures and luminosities of the binary star reported by Witt et al.[d] for calculating the emission. However, radio to far infrared radiation emanating directly from the stars is far too weak to produce ETR, and we rely on stellar heating of the environment. We find that radiative temperatures on the order of 1000 K are sufficient to explain the largest tail and red-shifted peak at the smallest angular offset. We believe the molecules causing $\lambda 5797$ DIB[e] and $\lambda 5800$ RRB are identical.

---

[a] Schmidt, G. D., & Witt, A. N. 1991, ApJ, 383:698
[b] Sharp, R. G., Reilly, N. J., Kable, S. H., & Schmidt, T. W. 2006, ApJ, 639:194
[c] Oka, T., Welty, D. E., Johnson, S., York, D. G., Dahlstrom, J., & Hobbs, L. M. 2013, ApJ, 773:42
[d] Witt, A. N., Vijh, U. P., Hobbs, L. M., Aufdenberg, J. P., Thorburn, J. A., & York, D. G. 2009, 693:1946
[e] Huang, J. & Oka, T. 2015, Mol. Phys. 113, 15

CENTRAL 300 PC OF THE GALAXY PROBED BY THE INFRARED SPECTRA OF H_3^+ AND CO PART II. MORPHOLOGY AND DYNAMICS OF THE GAS

TAKESHI OKA, *Department of Astronomy and Astrophysics and Department of Chemistry, The Enrico Fermi Institute, University of Chicago, Chicago, IL, USA*; THOMAS R. GEBALLE, , *Gemini Observatory, Hilo, HI, USA*; MIWA GOTO, , *University Observatory Munich, Munich, Germany*; TOMONORI USUDA, , *National Astronomical Observatory of Japan, Tokyo, Japan*; BENJAMIN J. McCALL, *Departments of Chemistry and Astronomy, University of Illinois at Urbana-Champaign, Urbana, IL, USA*; NICK INDRIOLO, *Space Telescope Science Institute, Baltimore, MD, USA*.

Velocity-resolved spectra of infrared lines of H_3^+ at 3.7 μm and CO at 2.3 μm have been obtained toward $\sim$40 stars in the Central Molecular Zone (CMZ), a region of radius $\sim$150 pc centered on the Galactic center. Although the coverage of the region is limited by the available number of suitable stars for absorption spectroscopy, the rich Doppler profiles of the H_3^+ lines in warm ($T\sim$250 K) and diffuse ($n \leq 100$ cm^{-3}) clouds[a] have allowed us to draw a longitude-velocity (l-v) plot to reach the following conclusions.

(1) Based on the blue-shifted profiles of H_3^+ absorption lines, which are dominantly in the velocity range from -200 km s^{-1} to 10 km s^{-1}, the warm diffuse gas is moving outward from the center.

(2) Although limited in uniformity of longitudinal coverage, the observed (l-v) plot for H_3^+ suggests that the outer surface of the expanding gas forms a ring of radius of $\sim$140 pc and has a velocity of expansion of $\sim$140 km s^{-1}. This finding revives the idea of the expanding molecule ring proposed by Kaifu et al.(1972)[b] and Scoville (1972)[c] which contrasts with a more recent interpretation of the overall gas kinematics as due to a barred gravitational potential (Binney et al. 1991).[d]

(3) The results revive the idea of an explosion or overall expulsion of gas from the center within the last few million years. Unlike the original proposals that the EMR is also rotating, the H_3^+ l-v plot indicates purely expanding gas.

---

[a] Oka, T., Geballe, T.R., Goto, M., Usuda, M., McCall, B.J., Indriolo, N. to be submitted (2018)
[b] Kaifu, N., Kato, T. Iguchi, T., 1972, Nature, 238, 105
[c] Scoville, N.Z. 1972, ApJ, 175, L127
[d] Binney, J., Gerhard, O.E., Stark, A.A., Bally J., Uchida, K.I. 1991, Mon. Not. R. astr. Soc. 252, 210

FA. Vibrational structure/frequencies

Friday, June 22, 2018 – 8:30 AM

Room: 116 Roger Adams Lab

Chair: G. S. Grubbs II, Missouri University of Science and Technology, Rolla, MO, USA

FA01   8:30–8:45

THE 103 - 360 GHZ ROTATIONAL SPECTRUM OF BENZONITRILE, THE FIRST INTERSTELLAR BENZENE DERIVATIVE DETECTED BY RADIOASTRONOMY

MARIA ZDANOVSKAIA, *Department of Chemistry, University of Wisconsin–Madison, Madison, WI, USA*; BRIAN J. ESSELMAN, *Department of Chemistry, The Univeristy of Wisconsin, Madison, WI, USA*; HUNTER SINGH LAU, DESIREE M. BATES, *Department of Chemistry, University of Wisconsin–Madison, Madison, WI, USA*; R. CLAUDE WOODS, ROBERT J. McMAHON, *Department of Chemistry, The Univeristy of Wisconsin, Madison, WI, USA*; ZBIGNIEW KISIEL, *ON2, Institute of Physics, Polish Academy of Sciences, Warszawa, Poland.*

Benzonitrile (C_7H_5N, C_{2v}, μ_a = 4.5 D) has recently been detected in the interstellar medium (ISM), specifically in the Taurus Molecular Cloud 1 (TMC-1), using both the technique of composite averages[1–2] and by nine hyperfine-resolved rotational transitions[2] under 50 GHz. While benzonitrile has been thoroughly studied using infrared and cm-wave spectroscopy, no former studies have examined the rotational spectrum above 160 GHz. Herein, we present the analysis and assignment of the mm-wave rotational spectrum of benzonitrile (vibrational ground state) in the 103 – 350 GHz frequency range, which should assist in future astronomical searches. Additionally, we have completed a two-state least-squares fit of the hitherto unreported, Coriolis-coupled dyad of benzonitrile's two lowest frequency vibrational modes: ν_{22} (141 cm^{-1}) and ν_{33} (163 cm^{-1}), resulting in approximately 3000 transitions per state fit to within experimental accuracy. The two-state fit accounts for many resonances between the two states and 11 nominal interstate transitions. As a result, we have determined the energy gap between the vibrational states ($\Delta E_{22,33}$ = 19.108187(7) cm^{-1}) and the Coriolis coupling value ($\zeta_{22,33}{}^a$ = 0.841(5)). The study demonstrates that the lowest energy fundamentals of benzonitrile follow the previously described pattern of this molecular class.

1. Kalenskii, S. V., Possible Detection of Interstellar Benzonitrile. Proceedings of the Russian-Indian workshop on radio astronomy and star formation, October 10-12, 2016 (eds. I. Zinchenko & P. Zemlyanukha), Institute of Applied Physics RAS, p.43-50, 2017. 2. McGuire, B. A.; Burkhardt, A. M.; Kalenskii, S. V.; Shingledecker, C. N.; Remijan, A. J.; Herbst, E.; McCarthy, M. C. Science 2018, 359 (6372), 202-205.

FA02

INFRARED SPECTRUM OF CHLOROMETHYL HYDROPEROXIDE CH_2ClOOH PRODUCED FROM REACTION OF THE CRIEGEE INTERMEDIATE CH_2OO WITH HCl

<u>WEI-CHE LIANG</u>, *Department of Applied Chemistry, National Chiao Tung University, Hsinchu, Taiwan*; YUAN-PERN LEE, *Institute of Atomic and Molecular Sciences, Academia Sinica, Taipei, Taiwan.*

The Criegee intermediates, which are carbonyl oxides produced in ozonolysis of unsaturated hydrocarbons,[a] play important roles in the production of OH, aerosols and organic acids in the atmosphere. Criegee intermediates react readily with other atmospheric species such as NO_2, SO_2, $(H_2O)_2$ and HCl. The reaction of CH_2OO with HCl was reported to be rapid, with a rate coefficient of 4.6×10^{-11} cm^3 molecule^{-1} s^{-1}.[b] Quantum-chemical calculations indicate that the reaction CH_2OO + HCl proceeds through a barrierless association reaction to form chloromethyl hydroperoxide (CMHP, CH_2ClOOH), which was predicted to have three stable conformations, denoted as gauche-gauche CMHP, anti-CMHP and gauche-anti CMHP; the first two conformers were observed by microwave spectroscopy.[c]

In this work, a step-scan Fourier-transform spectrometer coupled with a multipass absorption cell was employed to record temporally resolved infrared (IR) absorption spectra of the reactants and products during the reaction of CH_2OO with HCl in a flow system. CH_2OO was produced from the reaction of O_2 with CH_2I, which was produced via photolysis of CH_2I_2 at 308 nm.[d] Time-resolved IR absorption spectra were recorded at resolution 0.25 cm^{-1}. The spectrum of gauche-gauche CMHP was characterized by the absorption bands at 1061.4, 1309.4, 1359.9 cm^{-1}, and the spectrum of gauche-anti CMHP was characterized by the absorption bands at 1054.8, 1310.3, 1359.3 cm^{-1}. The observed wavenumbers, rotational contours and relative intensities agree with those predicted with the B3LYP/aug-cc-PVTZ method; contributions of the hot bands from excited states of the low-lying torsional modes are significant.

---

[a] R. Criegee and G. Wenner, J. Liebigs Ann. Chem. 564, 9 (1949).
[b] Elizabeth S. Foreman, Kara M. Kapnas, Craig Murray, Angew. Chem. Int. Ed. 55, 1 (2016).
[c] C. Cabezas and Y. Endo, ChemPhysChem. 18, 1860 (2017).
[d] Y.-Y. Wang, C.-Y. Chung, and Y.-P. Lee, J. Chem. Phys. 145, 154303 (2016).

FA03

INFRARED SPECTRA OF C_2H_4BR AND C_2H_4I IN SOLID *PARA*-HYDROGEN: BRIDGED OR OPEN STRUCTURE?

<u>YU-HSUAN CHEN</u>, *Department of Applied Chemistry, National Chiao Tung University, Hsinchu, Taiwan*; YUAN-PERN LEE, *Applied Chemistry, National Chiao Tung University, Hsinchu, Taiwan, Institute of Atomic and Molecular Sciences, Academia Sinica, Taipei, Taiwan.*

The electrophilic addition of halogens to alkenes is important in organic synthesis. The structure of the reaction intermediate, haloalkyl radical, plays an important role in the stereo specificity of the addition reaction. The simplest intermediate, haloethyl radical C_2H_4X, has two possible geometries: an open (classical) structure and a bridged structure. However, which structure is more stable is controversial because several calculations yield varied results. Experiments using time-resolved X-ray diffraction[a] and UV absorption[b] indicated that C_2H_4I has a briedged structure, but our results of Cl + C_2H_4 in solid *para*-hydrogen (*p*-H_2) indicated that C_2H_4Cl has an open structure.[c] Even though experiments in noble-gas matrices suggest that C_2H_4Br is open,[d] whereas C_2H_4I is bridged,[e] the spectral evidence remains uncertain.

We took advantage of the diminished cage effect of *p*-H_2 to investigate the infrared (IR) spectra of C_2H_4Br and C_2H_4I by irradiating *p*-H_2 matrices containing C_2H_4 and Br_2 or I_2 with light at varied wavelengths. New spectral lines were grouped according to their behavior upon subsequent annealing and secondary photolysis. The assignments were derived on comparison with scaled vibrational wavenumbers and IR intensities calculated with the B3LYP/aug-cc-pVTZ-pp method. Our preliminary results indicate that lines at 676.9, 776.7, 1068.1, 1148.0, and 3126.8 cm^{-1} are assigned to the 2-bromoethyl radical (C_2H_4Br) in an open form, whereas those at 933.7, 1139.0, 1436.8, and 1609.0 cm^{-1} to the iodoethyl radical (C_2H_4I) in the bridged form. A small amount of $C_2H_4Br_2$ and $C_2H_4I_2$ was also observed; their stereochemistry will also be discussed.

---

[a] H. Ihee, M. Lorenc, T. K. Kim, Q. Y. Kong, M. Cammarata, J. H. Lee, S. Bratos, M. Wulff, *Science*. 2005, **309**, 1223.
[b] A. Kalume, L. George, A. D. Powell, R. Dawes, S. A. Ried, *J. Phys. Chem. A*. 2014, **118**, 6838.
[c] J. C. Amicangelo, B. Golec, M. Bahou, Y. P. Lee, *Phys. Chem. Chem. Phys*. 2012, **14**, 1014.
[d] A. Kalume, L. George, P. Z. El-Khoury, A. N. Tarnovsky, S. A. Reid, *J. Phys. Chem. A*. 2010, **114**, 9919.
[e] L. George, A. Kalume, S. A. Reid, *Chem. Phys. Lett*. 2012, **554**, 86.

FA04 9:21–9:36

A MULTIMODE-LIKE SELECTION OF CENTERS OF GAUSSIAN BASIS FUNCTIONS WHEN COMPUTING VIBRATIONAL SPECTRA USING COLLOCATION

SERGEI MANZHOS, *Department of Mechanical Engineering, National University of Singapore, Singapore, China*; XIAO-GANG WANG, TUCKER CARRINGTON, *Department of Chemistry, Queen's University, Kingston, ON, Canada.*

When computing vibrational spectra with collocation and localized basis functions such as Gaussian functions, one can choose the distribution of points and of function centers to increase the accuracy or decrease the cpu cost. Here, we compute vibrational energy levels using Gaussian basis functions whose centers are in slabs that include the lower-dimensional hyperplanes on which the Multimode approximation to the potential is based. We use more potential points than basis functions to increase the accuracy, i.e. we use rectangular collocation. The number of Gaussian basis functions is smaller than the number required using the best existing methods. For formaldehyde, the first 50 (100) levels we compute, using 30,000 Gaussians and 120,000 points, in 4D-like slabs, differ from numerically exact levels by 0.3 (0.6) cm-1 (mean absolute error). With 3D-like slabs, the mae for the first 50 (100) levels is 0.17 (0.47) cm-1 with 30,000 basis functions and 0.95 (2.06) cm-1 with 20,000 basis functions. Although we use a multimode-like idea to select Gaussian centers, we use a single point set and there is no need to write the potential in multimode form and no need to neglect high-order terms.

FA05 9:38–9:53

TRANSIENT RAMAN SPECTRA, STRUCTURE AND THERMOCHEMISTRY OF THE SELENOCYANATE DIMER RADICAL ANION IN WATER

IRENEUSZ JANIK, G. N. R. TRIPATHI, *Radiation Laboratory, University of Notre Dame, Notre Dame, IN, USA.*

Time-resolved resonance-enhanced Raman spectra of the selenocyanate dimer radical anion, $(SeCN)_2^{\cdot-}$, prepared by pulse radiolysis in water, have been obtained and interpreted in conjunction with theoretical calculations to provide detailed information on the molecular geometry and bond properties of the species. The structural properties of the radical are used to develop a molecular perspective on its thermochemistry in aqueous solution. Twelve Stokes Raman bands of the radical observed in the 100-3000 cm^{-1} region are assigned in terms of the strongly 140.5 cm^{-1}, weakly 550 cm^{-1}, and moderately 2095 cm^{-1}, enhanced fundamentals, their overtones and combinations. Calculations by range-separated hybrid density functionals (ωB97x and LC-ωPBE) support the spectroscopic assignments of the 140.5 cm^{-1} vibration to a predominantly SeSe stretching mode and the features at 550 cm^{-1} and 2095 cm^{-1} to SeC and CN symmetric stretching modes, respectively. The corresponding bond lengths are 2.957Å, 1.823Å and 1.157Å. A first order anharmonicity of 0.44 cm^{-1} determined for the SeSe stretching mode suggests a convergence of vibrational states at an energy 1.4eV, using the Birge–Sponer extrapolation. This value, estimated for the radical confined in solvent cage, compares well with the calculated gas-phase energy, 1.32eV, required for the radical to dissociate into $SeCN^{\cdot}$ and $SeCN^{-}$ fragments. The enthalpy of dissociation drops to 0.70eV in water when solvent dielectric effects on the radical and its dissociation products upon Se-Se bond scission are incorporated in the calculations. Our findings are compared to analogous symmetric and asymmetric hemibonded radical anions i.e. $(SCN)_2^{\cdot-}$ and $NCSOH^{\cdot-}$ to provide insights into relationship between their structure and properties.

FA06 9:55 – 10:10

MOLECULAR-SCALE INTERROGATION OF CATALYTIC INTERACTIONS BETWEEN OXYGEN AND COBALT PHTHALOCYANINE USING ULTRAHIGH VACUUM TIP-ENHANCED RAMAN SPECTROSCOPY

DUC NGUYEN, GYEONGWON KANG, GEORGE C. SCHATZ, RICHARD P. VAN DUYNE, *Department of Chemistry, Northwestern University, Evanston, IL, USA.*

Ultrahigh vacuum tip-enhanced Raman spectroscopy (UHV-TERS) is a powerful method combining the rich chemical information of vibrational spectroscopy with the ultrahigh spatial resolution of scanning tunneling microscopy (STM). However, despite of its potential, there is a lack of studies demonstrating the capability of UHV-TERS in investigating chemical reactions and molecular adsorptions under controlled environment. Herein, we use UHV-TERS to investigate adsorption of oxygen (O_2) with cobalt (II) phthalocyanine (CoPc) supported on Ag(111) single crystal surfaces under highly controlled environment, which is the initial step for the oxygen reduction reaction (ORR) using metal Pc catalysts. Two adsorption configurations are primarily observed, assigned as O_2/CoPc/Ag(111) and O/CoPc/Ag(111) based on STM imaging, TERS, isotopologue substitution, and density functional theory (DFT) calculations. Distinct vibrational features are observed for different adsorption configurations such as the ^{18}O-^{18}O stretching frequency at 1151 cm^{-1} for O_2/CoPc/Ag(111), and Co-^{16}O and Co-^{18}O vibrational frequencies at 661 cm^{-1} and 623 cm^{-1}, respectively, for O/CoPc/Ag(111). DFT calculations show vibrational mode coupling of O-O and Co-O vibrations to the Pc ring, resulting in different symmetries of oxygen-related normal modes for different isotopes. This study establishes UHV-TERS as a chemically sensitive tool for probing catalytic systems at the molecular-scale.

FB. Metal containing

Friday, June 22, 2018 – 8:30 AM

Room: 100 Noyes Laboratory

Chair: Leah C O'Brien, Southern Illinois University, Edwardsville, IL, USA

FB01   8:30–8:45

PROBING SELECTIVE BOND ACTIVATION IN ALKYLAMINES: LANTHANUM-MEDIATED C-H AND N-H BOND ACTIVATION STUDIED BY MATI SPECTROSCOPY.

SILVER NYAMBO, YUCHEN ZHANG, DONG-SHENG YANG, *Department of Chemistry, University of Kentucky, Lexington, KY, USA.*

In this work, La atom reactions with small alkylamines $R-NH_2$ and $H_2N-(CH_2)_n-NH_2$ (n=2, 3, 4) are carried out in a laser-ablation supersonic molecular beam source. Reaction products are observed with photoionization time-of-flight mass spectrometry and characterized by mass-analyzed threshold ionization (MATI) spectroscopy and DFT calculations. These reactions proceed via an exothermic H_2 loss from the ligand and form a metal complex. The $R-NH_2$ ligands favor the formation of acyclic metal complexes where La atom is doubly bonded to the N atom, whereas the $H_2N-(CH_2)_n-NH_2$ ligands prefer the formation of cyclic metal complexes where La atom is bonded to both N atoms. The reaction between La atom and CH_3NH_2 produces $LaNCH_3$ in two low-energy isomeric forms: an acyclic C_{3v} structure and a three-membered C_s ring. The C_{3v} isomer is formed by concerted elimination of two H atoms from the NH_2 group, whereas the C_s ring is formed by H atom elimination from both the alpha C and NH_2 group. Both isomers prefer a doublet ground state with a La $6s^1$-based electron configuration and a singlet ionic state by removing the 6s electron. The C_{3v} structure is calculated to be more stable than the C_s isomer but has a slightly lower adiabatic ionization energy (40022 (5) cm^{-1}) than the ring isomer (40399 (5) cm^{-1}) as measured from the MATI spectra. The MATI spectrum of the C_{3v} isomer is dominated by the origin band and La-Ligand and C-N stretching vibronic bands. The MATI signal of the C_s isomer is very weak and is only 10% of the C_{3v} MATI signal. This observation confirms that the C_{3v} isomer is more stable and its formation is more favorable than the C_s ring.

FB02   8:47–9:02

FTIR STUDY OF THE REACTIVITY OF HETERONUCLEAR SMALL TRANSITION METAL CLUSTER WITH CARBON MONOXIDE

MOHAMAD IBRAHIM, PASCALE SOULARD, *MONARIS, Sorbonne Université, CNRS, Paris, France*; ESMAÏL ALIKHANI, *MONARIS, Sorbonne Université, Paris, France*; BENOÎT TREMBLAY, *MONARIS, Sorbonne Université, CNRS, Paris, France*.

Bimetallic catalysts have attracted great research efforts in the past decades due to their chemical and physical properties different from the individual pure metals and promising applications in chemical conversion, energy technology and environmental protection. Numerous experimental and theoretical investigations have been focused on the reactions of transition metal atoms and small clusters with CO, and a variety of transition-metal carbonyl complexes have been characterized in gas phase and in solid argon. Since they are a very few studies on the transition metal heteronuclear dimer, we have studied in solid argon the reactivity of carbon monoxide with the heterodimer PdTi by infrared spectroscopy (FTIR). Heteronuclear cluster carbonyls, $PdTi(CO)_n$ (n=1-3) and $Pd_2Ti(CO)_2$, have been characterized on the basis of the isotopic substitution and irradiation effect. DFT calculations of the geometrical and electronic properties are also presented, and compared with the experimental values. An irradiation in visible leads to conversion between the isomers Pd-Ti-CO and Ti-Pd-CO distinguished by a large shift of the stretching frequency of the diatomic CO.

FB03 9:04 – 9:19

MAPPING THE INTRINSIC PHOTOCHEMISTRY OF PhotoCORMS VIA GAS-PHASE LASER SPECTROSCOPY

ROSARIA CERCOLA, JASON M. LYNAM, CAROLINE H. E. DESSENT, *Department of Chemistry, University of York, York, United Kingdom.*

We perform, for the first time, gas-phase laser photodissociation spectroscopy on a series of metal carbonyls that can lose CO upon irradiation. These molecules (PhotoCORMs) can be used for delivering and releasing CO molecules for medicinal purposes, such as in cancer therapy and as antimicrobials. Photodepletion (PD) and photofragmentation (PF) spectra of $[CpRu(Ph_3)_2CO]^+$ and $[CpRu(dppe)CO]^+$ were acquired between 230 and 400 nm, and the range 230-500 nm was explored for $[Mn(CO)_4Br_2]^-$. All the PhotoCORMs lose CO after irradiation, accessing different fragmentation channels when different excited states are populated. Indeed, while scanning the wavelength range in our laser-interfaced electrospray mass spectrometer, we observe the production spectra of the photofragments and can track the variation in the intensity of their production. $[Mn(CO)_4Br_2]^-$ loses 3 CO molecules in the key visible region and 4 COs in the UV. $[CpRu(Ph_3)_2CO]^+$ fragments into $[CpRuPh_3]^+$ via the loss of CO and Ph_3. This observation can be used to improve the design of new CO-releasing molecules, as we demonstrate in the $[CpRu(dppe)CO]^+$ system where we successfully observe only the CO loss across the whole explored wavelength range. Finally, solution-phase irradiation results are presented for 365 nm photoexcitation, showing comparable photofragmentation results to the ones obtained in the gas-phase.

FB04 9:21 – 9:36

TIME-RESOLVED RELAXATION DYNAMICS OF NEAR-INFRARED EXCITED ELECTRONIC STATES IN TRANSITION METAL COMPLEXES.

DARYA S. BUDKINA, *Department of Chemistry and Center for Photochemical Sciences, Bowling Green State University, Bowling Green, OH, USA*; SERGEY M. MATVEEV, *Chemistry, University of Illinois at Urbana-Champaign, URBANA-CHAMPAIGN, IL, USA*; CHRISTOPHER M. HICKS, *Department of Chemistry and Center for Photochemical Sciences, Bowling Green State University, Bowling Green, OH, USA*; VENIAMIN A. BORIN, *Physical Chemistry, The Hebrew University, Jerusalem, Israel*; ANDREY S. MERESHCHENKO, *Faculty of Chemistry, Saint-Petersburg State University, Saint-Petersburg, Russia*; ALEXANDER N TARNOVSKY, *Department of Chemistry and Center for Photochemical Sciences, Bowling Green State University, Bowling Green, OH, USA.*

Sub-100 fs time-resolved, broadband transient absorption spectroscopy was employed to investigate ultrafast radiationless relaxation dynamics of near-infrared, metal-centered (MC), electronic excited states of several d^5 and d^9 transition metal complexes (e.g., $CuCl_4^{2-}$, $CuBr_4^{2-}$, $IrBr_6^{2-}$, $IrCl_6^{2-}$, etc.) in acetonitrile solution. The results yield insights into the topology of the involved potential energy surfaces, Jann-Teller distortions, and the dynamics through conical intersections connecting the first excited and ground electronic states (energy gap, less than 8000 cm^{-1}). Furthermore, it was found that the addition of water to the solutions efficiently quenches the MC excited states via energy transfer.

FB05 9:38 – 9:53

DISCOVERY OF DATIVE BONDING OF BERYLLIUM FLUORIDE ANION BY PHOTOELECTRON VELOCITY MAP IMAGING SPECTROSCOPY

MALLORY THEIS, *Department of Chemistry, Emory University, Atlanta, GA, USA*; PEARL JEAN, *Chemistry Department, Emory University, Atlanta, Georgia, United States*; MICHAEL HEAVEN, *Department of Chemistry, Emory University, Atlanta, GA, USA.*

Beryllium can exhibit unusually strong attractive interactions under conditions where it is nominally a closed-shell atom. Two prominent examples are the Be_2 dimer and the He-BeO complex. Most recently, we examined the bonding of a similarly interesting molecule, the closed-shell Be-F^- anion. This molecule preserves the closed-shell character of the atoms as the electron affinity of F is high (328.16 kJ mol^{-1}) while that of Be is negative. Photoelectron velocity map imaging spectroscopy, in conjunction with coupled cluster electronic structure calculations, were used to determine the vibrational frequency for BeF^- and the electron affinity of BeF (approximately 8700 cm^{-1}). The latter has been used to determine a lower bound of 28480 cm^{-1}(343 kJ mol^{-1}) for the bond energy of BeF^-. The electronic structure calculations yielded predictions that were in good agreement with the observed data. A natural bond orbital analysis shows that BeF^- is primarily bound by a dative interaction.

Intermission

FB06   10:29 – 10:44

SPECTROSCOPY OF TiO SINGLET STATES

<u>DROR M. BITTNER</u>, PETER F. BERNATH, *Department of Chemistry and Biochemistry, Old Dominion University, Norfolk, VA, USA.*

TiO is a molecule of considerable astronomical importance. It is present in the atmospheres of oxygen-rich low-mass stellar objects. Three Fourier transform emission spectra have been used to determine improved and consistent spectroscopic constants of the $a^1\Delta$, $b^1\Pi$, $d^1\Sigma^+$, $c^1\Phi$ and $f^1\Delta$ states of TiO by fitting the $b^1\Pi$-$a^1\Delta$, $b^1\Pi$-$d^1\Sigma^+$, $c^1\Phi$-$a^1\Delta$ and $f^1\Delta$-$a^1\Delta$ systems. This analysis provides the most extensive fit of the TiO singlet states. New bands of the $b^1\Pi$-$a^1\Delta$ and $c^1\Phi$-$a^1\Delta$ systems have been measured and an extensive list of line positions will be published.

FB07   10:46 – 11:01

ROTATIONAL ANALYSIS OF SEVERAL VIBRATIONAL BANDS OF THE [7.7] $Y\ ^2\Sigma^+$ - $X\ ^2\Pi_i$ TRANSITION OF ^{63}CuO

<u>JACK C HARMS</u>, JAMES J O'BRIEN, *Chemistry and Biochemistry, University of Missouri, St. Louis, MO, USA*; LEAH C O'BRIEN, *Department of Chemistry, Southern Illinois University, Edwardsville, IL, USA.*

The [7.7] $Y\ ^2\Sigma^+$ - $X\ ^2\Pi_i$ transition of CuO was observed in emission from a Cu hollow cathode recorded with the FT-spectrometer associated with the McMath-Pierce Solar Telescope at Kitt Peak in 1994. In 1996, a rotational analysis of the (0,0) band of the Y - X transition was reported by O'Brien *et al.* In a recent analysis of the (0,0) and (1,1) bands of the [16.4] $A\ ^2\Sigma^-$ - $X\ ^2\Pi_i$ transition of CuO performed by the authors, improved centrifugal distortion constants for the $X\ ^2\Pi_i$ state were obtained. Line positions from the millimeter wave spectrum of CuO reported by Steimle *et al.* were successfully incorporated into the fit of the $A - X$ transition, however, a fit including the line positions of the Y - X transition reported by O'Brien *et al.* showed small, yet significant, deviations in the residuals of the fit. In this study, the FTS data from 1994 was accessed from the FTP archive available from the National Solar Observatory website to investigate these deviations. The calibration of the data was verified to ± 0.001 cm^{-1} using Ne reference lines from Sansonetti *et al.* Using PGOPHER simulations and the improved rotational constants for the $X\ ^2\Pi_i$ ground state, over 1000 additional line positions belonging to the (0,0) band of the Y - X transition were identified in the FTS data. Several previously unidentified vibrational bands were also observed and rotationally analyzed using PGOPHER, specifically the (2,0), (2,1), (2,2), (2,3), (1,0), (1,1), (1,2), (1,3), (0,0), (0,1), and (0,2) bands. A comprehensive fit of the data containing more than 10,000 line positions has been conducted using PGOPHER. The fit successfully incorporated the millimeter wave data for the $X\ ^2\Pi_i$ state from Steimle *et al.* and the intracavity laser absorption data for the A - X state from the authors. Results of this analysis will be presented.

FB08   11:03 – 11:18

ROTATIONAL ANALYSIS OF AN ELECTRONIC TRANSITION OF CuOH OBSERVED WITH INTRACAVITY LASER SPECTROSCOPY

<u>JACK C HARMS</u>, *Chemistry and Biochemistry, University of Missouri, St. Louis, MO, USA*; LEAH C O'BRIEN, *Department of Chemistry, Southern Illinois University, Edwardsville, IL, USA*; JAMES J O'BRIEN, *Chemistry and Biochemistry, University of Missouri, St. Louis, MO, USA*.

An electronic transition of CuOH has been observed in the red using Intracavity Laser Spectroscopy (ILS). The CuOH molecules were produced in the 0.60 A RF-discharge of a Cu hollow cathode using either 1 Torr of H_2 and 200 mTorr of O_2 as sputter gases to produce CuOH, or 800 mTorr of O_2 and 200 mTorr of D_2 to produce CuOD. The hollow cathode was located in the resonator cavity of a dye laser, enabling the enhancement of molecular absorption features through laser action. Using a generation time of 200 μsec and a 130 mm long cathode, the effective pathlength for the measurements was approximately 7 km. Each 6 cm^{-1} wide spectral segment was calibrated by fitting I_2 absorbance features from spectra collected from an extracavity I_2 cell at each monochromator position (with the plasma turned off) to the I_2 reference data of Salami and Ross. Deviations between the recorded and reference I_2 features were typically less than ± 0.002 cm^{-1}. Two strong red-degraded Q-bandheads are observed in the CuOH spectrum, located at 15,155 cm^{-1} and 15,093 cm^{-1}. These bands are consistent with an electronic transition briefly mentioned but not analyzed in a publication on the $A\ ^1A'$-$X\ ^1A'$ transition of CuOH by Jarman *et al.* from 1991. The isotopologue shifts of the CuOD Q-heads are +4.5 cm^{-1} and +5.0 cm^{-1}, respectively. The transitions for both isotopologues have P-, Q-, and R-branches with no observed splitting. Each branch is accompanied by features due to ^{65}CuOH that are approximately 40% of the intensity of the main features, consistent with natural abundances of 69.17% ^{63}Cu and 30.83% ^{65}Cu. Ground state constants for CuOH and CuOD are determined from the millimeter wave spectra of Whitham *et al.*, and the transitions will be fit using PGOPHER. Results of the analysis will be presented.

FC. Comparing theory and experiment
Friday, June 22, 2018 – 8:30 AM
Room: 1024 Chemistry Annex

Chair: Jennifer van Wijngaarden, University of Manitoba, Winnipeg, MB, Canada

FC01 8:30 – 8:45
THE ROTATIONAL SPECTRUM AND POTENTIAL ENERGY SURFACE OF AR-SIO: AN EXPERIMENTAL INVESTIGATION

MICHAEL C McCARTHY, *Atomic and Molecular Physics, Harvard-Smithsonian Center for Astrophysics, Cambridge, MA, USA*; RICHARD DAWES, *Department of Chemistry, Missouri University of Science and Technology, Rolla, MO, USA.*

The rotational spectra of five isotopic species of the Ar–SiO complex have been observed at high-spectral resolution between 8 and 18 GHz using chirped Fourier transform microwave spectroscopy and a discharge nozzle source; follow-up cavity measurements have extended these measurements to as high as 35 GHz. The spectra of the normal species is dominated by a strong progression of a-type rotational transitions arising from increasing quanta in the Si–O stretch. A rotational analysis of these lines and a hyperfine analysis of Ar–Si^{17}O suggest that the complex is a highly fluxional prolate symmetric rotor with a vibrationally-averaged structure close to T-shaped in which the oxygen atom lies closer to argon than the silicon atom, much like Ar–CO. Newly performed calculations of the rovibrational level pattern are in good agreement with the experimentally-derived rotational constants of normal and isotopic Ar–SiO up to $v=12$ ($\sim$14,500 cm^{-1}) in the Si–O stretch suggesting that the present theoretical treatment well reproduces the salient properties of the intramolecular potential.

FC02 8:47 – 9:02
THE ROTATIONAL SPECTRUM AND POTENTIAL ENERGY SURFACE OF AR-SIO: A THEORETICAL INVESTIGATION

RICHARD DAWES, *Department of Chemistry, Missouri University of Science and Technology, Rolla, MO, USA*; MICHAEL C McCARTHY, *Atomic and Molecular Physics, Harvard-Smithsonian Center for Astrophysics, Cambridge, MA, USA.*

The rotational spectra of five isotopic species of the Ar–SiO complex have been observed at high-spectral resolution, employing various techniques to obtain spectra between 8 and 35 GHz. Progressions of rotational transitions were recorded for a range of quanta in the Si-O stretch which correspond to resonance states of the complex since the vibrational frequency of the diatomic exceeds the binding energy of the complex. A complementary theoretical study was performed in which variational rovibrational calculations were performed using a series of potential energy surfaces (PESs) rep-

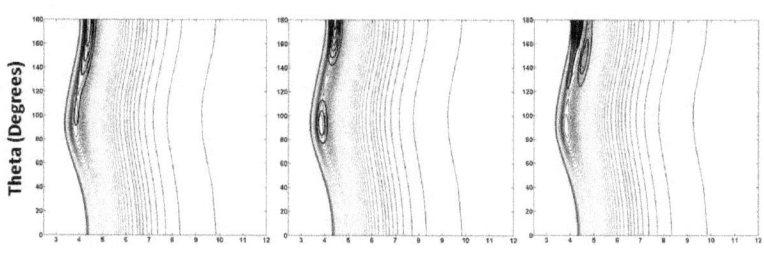

Probability density plots for some J=0 vdW vibrational states on the SiO(v=0)—Ar PES. (left) ground state (E= 0.0 cm^{-1}), (middle) first bend-excited state (E = 2.63 cm^{-1}), (right) first stretch-excited state (E = 33.57 cm^{-1}). Probability is delocalized along a low energy channel towards linearity.

resenting the SiO + Ar interaction and describing a range of vibrational quanta in the SiO(v=0,1...7) fragment. As seen in the Figure, the global minimum (V = -152.2 cm^{-1}) is nearly T-shaped, but a barrier of only 7.2 cm^{-1} leads to a second minimum in a long channel along the angular coordinate. The relative energy of the T-shaped minimum and the channel toward linearity, varies with the number of quanta in SiO (progressively favoring the more linear structure), and for SiO(v=7), the T-shaped structure is no longer the global minimum. To compute the rovibrational levels and wavefunctions, the RV3 three-atom variational code of Wang and Carrington was used.

FC03 9:04 – 9:19

HIGH RESOLUTION ROTATIONAL SPECTROSCOPY OF CH_3^+-He

MATTHIAS TÖPFER, THOMAS SALOMON, *I. Physikalisches Institut, Universität zu Köln, Köln, Germany*; OTTO DOPFER, *Institut für Optik und Atomare Physik, Technische Universität Berlin, Berlin, Germany*; KOICHI MT YAMADA, *Institute for Environmental Management Technology (EMTech), National Institute of Advanced Industrial Science and Technology (AIST), Tsukuba, Japan*; HIROSHI KOHGUCHI, *Department of Chemistry, Hiroshima University, Hiroshima, Japan*; STEPHAN SCHLEMMER, OSKAR ASVANY, *I. Physikalisches Institut, Universität zu Köln, Köln, Germany*.

Using an infrared and millimeter double resonance action spectroscopic scheme in a cryogenic ion trap, high resolution rotational transitions are recorded for the (near) symmetric top CH_3^+-He. Eighteen rotational transitions up to $J'' = 6$ are recorded in the range 60 - 410 GHz. Small unexpected splittings are resolved for $K = 1$. Advantages of this novel approach of high-resolution ion spectroscopy and potential future applications to spectra of cold cation-helium complexes are discussed.

FC04 9:21 – 9:36

SLOW PHOTOELECTRON VELOCITY-MAP IMAGING (SEVI) SPECTROSCOPY OF CRYO-COOLED ANIONS

MARISSA L. WEICHMAN, *JILA, National Institute of Standards and Technology and Univ. of Colorado Department of Physics, University of Colorado, Boulder, CO, USA*; JONGJIN B. KIM, JESSALYN A. DeVINE, DANIEL NEUMARK, *Department of Chemistry, The University of California, Berkeley, CA, USA*.

Slow photoelectron velocity-map imaging spectroscopy of cryogenically-cooled anions (cryo-SEVI) is a powerful technique for elucidating the vibrational and electronic structure of exotic neutral species. SEVI is a high-resolution variant of anion photoelectron imaging that yields spectra with energy resolution as high as 1 cm^{-1}. The preparation of cold anions eliminates hot bands and narrows rotational envelopes, enabling the acquisition of well-resolved photoelectron spectra for complex and spectroscopically challenging species.[1,2]

Recently, cryo-SEVI has been applied as a spectroscopic probe of transition state dynamics on neutral reactive surfaces, through photodetachment of a bound anion similar in geometry to the desired transition state. In the benchmark F + H$_2$ reaction, we probe the transition state region through detachment of FH$_2^-$ and directly observe new reactive resonances. Comparison to new theory allows for the assignment of resonances associated with quasi-bound states of the transition state and products.[3] We also report spectra of the F + CH$_3$OH hydrogen abstraction reaction through photodetachment of the CH$_3$OHF$^-$ van der Waals cluster. We gain insight into the energetics and vibrational structure of transient complexes along the reaction coordinate of this complex polyatomic system.[4]

Finally, we report a new cryo-SEVI study of vinylidene (H$_2$CC), a high energy isomer of acetylene, which is accessed directly through detachment of H$_2$CC$^-$. We find spectroscopic evidence that the isomerization of vinylidene to acetylene is highly state-specific, with excitation of the ν_6 in-plane rocking mode resulting in appreciable tunneling-facilitated mixing with highly vibrationally excited states of acetylene.[5]

---

[1]Hock *et al. JCP* **137**, 244201 (2012); [2]Weichman *et al. PNAS* **113**, 1698 (2016); [3]Kim *et al. Science* **349**, 510 (2015); [4]Weichman *et al. Nat. Chem.* **9**, 950 (2017); [5]DeVine *et al. Science* **358**, 336 (2017)

Intermission

FC05 10:12 – 10:27

DFT CALCULATION OF TORSIONAL LEVELS OF ETHANOL MOLECULE INCLUDING NONCOVALENT INTERACTIONS.

ULADZIMIR SAPESHKA[a], *Physics, University of Illinois at Chicago, Chicago, IL, USA*; GEORGE PITSEVICH, DOROZHKIN NIKOLAY, *Physics, Belarusian State University, Minsk, Belarus*; VYTAUTAS BALEVICIUS, *Physics, Vilnius University, Vilnius, Lithuania*.

Ethanol is the simplest molecule with two internal rotors, two stable configurations, which able to form H-bonded clusters of different sizes. In [1] 2D potential energy surface (PES) was calculated using MP4/cc-pVTZ levels of theory, however, geometry optimizations were performed at MP2/cc-pVTZ level. Results are in good agreement with experimental frequencies for HO group, but difference for the CH_3 group was essential. Similar calculations were performed [2,3] at B3LYP/cc-pVQZ and B3LYP/acc-pVQZ levels of theory using a denser grid for calculating 2D PES. Obtained results agree more with experiment, but problem with HO top still exists. It can be assumed that basis set too incomplete in [1], whereas B3LYP hybrid functional used in [2,3] cannot estimate the noncovalent interactions between atoms. To overcome these limitations we chose two DFT methods (wB97XD and M06-2X) and acc-pVQZ basis set. The first [4] refers to the long-range corrected hybrid density functionals, the second [5] is well suited for calculation of the main-group thermochemistry properties, barrier heights, noncovalent and "dispersion like" interactions. [6-8] show these hybrid fuctionals are good for both complexes and covalent systems.

The torsional coordinates of the methyl and hydroxyl groups were denoted as ϕ and γ with periodicity $2\pi/3$ and 2π respectively. The energies were found using wB97XD/acc-pVQZ and M06-2X/acc-pVQZ levels of theory. The kinematic coefficient matrix was calculated using Wilson's s vectors [2]. 2D vibrational Schrödinger equation was solved using DVR method.

[a][1] M.L. Senent, Y.G. Smeyers, R. Dominguez-Gomez, M. Villa, J.Chem.Phys., 112, 5809 (2000). [2] G. Pitsevich, A. Malevich, J.Appl.Spectr. 82, 505 (2015). [3] G. Pitsevich, A. Malevich, E. Kozlovskaya, A. Kapskaya, Sci.Adv.Today. 2, 25238 (2016). [4] J.-D. Chai, M. Head-Gordon, Phys.Chem.Chem.Phys. 10, 6615 (2008). [5] Y. Zhao, D.G. Truhlar, Theor.Chem.Account, 120, 215 (2008). [6] G.J. Jones, A. Robertazzi, J.A. Platts, J.Phys.Chem. B, 117, 3315 (2013). [7] Y. Liu, J. Zhao, F. Li, Z. Chen, J.Comp.Chem., 34, 121 (2013) [8] S.A. Smith, K.E. Hand, M.L. Love, G. Hill, D.H. Magers, J.Comp.Chem., 34, 558 (2013).

FC06 10:29 – 10:44

IMPROVEMENT OF THE DISSOCIATION ENERGY OF THE HYDROGEN MOLECULE (PART ONE)

JOËL HUSSELS, CUNFENG CHENG, MING LI NIU, HENDRICK BETHLEM, K.S.E. EIKEMA, EDCEL JOHN SALUMBIDES, WIM UBACHS, *Department of Physics and Astronomy, VU University, Amsterdam, Netherlands*; MAXIMILIAN BEYER, NICOLAS HOELSCH, JOSEF A. AGNER, FREDERIC MERKT, *Laboratorium für Physikalische Chemie, ETH Zurich, Zurich, Switzerland*; LEI-GANG TAO, SHUI-MING HU, *Hefei National Laboratory for Physical Science at Microscale, University of Science and Technology of China, Hefei, China*; CHRISTIAN JUNGEN, *Laboratoire Aimé Cotton, CNRS, Orsay, France*.

The dissociation energy (D_0) of ortho H_2 is a benchmark value in quantum chemistry, with recent QED calculations now approaching accuracies achievable in simple atoms. A precision measurement of the GK-X molecular transition, in combination with other precision measurements (see also part two), provides an improved value for D_0.[1, 2] The GK-X transition is excited through Doppler-free two-photon spectroscopy using 179-nm radiation, based on frequency up-conversion using a special KBBF crystal. The optical frequency of the fundamental (716 nm), which is the output of a narrowband pulsed Ti:Sa laser system, is locked to a frequency comb. This enables accuracies at the sub-MHz level, leading to an order-of-magnitude improvement for D_0 to the 10^{-9} level of accuracy. The comparison of this accurate experimental result with the best calculations may provide a test of the Standard Model of Physics.[3]

[1] D. Sprecher, Ch. Jungen, W. Ubachs and F. Merkt, Faraday Discussions 150, 51-70 (2011)
[2] W. Ubachs, J.C.J. Koelemeij, K.S.E. Eikema and E.J. Salumbides, J. Mol. Spectr. 320, 1-12 (2016)
[3] J. Liu, E. J. Salumbides, U. Hollenstein, J. C. J. Koelemeij, K. S. E. Eikema, W. Ubachs and F. Merkt, J. Chem. Phys. 130 (17), 174306 (2009)

FC07 10:46 – 11:01

IMPROVEMENT OF THE DISSOCIATION ENERGY OF THE HYDROGEN MOLECULE (PART TWO)

MAXIMILIAN BEYER, <u>NICOLAS HOELSCH</u>, JOSEF A. AGNER, FREDERIC MERKT, *Laboratorium für Physikalische Chemie, ETH Zurich, Zurich, Switzerland*; CUNFENG CHENG, JOËL HUSSELS, MING LI NIU, HENDRICK BETHLEM, K.S.E. EIKEMA, EDCEL JOHN SALUMBIDES, WIM UBACHS, *Department of Physics and Astronomy, VU University, Amsterdam, Netherlands*; CHRISTIAN JUNGEN, *Laboratoire Aimé Cotton, CNRS, Orsay, France*.

The dissociation energy D_0 of ortho H_2 is a benchmark quantity in quantum chemistry, with recent QED calculations now approaching accuracies achievable in simple atoms. In the light of recent discrepancies between experiment and theory [1], a combined effort (see also part one) has been undertaken to provide an improved experimental value for D_0.

We report the transition frequency from the $GK\ ^1\Sigma_g^+$ ($v = 1, N = 1$) state to the 56p ($N = 1, S = 0, F = 0 - 2$) Rydberg state belonging to the series converging on the $X^+\ ^2\Sigma_g^+$ ($v^+ = 0, N^+ = 1$) ground state of ortho H_2^+. A resonant three-photon excitation scheme was employed, using pulsed VUV and VIS laser sources to reach the intermediate GK state and a continuous-wave near-infrared (NIR) laser source for the transition to the Rydberg state. To reach the desired accuracy, the procedure involved [2]: (i) minimizing the Doppler width through the use of a doubly skimmed, supersonic molecular beam produced by a cryogenic pulsed valve, (ii) minimizing stray electric and magnetic fields, (iii) cancelling the first-order Doppler shift using two counterpropagating laser beams, (iv) calibrating the NIR-laser frequency using a frequency comb referenced to an atomic clock.

The ionization energy of the intermediate GK state was obtained by adding the binding energy of the Rydberg state determined previously by millimeter-wave spectroscopy and multichannel quantum-defect theory [3]. In combination with the $GK\ ^1\Sigma_g^+$ ($v = 1, N = 1$) ← $X\ ^1\Sigma_g^+$ ($v = 0, N = 1$) transition frequency presented in part one, an order-of magnitude improvement for D_0 at the 10^{-9} level of accuracy has been achieved, while remaining consistent with the previously most precise determination [4].

[1] M. Puchalski et al., Phys. Rev. A 95, 052506 (2017)
[2] M. Beyer et al., Phys. Rev. A 97, 012501 (2018)
[3] D. Sprecher et al., J. Chem. Phys. 140, 104303:1-18 (2014)
[4] J. Liu et al., J. Chem. Phys. 130 (17), 174306 (2009)

FC08 11:03 – 11:18

BENCHMARK CALCULATION OF K-EDGE IONIZATION ENERGIES USING EXACT-TWO-COMPONENT COUPLED-CLUSTER METHODS

<u>JUNZI LIU</u>, *Department of Chemistry, Johns Hopkins University, Baltimore, MD, USA*; DEVIN A. MATTHEWS, *Institute for Computational Engineering and Sciences, The University of Texas at Austin, Austin, TX, USA*; LAN CHENG, *Department of Chemistry, Johns Hopkins University, Baltimore, MD, USA*.

With the recent efficient implementation of coupled-cluster techniques with the inclusion of quadruple excitations [1] as well as the development of relativistic exact two-component theory [2], calculations of core-ionized/excited states aiming at high accuracy have become feasible [3]. Here we extend a preliminary study presented during the last ISMS to an extensive benchmark study involving twenty-six K-edge ionization energies of first-row elements (C, O, N, F) in fifteen molecules. Core-valence separation [4] has been used to facilitate the convergence of the equation-of-motion coupled-cluster equation for the core-ionized states. Effects from high-level correlation, geometrical relaxation, and vibrational corrections are critically analyzed, and the computed results are carefully compared with the corresponding experimental data.

References

[1] D. Matthews, and J. F. Stanton, J. Chem. Phys. **142**, 064108 (2015).

[2] K. G. Dyall, J. Chem. Phys. **106**, 9618 (1997).

[3] For example, see R. H. Myhre, T. J. A. Wolf, L. Cheng, S. Nandi, S. Coriani, M. Gühr and H. Koch, J. Chem. Phys. **148**, 064106 (2018).

[4] S. Coriani, and H. Koch, J. Chem. Phys. **143**, 181103 (2015).

FD. Fundamental interest

Friday, June 22, 2018 – 8:30 AM

Room: 217 Noyes Laboratory

Chair: Shui-Ming Hu, University of Science and Technology of China, Hefei, China

FD01   8:30 – 8:45

HAVING A BALL! MICROWAVE SPECTRUM OF THE (NEARLY) SPHERICAL TOP TEFLIC ACID

<u>SVEN HERBERS</u>, DANIEL A. OBENCHAIN, PETER KRAUS, JENS-UWE GRABOW, *Institut für Physikalische Chemie und Elektrochemie, Gottfried-Wilhelm-Leibniz-Universität, Hannover, Germany*.

The microwave spectrum of teflic acid (TeF_5OH) in the range of 3 to 25 GHz was analyzed. Though teflic acid is an asymmetric top in its equilibrium structure, it behaves like a symmetric top because of the OH group internal rotation. The strongest transitions in the spectrum originate from the $^{130}TeF_5OH$, $^{128}TeF_5OH$ and $^{126}TeF_5OH$ species as shown in the figure below. The $TeF_5{}^{18}OH$, TeF_5OD and $TeF_5{}^{18}OD$ isotopologues were also analyzed. From the rotational constants of the different isotopologues and with help of quantum chemical calculations a semi empirical equilibrium structure of teflic acid was determined. The Te-O equilibrium bondlength was determined with accuracy to the hundredth of an angstrom.

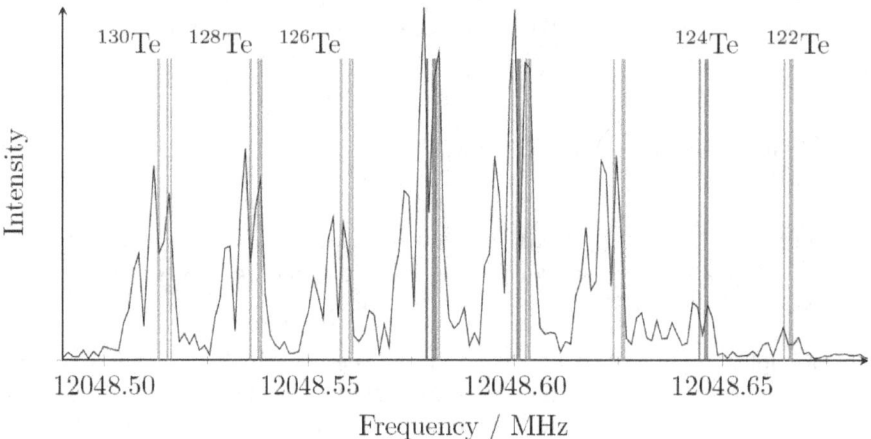

FD02   8:47 – 9:02

HIGH PRECISION SPECTRUM OF THE SECOND OVERTONE OF $^{12}C^{16}O$

<u>AN-WEN LIU</u>, JIN WANG, YU ROBERT SUN, SHUI-MING HU, *Hefei National Laboratory for Physical Science at Microscale, University of Science and Technology of China, Hefei, China*.

CO is the second most abundant molecule in the universe. Precise spectrum of the carbon monoxide molecule is great importance in astrophysical observation and in the test of the quantum chemistry model. Thirty-nine ro-vibrational transitions of $^{12}C^{16}O$ in the second overtone band were measured by a comb-locked cavity ring-down spectrometer. The line positions were determined with kHz accuracy, or relatively 10^{-12} level, which is over two orders of magnitude more accurate than previous Doppler-limited studies. Comparisons of the line positions determined in this work with literature experimental values and the calculations by the Dunham-Watson model are given. The bond length in the second overtone vibrational state was determined for carbon monoixde. The calculated pure rotational line positions agree with the experimental values recorded by lamb-dip spectrometer within the experimental uncertainties.

FD03 9:04 – 9:19

IMPLEMENTING THE NEW KELVIN BY MOLECULAR PRECISION SPECTROSCOPY

ELIAS MOUFAREJ, *Laboratoire de Physique des Lasers, CNRS, Université Paris 13, Sorbonne Paris Cité, Villetaneuse, France*; OLGA KOZLOVA, CATHERINE MARTIN, STEPHAN BRIAUDEAU, *Laboratoire Commun de Métrologie, LNE-CNAM, La Plaine Saint-Denis, France*; BENOIT DARQUIE, CHRISTOPHE DAUSSY, *Laboratoire de Physique des Lasers, CNRS, Université Paris 13, Sorbonne Paris Cité, Villetaneuse, France.*

Next autumn, the redefinition of the international System of Units (SI) will be accepted by the General Conference on Weights and Measures. This major reform will establish a new definition of units in terms of a set of 7 defining constants. The temperature unit, the kelvin, the definition of which is currently based on the triple point of water (TPW), will for example be redefined by fixing the value of the Boltzmann constant k_B. For the implementation of the new kelvin, various primary thermometry methods are currently being developed such as Doppler-broadening thermometry (DBT), acoustic gas thermometry, Johnson noise thermometry, dielectric constant gas thermometry, etc... We have previously proposed and developped the DBT method and have used it to demonstrate an accurate determination of $k_B{}^{a,b}$. Once k_B fixed in the new SI, DBT will become a primary spectroscopic method for thermodynamic temperature measurements and thus for the implementation of the new kelvin. DBT is based on the precise measurement of the Doppler broadening of absorption line of a gaz phase atomic or molecular species, an ammonia ro-vibrational transition in the mid-infrared range ($\sim 10\mu m$) in our case, combined with some highly accurate modeling of the line profile. We are currently developing DBT in the temperature range 300-430K, in order to demonstrate its potential and study its limitations beyond the temperature of the TPW. We will present our progress towards temprature measurements with uncertainties at the 25ppm level. The existing set-up (previously used for the determination of k_B) has been upgraded. We have placed the spectroscopic cell in a variable thermostat, the temperature stability and gradient of which has been characterized. We have also improved our mid-infrared spectrometer to investigate, in particular the influence of the line-mixing on the temperature measurement accuracy.

[a] C. Daussy et al., *First direct determination of the Boltzmann constant by an optical method*, Phys. Rev. Lett. **98**, 250801 (2007).
[b] S. Mejri et al., *Measuring the Boltzmann constant by mid-infrared laser spectroscopy of ammonia*, Metrologia **52**, S314 (2015).

FD04 9:21 – 9:36

TESTING THE PARITY SYMMETRY IN COLD CHIRAL MOLECULES USING VIBRATIONAL SPECTROSCOPY

MATTHIEU PIERENS, LOUIS LECORDIER, ANNE COURNOL, MATHIEU MANCEAU, SEAN TOKUNAGA, ALEXANDER SHELKOVNIKOV, OLIVIER LOPEZ, CHRISTOPHE DAUSSY, ANNE AMY-KLEIN, CHRISTIAN CHARDONNET, *Laboratoire de Physique des Lasers, CNRS, Université Paris 13, Sorbonne Paris Cité, Villetaneuse, France*; PIERRE ASSELIN, YANN BERGER, *CNRS, De la Molécule aux Nano-Objets: Réactivité, Interactions, Spectroscopies, MONARIS, Sorbonne Université, PARIS, France*; THERESE R. HUET, L. MARGULÈS, R. A. MOTIYENKO, *UMR 8523 - PhLAM - Physique des Lasers Atomes et Molécules, University of Lille, CNRS, F-59000 Lille, France*; RICHARD J. HENDRICKS, *National Physical Laboratory, NPL, Teddington, United Kingdom*; THOMAS WALL, MICHAEL TARBUTT, *Centre for Cold Matter, Imperial College London, London, United Kingdom*; BENOIT DARQUIE, *Laboratoire de Physique des Lasers, CNRS, Université Paris 13, Sorbonne Paris Cité, Villetaneuse, France.*

Parity violation (PV) has never been observed in chiral molecules. Caused by the weak nuclear force, PV should lead to a tiny energy difference between the enantiomers of a chiral molecule, and in turn to frequency differences in the rovibrational spectra of the two enantiomers of a chiral molecule. A successful PV measurement will shed some light on the origins of biomolecular homochirality. It can also constitute a test of the standard model in the low-energy regime and a probe of physics beyond it, and serve as a stringent benchmark in relativistic quantum chemistry calculations.

We present our ongoing work towards developing the technologies needed for measuring PV in chiral molecules via Ramsey interferometry in the mid-infrared. This includes amongst other things developing frequency stabilised quantum cascade lasers calibrated against primary standards and a buffer-gas source of organo-metallic species of interest for a PV measurement formed using laser ablation of solid-state molecules in a cryogenic cell containing gaseous helium at 4 K.

We also present the results of preliminary spectroscopic investigations conducted on various species, in particular methyltrioxorhenium (MTO), an achiral test molecule from which promising chiral derivatives have recently been synthesized. We report on the high-resolution spectroscopy of MTO, both in cells and in molecular beams, at various temperatures and resolutions.

FD05 9:38 – 9:53

VIBRATIONAL MODE MIXING FERMI RESONANCE AND VIBRATIONAL RELAXATION IN THE EXCITED STATE OF HYDROGEN BONDED COMPLEXES OF A PHENOLIC CHROMOPHORE

<u>DEB PRATIM MUKHOPADHYAY</u>, SOUVICK BISWAS, TAPAS CHAKRABORTY, *Physical Chemistry, Indian Association for the Cultivation of Science, Kolkata, India.*

We report here Laser-induced fluorescence (LIF) study of hydrogen-bonded 1:1 complexes of p-fluorophenol with common solvents e.g. water and methanol in supersonic jet expansion condition. We successively excite these binary complexes with different amount of vibronic energies over electronic origin transitions and monitor the energy-dependent photophysics by measuring the fluorescence emission from each excited level. Though the two complexes have very similar hydrogen bonding, O-H•••O type, but the lack of C_s symmetry going from water to methanol complex has enormous effects on electronic transitions. All intermolecular bands with their excited state counterparts are assigned with the help of quantum chemical calculations. Fermi type interactions between inter and intramolecular vibrations are observed for both the complexes. In water complex, the excitation energy is selectively transferred to another isoenergetic level when excited to an intramolecular ring mode (422 cm^{-1}) but the complete relaxation of the excitation energy is observed when ring breathing mode at 822 cm^{-1}is excited. However, this energy threshold for complete vibrational energy relaxation is even lower for p-fluorophenol-methanol complex due to increase of density of states and absence of symmetry restriction for mode coupling in S_1

FD06 9:55 – 10:10

2-METHYl-1-HEXEN-3-YNE AND 3-HEXYN-2-ONE ADVENTURES IN METHYL GROUP INTERNAL ROTATION

<u>SUSANNA L. STEPHENS</u>, ZAIN KHANNA, *Department of Chemistry, Wesleyan University, Middletown, CT, USA*; ROBERT KARL BOHN, *Department of Chemistry, University of Connecticut, Storrs, CT, USA*; STEWART E. NOVICK, *Department of Chemistry, Wesleyan University, Middletown, CT, USA*; S. A. COOKE, *Natural and Social Science, Purchase College SUNY, Purchase, NY, USA.*

The two titled molecules have been studied in the past by one of us (RKB) and coworkers. Due to unassigned splittings of the methyl internal rotations, and inconsistencies in the spectroscopic fits the studies were deemed incomplete for a manuscript but were previously presented in isolation at ISMS in Columbus.[a,b] By re-measurement we have identified new small splittings and added breadth the spectral fits to allow improved fitting. Parameters characterizing internal rotation have been determined for 2-methyl-1-hexen-3-yne for the first time. This work is part of a larger series of work which attempts to rationalize how molecules undergo structural and dynamical changes when altering one moiety, for example how an environment surrounding a double bond hinders the free rotation of an adjacent methyl group.

---

[a]Microwave study of 2-Methyl-Hexene-3-yne; Ground state and tortionally excited state; Yeager. Joseph, Bohn. Robert.K ISMS 2007
[b]Rotational spectrum and structure of 3-Hexyn-2-one; Bohn. Robert.K ISMS 1997

Intermission

FD07 10:46 – 11:01

ROTATIONAL-PREDISSOCIATION DOUBLE RESONANCE SPECTROSCOPY OF THE He-HCO$^+$ COMPLEX

THOMAS SALOMON, MATTHIAS TÖPFER, PHILLIP SCHREIER, *I. Physikalisches Institut, Universität zu Köln, Köln, Germany*; HIROSHI KOHGUCHI, *Department of Chemistry, Hiroshima University, Hiroshima, Japan*; LEONID SURIN, *Molecular Spectroscopy, Institute of Spectroscopy, Troitsk, Moscow, Russia*; <u>STEPHAN SCHLEMMER</u>, OSKAR ASVANY, *I. Physikalisches Institut, Universität zu Köln, Köln, Germany.*

Cation-Helium complexes are interesting spectroscopic systems due to the floppy bond of the helium atom. He-HCO$^+$ is a particularly interesting test system, as it is linear and has a $^1\Sigma$ ground state. So far experimental data have been limited to infrared studies on the ν_1 C-H stretching mode of He-HCO$^+$. In order to obtain high-resolution rotational data the recently developed rotational-predissociation double resonance method has been applied to this complex. Accurate molecular parameters will be presented and discussed.

FD08 11:03 – 11:18

AMMONIA AT 10^6 V/CM IN AN 8K ARGON MATRIX: POLARIZATION, ORIENTATION, AND PENDULARIZATION

YOUNGWOOK PARK, *Chemistry, Seoul National University, Seoul, South Korea*; ROBERT W FIELD, *Department of Chemistry, MIT, Cambridge, MA, USA*; HEON KANG, *Chemistry, Seoul National University, Seoul, South Korea*.

The ν_2 band of NH_3 and ND_3 in solid Ar at 8K inside an ice-film nanocapacitor is observed by reflection-absorption infrared spectroscopy (RAIRS). Ammonia is an almost-free rotor in the Ar matrix. As the electric field is increased up to 1×10^6 V/cm, the spectrum undergoes a sequence of (reversible) changes driven by two kinds of Stark effect: mixing of inversion doublet components (for $K_c \geq 1$) and c-dipole type $\Delta K=0$, $\Delta J=\pm 1$ J-mixing. Mixing of inversion doublets results in quenching of the inversion dynamics and J-mixing leads toward pendularization. H vs. D nuclear permutation symmetry effects are clearly visible in the spectrum. At 10^6 V/cm and 8 K, the Stark interaction energy, $\mu E = 25$ cm^{-1}, is larger than kBT = 5.6 cm^{-1} and intermediate between the inversion splittings in the NH_3 v=0 and v_2=1 levels, 0.79 and 36.5 cm^{-1}. This frequency domain spectrum in a scanned DC electric field encodes a more complete dynamical picture than experiments in which an extremely high electric field is generated by a focused ultrafast laser pulse.

Samsung Science and Technology Foundation (SSTF-BA1301-04)

FD09 11:20 – 11:35

QUANTUM CASCADE LASER SPECTROSCOPY OF CARBONYL SULFIDE AND METHANOL ISOTOPOLOGUES IN HELIUM NANODROPLETS

ISAAC JAMES MILLER, TY FAULKNER, PAUL RASTON, *Chemistry and Biochemistry, James Madison University, Harrisonburg, VA, USA*.

Superfluid helium nanodroplets present a unique environment for the investigation of the coupling of solvent density to the rotation of embedded molecules [1]. This coupling results in a reduction of the gas phase rotational constant, B_{gas}, by an amount that depends on both the gas phase rotational velocity and the anisotropy of the helium-rotor interaction potential [2]. We can gain insight into the dependence of B_{gas} on the coupling by investigating different isotopologues of a given molecule, such as HCN/DCN [3] (since the interaction potential is approximately the same between them). With this in mind, we recorded the high-resolution infrared spectra of carbonyl sulfide and methanol isotopologues from 4.7 to 5.0 μm, using a newly built spectrometer. This spectral region allows for coverage of the CO stretching and third overtone bending bands of carbonyl sulfide, and the symmetric CD_3 stretching band of methanol. For both systems, we find that the heavier isotopologues couple to more helium density, and explore the connection between the two molecules in terms of their dependence of B_{gas} on the amount of coupled helium density.

[1] S. Grebenev, J. P. Toennies, and A. F. Vilesov, Science 279, 2083 (1998).
[2] S. Paolini, S. Fantoni, S. Moroni, and S. Baroni, J. Chem. Phys. 123, 114306 (2005).
[3] A. Conjusteau, C. Callegari, I. Reinhard, K. K. Lehmann, and G. Scoles, J. Chem. Phys. 113, 4840 (2000).

FE. Ions

Friday, June 22, 2018 – 8:30 AM

Room: B102 Chemical and Life Sciences

Chair: Caroline Chick Jarrold, Indiana University, Bloomington, IN, USA

FE01   8:30–8:45

CHARACTERIZING PEPTIDE ALPHA HELICES VIA COLD ION SPECTROSCOPY OF MODEL COMPOUNDS

JOHN T LAWLER, CHRISTOPHER P HARRILAL, *Department of Chemistry, Purdue University, West Lafayette, IN, USA*; TIMOTHY HILL, DAVID FAIRLIE, *Institute for Molecular Bioscience, The University of Queensland, Brisbane, Australia*; SCOTT A McLUCKEY, TIMOTHY S. ZWIER, *Department of Chemistry, Purdue University, West Lafayette, IN, USA.*

Tethered peptides are synthetic peptides in which a chemical linkage between two remote sites in the peptide sequence bind these sites together. The Fairlie group has devised a tether that locks the pentapeptide 'core' into a single turn of an alpha helix, robust to large swings in pH, temperature, and denaturant. By changing the chirality of the amino acids, left- (D) and right-handed (L) helices can be exclusively formed. Catenating these sub-units leads to α-helices of greater length. This talk describes the propensity of these tethered peptides to maintain their alpha helical nature upon the removal of solvent and transition into the gas phase as protonated ions. To this end we explore the structures of three tethered peptides linked together through the formation of a lactam between the lysine and aspartic acid residues, Y[KAAAD]-NH$_2$, F[KAAAD]-NH$_2$, and YR[KAAAD]-NH$_2$ (tether denoted by brackets). UV photofragment spectroscopy and IR-UV double resonance methods will be carried out on the cryocooled, protonated ions to probe the hydrogen bonding patterns of these molecules with the goal of elucidating the unique spectroscopic signatures of isolated single-turn alpha helices. The effect of the protonation site and handedness of the helix on the H-bonds in the single turn helices will also be described.

FE02   8:47–9:02

INVESTIGATING ELECTRONIC AND STRUCTURAL CHANGES IMPOSED BY ZWITTERIONIC PARING IN MODEL PEPTIDE SYSTEMS USING IR-UV-IR TRIPLE RESONANCE SPECTROSCOPY

CHRISTOPHER P HARRILAL, *Department of Chemistry, Purdue University, West Lafayette, IN, USA*; ANTHONY PITTS-MCCOY, *Chemistry, Purdue University, West Lafayette, IN, USA*; SCOTT A McLUCKEY, TIMOTHY S. ZWIER, *Department of Chemistry, Purdue University, West Lafayette, IN, USA.*

Strong electrostatic interactions such as zwitterionic pairing between oppositely charged amino acids are common in the condensed phase at neutral pH and can play a large role in determining the conformational landscape of peptides and proteins. Whether such interactions are possible in the absence of solvent however, has been previously debated. Growing experimental evidence suggests that these interactions are indeed possible in isolated gas phase ions and may give rise to unique fragmentation upon UV irradiation. In this study we use a series IR-UV-IR triple resonance techniques performed at 10 K to investigate the influences of these electrostatic interactions on the electronic and structural properties of model YGRXR (X = gly, asp) pentapeptide systems and their methyl ester counterparts. The initial electronic spectra, under single UV photon conditions, of model systems which may possess zwitterionic pairing hardly show discrete electronic transitions, rather a broad absorbtion which mainly gives rise to tyrosine side chain cleavage is observed. Upon methylation of the carboxylate functional groups, which prevents zwitterionic interactions, the cold action spectra become well resolved such that sharp electronic transitions due to the $\pi\pi^*$ transition of the tyrosine aromatic ring are observed. Using an UV-IR double resonance scheme it is possible to enhance the tyrosine side chain cleavage after an initial UV excitation, provided that the IR laser is fixed on a vibrational. Under these conditions the Franck Condon progressions for the non-methyl esterified systems become clearly observable. These initial results suggest that local excitation of the chromophore may couple to the autoionizing state responsible for electron detachment, similar to the mechanism postulated for photoinduced electron detachment from gas phase anions. Using IR-UV-IR triple resonance, conformer specific IR spectra can be taken for zwitterionic systems despite the large "off-resonance" absorbtion. A comparison of the IR spectra reveal that the +1 charge states are more prone to form zwitterionic interactions than the +2. Harmonic-level vibrational frequency calculations will be performed on candidate structures and compared to experimental spectra such that the influences of zwitterionic ionic pairing on the 3-dimensional structure can be directly compared to conformations without such parings.

FE03
9:04 – 9:19

MASS-ANALYZED THRESHOLD IONIZATION SPECTROSCOPY OF P-CHLOROANISOLE

SHEN-YUAN TZENG, WEN-BIH TZENG, *Institute of Atomic and Molecular Sciences, Academia Sinica, Taipei, Taiwan.*

We applied the two-color resonant two-photon photoionization efficiency and mass-analyzed threshold ionization (MATI) spectroscopic techniques to record the cation spectra of p-chloroanisole. In particular, several vibronic states were used as the intermediate levels to record the MATI spectra to investigate whether a significant change in molecular geometry upon ionization and to obtain more information about the active cation vibrations. The adiabatic ionization energy of this molecule has been precisely measured to be $66\,100 \pm 5$ cm-1. These experimental data suggest that the molecular geometry of p-chloroanisole in the cationic ground D0 state resembles that in the electronically excited neutral S1 state. Most of the observed distinct MATI bands result from the active vibrations involving in-plane ring deformation of the p-chloroanisole cation.

FE04
9:21 – 9:36

PHOTODETACHMENT AND RESONANT PHOTOELECTRON SPECTROSCOPY OF CRYOGENICALLY-COOLED PHENOXIDE ANIONS VIA DIPOLE-BOUND EXCITED STATES

CHEN-HUI QIAN, GUO-ZHU ZHU, LAI-SHENG WANG, *Department of Chemistry, Brown University, Providence, RI, USA.*

The phenoxide anion ($C_6H_5O^-$), with a large dipole moment of 4.0 D for its neutral core, can support an excited dipole-bound state (DBS) 97 cm^{-1} below the electron detachment threhold [Liu et al. Angew. Chem. Int. Ed. 52, 8976-8979 (2013)]. The vibrational mode-specific autodetachments from the vibrational levels of the DBS were first discovered in this system by high-resolution resonant photoelectron (PE) imaging. Here, we report a photodetachment spectrum of the phenoxide anion up to 2000 cm^{-1} above the detachment threshold. Several new vibrational peaks are observed due to autodetachment from combinational levels of the DBS. The corresponding resonant PE spectra show a clear peak due to the lowest-frequency bending mode, suggesting the possibility of a slight nonplanar structure for the phenoxy radical.

FE05
9:38 – 9:53

PHOTOINDUCED CHARGE TRANSFER IN CATION-π COMPLEXES STUDIED WITH VMI

BRANDON M. RITTGERS, DANIEL LEICHT, MICHAEL A DUNCAN, *Department of Chemistry, University of Georgia, Athens, GA, USA.*

The photodissociation charge transfer processes of Ag^+ cation-π complexes with small aromatics were studied using velocity map imaging. Ions formed by laser vaporization are pulse extracted and mass-selected in a linear time-of-flight mass spectrometer. The ion beam is then intersected with a UV laser causing dissociation, and the ions are detected using a fast-phosphor screen. The detector has spatial resolution which allows us to extract the total kinetic energy release of the dissociation process, which gives us information on the binding energy of the ion. Excitation with 355 nm lead to the dissociative charge transfer of Ag^+-toluene and Ag^+-furan, as seen previously with the Ag^+-benzene complex.

FE06   9:55 – 10:10

INFRARED SPECTROSCOPY OF Zn(ACETYLENE)$_{1-5}^+$: EVIDENCE OF ACETYLENE ACTIVATION BY A METAL RADICAL

<u>JOSHUA H MARKS</u>, TIMOTHY B WARD, MICHAEL A DUNCAN, *Department of Chemistry, University of Georgia, Athens, GA, USA.*

Zinc cation is studied as a model system for single atom catalysis in the gas phase with infrared photodissociation spectroscopy. Zn(C$_2$H$_2$)$_n^+$ (n = 1–5) clusters are produced via laser vaporization of zinc in a supersonic expansion of acetylene and argon. Clusters are mass-selected and studied with infrared photodissociation spectroscopy in the C–H stretching region. Smaller clusters (n = 1–3) are studied with the use of a weakly bound argon tag. These spectra are assigned with B3LYP/Def2TZVP computational studies. Zn(C$_2$H$_2$)$^+$ is found to consist of a C$_{2v}$ three membered metallacycle, where zinc is equidistant from both carbon atoms of acetylene. Zn(C$_2$H$_2$)$_2^+$ does not contain a metallacycle, but features zinc binding more closely to one of the carbon atoms of each acetylene in a C$_2$ configuration. The three-coordinate cluster is predicted to be lowest in energy as a π-bound D$_{3h}$ structure, with a low energy C$_3$ structure. When the spectrum of this cluster is measured with argon tagging the D$_{3h}$ isomer is most abundant. When measured without the tag the C$_3$ isomer is found to be in abundance. The spectra of the four and five coordinate clusters are found to contain a feature 160 cm^{-1} to the red of the acetylene C–H asymmetric stretch. This is attributed to a fourth acetylene ligand forming a metal vinyl radical, accompanied by formation of Zn(II). This transfer of the radical center from zinc to a ligand activates the acetylene, and could be the first step in single atom catalysis by zinc.

Intermission

FE07   10:46 – 11:01

SINGLE ATOM CATALYTIC CYCLOTRIMERIZATION OF V(ACETYLENE)$_3^+$ STUDIED WITH INFRARED SPECTROSCOPY

<u>JOSHUA H MARKS</u>, TIMOTHY B WARD, MICHAEL A DUNCAN, *Department of Chemistry, University of Georgia, Athens, GA, USA.*

Vanadium cation is studied as a model system for single atom catalysis in the gas phase with infrared photodissociation spectroscopy. Intermediates in the cyclotrimerization of acetylene to form benzene are observed. V(C$_2$H$_2$)$_n^+$ clusters are produced via laser vaporization of vanadium in a supersonic expansion of argon containing acetylene. Clusters of V(C$_2$H$_2$)$^+$, V(C$_2$H$_2$)$_2^+$, and V(C$_2$H$_2$)$_3^+$ are studied with infrared photodissociation spectroscopy with the aid of argon tagging in the C–H stretching region. These spectra are assigned on the basis of B3LYP computations. V(C$_2$H$_2$)$^+$ is a three membered metallacycle, where the hydrogens bend away from the vanadium. V(C$_2$H$_2$)$_2^+$ is a bimetallacycle where both acetylene ligands interact with vanadium symmetrically through their π-bonds in a C$_{2v}$ configuration. The structure of V(C$_2$H$_2$)$_3^+$ is found to vary with the concentration of acetylene in the supersonic expansion. At low concentrations of acetylene two isomers of V(C$_2$H$_2$)$_3^+$ are observed, a trimetallacycle, and a bimetallacycle which includes a five membered ring and a three membered ring. As the concentration of acetylene is increased past 5% the trimetallacycle decreases in abundance. Benzene is observed at yet higher concentration of acetylene. An expansion gas consisting of 15% acetylene in argon results in the exclusive formation of V(Bz)$^+$.

FE08 11:03–11:18

THRESHOLD IONIZATION SPECTROSCOPY AND SPIN-ORBIT COUPLING OF LnNH (Ln = La and Ce) FORMED BY Ln REACTIONS WITH AMMONIA

YUCHEN ZHANG, SILVER NYAMBO, DONG-SHENG YANG, *Department of Chemistry, University of Kentucky, Lexington, KY, USA.*

Ln (Ln = La and Ce) atom reactions with ammonia are carried out in a pulsed laser vaporization supersonic molecular beam source, and metal-containing species are observed with time-of-flight mass spectrometry and characterized by mass-analyzed threshold ionization (MATI) spectroscopy. The MATI spectrum of LaNH exhibits a single vibronic band system with a strong origin band and two weak vibronic progressions, whereas the spectrum of CeNH shows two band systems separated by 80 cm^{-1}, with each being similar to the LaNH spectrum. By comparing with theoretical calculations, both LaNH and CeNH are identified as linear molecules with symmetry, and the two vibronic progressions are attributed to excitations of Ln-N stretching and LnNH bending modes in the ions. The additional band system observed for CeNH is due to spin-orbit splitting from interactions of a pair of nearly degenerate triplets and a pair of nearly degenerate singlets. The ground valence electron configurations of LaNH and CeNH are La $6s^1$ and Ce $4f^1 6s^1$, and ionization of each species removes the Ln $6s^1$ electron. The remaining two electrons that are associated with the isolated Ln atoms or ions are spin paired in a molecular orbital that is a bonding combination between a Ln 5d orbital and a N π^* antibonding orbital.

FE09 11:20–11:35

TWO-PHOTON IONIZATION STUDY OF THE LOW LYING STATES OF UN^+

ROBERT A. VANGUNDY, THOMAS D PERSINGER, MICHAEL HEAVEN, *Department of Chemistry, Emory University, Atlanta, GA, USA.*

The electronic structures of UN and UN^+ are of interest for the testing and development of relativistic quantum chemistry methods. The ground state UN was probed by Matthew and Morse[1], who found that the electronic configuration ($5f^2 7s$) differed from that of the isoelectronic UO^+ cation ($5f^3$). In the present study we examine the ionization energy of UN and the low energy states of UN^+ by means of pulsed-field ionization zero kinetic energy photoelectron spectroscopy (PFI-ZEKE). Resonantly enhanced two photon ionization (R2PI) coupled with a time of flight mass spectrometer was used to confirm production of the UN molecule and locate suitable electronically excited states for subsequent access to UN^+ via high-n Rydberg states. The results will be compared to the predictions from ligand field theory and high-level ab ignition calculations.

1. D. J. Matthew and M. D. Morse, J. Chem. Phys. 138, 184303 (2013)

FE10 11:37–11:52

CESIUM IONIZATION AND RECOMBINATION

SEAN MICHAEL BRESLER, *Physical Chemistry, Emory University, Atlanta, GA, USA*; MICHAEL HEAVEN, *Department of Chemistry, Emory University, Atlanta, GA, USA.*

Diode Pumped Alkali Lasers (DPALS) are promising candidates for high-power directed energy applications including data transmission and ballistics defense. Selection of an appropriate buffer gas for the gain media requires absence of undesirable chemical reactions while still meeting the kinetic requirements of the system. Small hydrocarbons have been investigated as a potential buffer gas, and while these meet many of the kinetic requirements of the system, they produce unwanted side products, depleting the gain media. Recent measurements including Laser Induced Fluorescence (LIF), dispersed fluorescence, and ion lifetime measurements indicate that the dominant pathway to the products involves ions and highly excited alkali atoms with very long relaxation times.

FF. Atmospheric science
Friday, June 22, 2018 – 8:30 AM
Room: 2079 Natural History

Chair: Jacob Stewart, Connecticut College, New London, CT, USA

FF01 8:30–8:45

POSITIONS, INTENSITIES AND AIR-BROADENED LINE SHAPE PARAMETERS FOR THE 1←0 BANDS OF CO ISOTOPOLOGUES

V. MALATHY DEVI, D. CHRIS BENNER, *Department of Physics, College of William and Mary, Williamsburg, VA, USA*; KEEYOON SUNG, TIMOTHY J. CRAWFORD, *Jet Propulsion Laboratory, California Institute of Technology, Pasadena, CA, USA*; GANG LI, *PTB, Physikalisch-Technische Bundesanstalt, Braunschweig, Germany*; ROBERT R. GAMACHE, *Department of Environmental, Earth, and Atmospheric Sciences, University of Massachusetts, Lowell, MA, USA*; MARY ANN H. SMITH, *Science Directorate, NASA Langley Research Center, Hampton, VA, USA*; IOULI E GORDON, *Atomic and Molecular Physics, Harvard-Smithsonian Center for Astrophysics, Cambridge, MA, USA*; ARLAN MANTZ, *Department of Physics, Astronomy and Geophysics, Connecticut College, New London, CT, USA.*

High-resolution spectra recorded with Fourier transform spectrometers (FTS) have been analyzed to determine line positions and intensities for transitions in the 1←0 bands of $^{12}C^{16}O$, $^{13}C^{16}O$, $^{12}C^{18}O$ and $^{13}C^{16}O$, and air-broadened half-width and shift coefficients, their temperature dependences, line mixing and speed dependence parameters were measured for $^{13}C^{16}O$ and $^{12}C^{18}O$ transitions. These parameters were retrieved from two multispectrum fittings (1940-2260 cm^{-1}) of a data set that included two room-temperature spectra of a natural sample of CO recorded with the Kitt Peak FTS and self- and air-broadened spectra (up to 626 Torr) of ^{13}C-enriched and ^{18}O-enriched CO samples between 150 K and room temperature recorded with the JPL Bruker IFS-125HR FTS. Sample cells with path lengths of about 0.5, 1.1, 4.3 and 20.4 cm were used, and all but the shortest cell were temperature controlled. The retrieved 1←0 band strengths of $^{12}C^{16}O$, $^{12}C^{18}O$ and $^{13}C^{18}O$ are very close to the HITRAN2012[a] values, but for $^{13}C^{16}O$ the band strength is ∼4.5% larger than the HITRAN2012 value and 2.6% higher than the HITRAN2016[b] value.[c]

[a]L. S. Rothman et al.,*JQSRT* **130** (2013) 4-50.
[b]I. E. Gordon et al.,*JQSRT* **203** (2016) 3-69.
[c]Research described in this talk was performed at Connecticut College, the College of William and Mary, Langley Research Center and the Jet Propulsion Laboratory, California Institute of Technology, under contracts and cooperative agreements with NASA.

FF02 8:47–9:02

SPECTROSCOPIC STUDY OF SELF- AND AIR-BROADENED METHANE IN THE 4100-4300 cm^{-1} REGION

ADRIANA PREDOI-CROSS[a], *Department of Physics and Astronomy, University of Lethbridge, Lethbridge, Canada*; V. MALATHY DEVI, *Department of Physics, College of William and Mary, Williamsburg, VA, USA*; KEEYOON SUNG, *Jet Propulsion Laboratory, California Institute of Technology, Pasadena, CA, USA*; ANDREI V. NIKITIN, *Atmospheric Spectroscopy Div., Institute of Atmospheric Optics, Tomsk, Russia*; MARY ANN H. SMITH, *Science Directorate, NASA Langley Research Center, Hampton, VA, USA.*

The line parameters of self- and air-broadened methane in the $\nu_1+\nu_4$ and $\nu_3+\nu_4$ bands are determined using a nonlinear least-squares multispectrum fitting technique. We have analyzed a set of 14 laboratory spectra of pure methane and lean mixtures of methane in air which were recorded using a high-resolution Fourier Transform Spectrometer (FTS) at the Jet Propulsion Laboratory, California, employing a coolable sample cell with optical path length 20.38 cm. The line parameters determined in this analysis include line positions, intensities, self- and air-broadened line widths and pressure-induced shifts along with their temperature dependences, assuming a Speed-Dependent Voigt Profile (SDVP). The line mixing coefficients are quantified via the off-diagonal relaxation matrix element formalism. The broadening and shift parameters show good agreement with literature values and spectroscopic database entries. The observed line positions and intensities also agree fairly well with theoretically calculated results and values found in the spectroscopic databases. Spectroscopic parameters are also determined for some transitions of the $\nu_2+2\nu_4$, $2\nu_2+\nu_4$ and $3\nu_4$ bands of methane in the spectral range 4100-4300 cm^{-1}.[b]

[a]Current address: 512 Silkstone Crescent West, Lethbridge AB T1J4C1 Canada.
[b]Research described in this talk was performed at the College of William and Mary, Langley Research Center and the Jet Propulsion Laboratory, California Institute of Technology, under contracts and cooperative agreements with NASA.

FF03 9:04 – 9:19

A SPECTROSCOPIC PERTURBATION ORIGIN FOR SULFUR MASS INDEPENDENT FRACTIONATION VIA THE B-X SYSTEM OF S_2

ALEXANDER W HULL, *Department of Chemistry, MIT, Cambridge, MA, USA*; SHUHEI ONO, *Earth, Atmospheric, and Planetary Sciences, MIT, Cambridge, MA, USA*; ROBERT W FIELD, *Department of Chemistry, MIT, Cambridge, MA, USA.*

The Great Oxygenation Event (GOE), the introduction of O_2 into the Earth's atmosphere approximately 2.4 billion years ago, is a critical signpost in the development of life on Earth. The vanishing of sulfur isotope anomalies, called Sulfur Mass-Independent Fractionation (S-MIF), in the rock record is thought to be correlated with oxygenation of the early atmosphere. However, the mechanism for the generation of S-MIF in an anoxic atmosphere is unknown. Here, I propose a mechanism that involves spectroscopic perturbations in the B-X UV band system of S_2. This proposal is based on a global deperturbation analysis done by Green and Western[a][b] and work that I presented previously at this conference in 2015 (MG12) and 2016 (MG08). Specifically, perturbations of the "bright" B state by a "dark" B" state cause some isotopologues to have longer average excited state lifetimes than others. I demonstrate a difference between the shorter-lifetime symmetric (e.g. ^{32}S-^{32}S) isotopologues of S2, for which nuclear permutation symmetry causes half of the rotational lines to be missing, and the longer-lifetime asymmetric isotopologues (e.g. ^{33}S-^{32}S). I also comment on general features of the B/B" system of S_2 that make it uniquely well-suited to generate a large MIF isotope effect.

[a]M.E. Green, C.M. Western, A deperturbation analysis of the B $^3\Sigma_u^-$ (v' = 0-6) and the B" $^3\Pi_u$ (v' = 2-12) states of S_2, J. Chem. Phys. 104 (3) (1996) 848-864.
[b]M.E. Green, C.M. Western, Upper vibrational states of the B" $^3\Pi_u$ state of $^{32}S_2$, J. Chem. Soc., Faraday Trans. 93 (3) (1997) 365-372.

FF04 9:21 – 9:36

A POSSIBLE MECHANISM FOR SULFUR MASS INDEPENDENT FRACTIONATION IN THE B-X SYSTEM OF S_2

ALEXANDER W HULL, *Department of Chemistry, MIT, Cambridge, MA, USA*; SHUHEI ONO, *Earth, Atmospheric, and Planetary Sciences, MIT, Cambridge, MA, USA*; ROBERT W FIELD, *Department of Chemistry, MIT, Cambridge, MA, USA.*

Here, I continue my discussion of Sulfur Mass Independent Fractionation (S-MIF) in the S_2 B-X UV band system, relevant to the geologic signature of the anoxic atmosphere that existed prior to the Great Oxygen Event 2.4 billion years ago. To test a possible mechanism for the isotope effect, I implement a steady state, master equation kinetic model for every bound rovibronic state in the B/B" system. This model incorporates both rotationally inelastic and electronically inelastic collisions. The output of the model suggests that such collisions have negligible impact on average excited state lifetimes, indicating that the isotope effect is primarily spectroscopic in nature. The steady state populations that are output from the deperturbation/master equation model are useful in identifying which bright/dark state crossings are most proficient at populating long lifetime states, and thereby generating an S-MIF signature. A major conclusion from this analysis is that only a small minority of level crossings have such a capability, and, consequently, these have a dominant influence on the isotope effect averaged over the total system.

FF05 9:38 – 9:53

NEAR-GLOBAL ATMOSPHERIC DISTRIBUTIONs OF CARBONYL SULFIDE (OCS) ISOTOPOLOGUES

MAHDI YOUSEFI, *Department of Physics, Old Dominion University, Norfolk, VA, USA*; PETER F. BERNATH, *Department of Chemistry and Biochemistry, Old Dominion University, Norfolk, VA, USA*; CHRIS BOONE, *Department of Chemistry, University of Waterloo, Waterloo, ON, Canada.*

The distributions of the three most abundant isotopologues of carbonyl sulfide (OCS, $O^{13}CS$, and $OC^{34}S$) have been measured in the Earth's stratosphere by infrared remote sensing with the Atmospheric Chemistry Experiment (ACE) Fourier transform spectrometer. These satellite observations have provided a near-global picture of OCS isotopic fractionation. The ACE data indicate a different enrichment trend with altitude for the $O^{13}CS$ and $OC^{34}S$ isotopologues. The seasonal variation of the isotopologue enrichment was also studied using the ACE data. The Whole Atmosphere Community Climate Model (WACCM) has been used to model OCS and its isotopologue distributions in the stratosphere.

Intermission

FF06   10:29 – 10:44

INFRARED SPECTROSCOPIC CHARACTERIZATION OF THE STRUCTURES OF SULFURIC ACID/AMINE/WATER CLUSTERS

YI YANG, SARAH WALLER, ELEANOR CASTRACANE, EMILY E. RACOW, JOHN J. KREINBIHL, KATHLEEN A. NICKSON, CHRISTOPHER J JOHNSON, *Department of Chemistry, Stony Brook University, Stony Brook, NY, USA.*

It is estimated that ~50% of climatically relevant atmospheric aerosols arise from new particle formation (NPF), the process by which trace atmospheric gases such as sulfuric acid and ammonia cluster and grow. Amines are expected to enhance NPF, with greater enhancement from larger amines. Using cryogenic ion vibrational predissociation (CIVP) spectroscopy, we studied the structural evolution of clusters with up to 3 sulfuric acids. It is shown that substitution of amines for ammonia can induce structural rearrangement, which is driven by the ability of the alkylamines (MA, DMA, TMA) to form hydrogen bonds, and can lead to direct bisulfate-bisulfate hydrogen bonds. This direct interaction between formal anions indicates that hydrogen bonding can compete with Coulombic force in determining cluster structure. From these observations we have developed a model to predict when these arrangements may arise in ionic and neutral clusters with a variety of compositions. This structural motif is correlated with the fastest growing amines, and could play a role in the mechanism of NPF.

FF07   10:46 – 11:01

RATE CONSTANTS AND MECHANISM FOR THE REACTION OF ALKANES WITH ELECTRONICALLY EXCITED SO_2

JAY A KROLL, VERONICA VAIDA, *Department of Chemistry and Biochemistry, University of Colorado, Boulder, CO, USA.*

Sulfur compounds have been observed in a number of planetary atmospheres throughout our solar system. While our current understanding of sulfur chemistry explains much of what we observe in Earth's atmosphere, several discrepancies between modeling and observations of the Venusian atmosphere show there are still problems in our fundamental understanding of sulfur chemistry. Recent work in the Vaida lab has shown that electronically excited sulfur dioxide is incredibly reactive with a wide range of molecules including saturated alkanes. Using Infrared spectroscopic techniques, we have undertake a study to measure the rate constants for the reaction of electronically excited sulfur dioxide with a series of alkanes ranging from methane to n-nonane. We will present on the effect of chain length on the reaction rate and the effect of branched and ringed structures of alkanes on reaction with electronically excited sulfur dioxide.

HIGH RESOLUTION MICROWAVE SPECTROSCOPY IN A CRYOGENIC BUFFER GAS CELL: BRANCHING RATIOS AND REACTIVE INTERMEDIATES IN THE OZONOLYSIS OF ISOPRENE

JESSIE P PORTERFIELD, *AMP Division, Harvard-Smithsonian Center for Astrophysics, Cambridge, MA, USA*; SANDRA EIBENBERGER, DAVID PATTERSON, *Department of Physics, Harvard University, Cambridge, MA, USA*; MICHAEL C McCARTHY, *Atomic and Molecular Physics, Harvard-Smithsonian Center for Astrophysics, Cambridge, MA, USA*.

A new method to quantify reaction product ratios using high resolution microwave spectroscopy in a cryogenic buffer gas cell has been developed. We demonstrate its power with product ratio quantification in the ozonolysis of isoprene, $CH_2=C(CH_3)-CH=CH_2$, the most abundant, non-methane hydrocarbon emitted into the atmosphere by vegetation. Purified O_3 and isoprene were mixed for approximately 10 s under dilute (1.5-4% in argon) continuous flow conditions in an alumina tube held at 298 K and 5 Torr. Products exiting the tube were rapidly slowed and cooled within the buffer gas cell by collisions with cryogenic (4-7 K) He. High resolution chirped pulse microwave detection between 12 and 26 GHz was used to achieve isomer-specific product quantification with ppb sensitivity. We determined a ratio of MACR to MVK of 2.1 $\pm$ 0.4 under 1:1 ozone to isoprene conditions and 2.1 $\pm$ 0.2 under 2:1 ozone to isoprene conditions, a finding which is consistent with previous experimental results. The potential to perform a complete branching ratio analysis is discussed using ^{13}C isotopic substitution of isoprene. We also discuss the prospects for detecting the proposed Criegee intermediates in this reaction, methacrolein-oxide and methyl vinyl ketone-oxide, neither of which have ever been observed.

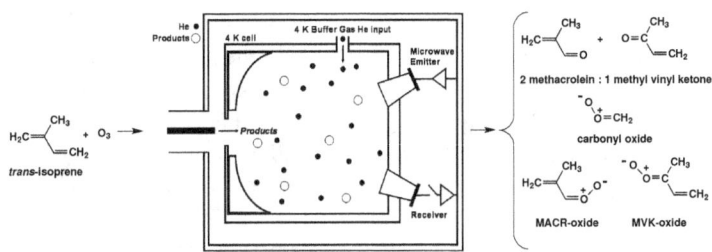

AUTHOR INDEX

A

Abdul-Munaim, Ali Mazin – TH10
Abeysekera, Chamara – WK02
Abgrall, Michel – TG02
Abplanalp, Matthew James – WL01
Adamovich, Igor V. – RH01, RH02
Adams, Keir – RL10
Adkins, Erin M. – MI05
Agner, Josef A. – RG08, FC06, FC07
Ahmed, Musahid – MJ10
Aidas, Kestutis – ML12
Alam, Jahangir – MJ11
Alcaraz, C. – WB07
Alden, Caroline – WA01
Alekseev, E. A. – TK06, TK08, WI07, WI08
Alikhani, Esmaïl – WF03, FB02
Allison, Thomas K – TG04, WG03
Allmendinger, Pitt – TG03
Almond, Matthew – RH09
Alonso, Elena R. – ML01, ML02, TJ08, TK12, WJ01
Alonso, José L. – ML01, ML02, TJ08, TK12, WJ01, RK02
Alsaif, Bidoor – TG05
Altmann, R.K. – MG01
Amicangelo, Jay C. – MJ02, TE11
Amy-Klein, Anne – TG02, FD04
Anderson, David T. – TD03
Annesley, Christopher – WH04
Antonov, Ivan – RG06, RG09
Appadoo, Dominique – WL08
Araki, Mitsunori – RL05
Arunan, Elangannan – WK11
Ashford, James R. – WH05
Asselin, Pierre – WC07, FD04
Asvany, Oskar – TH06, TL01, TL09, RA01, FC03, FD07
Axner, Ove – TG08

B

Baba, Masaaki – WG04
Baig, Nameera F. – RH08
Bailey, Josiah R – ML03
Bailleux, S. – RI06
Bailleux, Stephane – TK06
Balabanoff, Morgan E. – TD03
Balevicius , Vytautas – ML12, FC05
Banerjee, Pujarini – MK09
Bao, Yiliang – TA05
Baraban, Joshua H – MJ10, TK05, TL10, WC08
Barclay, A. J. – MK04, MK05, TE04
Barnum, Timothy J – MH06
Barone, Vincenzo – TL03, WB05
Basenback, Brittany – WE10
Basom, Edward – TD09
Bates, Desiree M. – FA01
Bauerecker, Sigurd – TK08
Baumann, Esther – TA09, WA01
Bechtel, Hans A – MI03
Becker, Jörg August – TK01
Behera, Bedabyas – TE06
Belloche, Arnaud – WI08, RI06
Belov, Sergey – TK06
Benke, Kristin – TD10
Benner, D. Chris – FF01
Berden, Giel – WI03
Berger, Yann – FD04
Bermúdez, Celina – TK03
Bernal, Jacob – RL06
Bernath, Peter F. – TF04, WB02, WL07, WL08, FB06, FF05
Bethlem, Hendrick – FC06, FC07
Bevan, John W. – MK08
Beyer, Maximilian – MG04, FC06, FC07
Bhattacharya, Indrani – MK09
Bieniewska, Julia – RG05
Billinghurst, Brant E. – WC04, WH10, WH11, WL08
Bird, Ryan G – ML03
Biswas, Souvick – WK07, FD05
Bittner, Dror M. – WL07, FB06
Blackstone, Christopher C – ML08
Blanco, Susana – WE02, WE05, WE07, WE09, RK01, RK05, RK06, RK09
Bloch, Daniel – TI07
Blodgett, Karl N. – ML09, ML10, WJ09
Bocquet, Robin – TK02, WJ02
Bocwinski, Rachel – RH06
Bogdanowicz, Robert – TH09
Bohlen, Matthias – TE02
Bohn, Paul W. – RH08
Bohn, Robert Karl – WJ06, FD06
Bojarski, Piotr – TH09
Boone, Chris – TF04, FF05
Borin, Veniamin A. – FB04
Borisov, Kirill – WJ04
Bose, Maitrayee – RL06
Boudon, Vincent – TF01, TF02, TF03, TJ02, WC02, WC05, WH08
Boulos, Victoria M. – WJ07, WJ08
Bounds, James R – WA05, RH03

Bowden, Hannah – RH09
Brackertz, Stefan – TH06
Bray, Cédric – WJ02
Breier, Alexander A. – TL06, WL09
Bresler, Sean Michael – FE10
Briaudeau, Stephan – FD03
Brice, Joseph T. – MJ01
Bridgmohan, Chelsea N – WJ10
Brown, Alaina R. – MJ01
Brown, Glenna J. – MJ03
Brünken, Sandra – RA01
Bteich, Sabath – TK02, WC07
Bucchino, Matthew – WD05
Büchling, Thomas – TL06
Budkina, Darya S. – TD05, FB04
Bui, Thinh Quoc – WA06
Bunn, Hayley – WL03
Buoniconti, Patrick – TC08
Burghardt, Irene – MA02
Burhani, Muffaddal – TD10
Burkhardt, Andrew M – WL11, RL03, RL04
Burton, Mark – WD05, WF02, RL02
Butler, Thomas – WG02

C

Cabezas, Carlos – ML02, TJ08, TK03, WJ01
Calabrese, Camilla – WK08
Caminati, Walther – WK09
Campargue, Alain – MG07, MI04, TF06, TG05, TG06, TJ01
Cao, Tianyuan – RH08
Carlson, Michaela – TD10
Carrigan-Broda, Theodore – TJ10
Carrington, Tucker – MH03, MH04, MH10, WB03, FA04
Carroll, Brandon – RK10
Casado, Mathieu – TG06
Caselli, Paola – TD01, RL01
Castracane, Eleanor – FF06
Castrillo, Antonio – WG05
Cercola, Rosaria – FB03
Cermak, Micheal – WA04
Čermák, Peter – TJ01
Cernicharo, Jose – WL02,RI06
Cerullo, Giulio – MG08
Chakraborty, Tapas – MK09, WK07, FD05
Chang, M.-C. Frank – TI10, WI06
Chang, Yih-Chung – RJ10
Chang, Ziqiao – TI05
Changala, Bryan – MH05, TD06, TK05, TL10, WA06, WC08,

Charczun, Dominik – TG09, TG10, WG06
Chardonnet, Christian – FD04
Chen, Jie – TA06, WA02
Chen, Junha – WK05, WK06
Chen, Kun – TA08
Chen, Ting-Yu – TD04
Chen, Weidong – TI03
Chen, Yu-Hsuan – FA03
Chen, Yuning – TG04, WG03
Chen, Zaijun – TA02
Chen, Zhihao – TI08
Cheng, Cunfeng – FC06, FC07
Cheng, Lan – WF07, WF08, FC08
Cheramy, Joseph – TI08
Chernolevska, Yelyzaveta – ML12
Cheung, Allan S.C. – RJ02
Cheung, Ling Fung – TE03
Chick Jarrold, Caroline – MA03, TE08, WD09, WD10, WF01
Chittari, Supraja – TC10
Chiu, Randall – MI08
Chowdhury, Papia – RH05
Chrayteh, Mhamad – WE01
Cich, Matthew J. – WA03
Clark, Andrew – RJ02
Clouthier, Dennis – WD06, WD07
Coburn, Sean – WA01
Coddington, Ian – TA09, WA01, WA04
Coeur, Cécile – TI03
Colditz, Sebastian – RL07
Cole, Ryan K. – WA03
Coluccelli, Nicola – WA08
Conway, Eamon K – TF07
Cooke, S. A. – TK10, TK11, WJ06, FD06
Cooper, Graham A. – WK10
Corby, Joanna F. – RL04
Cossel, Kevin C – WA01, WA04
Coudert, L. H. – WB07, RI04
Cournol, Anne – FD04
Cozijn, F.M.J. – MG10
Crabtree, Kyle N. – TI12, TL04, RJ10
Crawford, Timothy J. – RI02, FF01
Császár, Attila – MH02
Cuisset, Arnaud – TK02, WC03, WI02, WJ02, WJ03
Cygan, Agata – TG09, TG10
Czekner, Joseph – TE03

D

Daëron, Mathieu – TG06
Dagys, Laurynas – ML12
Daily, John W – MJ10
Dale, Harvey J A – RA02
Darquie, Benoit – TG02, TI07, RG05, FD03, FD04
Das, Puspendu Kumar – TE06
Dattani, Nikesh S. – WB04, WB08
Daunt, Stephen J. – WH10, WH11
Daussy, Christophe – FD03, FD04
Dawadi, Mahesh B. – TB05
Dawes, Richard – TD07, RJ05, RJ08, RJ09, FC01, FC02
De Natale, Paolo – TG01
de Oliveira, Nelson – MI09
de Vicente, Pablo – WL02
Debus, Michael – TG09
DePalatis, Michael – MG05
Dessent, Caroline H. E. – FB03
Devi, V. Malathy – FF01, FF02
DeVine, Jessalyn A. – FC04
Dhont, Guillaume – WC03, WJ03
Diddams, Scott – WG03
Dieter, Thomas S – MI02
Dittman, James – TC08
Djuricanin, Pavle – MK10
Dobulis, Marissa – WD10
Dodson, Leah G – RA03, RG01, RG02
Doizi, Denis – WC02, WH08
Doménech, José Luis – TL09, WL02
Dopfer, Otto – FC03
Dore, Luca – TL03
Doroshenko, Iryna – ML12
Dorovskaya, Olga – TK08, RI06
Douberly, Gary E. – MJ01, RG03, RG04
Draper, Anthony D. – WA03
Dréan, Pascal – WE01
Dreissen, L.S. – MG01
Drewsen, Michael – MG05
Drouin, Brian – TI10, WA03, WC06, WI06, RI01, RI02, RI08
Duan, Chuanxi – MK03
Dubernet, Marie-Lise – TF01
Dubey, Pankaj – WJ11
Duerden, Amanda Jo – TH03, TJ10
Duncan, Michael A – ML11, TH08, WK03, FE05, FE06, FE07
Duong, Chinh H. – TH04, TH05
Dupont, Jinbum – RH06
Dupré, Patrick – MG10

E

Eibenberger, Sandra – FF08
Eidelsberg, Michele – MI09
Eikema, K.S.E. – MG01, MG10, FC06, FC07
Eliet, Sophie – WI02, WI04
Ellis, Martha – MJ03
Ellison, Barney – MJ10, WC08
Elvir, Brayan R. – ML10
Embly, Caitlin – TC06, TC08
Encalada, Frankie – RL07
Encrenaz, Thérèse – TJ02
Endo, Yasuki – TK03, WD04
Endres, Christian – TD01
Enriquez, Lourdes – TJ09
Esposito, Anne Marie – MG09, MI02
Esselman, Brian J. – FA01
Esteki, Koorosh – TE04
Evangelisti, Luca – TC01, TC06, TC07, TC08, TC11, WK04, WK08, WK09

F

Fábri, Csaba – MH02
Fadda, Dario – RL07
Fairlie, David – FE01
Faist, Jerome – TG03
Farooq, Aamir – TG05
Fast, Arthur – MG03
Faulkner, Ty – FD09
Favero, Laura B. – WK08
Favier, M.G.J. – MG01
Faye, Mbaye – TF02, TF03, WC02, WC05
Federman, Steven – MI09, RL08
Fellinger, Jakob – TA07
Feng, Gang – TJ05, WK05, WK06, RK07
Fermann, Martin – TG05, WG06
Fernando, Anton Madushanka – TF04
Fertein, Eirc – TI03
Field, Robert W – MH06, FD08, FF03, FF04
Filipsson, Anna – TG08, WG07
Fineman, Liam – RH06
Fischer, Christian – RL07
Fischer, Joshua L. – ML09, ML10, TE09, WJ09
Fisher, Karin – MG05
Flaud, Jean-Marie – WH10, WH11
Fleisher, Adam J. – TA05, WA07
Flores, Jonathan – MJ07
Foguel, Lidor – RJ01
Foley, Aisling – RH06
Foltynowicz, Aleksandra – TG08, WG07
Fontanari, Daniele – WC03, WJ02, WJ03
Franke, Peter R. – MJ01, RG03, RG04
Frederickson, Kraig – RH01, RH02
Freund, Jens – MH08
Frigge, Robert – WL01
Fritz, Sean – TI11, WK01

Fuchs, Guido W – TL06, WL09
Fujii, Masaaki – TE05
Fukushima, Masaru – MJ09, TB01, WD02
Fuller, Tyler G. – TE10
Furneaux, John – MG03

G

Gaida, Christian – WG02
Gallego, Juan Daniel – WL02
Galleguillos Kempf, Sarah Caroll – MH08
Galzerano, Gianluca – WA08, WG05
Gamache, Robert R. – FF01
Gambetta, Alessio – WA08
Gans, B. – WB07
Gao, Xiaoming – TI03
Garcia, G. A. – WB07
Garner, Scott M. – TB06, TB07
Garrod, Robin T. – WI08
Gatti, Davide – MG08, TG05, WG05
Gatti, Fabien – TD07
Gauss, Jürgen – TL06
Gavilan, Lisseth – MI09
Gazali, Zainab – RH07
Geballe, Thomas R. – RL11
Gebhardt, Martin – WG02
Geis, Norbert – RL07
Georges, Robert – WC07
Gerz, Daniel – WG02
Gharaibeh, Mohammed – WD06
Gharib-Nezhad, Ehsan – MI03
Gheorghe, Andrei – RG10
Ghysels, Mélanie – MI05
Gianfrani, Livio – WG05
Gibson, Stephen T – MJ06, WH01
Giesen, Thomas – TL06, WL09
Gillcrist, David Joseph – TJ10
Giorgetta, Fabrizio – TA09, WA04
Gjuraj, Daniel – WH11
Gnanasekar, Sharon Priya – WK11
Golubiatnikov, G Yu – TK06
Gordon, Iouli E – MI07, FF01
Gore, Jay P – TI05
Gorman, Jason J – TA05
Gorman, Maire N. – WH05
Goto, Miwa – RL11
Gotti, Riccardo – MG08, TG07, WG05
Gou, Qian – TJ05, WK05, WK06, RK07
Goubet, Manuel – TK02, TK08, WC07, RI04
Gougoula, Eva – WK10
Grabow, Jens-Uwe – TK01, WJ04, RK03, FD01

Graneek, Jack B. – ML04
Grigoryan, Tigran – TJ02
Grilli, Roberto – TF06
Grimes, David – RG10
Groner, Peter – TK09, TK10
Gross, Eisen C. – MJ05
Grubbs II, G. S. – TC09, TH03, TJ10
Gruebele, Martin – TI01
Gruet, Sébastien – TL08, WE08, RI04
Grzywacz, Robert – WH10, WH11
Guillemin, J.-C. – TK03, TL11, RI06
Guirgis, Gamil A – TJ10
Gunthardt, Carolyn E. – WH02
Guo, Hairun – WG01
Gupta, Harshal – TL01, TL10
Gupton, B Frank – TC01, TC02
Gurusinghe, Ranil – WE10
Gutierrez, Imanol Usabiaga – WK08

H

Haghtalab, Golara – TC08
Halfen, DeWayne T – TJ04, TL07, RL02
Hall, Gregory – MI01, MI10, MJ05, WH02
Han, Dong – TI05
Hänsch, Theodor W. – TA02
Harms, Jack C – FB07, FB08
Harper, Nathan – WE10
Harren, Frans – WG10
Harrilal, Christopher P – FE01, FE02
Harris, Robert J – RL09
Harrison, Jeremy J. – TF05
Harrison, Rachel E. – RL09
Hartl, Ingmar – WG08
Hausmaninger, Thomas – TG08
Hayes, Wayne – RH09
Hays, Brian M – TI09, WK01, WL03
Hazrah, Arsh – WE04
Heaven, Michael – RJ04, FB05, FE09, FE10
Heays, Alan – MI09
Heays, Alan N – MI03
Heckl, Oliver H – TA07
Heger, Matthias – TI08
Hemberger, Patrick – MJ10
Hendricks, Richard J. – FD04
Herbers, Sven – TK01, RK03, FD01
Herbst, Eric – WL04, WL11, RL03
Herman, Daniel I. – TA09
Hermanns, Marius – WJ04, WJ05
Hernandez, Rodrigo – WB03
Hernandez-Castillo, Alicia O. – TI11, WK01, WK02
Herrero, Victor Jose – WL02
Hesselius, Daniel – WA04

Heuermann, Tobias – WG02
Hewett, Daniel M. – WJ07, WJ08
Heyl, Christoph – WG08
Hicks, Christopher M. – FB04
Hill, Christian – MI07
Hill, Timothy – FE01
Hindle, Francis – TK02, WI02, WJ02
Hirano, Tsuneo – MH08, MH09, TB02, TB03, TB04
Hoadley, Tyler – TF08
Hodges, James Neil – WL07
Hodges, Joseph T. – MI05, TA05, WH09
Hodyss, Robert – RI03
Hoelsch, Nicolas – MG04, FC06, FC07
Hofer, Christina – WG02
Holdren, Martin S. – TC07, TC08, WE07
Holland, Torrey E. – TH10
Holzmeier, F. – WB07
Hönle, Rainer – RL07
Hougen, Jon T. – TK06, TK07, TK08
Hoyt, Sasha – RH06
Hu, Ming-Guang – RG10
Hu, Shui-Ming – MI06, FC06, FD02
Hua, Tian-Peng – MI06
Huang, Xinchuan – WC09
Huet, Therese R. – WC07, WE01, RK08, FD04
Huff, Anna – WE06, WH06, WH07, RK04
Hugi, Andreas – TG03
Huke, Philipp – TG09
Hull, Alexander W – FF03, FF04
Hussels, Joël – FC06, FC07
Hutzler, Nicholas R – WF04

I

Ibrahim, Mohamad – FB02
Ilyushin, V. – TK08, WI08, RI06
Indriolo, Nick – RL11
Iranpour, Michael Cyrus – TF08
Iserlohe, Christof – RL07
Ishikawa, Haruki – MK01, MK02, MK06, MK07
Ishiwata, Takashi – MJ09, TB01, WD02
Iwakuni, Kana – WA06
Iwasaki, Hikari – MH07

J

Jabri, Atef – WJ02, WJ03
Jackson, William M. – RJ10

Jäger, Wolfgang – TC04, TI08, WE04
Janik, Ireneusz – FA05, FD10
Jans, Elijah R – RH01
Jansen, Paul – MG02, RG07, RG08
Jaraiz, Martin – TJ09
Jean, Pearl – FB05
Jensen, Per – MH08, MH09, TB02, TB03, TB04
Jiang, Jun – MH06
Jin, Jiaye – TE01
Johansen, Sommer L. – TI12, TL04
Johanssson, Alexandra C – TG08, WG07
Johnson, Christopher J – FF06
Johnson, Dylan – TE11
Johnson, Erika – TI11, RK02
Johnson, Mark – TH04, TH05
Johnson, Sarah – RH06
Jones, Grier – TJ10
Juanes, Marcos – TJ09, WK09
Jungen, Christian – MG04, FC06, FC07
Jurak, Jessica C – TI02
Jusko, Pavol – TL09, RA01

K

Kaiser, Catherine – TE11
Kaiser, Ralf Ingo – WL01
Kalenskii, Sergei – WL11, RL03
Kamegai, Kazuhisa – RL05
Kang, Gyeongwon – FA06
Kang, Heon – FD08
Kang, Peng – MI06
Kannangara, Prashansa – TH02, WE03
Karabaeva, Kanykey E. – TD05
Karczewski, Jakub – TH09
Karlovets, Ekaterina – TJ01
Karns, Joshua – MI07
Karunanithy, Robinson – TH10
Kasahara, Yasutoshi – MK01, MK02, MK06, MK07
Kassi, Samir – MG07, MI04, TF06, TG06, TG07, TJ01
Kato, Ryota – MK01, MK02
Kawashima, Yoshiyuki – ML05, ML06
Ke, Hanzhang – TF02
Kelleher, Patrick J – TH04, TH05
Keogh, John P – TJ04
Khanna, Zain – FD06
Khlopenkov, Alexander – RH06
Kim, Jongjin B. – FC04
Kim, Junggil – MI11, WK12
Kim, Sang Kyu – MI11, WK12
Kim, So-Yeon – WK12
Kim, Yanghyo – TI10, WI06

Kippenberg, Tobias J. – WG01
Kirmess, Kristopher M – WJ10
Kisiel, Zbigniew – WE02, RK09, FA01
Klein, Randolf – RL07
Kleiner, Isabelle – TK02, TK07, TK08, WJ03
Klocke, Jessica – TG03
Kobayashi, Kaori – WC04, WH03
Kochanov, Roman V – MI07
Kocheril, G. Stephen – TE03
Kocheril, Philip A. – MG09, MI02, TL05
Kohguchi, Hiroshi – FC03, FD07
Kohn, Alexander W. – TB08
Kokkin, Damian L – MJ07
Kolesniková, Lucie – WJ01, RK02
Kolomenskii, Alexander – WA05, RH03
Konefal, Magdalena – MG07
Kortyna, Andrew – MJ04, MJ08, WD01, WD11, WL06
Kottke, Tilman – TG03
Kowligy, Abijith S – WG03
Kowzan, Grzegorz – TG09, TG10
Koyama, Daisuke – RA02
Kozlova, Olga – FD03
Krabbe, Alfred – RL07
Kramer, Kristen M. – RH08
Krapivin, Igor – TK06, WI08
Kraus, Peter – RK03, FD01
Krausz, Ferenc – WG02
Krauth, J. – MG01
Kreinbihl, John J. – FF06
Krems , Roman – WB03
Krin, Anna – TC05
Kristinaityte, Kristina – ML12
Kroll, Jay A – FF07
Kubasik, Matthew A. – ML10
Kuper, Henning – TK01
Küpper, Jochen – WG08
Kurusu, Itaru – MK01
Kuze, Nobuhiko – ML05, ML06, RL05
Kwabia Tchana, F. – TJ02
Kwon, Woojin – RL09
Kyuberis, Aleksandra A. – TF07, RJ07

L

L'Ecuyer, Tristan S – RI01
Lafferty, Walter – WH10, WH11
LaForge, Aaron – TE02
Laliotis, Athanasios – TI07
Lamperti, Marco – MG08, TG05
Lampin, Jean-François – WI01, WI02, WI04

Landais, Amaelle – TG06
Lapinov, Alexander – TK06
Laporta, Paolo – TG05, WA08, WG05
Lattanzi, Valerio – TJ03, RL01
Lau, Hunter Singh – FA01
Lawler, John T – FE01
Laws, Benjamin A – MJ06, WH01
Le, Anh T. – MJ05, WF08
Le Coq, Yann – TG02
Le Gal, Romane – WL04
Le Targat, Rodolphe – TG02
LeBrun, Thomas W – TA05
Lechevallier, Loïc – TF06
Lecordier, Louis – FD04
Lee, Heesung – WK12
Lee, Kelvin – ML07, TJ03, TL01, WL10, WL12
Lee, Kevin – WG06
Lee, Sang – WD03
Lee, Timothy – WC09
Lee, Won-Kyu – TG02
Lee, Yuan-Pern – MA04, MJ02, TD04, WD04, RG04, FA02, FA03
Lees, Ronald M. – TL02, WI08
Lefebvre-Brion, H. – RJ11
Lei, Juncheng – RK07
Leicht, Daniel – FE05
Lemaire, Jean Louis – MI09
Lemus, Renato – MH01
Lengsfeld, Kevin G. – TK01
León, Iker – ML01, ML02, TJ08, TK12, WJ01, WK09, RK02
Leopold, Ken – WE06, WH06, WH07, RK04
Lesarri, Alberto – TJ09, WK09
Lesko, Daniel – MJ04, MJ08, WD01, WD11, WL06
Leung, Helen O. – TC03, TC04, TC11, TJ06, TJ07
Lewen, Frank – TL02, WI08, WJ04, WJ05
Lewkowicz, Aneta – TH09
Li, Gang – FF01
Li, Heng Ying – WH05
Li, Hui – TB04
Li, Ting – TA06, WA02
Li, Weixing – WK04, WK08, WK09
Li, Xiaolong – TJ05
LI, Yan – TA08
Li, Zhi-Yun – RL09
Liang, Wei-Che – FA02
Licari, Daniele – WB05
Liebermann, H. P. – RJ11
Lim, Jean Sun – MI11, WK12
Limpert, Jens – WG02
Lin, Ming-Fu – TD10
Lin, Paotai – WA05

Lin, Zhou – MH07, TB08
Lind, Alex – WG03
Line, Michael R – MI03
Linton, Colan – WF09
Lisak, Daniel – TG09, TG10
Liu, An-Wen – MI06, FD02
Liu, Gu-Liang – MI06
Liu, Jinjun – MJ11, MJ12, WF05, WF06
Liu, Junzi – FC08
Liu, Yu – RG10
Lockhart, James – MI10
Lockwood, Schuyler P – TE10
Loete, Michel – TF03
Loison, J.-C. – WB07
Long, David A. – MI05, TA05, WH09
Looney, Leslie – RL07, RL09
Loparo, Zachary E – TA03
Lopez, Juan Carlos – WE02, WE05, WE07, WE09, RK01, RK05, RK06, RK09
Lopez, Olivier – TG02, FD04
Loru, Donatella – TL08
Lowe, Albyn – TI05
Lu, Tao – WK05, WK06
Lucchese, Robert R. – MK08
Lucht, Robert P. – TI05
Lukusa Mudiayi, Junior – TI07
Luo, Pei-Ling – WD04
Lyding, Joseph – TI01
Lynam, Jason M. – FB03
Lyons, James R – MI03, MI09
Lyulin, Oleg – TG05

M

Macario, Alberto – WE09, RK01, RK05
Mackenzie, Becca – WE06, WH06
Makhnev, Vladimir Yu. – RJ07
Malaska, Michael – RI03
Manceau, Mathieu – FD04
Manceron, Laurent – TF02, TF03, TJ02, WC02, WC05, WH08
Mandrell, Christopher – TI02
Mangold, Markus – TG03
Mann, Jennifer – WD09, WD10
Mantz, Arlan – FF01
Manzhos, Sergei – WB03, FA04
Marangoni, Marco – MG08, TG05, TG07, WG05
Margulès, L. – TK03, TK06, TK08, TL11, WI04, WI07, RI06, FD04
Maris, Assimo – WK04, WK08
Marks, Joshua H – FE06, FE07
Marks, Julia – RH06

Markus, Charles R. – MG09, MI02, TL05
Marsalka, Arunas – ML12
Marshall, Frank E – TC09, TH03, TJ10
Marshall, Mark D. – TC03, TC04, TC11, TJ06, TJ07
Martens, Jonathan K – WI03
Martin, Catherine – FD03
Martin-Drumel, Marie-Aline – TJ03, TL10, WC03, WC08, RI04
Maslowski, Piotr – MG08, TG09, TG10, WG07
Mason, Jarrett – WF01
Masri, Assaad R – TI05
Mata, Santiago – ML01, ML02, TJ08, RK02
Matsumoto, Yoshiteru – MK06, MK07
Matsushima, Fusakazu – WH03
Matt, Wyatt – MI07
Matthews, Devin A. – FC08
Matveev, Sergey M. – FB04
Maul, Christof – TK08
Maurin, Isabelle – TI07
Mayer, Kevin J – TC06, TC07, TC08, TC10
McCabe, Morgan N – TI09, WL03
McCall, Benjamin J. – MG09, MI02, TL05, RL11
McCarthy, Michael C – ML07, TJ03, TK05, TL01, TL10, WC08, WL10, WL11, WL12, RK10, RL03, FC01, FC02, FF08
McCunn, Laura R. – MJ03
McDivitt, Lindsey M – ML03
McDonald II, David C – ML11, TH08, WK03
McGuire, Brett A. – TJ03, WL11, WL12, RI06, RL03, RL04
McKellar, Bob – MK04, TE04
McLuckey, Scott A – FE01, FE02
McMahon, Robert J. – FA01
McMahon, Timothy J – ML03
McRobbie, Stuart – RH09
Medcraft, Chris – WK10
Meek, Samuel – MG03
Meerts, W. Leo – TK02, WC03
Meiser, Jana – WE08
Melandri, Sonia – WK04, WK08
Mellau, Georg Ch. – TK08
Melosso, Mattia – TL03
Menten, Karl M. – TL02, WI08
Mereshchenko, Andrey S. – FB04
Merkt, Frederic – MG02, MG04, RG07, RG08, FC06, FC07
Meyer, Steffen – MG05
Michaelian, K. H. – MK05

Micica, Martin – WI04
Mikhonin, Alex – RH04
Milam, Stefanie N – WL05
Miliordos, Evangelos – RJ06
Miller, Isaac James – FD09
Miller, Terry A. – TB06, TB07, WF05, RH02
Minami, Yoshiaki – RL05
Mishra, Hirdyesh – WB06
Mishra, Piyush – TE09
Misono, Masatoshi – WG04
Mitchell, Brianna – TC08
Mo, Yirong – WH07
Moazzen-Ahmadi, Nasser – MK04, MK05, TE04
Modi, Alysa – TC10
Momose, Takamasa – MK10
Mondelain, Didier – MI04, TF06, TJ01
Mońka, Michał – TH09
Moon, Nicole – TH03, TJ10
Moore, Brendan – MK10
Morales-Soto, Nydia – RH08
Moreau, Nicolas – TF01
Moretti, Luigi – WG05
Moriwaki, Yoshiki – WH03
Motiyenko, R. A. – TK03, TK06, TK08, TL11, WI04, WI07, RI06, FD04
Moufarej, Elias – FD03
Mouret, Gaël – TK02, WI02, WJ02, WJ03
Muckle, Matt – TC02, RH04
Mukhopadhyay, Deb Pratim – MK09, WK07, FD05
Müller, Holger S. P. – TL02, WI08, WJ04, WJ05
Munshi, Musleh Uddin – WI03
Muraviev, Andrey – TA03

N

Nagashima, Umpei – MH08, MH09, TB02, TB03
Namekata, Takumi – WH03
Nava, Matthew – WC08
Ndengue, Steve Alexandre – TD07, RJ05
Neeman, Elias M. – RK08
Neill, Justin L. – TC01, TC02, TH07, RH04, RH06
Nemchick, Deacon J – TI10, WI06
Nemes, Coleen T. – WI05
Nesbitt, David – MJ04, MJ08, WD01, WD11, WL06
Neu, Jens – WI05
Neumark, Daniel – FC04

Newbury, Nathan R. – TA09, WA04
Newby, Josh – TE10
Ng, Cheuk-Yiu – RJ10
Ng, Yuk Wai – RJ02
Nguyen, Duc – TI01, FA06
Nguyen, Duc-Trung – WF08, WF09, RJ03
Nguyen, Ha Vinh Lam – WJ03
Nguyen, Thanh Lam – TK05
Ni, Kang-Kuen – RG10
Nickson, Kathleen A. – FF06
Niedermeyer, Justin – WA06
Nikitin, Andrei V. – WB02, FF02
Nikodem, Michal – TI06
Nikolay, Dorozhkin – FC05
Nimlos, Mark R – MJ10
Nishiyama, Akiko – TG09, WG04
Niu, Ming Li – FC06, FC07
North, Simon – WH02
Novick, Stewart E. – TK10, TK11, WJ06, FD06
Nyambo, Silver – WD08, FB01, FE08

O

O'Brien, James J – FB07, FB08
O'Brien, Leah C – FB07, FB08
Obenchain, Daniel A. – TK01, WJ06, RK03, FD01
Odermatt, Eric – TC10
Odom, Brian C. – RG06, RG09
Ohashi, Nobukimi – WC04
Ohyama, Ryo – WC04
Oka, Takeshi – RL10, RL11
Okuda, Shoko – MG06, WG09
Ono, Shuhei – FF03, FF04
Oomens, Jos – WI03
Orellana, W. – TK10, TK11
Orito, Masataka – MK01
Ormond, Thomas – MJ10
Orr-Ewing, Andrew – RA02
Oyama, Takahiro – RL05
Ozeki, Hiroyuki – RI05

P

Pardo, Juan R. – WL02
Park, Youngwook – FD08
Pate, Brooks – TC01, TC02, TC06, TC07, TC08, TC10, TC11, TH02, TH07, WE03, WE07, RH06, RI03
Patrick, Link – TI04
Patros, Kellyn M. – WD09, WD10
Patterson, David – FF08
Paul, Anam C. – MJ11, MJ12, WF05, WF06
Payagala, Yudhishtara – TE11
Pearson, Jeniveve – WC06, RI08
Pearson, John – WC06, RI03, RI08
Peebles, Rebecca A. – TH02, WE03
Peebles, Sean A. – TH02, WE03
Peláez, Ramón J. – WL02
Pereira, Marcus Vinicus – TF08
Perevalov, Valery – TJ01
Perez, Cristobal – TL08, WE05, WE08
Perry, David S. – TB05
Persinger, Thomas D – FE09
Philipot, Florian – TJ02
Picqué, Nathalie – TA02
Pienkina, A. – WI04
Pierens, Matthieu – FD04
Pietropolli Charmet, Andrea – MK05
Pilgram, Nickolas – WF04
Pinacho, Pablo – WE02, WE09, RK06, RK09
Pinacho, Ruth – TJ09, WK09
Pirali, Olivier – TK08, TL10, WC05, WC06, WC07, WC08, WI02, RI04, RI08
Pischer, Anna L – TI12
Pitsevich, George – ML12, FC05
Pitts-McCoy, Anthony – FE02
Plusquellic, David F. – TA05
Poglitsch, Albrecht – RL07
Pogorelov, Valeriy – ML12
Polli, Dario – MG08
Polyansky, Oleg L. – MI08, TF07, RJ07
Porterfield, Jessie P – MJ10, TK05, WC08, FF08
Postava, Kamil – WI04
Pottie, Paul-Eric – TG02
Powers, Carson Reed – TI09, WL03
Prasad, Kuldeep – WA01
Predoi-Cross, Adriana – FF02
Prevedelli, Marco – TG07
Pringle, Wallace C. – TK10
Prozument, Kirill – TH01
Pullen, Gregory T. – RG04
Pupeza, Ioachim – WG02
Puzzarini, Cristina – TL03, WB05
Pyatenko, Elizaveta – WH05

Q

Qian, Chen-hui – FE04
Quesada-Moreno, María Mar – TC05
Quintas Sánchez, Ernesto – RJ05, RJ08

R

Raab, Walfried – RL07
Racow, Emily E. – FF06
Raghavachari, Krishnan – TE08
Rai, Awadhesh Kumar – RH07
Ramos, Sashary – TD09
Raston, Paul – FD09
Ray, Manisha – TE08
Reck, Theodore J – RI03
Redlich, Britta – RA01
Reed, Zachary – WH09
Regan, Kevin P. – WI05
Reilly, Neil J – MJ07
Remijan, Anthony – WL11, RI06, RL03, RL04
Rey, Michael – WB02
Reymond-Laruinaz, Sébastien – WC02, WH08
Reza, Md Asmaul – MJ11, MJ12, WF05, WF06
Rice, Corey – TE02
Rice, Johnathan S – RL08
Richard, Cyril – TF01, TF02, TF03, TJ02
Richard, Lucile – TF06
Rieker, Gregory B – WA01, WA03, WA04
Riffe, Erika – TI11
Rios Leite, Jose Roberto – TI07
Rittgers, Brandon M. – FE05
Rivera-Rivera, Luis A. – MK08, TD08
Rizopoulos, Athena – TJ02
Robichaud, David – MJ10
Roenitz, Kevin – TI09
Romanini, Daniele – TF06, TG07
Ross, Sederra D – MJ07
Ross, Stephen Cary – WH03
Roth, C. – MG01
Rothman, Laurence S. – MI07
Rothschopf, Gretchen K – WD07
Roucou, Anthony – TK02, WC03, WJ02, WJ03
Roy, P. – TF03, WI02
Rubio, José Emiliano – TJ09, WK09
Russ, Benjamin – WD05
Rutkowski, Lucile – MG08, TG08, WG07
Rybchuk, Alex – WA01
Ryland, Elizabeth S – TD10

S

Salomon, Thomas – FC03, FD07
Salumbides, Edcel John – MG01, MG10, FC06, FC07
Sanov, Andrei – ML08
Santagata, Rosa – TG02

Sapeshka, Uladzimir – ML12, FC05
Saragi, Rizalina Tama – TJ09
Sasada, Hiroyuki – MG06, WG09
Satija, Aman – TI05
Sato, Hikaru – MK01, MK02
Sauer, B. E. – RG05
Savino, Brian – WL03
Savoia, Annunziata – WE01
Schatz, George C. – FA06
Schaugaard, Richard N – TE08
Scheer, Adam M – MJ10
Schelling, Rachel – RH06
Schlemmer, Stephan – TD01, TH06, TL01, TL02, TL09, WI08, WJ04, WJ05, RA01, FC03, FD07
Schmidt, Miroslaw R. – RI07
Schmitz, Joel R – RJ04
Schmuttenmaer, Charles A. – WI05
Schmutz, Hansjürg – RG08
Schnell, Melanie – ML04, TC05, TL08, WE05, WE08, RI04
Schrader, Alex W – MG09, MI02
Schreier, Phillip – FD07
Schroeder, Paul James – WA03
Schuessler, Hans A – WA05, RH03
Schwenke, David – WC09
Scolati, Haley N. – TI12
Scrape, Preston G. – MJ04, MJ08, WD01, WD11, WL06
Scribano, Yohann – TD07
Sears, Trevor – MI01, MI10, MJ05, RG05
Sedlacek, Ivan – TJ10
Sedo, Galen – TC09
Seifert, Nathan A. – TC04, WE04
Semeria, Luca – MG02, RG07, RG08
Sharma, Ketan – TB06, TB07, WF05
Shelkovnikov, Alexander – TI07, FD04
Sheridan, Phillip M. – WD05
Shields, George C – WE08
Shiery, Caleb B – ML03
Shimizu, Takutoshi – MK06, MK07
Shingledecker, Christopher N – WL04, WL11, RL03
Shipman, Steven – TD02, TI09, TI11, TK04, TK05, RK02
Shrout, Joshua D. – RH08
Shuvra, Pranoy Deb – WF06
Sibert, Edwin – TE09
Signore, Joshua A. – WJ06
Silfies, Myles C – TG04, WG03
Sivakumar, P – TH10, TI02
Smart, Taylor – TC07
Smith, CJ – WH06, WH07, RK04
Smith, Houston H – TI09, WL05

Smith, Mary Ann H. – FF01, FF02
Smith, Tony – WD06, WD07
Smolski, Viktor O – TA03
Soboń, Grzegorz – TG09
Solaro, Cyrille – MG05
Sonstrom, Reilly E. – TC01
Soulard, Pascale – WC07
soulard, pascale – FB02
Souvi, Sidi M.O. – WF03
Spada, Lorenzo – TL03, WB05
Spivey, Charles – TC10
Stachowiak, Dorota – TI06
Stahl, Wolfgang – WJ03
Stanton, John F. – MH05, TB06, TB07, TD06, TL10, WC08, WK02
Stark, Glenn – MI03, MI09
Stashower, Julian – TC10
Steber, Amanda – TL08, WE08
Steele, Ryan P – TH04, TH05
Steimle, Timothy – WF04, WF08, WF09, RJ03
Stephens, Ian – RL09
Stephens, Susanna L. – TK10, TK11, WJ06, FD06
Sterczewski, Lukasz A. – TA04
Stewart, Jacob – TF08
Stienkemeier, Frank – TE02
Stollenwerk, Patrick R – RG06, RG09
Stoltmann, Tim – MG07, TG06
Strom, Aaron I. – TD03
Sukegawa, Takashi – WA06
Sun, Dewei – ML09, WJ09
Sun, Yu Robert – FD02
Sung, Keeyoon – WC06, RI02, RI08, FF01, FF02
Sur, Sangeeta – RJ05
Surin, Leonid – FD07
Swann, William C – WA04
Sydow, Christian – TK08
Synak, Anna – TH09
Szczodrowski, Karol – TH09

T

Tabata, Mizuki – TE05
Takano, Shuro – RL05
Talbot, Justin J – TH04, TH05
Tan, Yan – MI07
Tanarro, Isabel – WL02
Tang, Adrian – TI10, WI06
Tani, Iori – WH03
Tao, Lei-Gang – FC06
Tarbutt, Michael – RG05, FD04
Tarnovsky, Alexander N – TD05, FB04
Tasinato, Nicola – WB05

Telfah, Hamzeh – MJ11, MJ12
Temelso, Berhane – WE08
Tennis, Jessica D. – WL04
Tennyson, Jonathan – TF07, WH05, RJ07
Thakur, Surya Narayan – RH07
Thapaliya, Bishnu P. – TB05
Theis, Mallory – FB05
Thibault, Franck – MG08
Thielges, Megan – TD09
Thomas, Javix – RK01
Thomas, Levi Michael – TI05
Thomas, Phillip – MH03
Thompson, Michael C – RA03
Thorwirth, Sven – TL01, TL10, WC03
Timmers, Henry – WG03
Toh, Shin Yi – MK10
Tokaryk, Dennis W. – WC04
Tokunaga, Sean – TG02, TI07, FD04
Tomaszewska, Dorota – TG09
Töpfer, Matthias – FC03, FD07
Topolski, Josey E – TE08, WF01
Tran, Dang Bao An – TG02
Tran, Minh Nhat – TF08
Trawiński, Ryszard S. – TG09, TG10
Tremblay, Benoît – WF03, FB02
Tripathi, G. N. R. – FA05
Troy, Tyler – MJ10
Truong, Gar-Wing – WA04
Tsukiyama, Koichi – RL05
Tubergen, Michael – WE10
Turner, Andrew Martin – WL01
Turut, Joan – WI02
Twagirayezu, Sylvestre – MI01, RH04
Tyuterev, Vladimir – WB02
Tzeng, Shen-Yuan – FE03
Tzeng, Wen-Bih – FE03

U

Ubachs, Wim – MG01, MG10, MI09, FC06, FC07
Uchida, Masaaki – MK06, MK07
Usuda, Tomonori – RL11

V

Vacca, William – RL07
Vaccaro, Patrick – RJ01
Vaida, Veronica – FF07
Valeviciene, Nomeda Rima – ML12
Vamos, Lenard – WG02
Van Duyne, Richard P. – FA06
Van Voorhis, Troy – MH07, TB08
Vander Auwera, Jean – TJ02, WC01, WH08

VanGundy, Robert A. – FE09
Vanwolleghem, Mathias – WI04
Vasilchenko, Semyon – TF06
Vaskivskyi, Yevhenii – ML12
Vázquez, Gabriel J. – RJ11
Vealey, Zachary – RJ01
Venkataramanababu, Sruthi – RG06, RG09
Venkatraman, Ravi Kumar – RA02
Ventrillard, Irene – TF06
Verkamp, Max A – TD10, TD11
Verma, Kanupriya – TE07
Vicentini, Edoardo – WA08
Virbila, Gabriel – TI10, WI06
Viswanathan, K S – TE07, WJ11
Vodopyanov, Konstantin L – TA03
Volkamer, Rainer – MI08
Vura-Weis, Josh – TD10, TD11

W

Wachsmuth, Dennis – RK03
Wagner, Albert F. – TD08
Wagner, J. Philipp – ML11, TH08, WK03
Walker, Nick – WK10
Wall, Thomas – RG05, FD04
Wallace, Colin J. – WH02
Waller, Sarah – FF06
Wang, Guanjun – TE01
Wang, Guishi – TI03
Wang, Haolu – TI08
Wang, Jin – FD02
Wang, Juan – TJ05
Wang, Lai-Sheng – TE03, FE04
Wang, Lichang – WJ10
Wang, Na – RJ02
Wang, Xiao-Gang – FA04
Wang, Yuchen – WA08
Ward, Meredith – MJ07
Ward, Timothy B – FE06, FE07
Waßmuth, Björn – TL06
Watanabe, Kyohei – WH03
Watanabe, Shinichiro – ML06
Watson, Dennis G. – TH10
Waxman, Eleanor – TA09, WA04
Wcislo, Piotr – MG08
Weber, J. Mathias – RA03, RG01, RG02

Wehres, Nadine – WJ04, WJ05
Wei, Haoyun – TA08
Weichman, Marissa L. – WA06, WG06, FC04
Welch, Bradley – RJ09
Weng, Wenle – WG01
West, Channing – TC07, TC11, WE07, RH06
Westberg, Jonas – TA04, TI04
Westerfield, J. H. – TD02, TK04, TK05, TL10
Widicus Weaver, Susanna L. – TI09, WL03, WL05
Wilkins, Olivia H. – WJ04
Williams, Michael R. C. – WI05
Wilson, R. Marshall – TD05
Winkler, Georg – TA07
Wishnow, Edward H – RI02
Witsch, Daniel – WL09
Wodraszka, Robert – MH10
Wojtewicz, Szymon – MG08
Wong, Andy – WB02, WL08
Wong, Ying-Tung Angel – MK10
Woods, R. Claude – FA01
Wörner, Hans Jakob – RA04
Wright, Robert – WA01
Wronkovich, Miles A. – TJ07
Wu, Arthur – TC08
Wu, Tao – TI03
Wu, Tao – TA08
Wu, Yuehan – TA06
Wysocki, Gerard – TA04, TI04

X

Xia, Jinbao – RH03
Xie, Fan – TI08
Xu, Dan – TG02
Xu, Li-Hong – TK06, TL02, WI08
Xu, Yunjie – TC04, TI08, RK01
Xu, Zhongxing – RJ10

Y

Yachmenev, Andrey – WG08
Yagi, Reona – MK01
Yamada, Koichi MT – FC03
Yang, Dong-Sheng – WD08, FB01, FE08
Yang, Guang – WG08
Yang, Nan – TH04, TH05
Yang, Yi – FF06
Yang, Yuan – TC01, TC02
Yao, Zijun – TA06
Yasui, Takeshi – TA01
Ycas, Gabriel – TA09
Ye, Hong-Zhou – MH07
Ye, Jun – MA01, WA06, WG06
Yi, Hongming – TI03
Yin, Siyao – WA02
Yocum, Katarina – WL05
Young, Justin W. – WH04
Yousefi, Mahdi – FF05
Yu, Shanshan – WC06, RI03, RI08
Yurchenko, Sergei N. – WB01, WH05

Z

Zagorec-Marks, Wyatt – RG01, RG02
Zak, Emil J – MH04
Zakharenko, Olena – TL02, WI08
Zaleski, Daniel P. – TH01
Zdanovskaia, Maria – FA01
Zhang, Huaiyu – WH07
Zhang, Kaili – TD10
Zhang, Sasa – RH03
Zhang, Weijun – TI03
Zhang, Yuchen – WD08, FB01, FE08
Zhao, Gang – TG08
Zhao, Weixiong – TI03
Zhao, Xin – TA06, WA02
Zheng, Yang – RK07
Zheng, Zheng – TA06, WA02
Zhou, Feng – TA05
Zhou, Junchao – WA05
Zhou, Mingfei – TE01
Zhu, Feng – WA05, RH03
Zhu, Guo-Zhu – FE04
Zhu, Xuayne – RH06
Zinga, Samuel – WL03
Ziurys, Lucy M. – TJ04, TL07, WD05, WF02, RL02, RL06
Zobov, Nikolay F. – RJ07
Zou, Wenli – RJ02
Zwier, Timothy S. – ML09, ML10, TE09, TI11, WJ07, WJ08, WJ09, WK01, WK02, FE01, FE02

Amplitude

Committed to science

 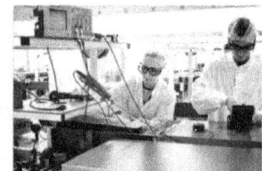 

Ti: Sapphire

- Most advanced PW class
- Unique kHz & hybrid solutions
- Versatile & reliable
- Modular solutions

ns systems

- Highest energy
- Reliable pump lasers
- Broadest portfolio
- Tailored solutions

Yb Ultrafast
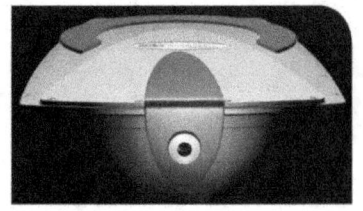

- Industrial & robust
- High repetition rate
- High energy
- Scaleable

www.amplitude-laser.com

Amplitude welcomes you to the

International Symposium on Molecular Spectroscopy

73rd Meeting

June 18-22
2018

Amplitude is a proud sponsor of the
Grad Student / Sponsor Mixer.

Please stop by and visit
our tabletop in the
exhibition
area.

www.amplitude-laser.com

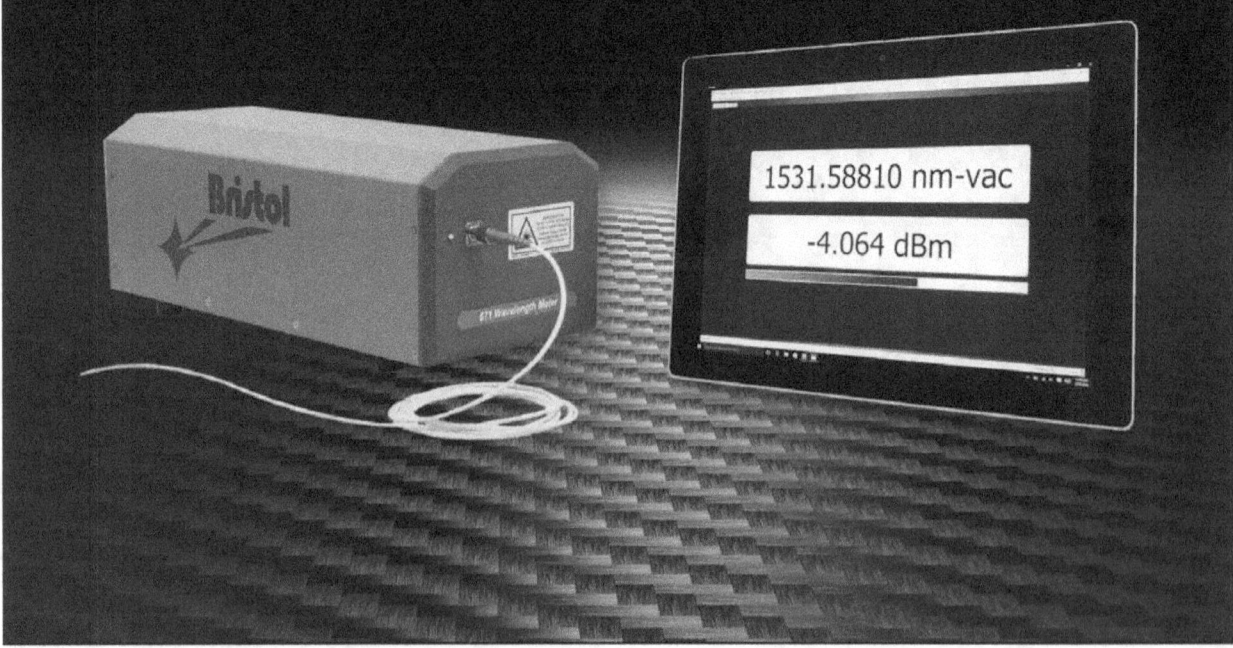

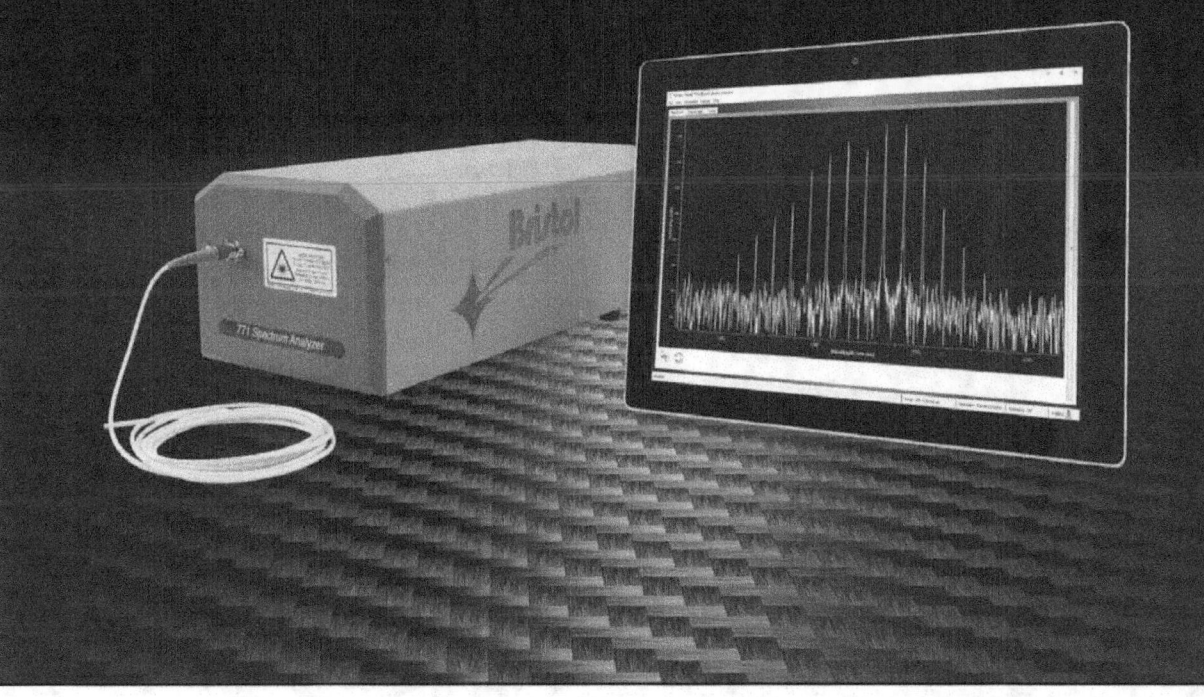

Volume 344, February 2018

ISSN 0022-2852

Journal of MOLECULAR SPECTROSCOPY

Editor in Chief
Terry A. Miller

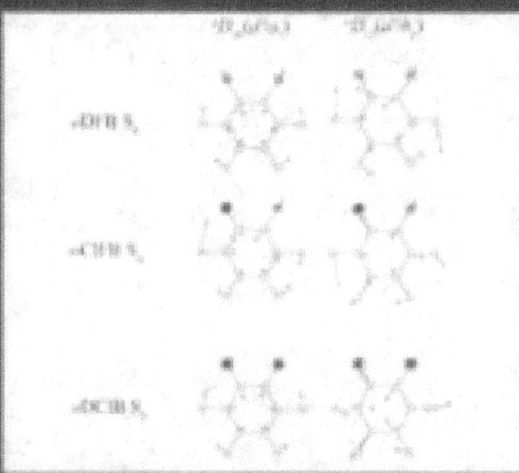

Cover Art: Consistent assignment of the vibrations of symmetric and asymmetric ortho-disubstituted benzenes (William D. Tuttle, Alexia M. Gardner, Anna Andrejeva, David J. Kemp, Jonathan C.A. Wakefield, Timothy G. Wright, J. Mol. Spectrosc. 344 (2018) 46–60.)

Available online at www.sciencedirect.com

ScienceDirect

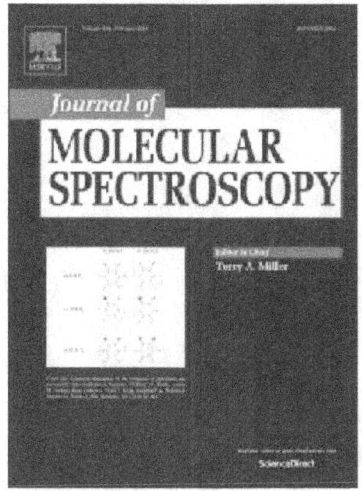

Journal of Molecular Spectroscopy
- established in 1957 -

Editor-in-Chief Terry A. Miller (USA)

Scope The *Journal of Molecular Spectroscopy* presents experimental and theoretical articles on all subjects relevant to molecular spectroscopy and its modern applications. An international medium for the publication of some of the most significant research in the field, the *Journal of Molecular Spectroscopy* is an invaluable resource for astrophysicists, chemists, physicists, engineers, and others involved in molecular spectroscopy research and practice.

Providing our authors a high quality and efficient peer review process, manuscript submitted to the *Journal of Molecular Spectroscopy* typically receive a first decision within 2-3 weeks and a final decision within 4-6 weeks.

Recent highlights from the *Journal of Molecular Spectroscopy*:

Special issue on Spectroscopy and Inter/Intramolecular Dynamics in Honor of Walther Caminati, guest edited by **Jens Uwe Grabow**, **Alberto Lesarri** and **Sonia Melandri**

Special issue on Molecular Spectroscopy in Traps, edited by **Stephan Schlemmer**, **Stefan Willitsch** and **Tim Steimle**

2017 Miller Prize:
Leah Dodson from University of Colorado at Boulder

2017 Rao Prize Winners:
- **Bryce Bjork** from University of Colorado at Boulder
- **Anna Huff** from University of Minnesota

Metrics
Impact Factor: 1.618*
5-Year Impact Factor: 1.482*
Cited half-life: >10 years*
SNIP: 0.742
SJR: 0.590

*Thomson Reuters journal citation reports 2016

For more info:
www.elsevier.com/locate/jms

Ideal Vacuum

Our products develop tomorrow's technologies

Ideal Vacuum Cubes

Manufactured and Built In Our Factory Here In The USA

6x6x6

6x6x12

Totally Modular
You Assemble with Simple Tools

Options
Plates, CF & KF Ports and Windows

Pressure to 1x10-6 Torr Dependent on Flange Type

Vacuum processes are used worldwide for research, development and in manufacturing of tomorrow's products. Here at Ideal Vacuum Products we are doing our part to better the future. We stock scientific equipment, vacuum pumps, pressure equipment, chambers, chillers, turbo pumps, blowers, leak detectors, oils and all accessories for your vacuum needs.

We have a highly qualified staff of technicians and engineers to answer any questions and the skill and knowledge to recommend the right vacuum products for your applications. We service the products we sell, know how they work and how you can get the best performance possible.

6x6x18

12x12x12

Ideal Vacuum Products, LLC. | Albuquerque, NM | 505.872.0037 | www.idealvac.com

Our products develop tomorrow's technologies

bellows valves

degas chambers

butterfly valves

All These Products Manufactured and Built In Our Factory Here In The USA

Vacuum processes are used worldwide for research, development and in manufacturing of tomorrow's products. Here at Ideal Vacuum Products we are doing our part to better the future. We stock scientific equipment, vacuum pumps, pressure equipment, chambers, chillers, turbo pumps, blowers, leak detectors, oils and all accessories for your vacuum needs.

We have a highly qualified staff of technicians and engineers to answer any questions and the skill and knowledge to recommend the right vacuum products for your applications. We service the products we sell, know how they work and how you can get the best performance possible.

nitrogen traps

vacuum cubes

Totally Modular
You Assemble
with Simple Tools
Options
Plates, CF & KF Ports
and Windows
Sizes
6X6X6, 6X6X12
6X6X18, 12X12X12

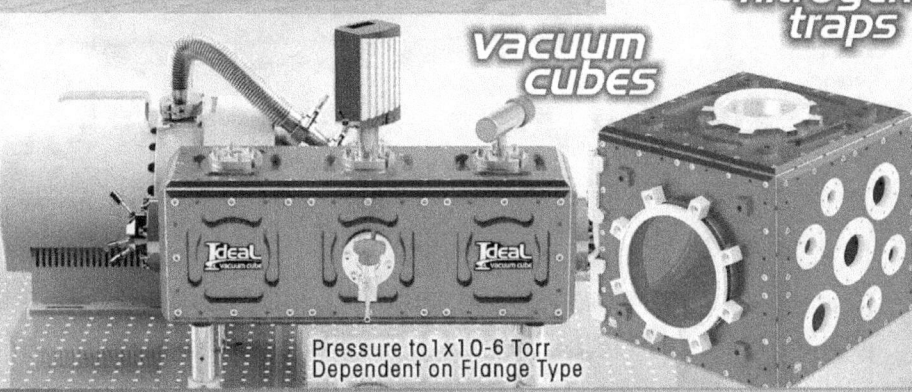

Pressure to 1×10^{-6} Torr
Dependent on Flange Type

Ideal Vacuum Products, LLC. | Albuquerque, NM | 505.872.0037 | www.idealvac.com

JPCA IMPACT FACTOR	JPCB IMPACT FACTOR	JPCC IMPACT FACTOR	JPCL IMPACT FACTOR
2.847	**3.177**	**4.536**	**9.353**

THE JOURNAL OF PHYSICAL CHEMISTRY

The leading journals in physical chemistry, with the most articles published and the largest worldwide readership

EDITOR-IN-CHIEF
George C. Schatz
Northwestern University

READ A RECENT VIRTUAL ISSUE ON
NMR DEVELOPMENTS & APPLICATIONS

connect.acspubs.org/NMR

DEPUTY EDITOR:
Anne B. McCoy, *University of Washington*

Journal Scope:
- Isolated Molecules, Clusters, Radicals, and Ions
- Environmental Chemistry, Geochemistry, and Astrochemistry
- Quantum Chemistry

DEPUTY EDITOR:
Joan-Emma Shea, *University of California, Santa Barbara*

Journal Scope:
Biophysical Chemistry, Biomaterials, Liquids, and Soft Matter

DEPUTY EDITOR:
Catherine J. Murphy, *University of Illinois at Urbana-Champaign*

Journal Scope:
Energy Conversion and Storage, Optical and Electronic Devices, Interfaces, Nanomaterials, and Hard Matter

DEPUTY EDITOR:
Gregory Scholes, *Princeton University*

Journal Scope:
Significant scientific advances in physical chemistry, chemical physics, and materials science

pubs.acs.org

North American ALMA Regional Center (NA-ARC)

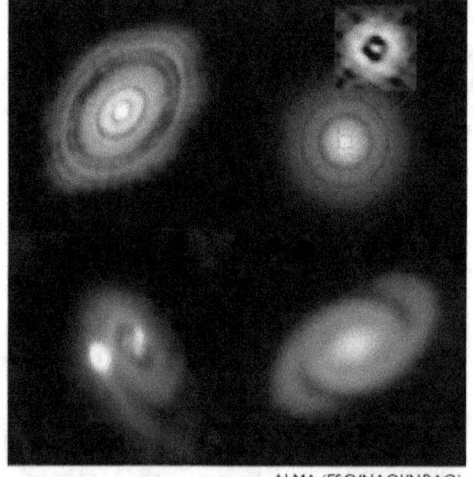
ALMA (ESO/NAOJ/NRAO)

The North American scientific community access to the **Atacama Large Millimeter/Submillimeter Array (ALMA)**

Headquartered at the **National Radio Astronomy Observatory (NRAO)** in Charlottesville, Virginia

Outreach services to inquisitive parties interested in ALMA science:

- **Data Reduction and Expert Analysis Assistance:** Travel and lodging support during visits to NRAO from investigators of successful ALMA programs or archival researchers.

- **Data Reduction Parties:** 10-12 PIs and their students visit the NA ARC for expert training.

- **Summer Student Program:** Introducing undergraduate/graduate students to innovative research.

- **Student Observing Support:** Funds graduate students working on eligible ALMA proposals.

- **Graduate Pre-Doctoral Program:** Conduct thesis research under the supervision of an NRAO scientist.

- **ALMA Ambassador Postdoctoral Fellows Program:** Provides training and $10,000 research grant to postdoctoral researchers interested in expanding their ALMA/interferomerty expertise and sharing that knowledge with their home institutions through ALMA proposal writing workshops.

To learn more about the NA-ARC and ALMA, visit:
science.nrao.edu/facilities/alma

National Radio Astronomy Observatory (NRAO)

NRAO/AUI/NSF

Founded in 1956, the NRAO provides state-of-the-art radio telescope facilities for use by the international scientific community. NRAO telescopes are open to all astronomers regardless of institutional or national affiliation. Observing time on NRAO telescopes is available on a competitive basis to qualified scientists after evaluation of research proposals on the basis of scientific merit, the capability of the instruments to do the work, and the availability of the telescope during the requested time.

www.nrao.edu

NRAO is a facility of the National Science Foundation operated by Associated Universities, Inc.

Pulsed Laser Solutions

Quantel Welcomes You to the
International Symposium on Molecular Spectroscopy
72nd Meeting – June 19-23, 2017

Sponsor of the Women's Networking Reception –
Wednesday June 21st

Quantel
2 bis, avenue du Pacifique
Z.A. de Courtaboeuf – BP 23
91941 Les Ulis Cedex – France
33 (0)1 69 29 16 45

www.quantel-laser.com

Quantel USA
49 Willow Peak Dr.
Bozeman, MT 59718
1-877-QUANTEL

The Coblentz Society – fostering understanding and application of vibrational spectroscopy

Call for Coblentz Award Nominations

The Coblentz Award is presented annually to an outstanding young molecular spectroscopist under the age of 40. This award is the Society's original award (first awarded in 1964), and is the complement of the 'Craver Award' that recognizes young spectroscopists for efforts in applied analytical vibrational spectroscopy. The candidate must be under the age of 40 on January 1 of the year of the award. The award comprises an honorarium, a plaque with a prism from the periscope of a World War II Navy submarine, and a travel allowance.

More information can be found at:
http://www.coblentz.org/awards/the-coblentz-award

Nominations for the 2019 Coblentz Award must include a detailed description of the nominee's accomplishments, a curriculum vitae or resume, and minimum of three supporting letters. Nominations for 2019 close on **July 15, 2018**. Files of candidates will be kept active for 3 years or until the age of eligibility is exceeded. Annual updates of candidate files are encouraged and will be solicited from the nomination source by the award's committee chair.

Please send nomination packages by email to nominations@coblentz.org

MenloSystems

Optical Frequency Combs

- 390 nm - 15 µm spectral range
- Unrivaled stability of 10^{-18}
- Portable optical frequency comb

Dual Comb Spectroscopy

- Customized solutions
- Acquisition rates on the kHz level
- Resolution down to Hz level

Optical Reference Systems

- Sub-Hz CW lasers in the visible and IR
- Stability of 10^{-15} in 1 second

www.menlosystems.com

Taking Center Stage.

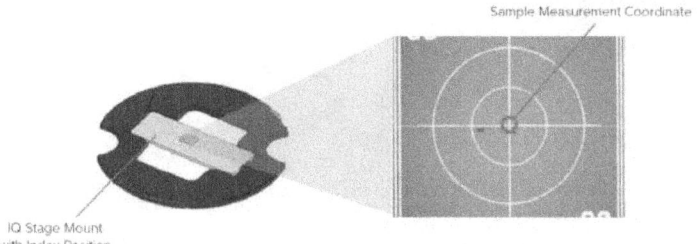

IQ STAGE™
For spectral correlation, move a sample between IR and Raman microscopes and measure exactly the same sample position using intelligent imaging technology

IRT SERIES
FTIR IMAGING MICROSCOPES
IQ Mapping™ for high speed IR mapping without moving the stage

NRS SERIES
CONFOCAL RAMAN MICROSCOPES
QRi™ for fast and high resolution confocal Raman imaging in 3D

PERFORMANCE • INNOVATION • RELIABILITY

ISMS MEETING VENUE INFORMATION

All contributed talks will be held in the Chemistry complex (and immediately adjoining buildings). The plenary talks will be held across the quad (about 600') in Foellinger Auditorium.

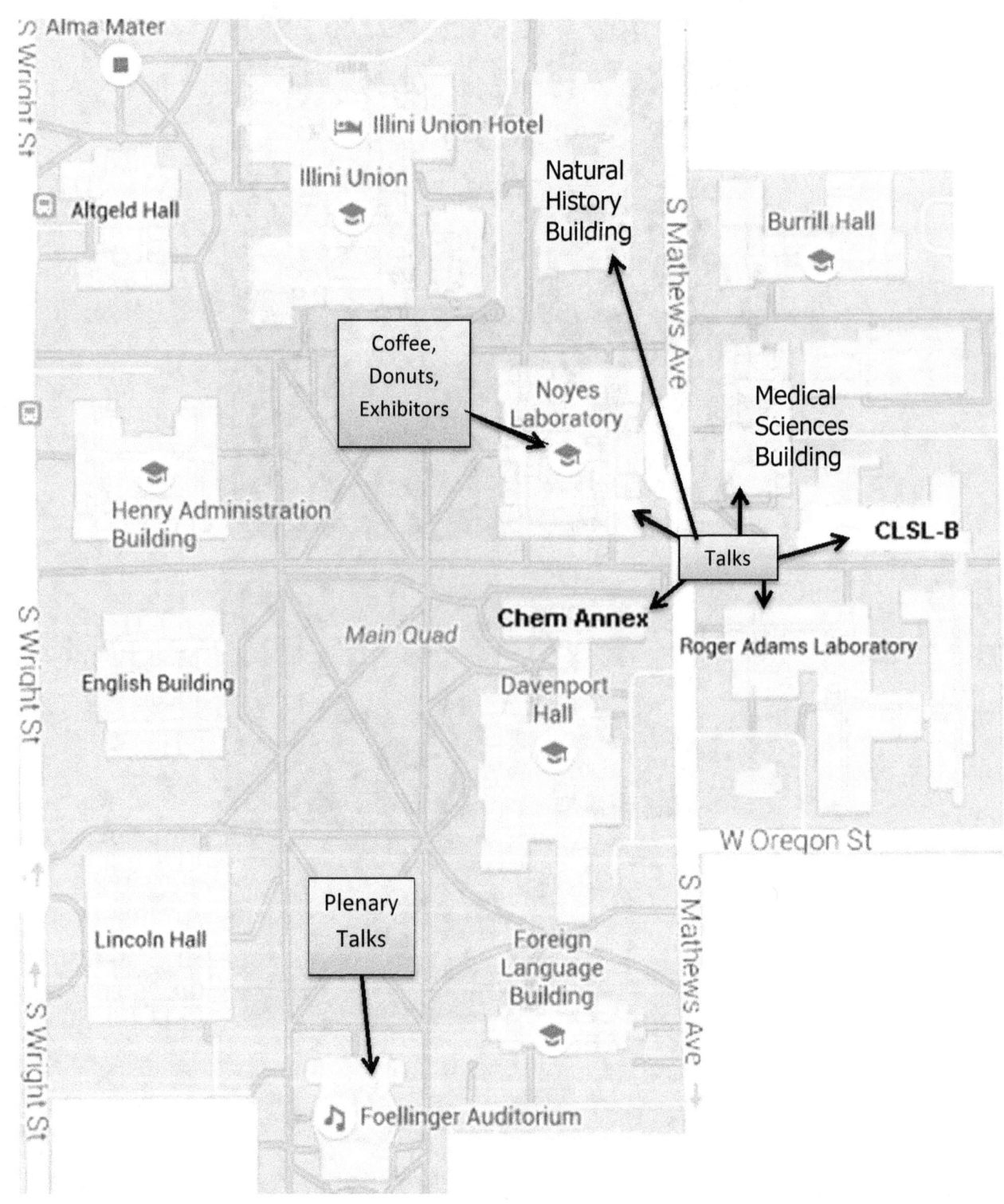

ACCESSIBLE ENTRANCES

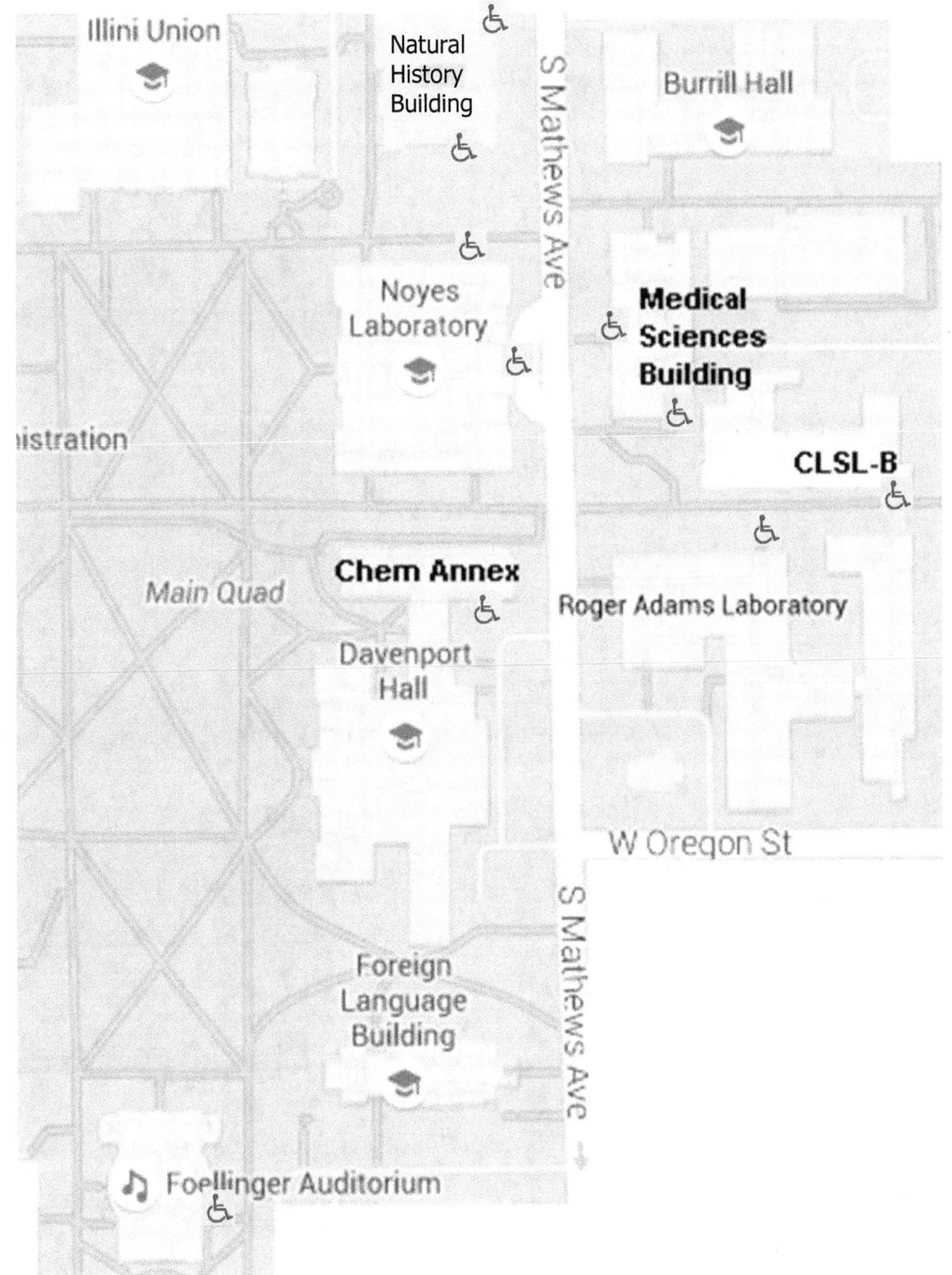

NOYES LABORATORY (NL)

Noyes Laboratory houses our Registration and Exhibitor/Refreshment Rooms (Chemistry Library), the Computer Lab (151), and two lecture halls (NL 100 and 217).

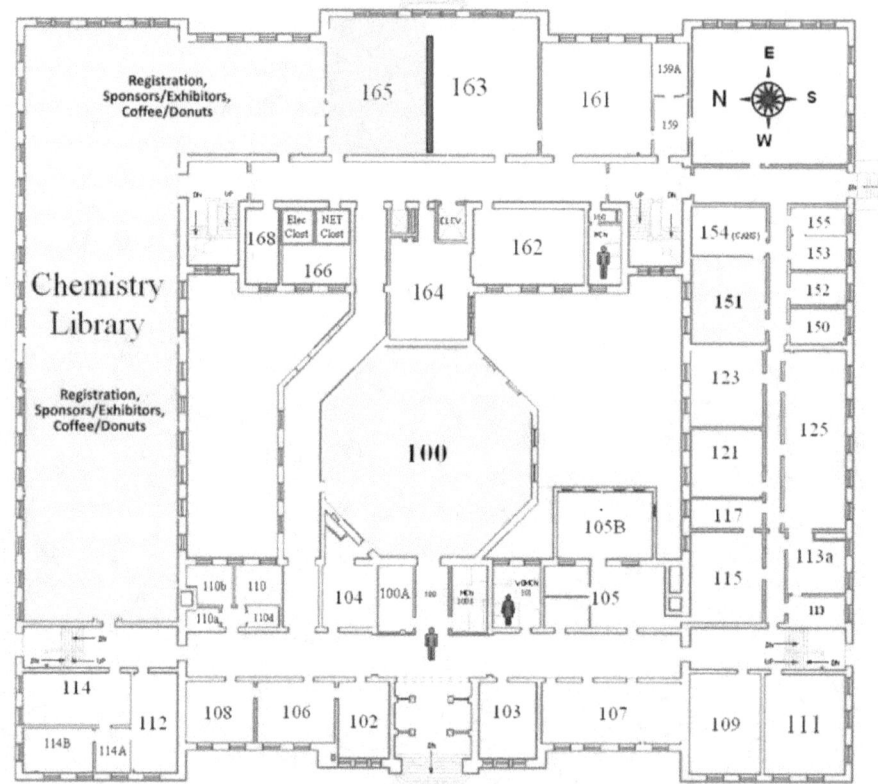

Noyes Laboratory - 1st Floor

CHEMISTRY ANNEX (CA)

Chemistry Annex is immediately to the south of Noyes Laboratory across a pedestrian walkway. It has one lecture hall (CA 1024)

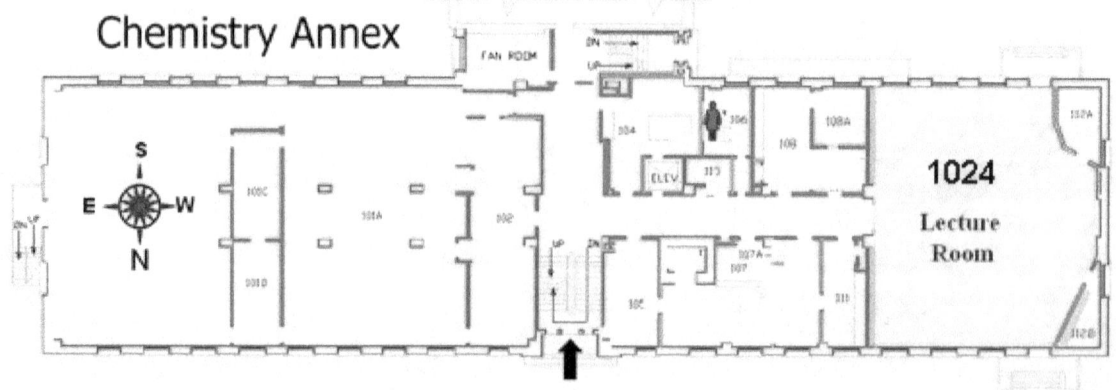

ROGER ADAMS LABORATORY (RAL)

Roger Adams Laboratory is across the street to the east of Chemistry Annex. It has one lecture hall (RAL 116). Please note that in Roger Adams Lab, the ground level is called "Ground" and the First Floor is equivalent to the Second Floor in the other buildings.

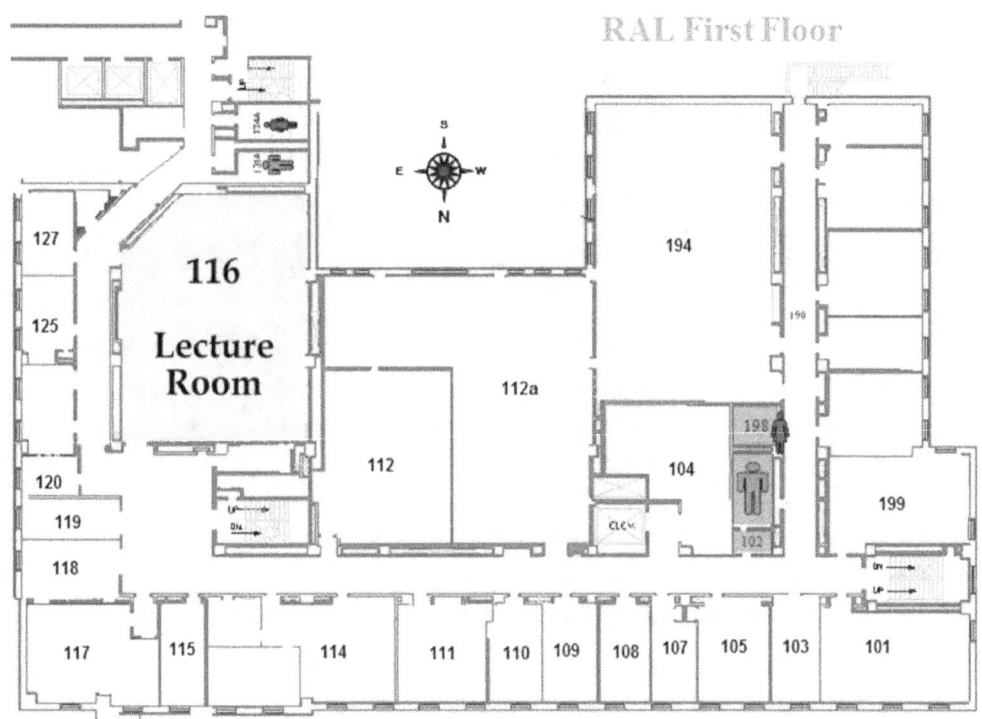

CHEMICAL AND LIFE SCIENCES (CLSL)

CLSL is a multi-wing building located across the street to the east of Noyes Laboratory. The lecture hall (CLSL B102) is in the B wing across the pedestrian walkway to the northeast of Roger Adams.

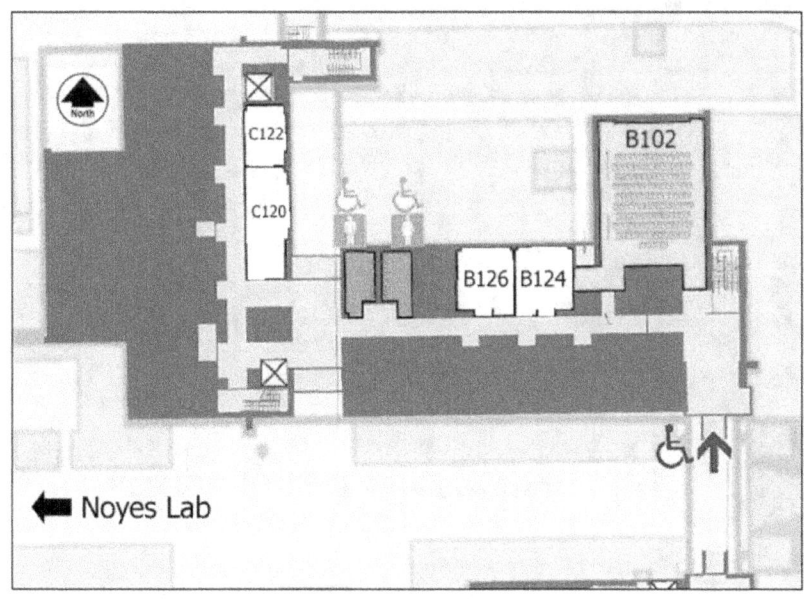

MEDICAL SCIENCES BUILDING (MSB)

Medical Sciences is across the pedestrian walkway to the north of RAL. It has one lecture hall (274).

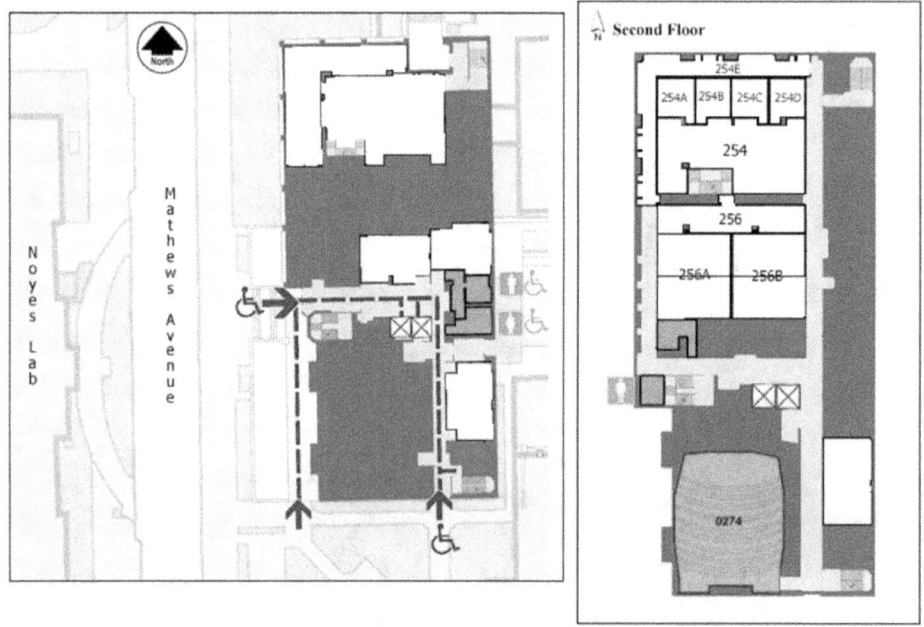

NATURAL HISTORY BUILDING

Natural History Building is immediately North of Noyes Lab. It has one lecture hall (NHB 2079).

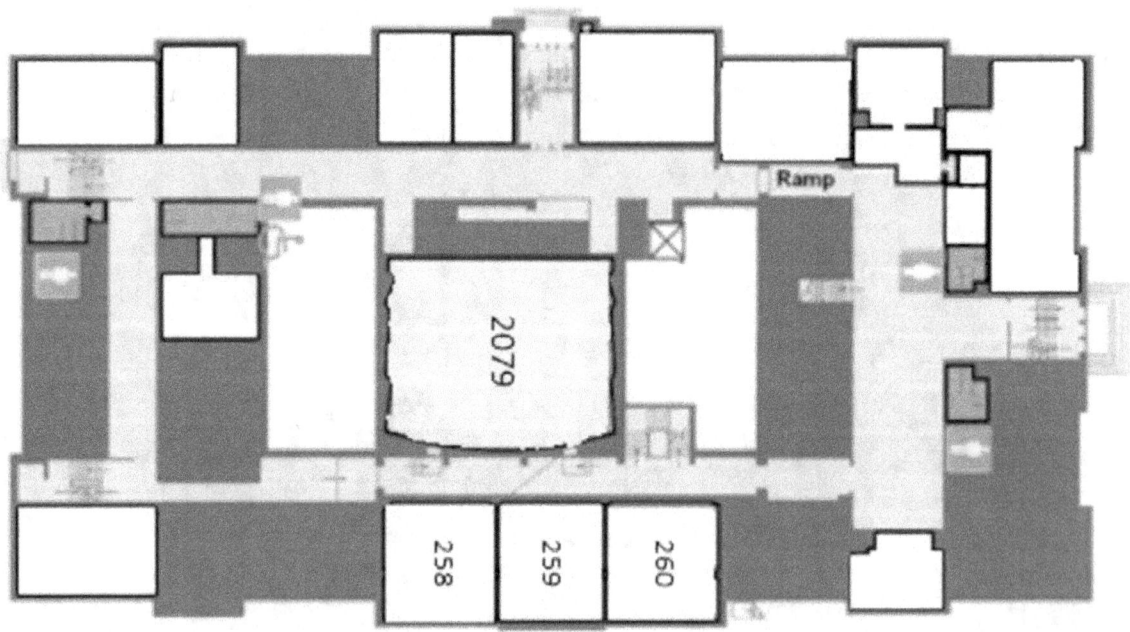

Foellinger Auditorium (Plenary and Intermission)

Foellinger Auditorium is located at the south end of the Quad. The main doors on the north (quad) side will open at 8:10 AM (the side ADA/wheelchair door will be open around 8:00 AM). There is seating on the main level and the upper balcony. There is no elevator in the building.

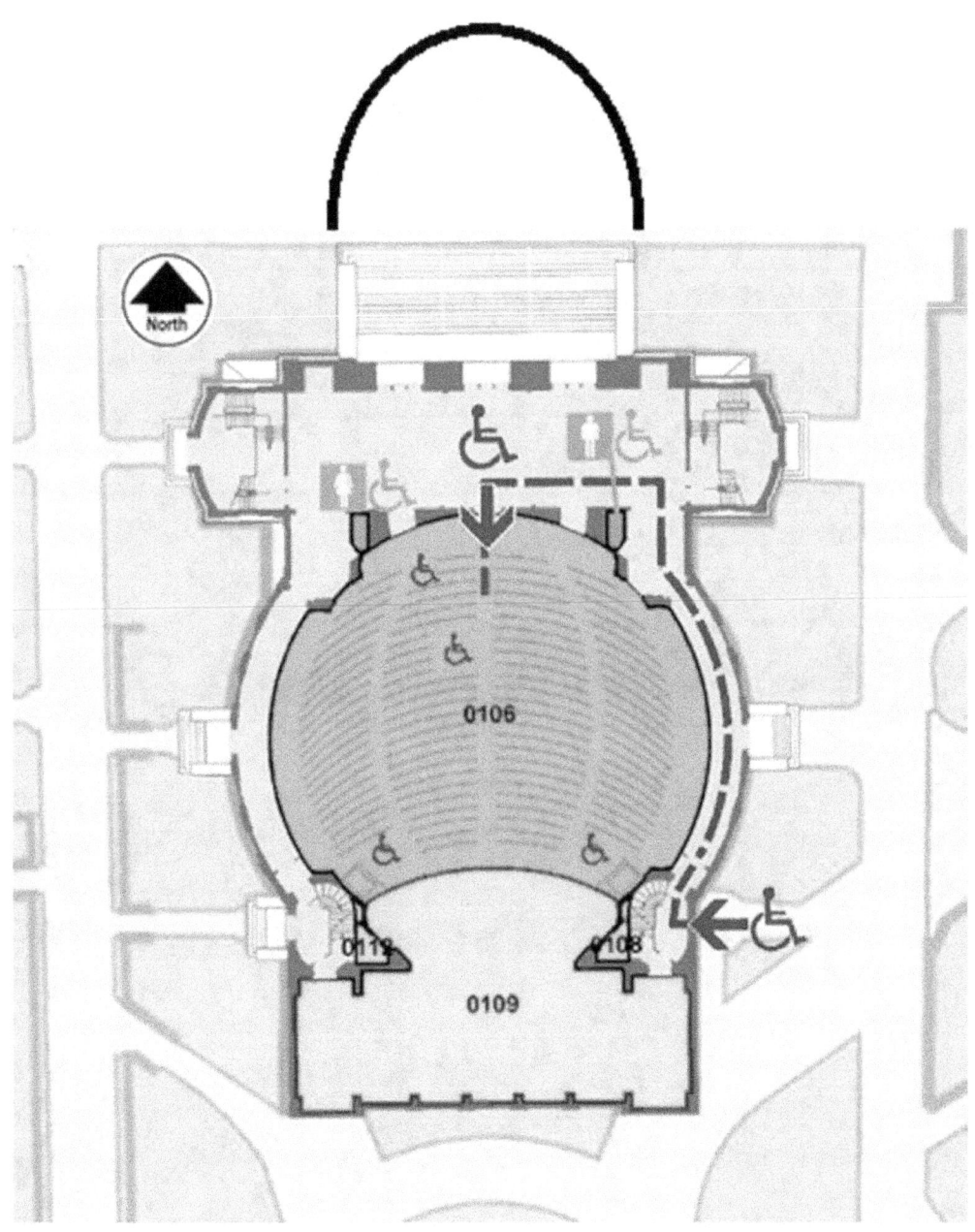

PARKING (E14) TO BOUSFIELD DORM

If you purchase a parking permit and are staying at the dorm, you will park in lot E14 (any spot). E14 is nearly due south of Bousfield Hall Dorm.

Parking enforcement begins at 6:00 AM on Monday, so you will need to have your car in lot E14 with your permit displayed before then. There are many parking meters on E. Peabody Drive (and in the lot across from Bousfield) if you wish to park closer for short periods (25 cents/15 minutes – generally between 6 AM and 6 PM, but check the meter because some go until 9 PM).

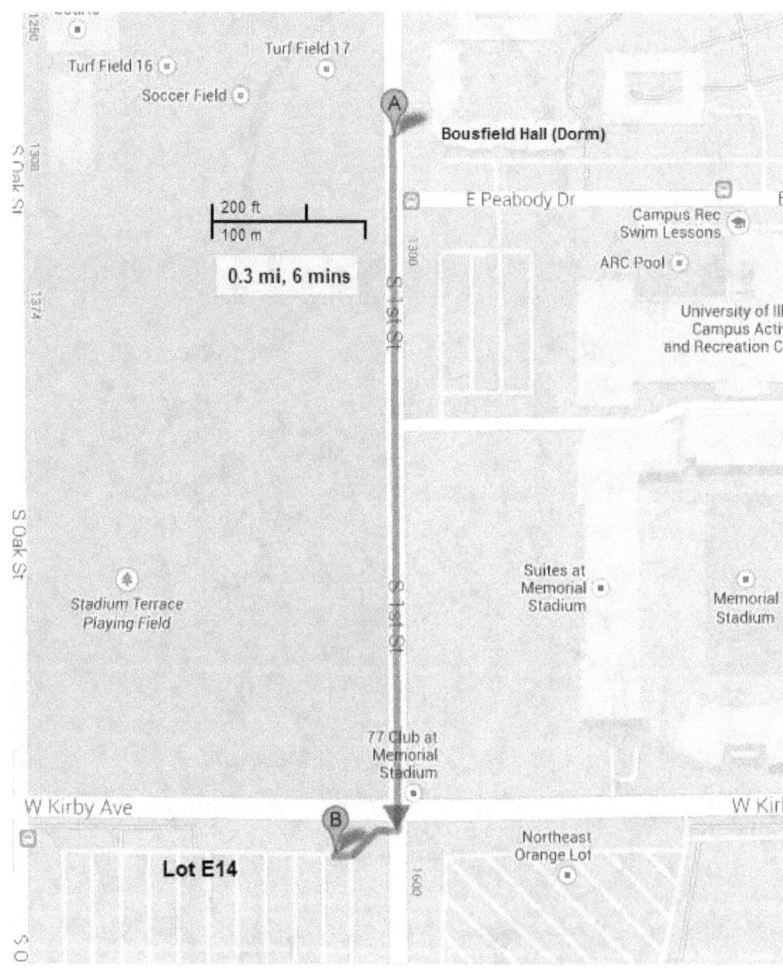

BOUSFIELD/WASSAJA DORM to MEETING VENUE (walking)

Bousfield & Wassaja Halls are just under a mile (15-20 minute walk) from the main symposium buildings

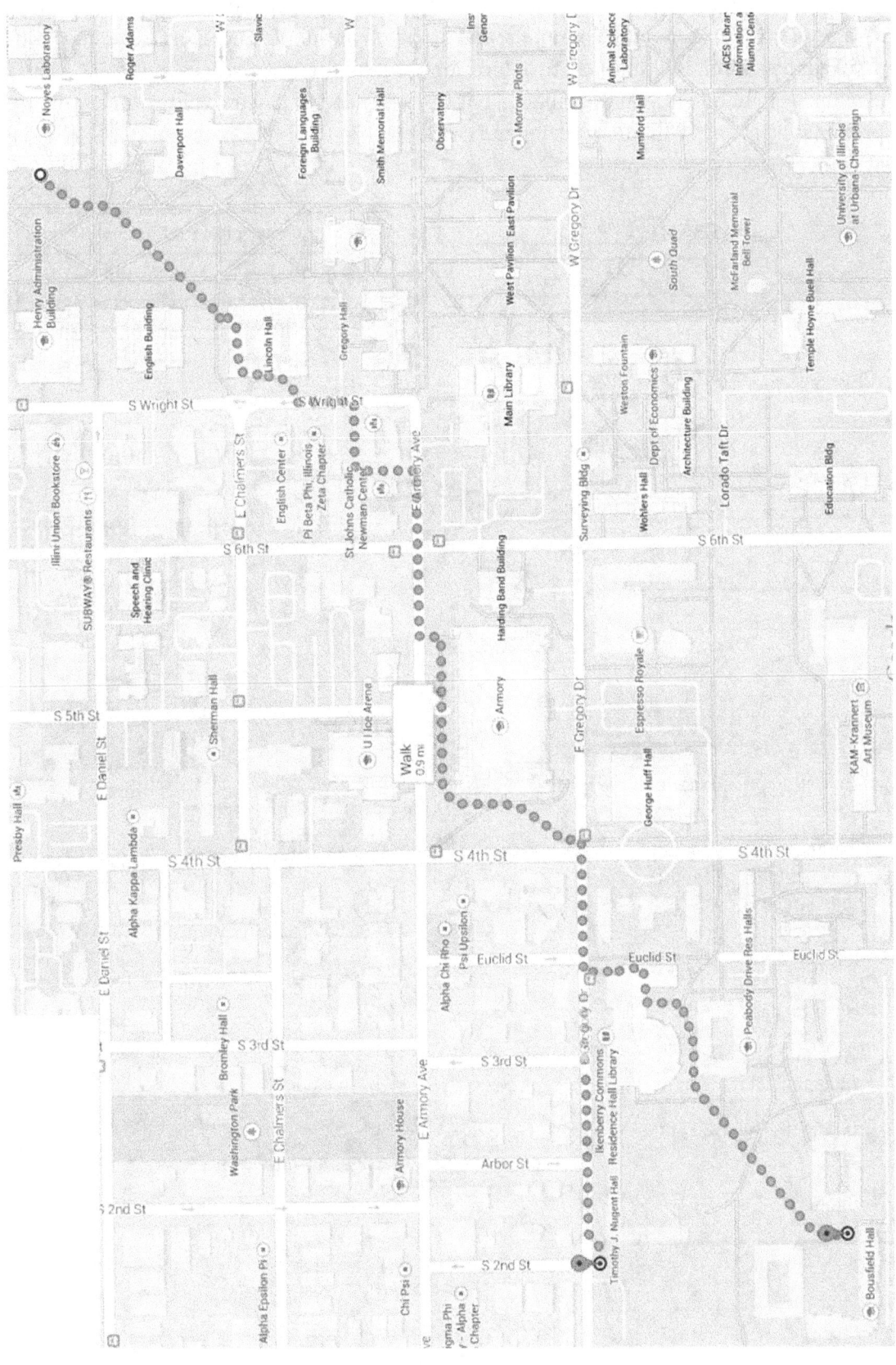

BOUSFIELD/WASSAJA DORM to MEETING VENUE (bus)

There is convenient and free bus service between Bousfield/Wassaja Dorms and 1 block from the meeting venue. The Yellow Line picks up on the corner of First and Peabody (Bousfield), and also on Gregory Drive (Wassaja) in front of Ikenberry Commons, and drops off at the Wright Street Terminal (just outside of the Henry Administration Building). Return locations are the same but across the street. The Yellow Line will also take you to downtown Champaign, but you will need to pay for your return (only iStops are free). Approximately every 10 minutes during the day.

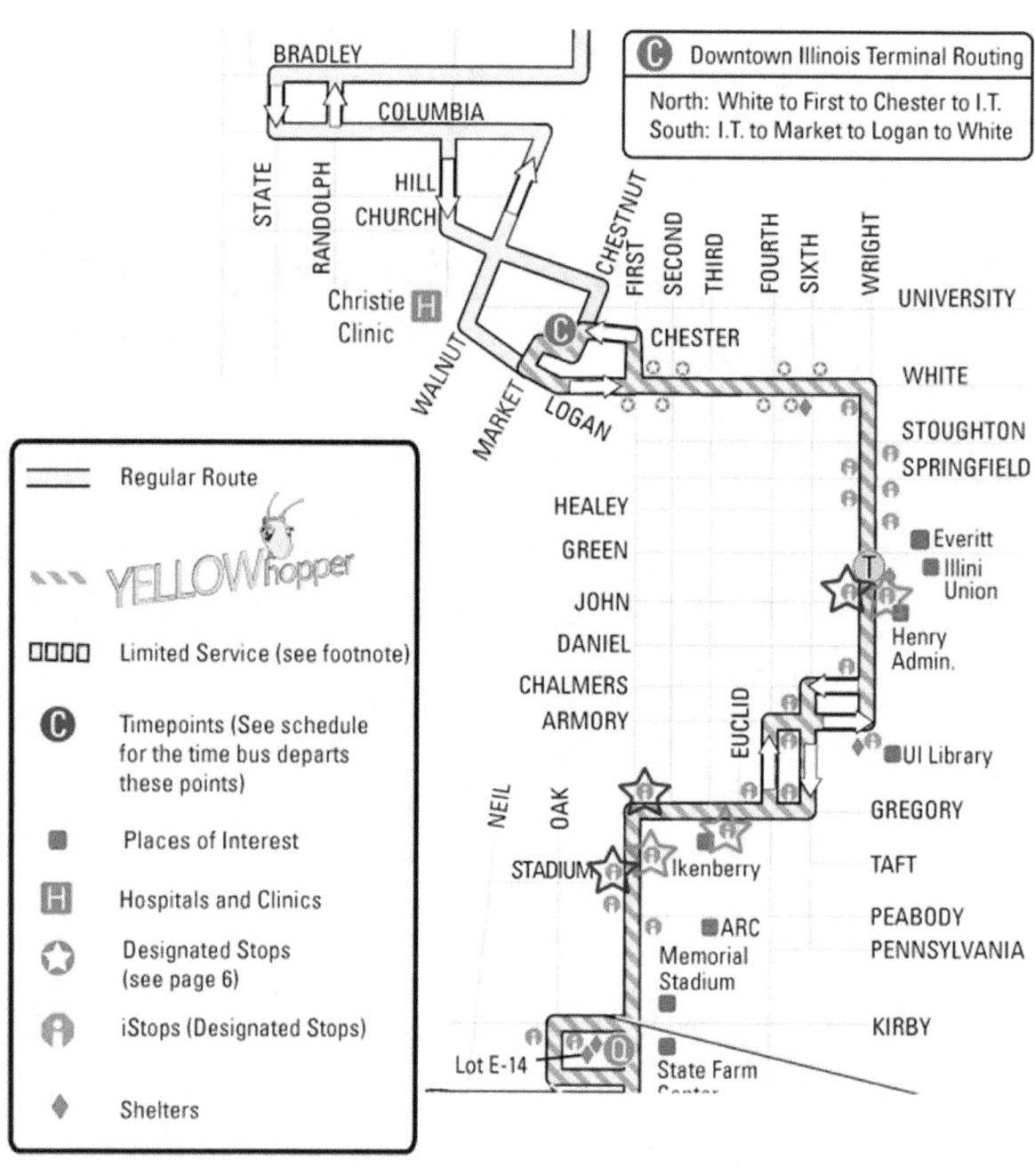

The Gold Line picks up on the corner of First and Peabody, and also on Gregory Drive in front of Ikenberry Commons and drops off at the Krannert Center (across the street from CLSL-B). Return locations are across the street. Runs every ~10 minutes during the day (offset from the Yellow Line by 5 minutes).

Bus Stops (Yellow Line = Left Arrow, Gold Line = Right Arrow, Foellinger Auditorium (Plenary) and Noyes Lab = Stars)

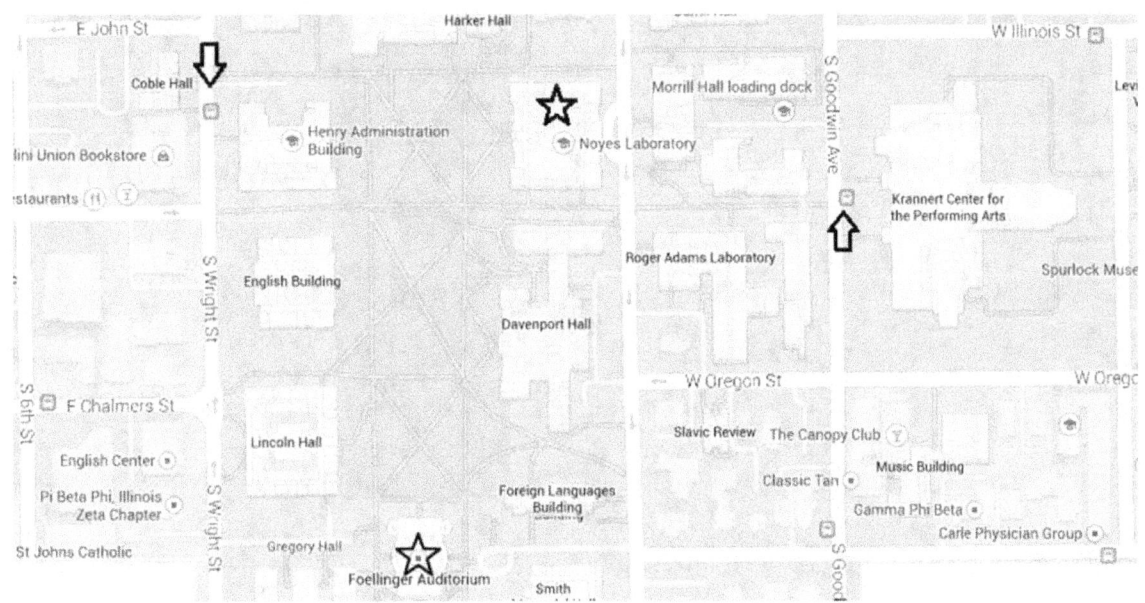

NOTES

NOTES

NOTES

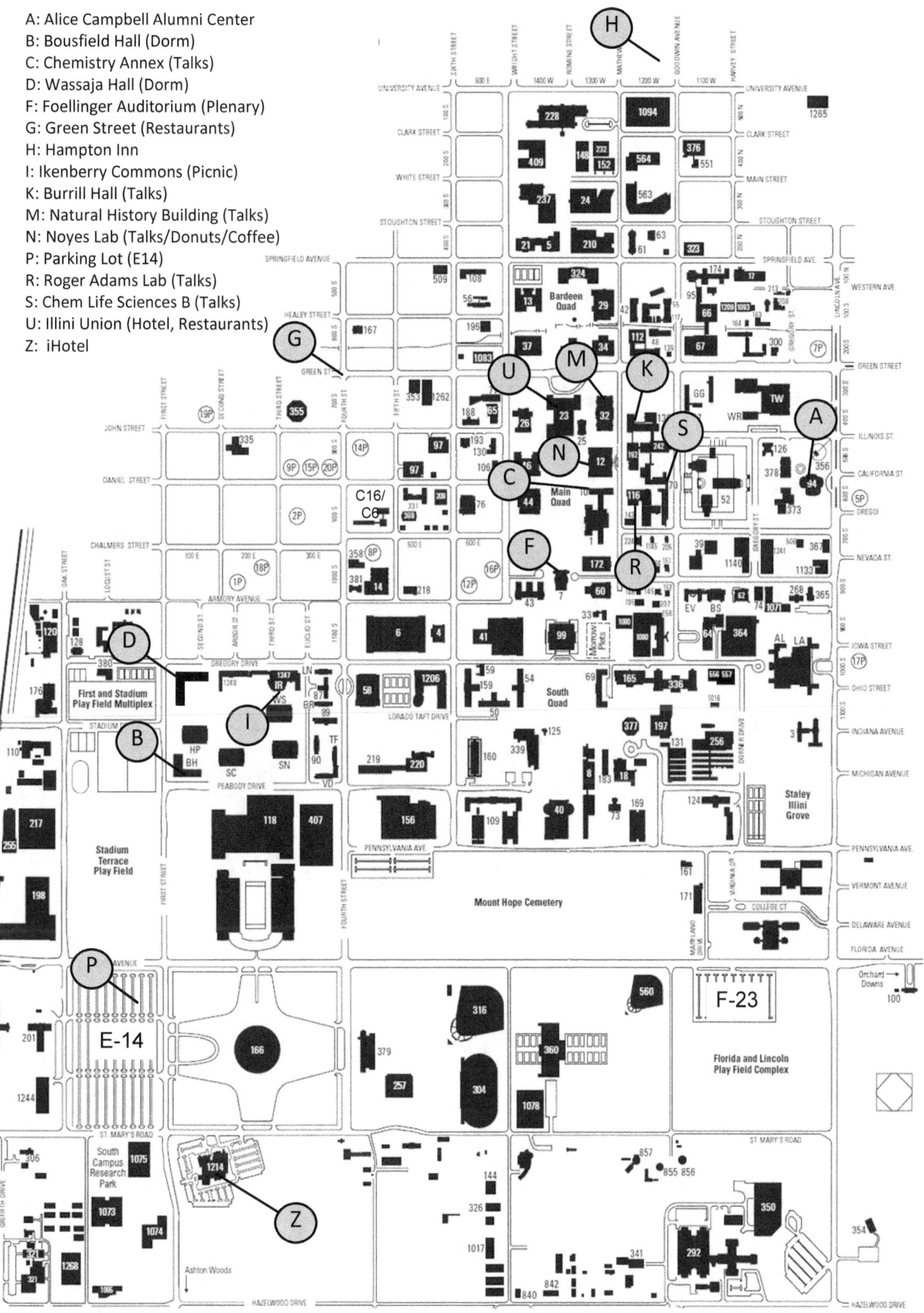

A: Alice Campbell Alumni Center
B: Bousfield Hall (Dorm)
C: Chemistry Annex (Talks)
D: Wassaja Hall (Dorm)
F: Foellinger Auditorium (Plenary)
G: Green Street (Restaurants)
H: Hampton Inn
I: Ikenberry Commons (Picnic)
K: Burrill Hall (Talks)
M: Natural History Building (Talks)
N: Noyes Lab (Talks/Donuts/Coffee)
P: Parking Lot (E14)
R: Roger Adams Lab (Talks)
S: Chem Life Sciences B (Talks)
U: Illini Union (Hotel, Restaurants)
Z: iHotel

www.ingramcontent.com/pod-product-compliance
Lightning Source LLC
Chambersburg PA
CBHW062349220526
45472CB00008B/1749